Pesticide Application Methods

Third Edition

G.A. Matthews

Professor in Pest Management
International Pesticide Application Research Centre
Imperial College, London, UK

Blackwell
Science

Editorial Offices:
Osney Mead, Oxford OX2 0EL
25 John Street, London WC1N 2BS
23 Ainslie Place, Edinburgh EH3 6AJ
350 Main Street, Malden
MA 02148 5018, USA
54 University Street, Carlton
Victoria 3053, Australia
10, rue Casimir Delavigne
75006 Paris, France

Other Editorial Offices:

Blackwell Wissenschafts-Verlag GmbH
Kurfürstendamm 57
10707 Berlin, Germany

Blackwell Science KK
MG Kodenmacho Building
7–10 Kodenmacho Nihombashi
Chuo-ku, Tokyo 104, Japan

First edition published by Longman Group UK Ltd 1979
Second edition published 1992
Third edition published by Blackwell Science Ltd 2000

Set in 10/12pt Times
by DP Photosetting, Aylesbury, Bucks
Printed and bound in Great Britain by
Biddles Ltd, Guildford and King's Lynn

DISTRIBUTORS

Marston Book Services Ltd
PO Box 269
Abingdon
Oxon OX14 4YN
(*Orders:* Tel: 01235 465500
Fax: 01235 465555)

USA
Blackwell Science, Inc.
Commerce Place
350 Main Street
Malden, MA 02148 5018
(*Orders:* Tel: 800 759 6102
781 388 8250
Fax: 781 388 8255)

Canada
Login Brothers Book Company
324 Saulteaux Crescent
Winnipeg, Manitoba R3J 3T2
(*Orders:* Tel: 204 837-2987
Fax: 204 837-3116

Australia
Blackwell Science Pty Ltd
54 University Street
Carlton, Victoria 3053
(*Orders:* Tel: 03 9347 0300
Fax: 03 9347 5001)

A catalogue record for this title is available from the British Library

ISBN 0-632-05473-5

Library of Congress
Cataloging-in-Publication Data is available

For further information on Blackwell Science, visit our website:
www.blackwell-science.com

Contents

Preface to third edition

At the start of the new millennium, crop protection is confronted with the public debate about genetically modified crops and demands for organic food. Pesticides are disliked due to the perception that residues are harmful, despite the knowledge that with their increased use over the last five decades, food quality has been vastly improved and life expectancy increased. Pesticides are also disliked due to the popular belief that toxic chemicals are overused and drift across the environment, adversely affecting water supplies and natural habitats.

The world's human population continues to increase, with greater demands for food of high quality, so there can be no return to growing crops without artificial fertilizers and some pesticide use. Genetically modified crops can provide a means of improving the quality of some crops by enhancing vitamin content or disease resistance. Unfortunately, the two GM crops most widely used initially were those expressing the *Bacillus thuringiensis* (Bt) toxin gene to check lepidopterous pests, and those with resistance to the herbicide glyphosate. Bt crops still require spray treatments to control other types of pests, notably sucking pests such as aphids, while the 'Roundup Ready' crops depend on using one particular herbicide – glyphosate, so they still require spray technology. These two types of GM crops have other problems, such as the strong possibility of insect pests becoming resistant to the Bt toxin and different weeds becoming a problem.

Biological and cultural controls are undoubtedly of great importance, but neither can respond rapidly to sudden outbreaks of pests, so pesticide use must form a key component of integrated crop management. Unfortunately, in many parts of the world the lack of infrastructure and trained personnel has resulted in misuse of pesticides. The challenge now is to spread the knowledge on safe use and correct application of pesticides beyond its present frontiers, so that higher yields of crops can be obtained in the developing countries. Pesticides are only one of the tools and can only protect crops with a high yield potential to justify the expense of their use. We know more about more precise application with fewer losses to the environment, but need to continue research so that new technologies can be incorporated to minimise pesticide use and improve the timing of applications. Since the last edition, development of hydraulic nozzles has provided droplet spectra less prone to drift beyond field boundaries, but care is need to maintain biological efficacy within fields. Legislation has led to greater safety in pesticide packaging and rules for maintaining equipment and minimising pollution, at least in some countries.

Pesticide application is a multidisciplinary subject, yet attracts little research funding compared with the inputs into chemistry and genetic engineering. Studies with the development of a biopesticide for locust control have demonstrated the need for careful integration of formulation and application technology research, to ensure that what is effective under laboratory conditions is also successful in the field. In preparing this edition, the opportunity has been taken to add two new chapters to discuss the application of biopesticides and the equipment needed in laboratory and field trial experiments. In updating chapters, the layout of some have not changed very significantly since the last edition, but some have been rearranged to reflect changing emphasis on certain types of equipment or techniques. It is hoped that this edition will continue to assist with training and improve the safety and efficiency of application.

Acknowledgements

As with the previous editions, I have been assisted by discussions with many of those involved in pesticides and their application. Much information has also been gained from visits to many countries to see how pesticides are being applied to crops and advise on integrated pest management in a range of agricultural environments.

In preparing the third edition, I have continued to receive generous support from many of those involved with pesticides and their application. In particular I wish to thank my colleagues at IPARC – Dr Roy Bateman, Evan Thornhill and Hans Dobson – for their assistance and discussions, and Miss Carole Collins for retyping much of the text of the second edition that had been lost. I would also like to thank the following for their contributions with updating information, supplying illustrations or reading drafts of the third edition: Carolyn Baeck, Tom Bals, Clive Barber, Martin Baxter, Clive Christian, John Clayton, John Crabtree, Mike de Lara, Bernd Dietrich, Andreas Deuble, Andrew Gilbert, Ken Giles, Andrew Hewitt, Christian Delcomyn Holst, Ivan Kirk, Mark Ledson, David McAuliffe, Frances McKim (BCPC), John Newton, Tim Nicholls, Malcolm Ogilvey (Chapter 16 in particular), Tim Sander (Chapter 13 in particular), Werner Stahl, John Tobutt, Stuart Wili. I owe a special thank you to Moira for her encouragement and support.

We are grateful to the following for permission to reproduce copyright material:

M. Abdalla for Fig. 2.16 (Abdalla, 1984); Academic Press and the author, Dr I. J. Graham-Bryce for Fig. 2.17; Academic Press and the respective authors for Figs 2.8 (Fryer, 1977) and 2.20 (Courshee, 1967); E. Allman & Company, Chichester for 7.2a and 16.3; E. Allman & Company and Hardi International for Fig. 7.1; Professor G. K. Batchelor for Fig. 4.8; Roy Bateman for Figs 5.2, 8.3, 10.14 and 17.2; BBA Germany for Fig. 5.20; Dr. S. Beernaerts for Fig. 15.1; BP Co. Ltd for Fig. 8.7; British Crop Protection Council for Figs 1.8 (Greaves & Marshall, 1987), 2.1a (Doble *et al.*, 1985), 4.3, 4.4, 5.6, 6.2, 6.6, 7.20 and 19.4; Dick Brown for Fig. 2.5; Burkard Manufacturing Co. for Figs 18.1–18.3; Capstan Ag

Systems Inc. USA for Fig. 7.17; Central Science Laboratory for Fig. 16.2; J. W. Chafer Ltd for Fig. 7.18; John Clayton (Micron Sprayers) for Fig. 2.4b; Cleanacres Ltd for Fig. 5.18; CP Products (USA) for Fig. 13.11; Collins, London and the author, H. A. Quantick for Fig. 13.20 (Quantick, 1985a); John Crabtree for Figs 18.5 and 18.6; Crop Protection Association for Fig. 16.7; Delavan USA for Figs 7.4, 7.5 and 7.7; Alan Dewar (IACR Brooms Barn) for Fig. 1.5; Professor N. Dombrowski for Fig. 5.1; Ellis Horwood and the authors, D. H. Bache and D. R. Johnstone for Fig. 4.2 (Bache & Johnstone, 1992); Farming Japan for Fig. 13.3; Food and Agricultural Organisation of the United Nations for Figs 2.10, 13.4, 13.5b, 13.6, 13.7, 13.19, 13.23; Professor Ken Giles (University of California, Davis) for Figs 5.16 and 5.17; Global Agricultural Technology Engineering, Florida for Fig. 6.3; Global Crop Protection Federation for Fig. 16.5; Graticules Ltd for Fig. 4.19; Hardi International for Figs 4.5, 5.19, 7.3 and 10.13; Dr C. A. Hart for Fig. 2.14; H D Hudson Manufacturing Company for Fig. 6.8; C. M. Himel for Fig. 4.18 (Himel, 1969a); Horstine Farmery for Figs 12. 4, 12.7, 12.8 and 12.9; Interscience and the author, R. P. Fraser for Figs 4.6, 10.4 and 10.5 (Fraser, 1958); D. R. Johnstone for Figs 2.18, 2.19 (Johnstone, 1972) and 4.9 (Johnstone *et al.*, 1974); M. Knoche for Fig. 10.16; Professor Ed Law (University of Georgia, Athens, USA) for Figs 9.7 and 9.8; Lechler GmbH for Fig. 7.12h; Lurmark Ltd for Fig 5.3; M. Morel Technoma France for Fig. 10.15c & d; Malvern Instruments for Fig. 4.13; Micron Sprayers for Figs 6.1, 8.1, 8.2, 8.4, 8.5, 8.6, 8.8–8.12, 10.11, 13.5d, 13.14–13.18; Motan GmbH for Figs 11.5 and 11.8c; Novartis for Fig. 3.1; NSW Agriculture for Fig.1.3; Overseas Development Natural Resources Institute (*Locust Handbook*) for Fig. 2.4a; Oxford Lasers for Fig. 4.16; Steve Parkin (Silsoe Research Institute) for Figs 2.2, 2.9, 4.14 and 4.15; Pulsfog GmbH for Figs 11.1 and 11.4; Satloc for Fig. 13.22; Silsoe Research Institute for Figs 1.10, 2.1b , 5.10, 5.11, 10.18b and 18.4; Simplex Manufacturing Co. (USA) for Fig. 13.8; Society of Chemical Industry for Fig. 2.15 (Ford & Salt, 1987); John Spillman for Figs 2.11, 2.12, 4.7, 13.5a, 13.5c and 13.10; Spraying Systems Co. (USA) for Figs 5.4, 5.5, 5.13, 7.8, 7.9, 7.11 and 7.12a–g; Stihl, USA for Fig. 10.8; Norman Thelwell for Fig. 13.12; Wisdom Systems for Figs 7.2b, e & f; WMEC for Fig. 16.6; World Health Organisation for Figs 5.7 and 6.9; Frank Wright for Figs 12.5, 14.4, 15.2, 15.3; Zeneca Agrochemicals (formerly ICI) for Figs 3.2, 4.10, 4.11, 9.5a and 12.1.

Conversion tables

	A	B	A→B	B→A
Weight	oz	g	× 28.35	× 0.0353
	lb	kg	× 0.454	× 2.205
	cwt	kg	× 50.8	× 0.0197
	ton (long)	kg	× 1016	× 0.000984
	ton (short)	ton (long)	× 0.893	× 1.12
Surface area	in^2	cm^2	× 6.45	× 0.155
	ft^2	m^2	× 0.093	× 10.764
	yd^2	m^2	× 0.836	× 1.196
	yd^2	acre	× 0.000207	× 4840
	acre	ha	× 0.405	× 2.471
Length	μm	mm	× 0.001	× 1000
	in	cm	× 2.54	× 0.394
	ft	m	× 0.305	× 3.281
	yd	m	× 0.914	× 1.094
	mile	km	× 1.609	× 0.621
Velocity	ft/s	m/s	× 0.305	× 3.281
	ft/min	m/s	× 0.00508	× 197.0
	mile/h	km/h	× 1.609	× 0.621
	mile/h	ft/min	× 88.0	× 0.0113
	knot	ft/s	× 1.689	× 0.59
	m/s	km/h	× 3.61	× 0.277
	cm/s	km/h	× 0.036	× 27.78
Quantities/area	lb/acre	kg/ha	× 1.12	× 0.894
	lb/acre	mg/ft^2	× 10.4	× 0.09615
	kg/ha	mg/m^2	× 100	× 0.01
	mg/ft^2	mg/m^2	× 10.794	× 0.093
	oz/yd^2	cwt/acre	× 2.7	× 0.37
	gal (Imp.)/acre	litre/ha	× 11.23	× 0.089
	gal (USA)/acre	litre/ha	× 9.346	× 0.107
	fl oz (Imp.)/acre	ml/ha	× 70.05	× 0.0143
	fl oz (USA)/acre	ml/ha	× 73.14	× 0.0137
	oz/acre	g/ha	× 70.05	× 0.0143
	oz/acre	kg/ha	× 0.07	× 14.27
Dilutions	fl oz/100 gal (Imp.)	ml/100 litres	× 6.25	× 0.16
	pint/100 gal (Imp.)	ml/100 litres	× 125	× 0.008
	oz/gal (Imp.)	g/litre	× 6.24	× 0.16
	oz/gal (USA)	g/litre	× 7.49	× 0.134
	lb/100 gal (Imp.)	kg/100 litre	× 0.0998	× 10.02

Density of water	gal (Imp.)	lb	× 10	× 0.1
	gal (USA)	lb	× 8.32	× 0.12
	lb	ft^3	× 0.016	× 62.37
	litre	kg	× 1	× 1
	ml	g	× 1	× 1
	lb/gal (Imp.)	g/ml	× 0.0997	× 10.03
	lb/gal (USA)	g/ml	× 0.1198	× 8.34
	lb/ft^3	kg/m^3	× 16.1	× 0.0624
Volume	in^3	ft^3	× 0.000579	× 1728
	ft^3	yd^3	× 0.037	× 27
	yd^3	m^3	× 0.764	× 1.308
	fl oz (Imp.)	ml	× 28.35	× 0.0352
	fl oz (USA)	ml	× 29.6	× 0.0338
	gal (Imp.)	gal (USA)	× 1.20	× 0.833
	gal (Imp.)	litre	× 4.55	× 0.22
	gal (USA)	litre	× 3.785	× 0.264
	cm^3	m^3	$\times 10^{-6}$	$\times 10^{6}$
	cm^3	μm^3	$\times 10^{12}$	$\times 10^{-12}$
Pressure	lb/in^2	kg/cm^2	× 0.0703	× 14.22
	lb/in^2	bar	× 0.0689	× 14.504
	bar	kPa	× 100	× 0.01
	lb/in^2	kPa	× 6.89	× 0.145
	kN/m^2	kPa	× 1	× 1
	N/m^2	kPa	× 0.001	× 1000
	lb/m^2	atm	× 0.068	× 14.696
Power	hp	kW	× 0.7457	× 1.341
Temperature	°C	°F	$\frac{9}{5}$ °C + 32	$\frac{5}{9}$ (°F–32)

Pesticide calculations

(1) To determine the quality (X) required to apply the recommended amount of active ingredient per hectare (A) with a formulation containing B percentage active ingredient.

$$\frac{A \times 100}{B} = X$$

Example: Apply 0.25 kg a.i./ha of 5 per cent carbofuran granules

$$\therefore \frac{0.25 \times 100}{5} = 5\,\text{kg granulates/ha}$$

(2) To determine the quantity of active ingredient (Y) required to mix with a known quantity of diluent (Q) to obtain a given concentration of spray.

$$Q \times \frac{\text{per cent concentration required}}{\text{per cent concentration of active ingredient}} = Y$$

(a) *Example:* Mix 100 litres of 0.5 per cent a.i., using a 50 per cent wettable powder

$$100 \times \frac{0.5}{50} = 1\,\text{kg of wettable powder}$$

(b) *Example:* Mix 2 litres of 5 per cent a.i. using a 75 per cent wettable powder

$$2000 \times \frac{5}{75} = 133\,\text{g of wettable powder}$$

Units, abbreviations and symbols

A	ampere
atm	atmospheric pressure
bar	baropmetric pressure
cd	candela
cm	centimetre
dB	decibel
fl oz	fluid ounce*
g	gram
g	acceleration due to gravity ($9.8\,m/s^2$)
gal	gallon*
h	hour
ha	hectare
hp	horsepower
kg	kilogram
km	kilometre
kN	kilonewton
kPa	kilopascal
kW	kilowatt
l	litre
m	metre
mg	milligram
ml	millilitre
mm	millimetre
µm	micrometre
N	newton
µP	micropoise
P	poise
p.s.i.	pounds per square inch
pt	pint
s	second
V	volt

* Volume measurements may be in Imperial or American units as indicated by (Imp.) or (USA).

Abbreviations and symbols

A	area
a	average distance between airstrip or water supply to fields
a.c.	alternating current
ADV	average droplet volume
AGL	above ground level
a.i.	active ingredient
AN	Antanov aircraft
BPMC	fenobucarb
C	average distance between fields
CDA	controlled droplet application
CFD	computional fluid dynamics
CU	coefficient of uniformity
D	diameter of centrifugal energy nozzle of opening of nozzle
d	droplet diameter
DCD	disposable container dispenser
'D' cell	a standard size dry battery
d.c.	direct current
DMI	demethylation inhibitor
DUE	deposit per unit emission
EC	emulsifiable concentrate
EDX	energy dispersive X-ray
EPA	Environmental Protection Agency (USA)
F	average size of field
FAO	Food and Agriculture Organization of the United Nations
FN	flow number
FP	fluorescent particle
GCPF	Global Crop Protection Federation

GIFAP Groupement International des Associations Nationales de Fabricants de Produits Agrochimiques. (International Group of National Associations of Manufacturers of Agrochemical Products)
GIS geographical information system
GPS global positioning system
GRP glass-reinforced plastic
H height
HAN heavy aromatic naphtha
HCN hydrogen cyanide
HLB hydrophile–lipophile balance
HP high power battery
HV high volume
Hz hertz
ICM integrated crop management
ID internal diameter
IGR insect growth regulator
IPM integrated pest management
IRM insecticide resistance management
ISA International Standard atmosphere
K, k constant
kV kilovolt
L length
LAI leaf area index
LD_{50} median lethal dose
LERAP local environmental risk assessment for pesticides
LIDAR light detection and range
LOK lever operated knapsack (sprayer)
LV low volume
MCPA 4-chloro-*o*-tolyloxyacetic acid
MRL maximum residue level
MV medium volume
N, n number of droplets
NMD number median diameter
NPV nuclear polyhedrosis virus
OES occupational exposure standard
P particle parameter
PDS pesticide dose simulator
PIC prior informed consent
PMS particle measuring system
PPE personal protection equipment
PRV pressure-regulating valve
PTFE polytetrafluoroethylene
p.t.o power take-off (tractor)
PVC polyvinyl chloride
Q application rate (litre/ha)
q application rate (litre/m^2)
Q_a volume of air
Q_f quantity of spray per load
q_n throughput of nozzle
Q_t volume applied per minute
rev revolution
r.p.m. revolutions per minute
S swath
s distance droplet travels
SC suspension concentrate
SP single power battery
SMV spray management values
SR stability ratio
T temperature
T_r time per loading and turning
T_w turn time at end of row
TDR turndown ratio
TER toxicity exposure ratio
U, u wind speed
UBZ unsprayed buffer zone
UCR unit canopy row
ULV ultralow volume
UR unsulfonated residue
UV ultraviolet light
V velocity
V_f velocity of sprayer while ferrying
V_s velocity of sprayer while spraying
VAD volume average diameter
VLV very low volume
VMD volume median diameter
VRU variable restrictor unit
W width
w angular velocity
WG water-dispersible granule
WHO World Health Organization
WP wettable powder
γ surface tension
η viscosity of air
ρ_a density of air
ρ_d density of droplet
$<$ is less than
$>$ is greater than

Multidisciplinary nature of pesticide application.

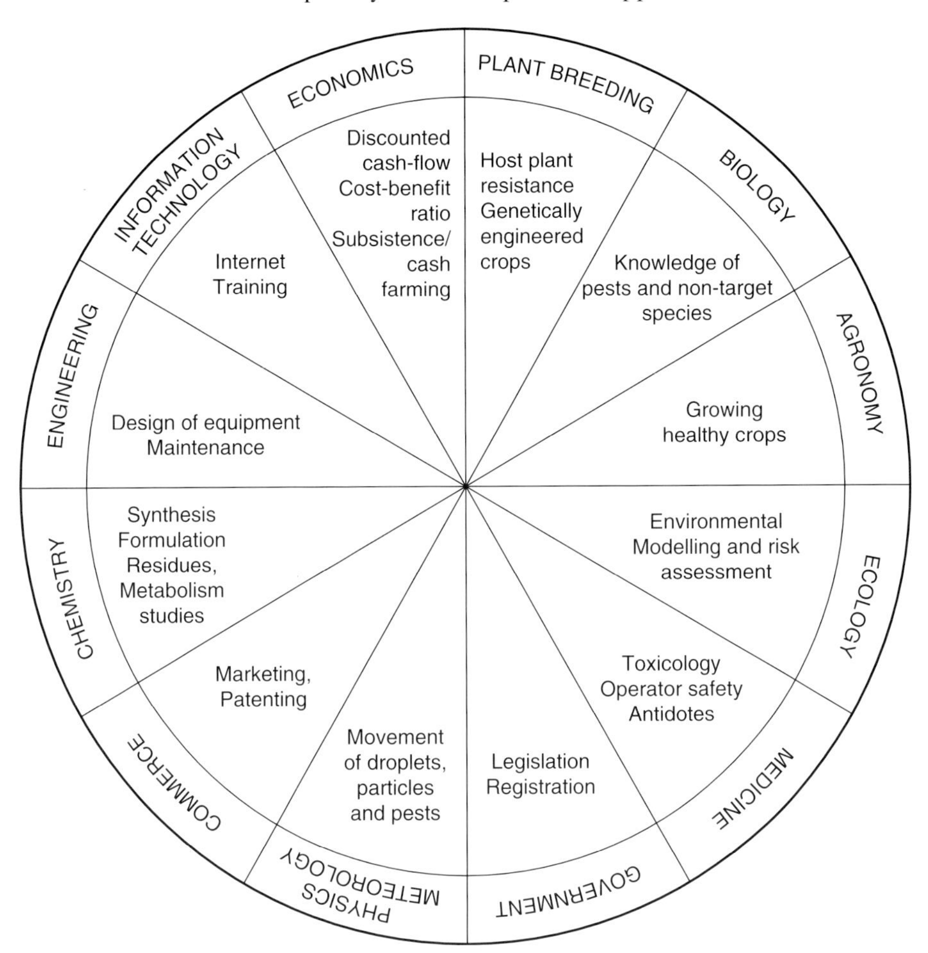

1

Chemical control in integrated pest management

Introduction

Over the last decade, global demand for pesticides has continued to grow (Fig. 1.1), although in Europe in particular some governments have adopted policies to limit their use. The restrictions have been in response to public perception of the risks associated with pesticide use in terms of residues in food and adverse effects on the environment. The perception is based erroneously on three false premises (Van Emden and Peakall, 1996) that good crops were obtained in an ideal pre-pesticide era, that chemicals like pesticides never occur in nature, and thirdly that these unnatural pesticides are causing an increase in cancer. In practice, plants contain many chemicals which are highly toxic. For example cyanide in cassava has to be removed by careful food preparation.

Without modern technology (including the use of pesticides) tripling world crop yields between 1960 and 1992, an additional 25–30 million square kilometres of land would have had to be cultivated with low-yield crops to feed the increased human population (Avery, 1997). Clearly, use of pesticides plays an important role in optimising yields. Modern technology is changing and many of the older pesticides, such the persistent organochlorine insecticides, are no longer registered for use as newer more active or selective chemicals take their place. At the same time, the agrochemical industry has invested in biotechnology and seed companies to exploit use of transgenic crops.

However, the growing of genetically modified crops has also aroused considerable public concern (Hill, 1998) and demands for legislation to control their use. While in many cases the transgenic crop is marketed on the basis that less pesticide will be used, other transgenic crops are associated with the application of particular herbicides, notably glyphosate used with 'Roundup Ready' crops. At present, single gene transfer often provides resistance to only one type of pest, thus the gene for *Bacillus thuringiensis* (Bt) toxin is effective against certain lepidopterous pests, but other insect groups may still have an

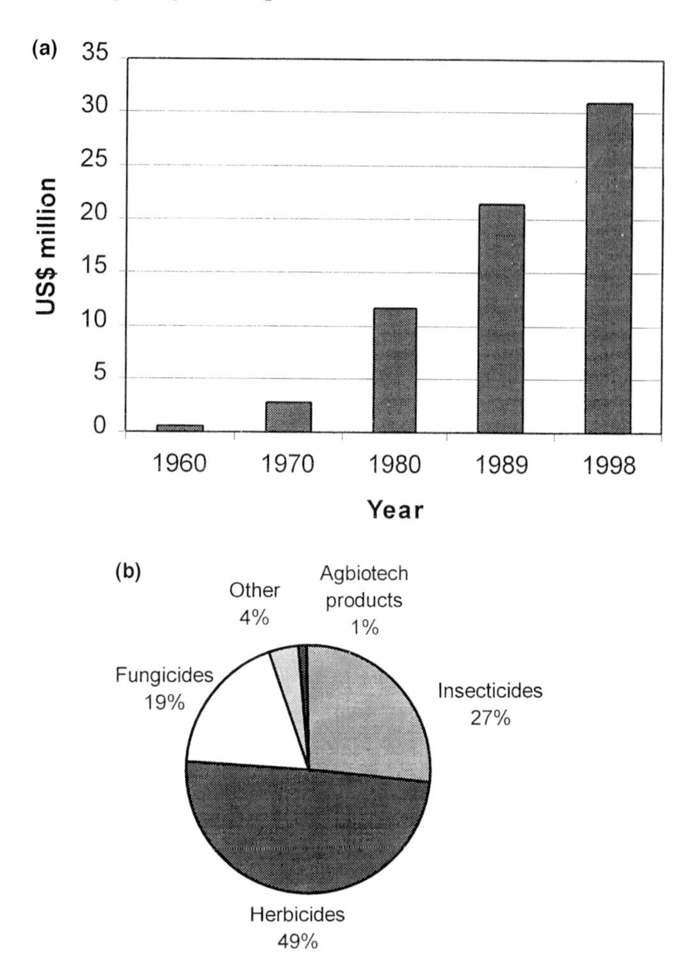

Fig. 1.1 (a) Increase in World Market for pesticides in US$ billion. (b) 1998 World Market for pesticides showing proportions of different types.

adverse effect on a crop and require an insecticide treatment (Hilder and Boulter, 1999). Furthermore, it has been quickly appreciated that pests resistant to the toxin in transgenic plants can be selected as occurs with overuse of a chemical pesticide, so the new varieties have been introduced with insecticide resistance management strategies (Merritt, 1998). The planting of genetically modified plants is therefore similar to use of new varieties from traditional plant breeding, and in relation to pest management their availability provides another tool to be integrated in the cropping programme.

Despite the criticisms of pesticide use, farmers will continue to need to apply them, because chemical control remains the most cost effective and rapid way of combatting the effects of weed competition and crop loss due to pathogens and insect pests. Our knowledge of the chemistry and suitability of an increasingly wide range of pesticides can now provide a more rational approach to their use and avoid the adverse outcomes associated with extensive use of the persistent organochlorines and the highly toxic organophosphate insecticides. International efforts have improved registration, and pesticides now commercially available have been rigorously evaluated with greater harmonisation of test procedures. Unfortunately, in many countries, especially in the less developed areas, farmers have inadequate training and too often use the least expensive pesticide, irrespective of its suitability for the pest situation; it is also frequently highly toxic, but the farmers do not have the appropriate protective clothing. In consequence, farmers in some areas have applied too many pesticide treatments and suffered economically and with poor health.

Modern farming practices have more intensive production of relatively few crops over large areas, while more traditional farming practices in tropical countries have a sequence of crops that provide a continual supply of food for polyphagous pests. Both these farming systems provide environments for pest populations to increase to such an extent that crop losses will occur unless control measures are implemented. Whereas these losses can be extremely serious and can result in total loss of a crop in some fields, for example the effect of an invasion of locusts or armyworms, the extent of damage is usually far less due to the intervention of natural enemies.

Considerable efforts have been put into training by means of farmer field schools, especially in relation to lowland irrigated rice production in South-East Asia in an attempt to get farmers to recognise the importance of natural enemies. The difficulty for the farmer is knowing when a pest population has reached a level at which economic damage will occur, so that preventative action can be taken. This decision should take into account the presence of natural enemies, but sampling for these can be quite time-consuming. Conservation of natural enemies is crucial in minimising the need for any chemical control, especially in the early vegetative stages of crop development. Areas with alfalfa or other fodder crops may provide a refuge for natural enemies, thus in Egypt berseem clover assists the overwintering survival of lacewings which are important predators of cotton pests. However, the farmer will need a pesticide when quick action must be taken to avoid economic crop loss. Various methods of assessing pest populations are used to assist farmers to determine when a pesticide may be applied as part of an integrated pest management programme.

Integrated pest management (IPM) utilises different control tactics (Fig. 1.2) in a harmonious manner to avoid, as far as possible, undesirable side effects on the environment. To many this means avoiding the use of any chemical pesticide and growing crops organically, but in many cases such a system is not sustainable where high yields are required. In some situations, the public will pay a premium for organic produce, but yields and quality are generally lower in comparison with crops receiving minimal intervention with chemical control.

Weeds are frequently the most important factor during crop establishment

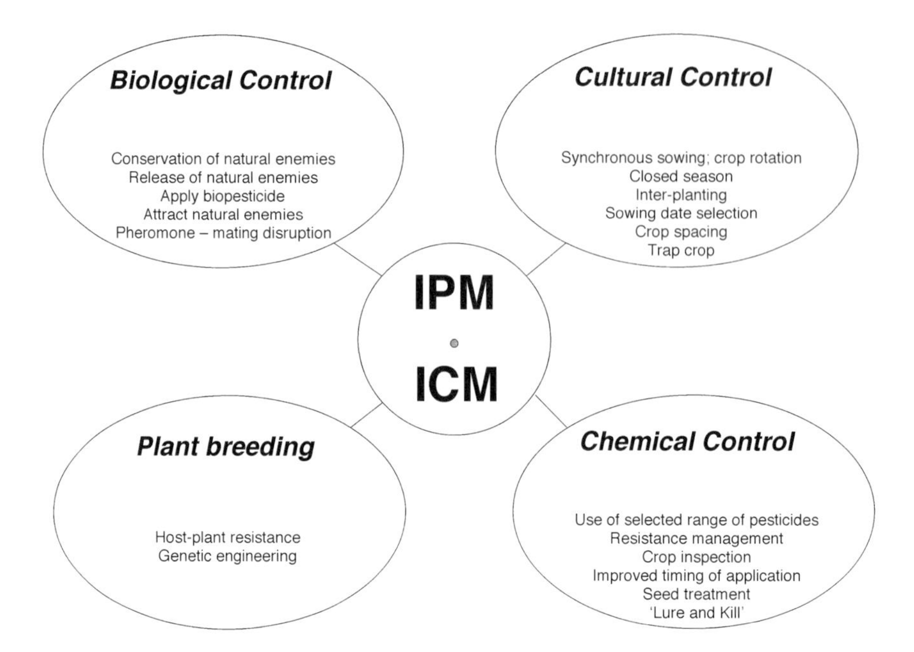

Fig. 1.2 IPM/ICM – the need to integrate different techniques.

at a time when demands for farm labour are high. Traditional hand weeding is very labour intensive and often not very effective, while general disturbance of soil by cultivation can increase erosion of some soils. Virtual weed-free conditions are possible with the range of herbicides now available, and on some well structured soils it is no longer necessary to plough every year as seed can be direct drilled after applying a broad action herbicide that is inactivated on contact with the soil.

Herbicide use has increased most where labour costs are high, there is a peak labour demand, or where mechanical hoeing will cause damage to the young crop. In conjunction with other agronomic practices such as tie ridging and planting along contours, herbicide use can reduce soil erosion by minimizing soil disturbance. Improved row weeding either by hand hoeing or by application of a herbicide increased yields by up to 35% in West Africa (Carson, 1987). With changes to direct seeding of rice and other factors, there is therefore an expectation that herbicide usage will increase in many crops in the tropics where traditional labour is no longer readily available for hand weeding or hoeing.

Wherever possible, farmers will select disease-resistant cultivars to reduce the need for fungicide treatments, but in some situations the farmer will continue to grow varieties which are susceptible to particular pathogens because of other qualities, such as taste and yield. The extensive damage to potato crops due to *Phytophthora infestans* that led to the Irish famine can be avoided by careful use of fungicides. The risk of selecting strains resistant to the fungicide can be reduced if the number of applications is restricted by monitoring climatic conditions so that treatments can be timed to coincide with periods favourable to the pathogen. Field application of fungicide will often improve fruit quality at harvest and allow longer storage.

The visibility of an insect is in no way related to the amount of damage and economic loss that can occur. Often farmers react to the presence of a low population of insects and may fail to distinguish between pest and beneficial species. The intervention of predators and parasitoids will often suppress a pest population such that economic damage is avoided. Thus, precipitate action with insecticides, especially those with a broad spectrum of activity, often disrupts this biological control too early in the crop, and in the absence of natural enemies, pest populations can increase dramatically. Furthermore, plants have evolved to withstand considerable damage due to insects by compensatory growth and production of chemicals toxic to the pests. Thus in integrated pest management programmes (Matthews, 1984; Van Emden and Peakall, 1996) pesticide use should always be confined to when a pest population has exceeded an economic threshold. The difficulty for the farmer is knowing when that economic threshold has been reached and then being able to take rapid action with minimal disruption of beneficial insects.

Pesticides

Thirty years ago Smith (1970) pointed out that despite intensive research into alternative methods of controlling pests, pathogens and weeds, pesticides remain our most powerful tool in pest management. This is particularly true when rapid action is needed. Southwood (1977) stressed the need to conserve pesticides as a valuable resource and reduce the the amount of chemical applied and the number of applications, to decrease the selection pressure for resistance, prolong the useful life of each pesticide and reduce environmental contamination. Over twenty years later, these comments remain true, even though much emphasis has been given to the development of transgenic crops. Pesticides will therefore continue to be an important part of integrated pest management programmes. There is, however, a greater realisation that pest management is only part of the wider requirement of integrated crop management, as investment in controlling pests can only be economic if there are sufficiently high potential yields. In practice, those marketing the produce (the supermarkets and food processing companies), are having a greater influence on pesticide use by insisting on specific management programmes.

Integrated crop management

Before the widespread availability of chemical pesticides, farmers had to rely first and foremost on the selection of cultivars resistant to pests and diseases. Unfortunately, not all resistant cultivars were acceptable in terms of the harvested produce due to bitter taste, poor yield or some other negative factor. Farmers therefore adopted various cultural techniques, including crop rotation, closed seasons with destruction of crop residues, intercropping and other practices, to mitigate pest damage. Biological control was also an important factor in supressing pest populations, but many of these basic techniques were forgotten due to the perceived convenience of applying chemical controls. The use of modern methods of manipulating genes in transgenic crops merely speeds up the process of selection of new crop cultivars. Whether they will provide a sustainable system of crop production has yet to be demonstrated. As indicated earlier, the introduction of the Bt toxin gene into plants will increase the mortality of certain lepidopterous pests, but it will not affect many other important insect pests and its effect on lepidoptera could be short-lived if insects resistant to Bt are selected.

Even partial plant resistance to a pest is important. As Van Emden (1972) pointed out, only half the dosage of the selective insecticide pirimicarb was required on plants with slight resistance to the cabbage aphid *Brevicoryne brassicae*. With the lower dosage of insecticide, the natural enemies were unaffected and controlled any of the pests that survived. In some crops, particularly those in glasshouses, the use of a low dosage of a non-persistent insecticide can be followed by release of natural enemies (GreatRex, 1998). A classic example is the application of resmethrin or the biopesticide containing the fungal pathogen *Verticillium lecani* to reduce whitefly *Trialeurodes vaporariorum* populations before the release of the parasitoid *Encarsia formosa*. This is important where light intensity and temperature are unfavourable to *Encarsia* early in the season (Parr *et al.*, 1976; Hussey and Scopes, 1985).

Area-wide IPM

Individual farmers can adopt an IPM programme, but increasingly many of the control tactics need to be implemented on a much larger scale. A farmer can choose a resistant cultivar, monitor the pest population and apply pesticides if the pest numbers reach economic significance, and subsequently destroy crop residues harbouring pests in the off-season. A good example has been in Central Africa, where cotton farmers grow a pubescent jassid resistant variety (Parnell *et al.*, 1949), time insecticide applications according to crop monitoring data (Tunstall and Matthews, 1961; Matthews and Tunstall, 1968), then uproot and destroy their cotton plants after harvest and bury crop residues by ploughing. Detailed recommendations were provided to farmers via a crop manual which has been updated frequently to reflect the availability of different varieties, and changes in insecticides. However, many tactics are only effective if all farmers within a defined area adopt them. A feature of the Central African programme has been a nationally accepted restricted list of recommended insecticides, discussed below in the next section. In Egypt, the use of pheromones was adopted on a national scale.

The selection of control techniques and their subsequent regulation throughout a given area or ecosystem, irrespective of county or national boundaries, is regarded as pest management. A distinction is made between the use of integrated control by individuals and pest management implemented co-operatively by everyone within the area. Pest management may give emphasis to one particular control technique, but in general there will be reliance on its harmonisation with other tactics. Furthermore, it must be a dynamic system requiring continual adjustment as information on the pest complex and control tactics increases. Modern information technology with computer databases, the internet and 'expert' systems can provide up-to-date information to farmers and their advisers.

Resistance to pesticides

The agrochemical industry has become more concerned about the impact of pesticide resistance and has recognised the role of IPM in reducing selection of resistant populations (Urech *et al.*, 1997). Efforts have been made to devise resistance management strategies, and to avoid disasters such as the cessation of cotton growing in parts of Mexico and Australia due to DDT resistance. Selection for resistance occurs if a particular chemical or chemical group is applied too frequently over a period to a given pest population. Initially, the impact of resistance was noted in glasshouses with a localised population, but resistance of red spider mite to organophosphates was also apparent on outdoor irrigated vegetable crops in the tropics where the same acaricide had been used throughout the year on different crops. Thus, resistance develops rapidly if most of a pest population is exposed to a specific pesticide, if the pest can multiply quickly, or if there is limited immigration of unexposed individuals. The user is tempted to increase either the dosage or the frequency of application, or both, if control measures are unsatisfactory, but this increases the selection for resistance.

Resistance selection is reduced if part of the pest population is on alternative host plants or other crops which are not treated with the same chemical. Thus, in introducing transgenic crops with the Bt toxin gene, a proportion of non-Bt crop is required as a refuge. Resistance to insecticides by the cotton bollworm *Helicoverpa armigera* has not been a serious problem in Africa, where large areas of maize and other host plants are untreated. However, in West Africa resistance to deltamethrin has now been reported, and this may be because farmers are using pyrethroids increasingly on vegetable crops in the same locality. Major problems of resistance in *H. armigera* have occurred in India and China where farmers have applied pyrethroids extensively with knapsack sprayers. Spray directed downwards from above the crop canopy was poorly deposited where the bollworms were feeding on buds, and in consquence lack of control led farmers to repeat treatments at frequent intervals. The continued exposure of larger larvae to pyrethroid deposits without significant mortality quickly led to resistant populations.

Unfortunately, the situation has been exacerbated by the availability of a number of generic insecticides with different trade names, but often based on the same or similar active ingredient; thus when the farmer thinks he may have changed to a different pesticide, in reality, the same pesticide is applied.

In Australia, the onset of pyrethroid resistance led to the introduction of a pragmatic resistance management strategy, which limited the application of any pyrethroid insecticide to a brief period each year irrespective of the crop. The original programmes have become more refined depending on the cropping practices in a given area, and generally there should be no more than two sequential sprays of any chemical group (Fig. 1.3) (Harris and Shaw, 1998). Where two bollworm species occur in the same area, a monoclonal antibody test has been used to check the percentage of *H. armigera* in the population which is likely to be resistant to pyrethroid insecticides.

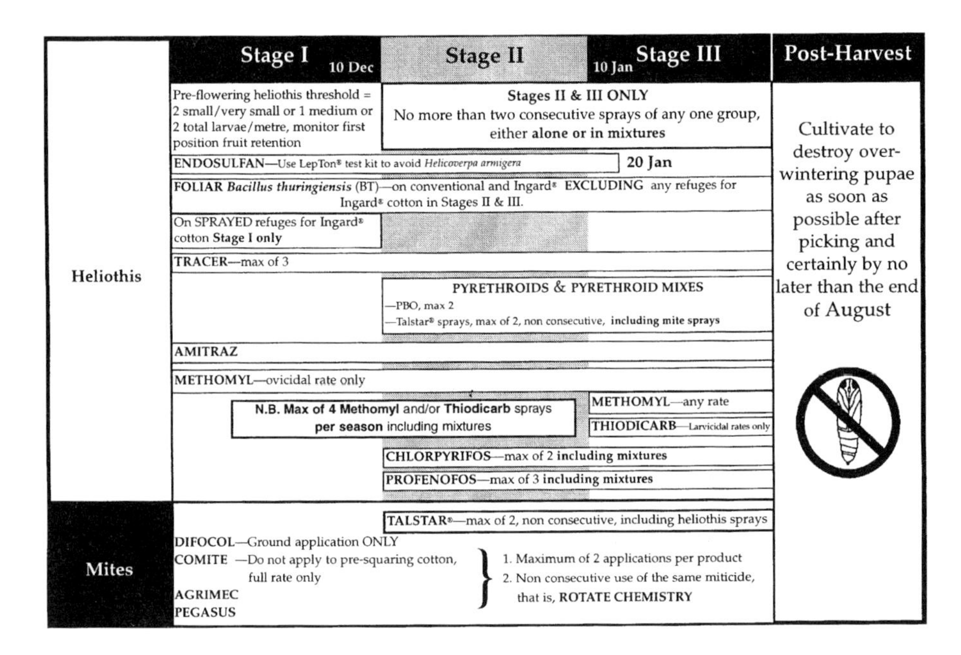

Fig. 1.3 Example of resistance management of insecticides on cotton in Australia.

Apart from the temporal control for pyrethroid insecticides, in Zimbabwe a spatial resistance management programme has been adopted for acaricides, whereby a particular type of acaricide may be used for only two seasons in one of three zones (Anon, 1998). The acaricides are rotated around the zones over a 6 year period (Fig. 1.4). In each of these resistance management programmes, the aim is to avoid a pest population being exposed too long to a particular pesticide. The alternative approach of using a mixture of pesticides, or adding a synergist such as piperonyl butoxide with pyrethroids, generally has limited value as insects are likely to become resistant to all the components

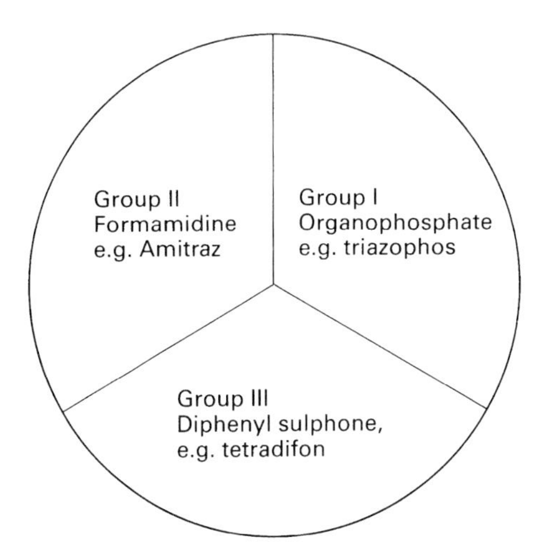

Fig. 1.4 Idealized acaricide rotation scheme based on system used in Zimbabwe (updated from Duncombe, 1973).

of the mixture. Whatever strategy is adopted, careful monitoring of levels of resistance in different localities is needed so that appropriate changes can be made to the strategy when needed.

Fungicide resistance

There is a similar problem with fungicides, in that if a chemical with a particular mode of action is used repeatedly, resistant strains of the fungi will be selected. Reduced dosages of fungicides showed significant selection for resistance to demethylation inhibitor (DMI) fungicides (Metcalfe *et al*., 1998), and that the strength of selection varied with fungicide, position of infection in the crop canopy and position on individual leaves. Clearly, with variations in deposits within a canopy and degradation of deposits, fungi will be exposed to low dosages of fungicide. Thus selection needs to be minimised by better disease forecasting so that fewer applications are required and those needed can be timed more accurately. Making sure the optimum dosage reaches where the infection is within the canopy is clearly most important. Assays have been developed for use in sensitivity monitoring schemes to help decide on future treatments; thus, Cooke *et al*. (1998) used a zoospore motility assay to test sensitivity of fluazinam to isolates of potato blight (*Phytophthora infestans*).

Herbicide resistance

Changes in the weed species often follow frequent use of a herbicide in one particular area, as the species tolerant to the chemical can grow without competition. This has resulted in the need for different and often more expensive herbicides or a combination of herbicides. Resistance to a particular herbicide may become evident more slowly than to insecticides or fungicides, because the generations of weeds overlap due to dormant seeds and there are fewer generations each year, but development of resistance to certain herbicides is already evident (Heap, 1997). In particular, there is resistance to the triazines, acetolactate synthase- or actyl CoA carboxylase-inhibitors due to mutated target sites (Schmidt, 1997). Some grass weeds have multiple resistance to herbicides with different modes of action. As an example, resistance of black grass (*Alopecurus myosuroides*), first detected in 1982, now affects over 700 farms in the UK (Moss *et al.*, 1999). Similarly, resistant wild oats was found on 65 farms. The problem has shown up on farms with many years of continuous winter wheat production (Orson and Harris, 1997).

Timing of spray application

One of the major problems of using pesticides is knowing in advance what pesticide and how much of it will be required during a season. To facilitate forward planning some farmers may prefer a prophylactic or fixed calendar schedule approach, but to minimise pesticide usage it is preferable to restrict treatments and only apply them when crop monitoring indicates a definite need. Forecasting pest incidence is an important means of improving the efficiency of timing applications, but is not always very accurate due to variations in weather conditions and survival of a pest population from the previous season. However, growers of sugarbeet in the UK havc benefited from the virus yellows warning scheme (Dewar, 1994). Modelling of the incidence of virus yellows has shown that over the last decade, up to five severe epidemics could have occurred since the major epidemic in 1974 (Fig. 1.5) if improved pest management practices had not been adopted (Werker *et al.*, 1998). Short term prediction of the potential for a disease outbreak based on weather forecasts can be useful for some diseases, for example where the temperature has to exceed a certain minimum coincident with high humidity and/or leaf wetness. Mini-meteorological stations can be set up to measure the conditions in crops sensitive to certain pathogens.

Economic thresholds

Ideally conservation of natural enemies would reduce the need for farmers to use any insecticides, but where climatic conditions and cropping practices result in a build-up of pest populations, quick action is needed to prevent economic losses. The actual loss of a crop will depend on when the pest infestation occurs during crop development and its severity. Often a crop can sustain some pest damage if there is sufficient time for plants to respond and compensate for the damage. The problem for farmers is deciding when action

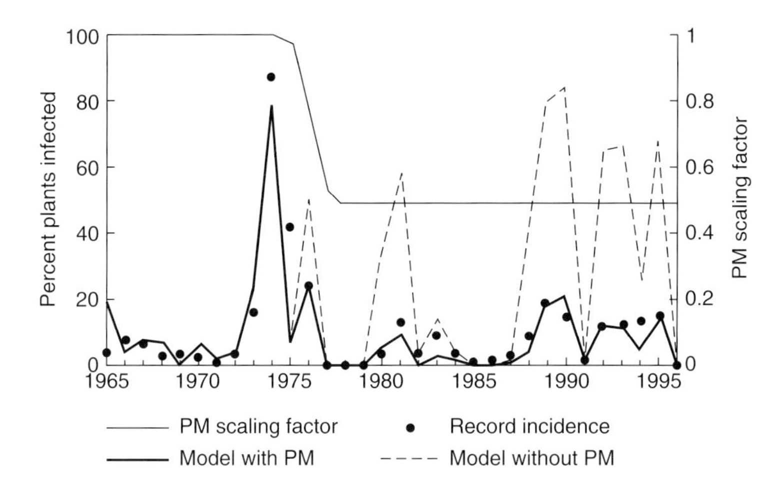

Fig. 1.5 Incidence of infected sugar beet and predicted levels as shown by model to indicate impact of integrated pest management.

has to be taken. One aspect of IPM is to use an economic threshold, defined as the population density at which control measures should be applied to prevent an increasing pest population from reaching the economic injury level. This economic injury level is the lowest population density that will cause economic damage (Stern, 1966; Onstad, 1987; Pedigo *et al.*, 1986). Changes in the market prices of crops make it very difficult to be precise about economic thresholds, so based on past experience farmers may have to follow a more pragmatic 'action threshold'. In some countries, farmers can employ independent crop consultants who will inspect fields and advise when chemical control is needed. However, in most situations it is the farmer who has to decide, so simple techniques of monitoring pest populations and/or damage are needed if the number of chemical treatments is to be minimised.

Timing of spray applications on cotton in relation to pest populations has been possible by using sequential sampling methods to reduce the time needed examining plants in the field (Fig. 1.6). The system allows a decision to spray if the population exceeds a set threshold even if the whole field has not been sampled, but generally requires sampling to continue if low populations are present. To simplify the crop monitoring, pegboards were developed (Beeden, 1972; Matthews, 1996a), the design of which has been adapted in different countries according to which pests are dominant and whether sampling considers the presence or absence of natural enemies. While it is important to avoid a spray treatment if large numbers of predators, such as lacewings, are present, natural enemies are generally less easy to detect.

In assessing whether to spray cotton, scouting for bollworm eggs (Fig. 1.7) has been advocated, as it is important to control first instar larvae, between the time eggs hatch and when larvae enter buds, to minimise the dosage of

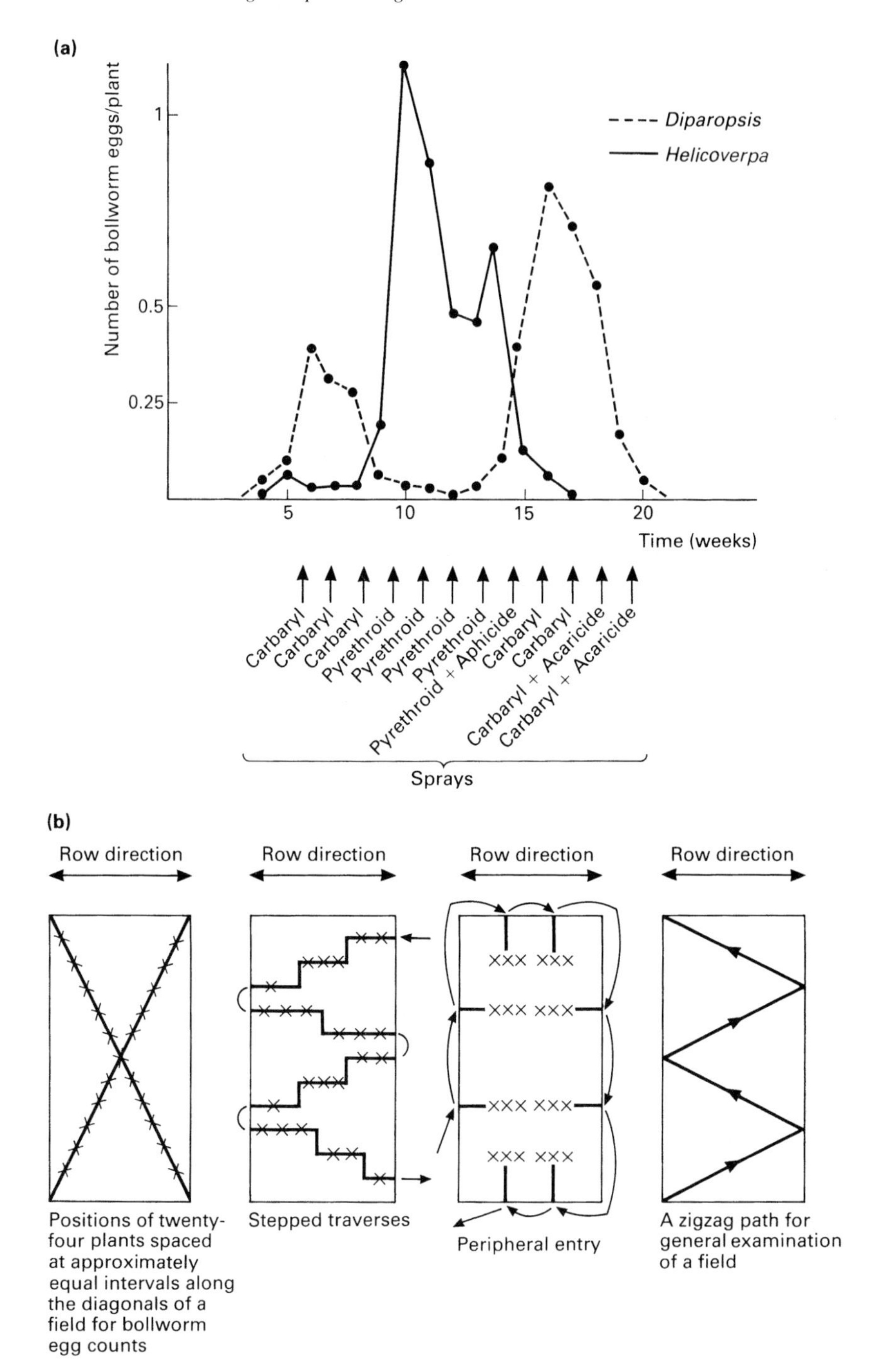

Fig. 1.6 (a) Timing of spray treatments on cotton, based on crop monitoring of bollworm eggs in Central Africa (updated from Matthews and Tunstall, 1968). (b) Sampling schemes for eggs in a cotton crop.

Fig. 1.7 Pegboard for small scale cotton farmer to record insect pests.

insecticide required. Once bollworm larvae are inside buds or bolls they are well protected from insecticide deposits, and to kill larger larvae requires a much larger amount of chemical. This same principle applies to most pests that attack the fruit and stems of crop plants. However, those advocating biological control prefer a delay until larvae are seen, as some eggs may not be viable or could be parasitised. This dilemma, whether to spray or wait, emphasises the importance of research in a particular area to assess the extent of biological control at different stages of crop development. Generally, if the 'action threshold' has been set correctly, insecticide is applied only when a pest infestation is no longer checked by natural controls and intervention is essential to avoid crop loss.

Other sampling systems have been devised depending on the crop and pest. Pheromone traps provide a selective and effective way of sampling low pest densities to determine whether an infestation is likely. At higher pest populations the trap data are less reliable, but trap use only indicates when pests are active and crops need to be monitored. Similar sticky traps, or traps with a food attractant, may be more appropriate for certain pests. Some scientists have suggested timing of treatments based on crop damage assessments, but it is likely that it is too late to justify an insecticide treatment when damage is observable. As an example, control of an insect vector of a viral disease requires action at very low pest populations, before the symptoms of disease can be seen, although reduction of further spread of an infection may be checked by a late treatment.

Application sites and placement

A key issue is the risk of 'spray drift' beyond the field boundary, especially if there is another crop susceptible to a herbicide, there is surface water or a ditch which could be contaminated by the pesticide (Croxford, 1998), or there are bees downwind of insecticide-treated fields. Protection of hedgerows around fields is also of crucial importance to avoid contaminating the habitat of important populations of natural enemies. Field boundaries are also important habitats for game birds and conservation of other wildlife (Oliver-Bellasis and Southerton, 1986; Forster and Rothert, 1998; Boatman, 1998) (Fig. 1.8). To minimise the risk of drift, some countries now have a legal requirement for a 'no-spray' or 'buffer' zone around fields or at least along the downwind edge of a field and to protect surface water (Van de Zande *et al.*, 2000) (Fig. 1.9). The width of the untreated buffer zone really depends on the spray droplet spectra, the height of release of the spray and wind conditions. To simplify the procedure some countries have fixed distances downwind from the field boundary; thus in the UK the unsprayed buffer zone (UBZ) has been set at 5 metres between the side of a ditch or watercourse and the edge of an arable crop, and 18 metres in orchards. However, following concern about the

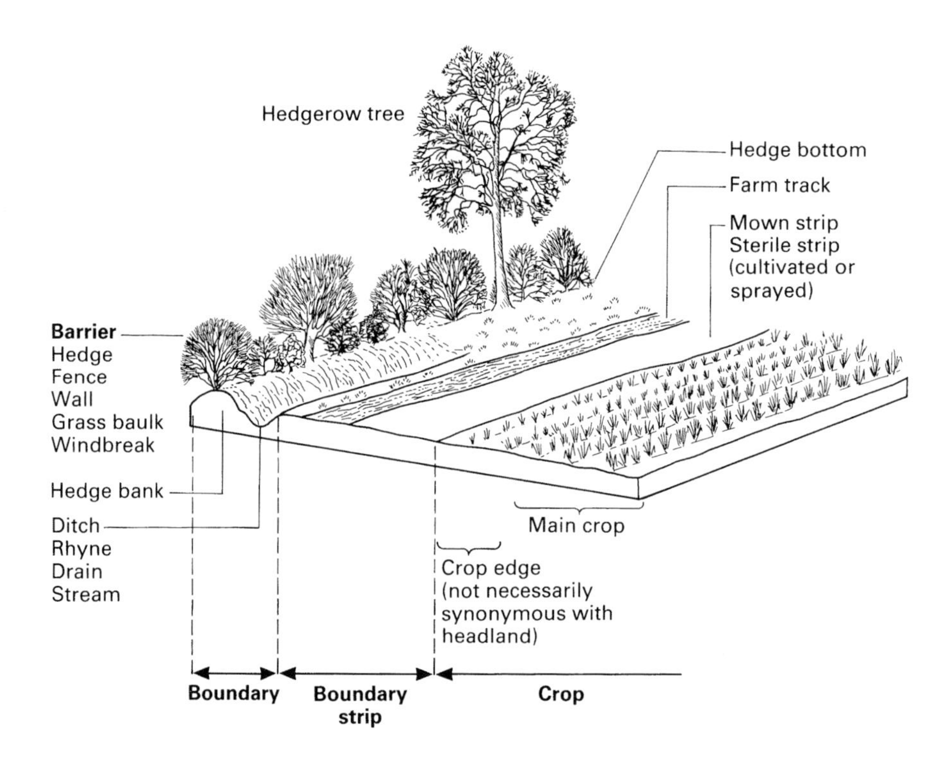

Fig. 1.8 Principal components of arable field margin (from Greaves and Marshall, 1987).

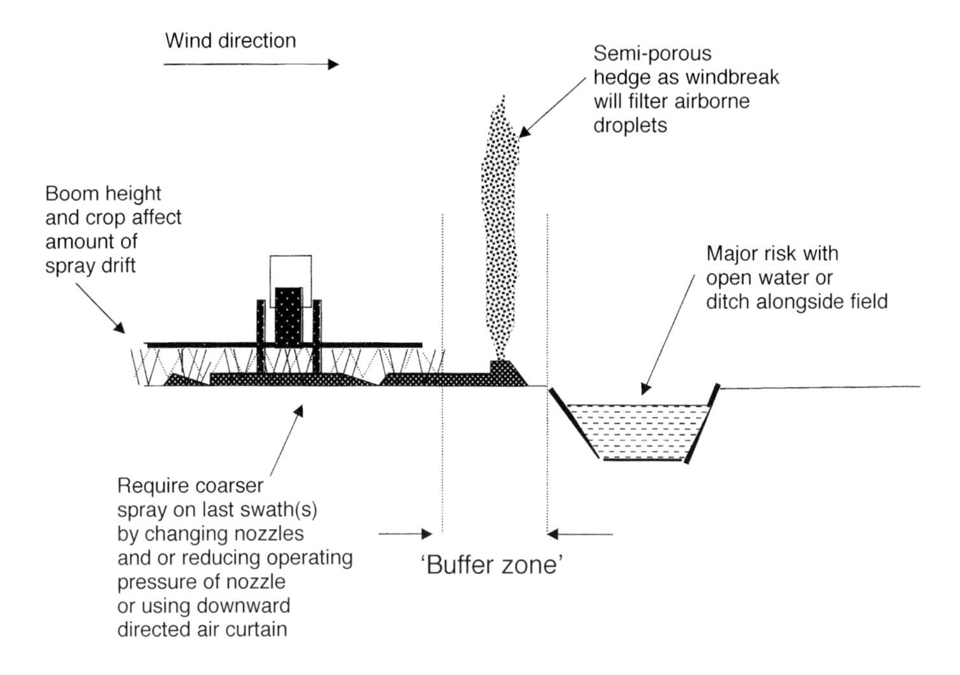

Fig. 1.9 Untreated buffer zone.

amount of crop area affected in the UK (Orson, 1998), this system has been modified by the introduction of a Local Environmental Risk Assessment for Pesticides (LERAP) where the UBZ can be reduced for ground based arable spray equipment from 5 metres to effectvely 1 metre from the top of the bank of a ditch if the spray method and equipment meets LERAP approval (Gilbert, 2000) (see also Chapters 4, 5 and 7). Longley and Southerton (1997) and Longley *et al.* (1997) examined the extent of drift into field boundaries and hedgerows and Miller *et al.* (2000) showed that differences in the plant structure will affect the extent of drift at field margins (Fig. 1.10). An established vegetative strip will significantly decrease drift compared with a cut stubble due to the filtration of the droplets (Miller, 1999). A grassed buffer strip, especially if sown perpendicular to the slope, will also restrict run-off of pesticide (Patty *et al.*, 1997). Heijne (2000) reported the use of artificial netting as an alternative to a hedge, which will take time to get established. The height and porosity of the netting determines the extent to which drift is reduced.

Crop monitoring for a pest may indicate a particular focus of infestation in a crop, and permit localised treatment to reduce the spread of the pest and avoid the cost of a treatment to the whole area. Some infestations may be initially at the edges of fields, e.g. pink bollworm may spread from villages if stalks have been stored for fuel. Many windborne insects collect on the lee side of hedges (Lewis, 1965) or other topographical feature. An isolated tree in a field can affect the initial distribution of red spider mites due to its effect on air

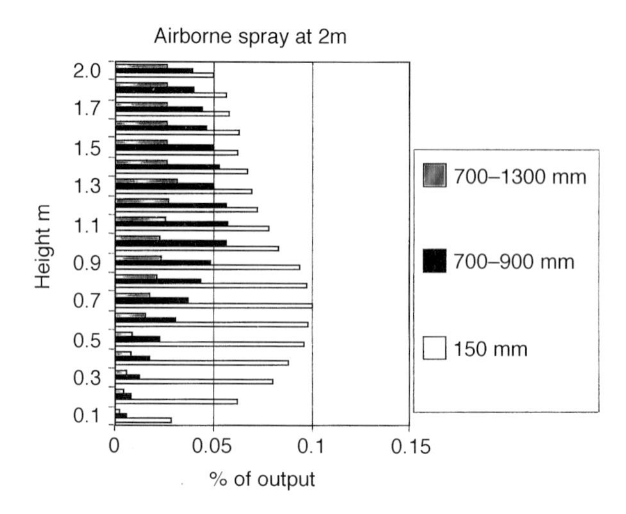

Fig. 1.10 Effect of vegetation in filtering spray droplets at field margin. Airborne drift at 2 m downwind recorded at different heights showing reduction if vegetation is 700 mm or higher.

movement across a field. If detected early, the initial patches of infestation can be treated with a knapsack sprayer to avoid treating the whole field.

Spatial differences within a field or crop canopy can also be exploited by using localised treatments to allow greater survival of natural enemies. Discrete droplets leaving areas untreated are generally more favourable than high volume treatments where all surfaces get wetted, when natural enemies inevitably are exposed to pesticides. Theoretically, some treatments can be localised by using an electrostatically charged spray, particularly to avoid pesticide fall-out on the soil and adversely affecting soil-inhabiting predators. However, this approach has not been exploited. Soil application of a systemic insecticide as granules or seed treatment will generally control sucking pests with less risk of direct effects on their natural enemies.

Conservation of natural enemies is especially important in perennial crops, so pesticide treatments may need to be separated in time. Thus, treatment of strips through an orchard with a non-persistent insecticide provides control of the pest, and natural enemies can re-establish from the untreated sections of the orchard which are treated several days later.

The importance of restricting pesticides as far as possible to the actual target is fundamental to good pest management and is considered in more detail in subsequent chapters.

2

Targets for pesticide deposition

Concern about the presence of pesticides in the environment has increased worldwide. As indicated in Chapter 1, regulatory authorities have introduced new controls to limit 'drift' of sprays outside treated areas, in particular by the introduction of 'buffer' or 'no-spray' zones. These regulations are aimed specifically at reducing the deposition of droplets immediately adjacent downwind of a treated field onto water surfaces and ditches that may have water flowing in them at some time during the year. However, a proportion of a spray in very small airborne droplets may be carried by air currents over much greater distances, sometimes several kilometres from the the site of application. Residues of persistent pesticides, such as the organochlorine insecticides, have been detected in meat and milk of cows grazing in pastures in the vicinity of treated areas. Such downwind drift can even occur after deposition, if the pesticide formulation is too volatile and the vapour is transported downwind. This has been most noticeable when certain herbicides, such as 2,4-D, were applied when susceptible plants showed signs of phytotoxicity. Changes to less volatile esters of the active ingredient and improved formulation have significantly reduced this particular problem.

To minimise drift and contamination of water, many farmers have applied coarser sprays, but this can lead to less efficient use of some pesticides. Larger droplets in coarse sprays may provide inadequate coverage to control pests. Much depends on the volume of spray applied, the properties of the pesticide, and the formulation in determining whether large droplets are collected on the foliage and whether subsequent redistribution of the pesticide compensates for inadequate coverage. Large droplets may bounce off difficult to wet foliage or fall between leaves to the soil surface. Increasing the volume, as many users do, may lead to coalescence of the droplets. The volume of liquid that can be retained on a leaf surface is limited, so once the leaf surface has been wetted, surplus liquid drips down to lower leaves and thence to the soil. Less liquid is retained on the leaf surface once 'run-off' has started, so the deposit achieved is proportional to the spray concentration, but independent of the volume applied. The amount of surfactant in the spray formulation or adjuvant mixed with a spray will affect spray retention, but run-off may start when as little as 100 litres/ha is applied to a low sparsely leaved crop (Johnstone, 1973a). A tree crop with dense foliage retains more spray, and in Australia run-off was significantly greater when more than 1500 litres/ha was applied (Cunningham and Harden, 1999).

As much as one-third of the spray applied to a crop may be lost to the soil at the time of application. This loss of pesticide within a treated field was referred to as 'endodrift' by Himel (1974) to differentiate it from the 'exodrift' outside the treated area. This pesticide may be adsorbed on the soil particles or subjected to microbial degradation, but certain chemicals are known to leach through the soil and may contaminate groundwater.

Pesticide deposited within the crop may be washed off later by rain, or in some cases by overhead irrigation. Some estimates have suggested that up to 80 per cent of the total pesticide applied to plants may eventually reach the soil (Courshee, 1960), where it can cause major changes in the populations of non-target species such as earthworms. Unfortunately, dosages are increased by some users to compensate for the losses due to drift, and farmers may repeat a treatment if rain falls soon after a spray has been applied.

Application of insecticides is very inefficient because much more is applied than the amount needed if it all reached the pests. Thus if 3×10^{-2} μg is required to kill an insect, only 30 mg need be applied to kill a population of 1 million insects, yet over 3000 times this amount is usually applied for effective control in the field (Brown, 1951). In practice, the foliage is the initial target for most insecticide applications so the efficiency is more like 30–40 per cent rather than the often quoted figures of less than 1 per cent. A similar level of efficiency applies to herbicides directed at weeds, but for soil surface applications, clearly most of the spray reaches its intended target, especially if a coarse spray is applied.

The volume of liquid in which a pesticide has to be applied is seldom indicated on the label except in general terms. This allows the farmer some flexibility in choosing an appropriate nozzle in relation to his equipment. However, in response to the concern about spray drift, the agrochemical industry is increasingly recommending the quality of the spray that should be applied. Most nozzles produce a range of droplet sizes, the smallest droplets being those most prone to exodrift. Thus, where drift of a particular product is likely to cause problems downwind of a treated area, the manufacturer can recommend on the label a specific nozzle which produces few fine droplets by using a code to define spray type, angle and output at a given pressure. Alternatively, the spray quality can be specified.

The spray quality assessments are based on data obtained by measuring the droplet spectra obtained with different nozzles by using a laser system (see Chapter 4) . The original scheme (Doble *et al.*, 1985) has been modified (Fig. 2.1) (Southcombe *et al.*, 1997) so that each category is clearly defined by selected reference nozzles. The spray quality scheme has now been adopted in several European countries and the USA (Hewitt *et al.*, 2000). Using hydraulic nozzles, there will be some small droplets liable to drift, even if spray is applied with a nozzle with a coarse quality (Fig. 2.2). As there are many more types of nozzle available to farmers (see Chapter 5), the spray quality scheme is being extended by taking into consideration wind-tunnel measurements of downwind movement of droplets to provide a drift-index (Fig. 2.1b). In Germany, where sprayer testing is compulsory, there is now equipment with the ability to reduce drift by between 50% and more than 90% (Ganzelmeier and Rautmann, 2000).

There can be a conflict between optimising the spray quality for efficient application of a pesticide and endeavouring to minimise the risk of drift. Each situation has to be judged in relation to the target for the pesticides and meteorological conditions. Ideally there is an optimum droplet size (Himel, 1969b) or spectrum which gives the most effect coverage of the target with minimum contamination of the environment. Greater attention is needed because of the trend to using smaller volumes of spray. The cost of collecting and transporting water to fields is significant, especially if weather conditions limit the time available to treat large fields. In the tropics, the use of low

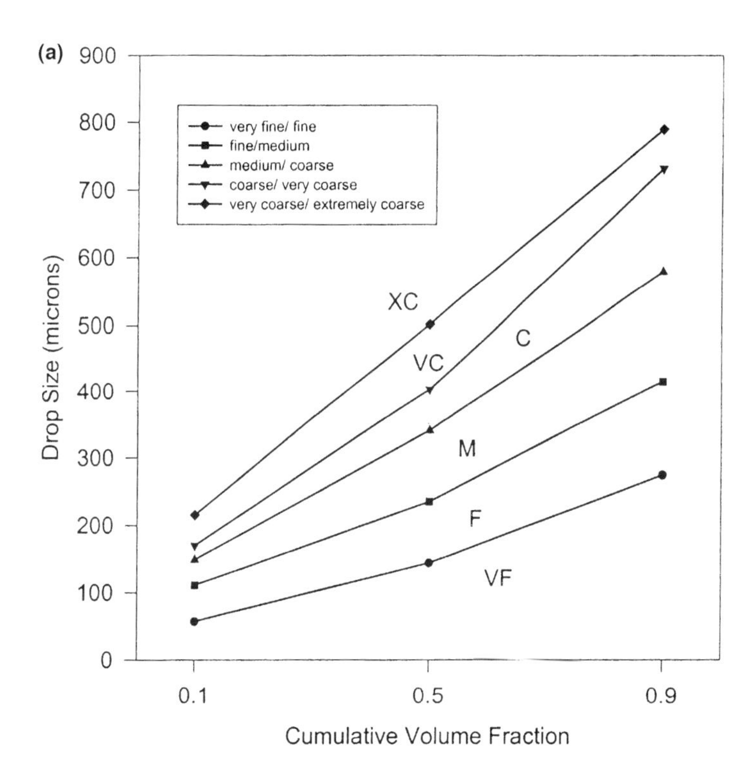

Fig. 2.1 (a) Spray quality chart for fan spray nozzles obtained by measuring droplet spectra of reference nozzles.

Fig. 2.1 (b) Measuring drift in a wind tunnel (photo: SRI).

volumes has been particularly important because the scarcity of water has been a major deterrent to farmers spraying to control their pests and weeds.

If pesticides are to be used more efficiently, the actual target needs to be defined in terms of both time and space. Furthermore, the proportion of emitted pesticide that reaches the target must be increased and in a form of deposit which is readily available to the pests. According to Hislop (1987), our objective is to place just sufficient of a selected active ingredient on the target to achieve a desired biological result with safety and economy. However, the process is quite complex (Fig. 2.3). Definition of the target requires a knowledge of the biology of the pest to determine at which stage it is most vulnerable to pesticides. Unfortunately, only a small proportion of a pest population may be at the most susceptible stage at any given time. Insects have several distinct stages during their life cycle; for example, adults, eggs and nymphs, or distinct larvae and pupae. These may not occupy the same habitat; for example, the larvae of mosquito which are vectors of malaria are aquatic while the adults are airborne and invade areas occupied by man. Similarly, with weeds, foliage may be affected by a herbicide while seeds can remain unaffected, enabling weeds to recolonise an area.

Difficulties such as these in defining the target have led to the use of persistent chemicals, but this has increased the risk of selecting populations resistant to a particular pesticide and then the risk of adverse effects in the environment. If users are to apply less persistent and more selective pesticides,

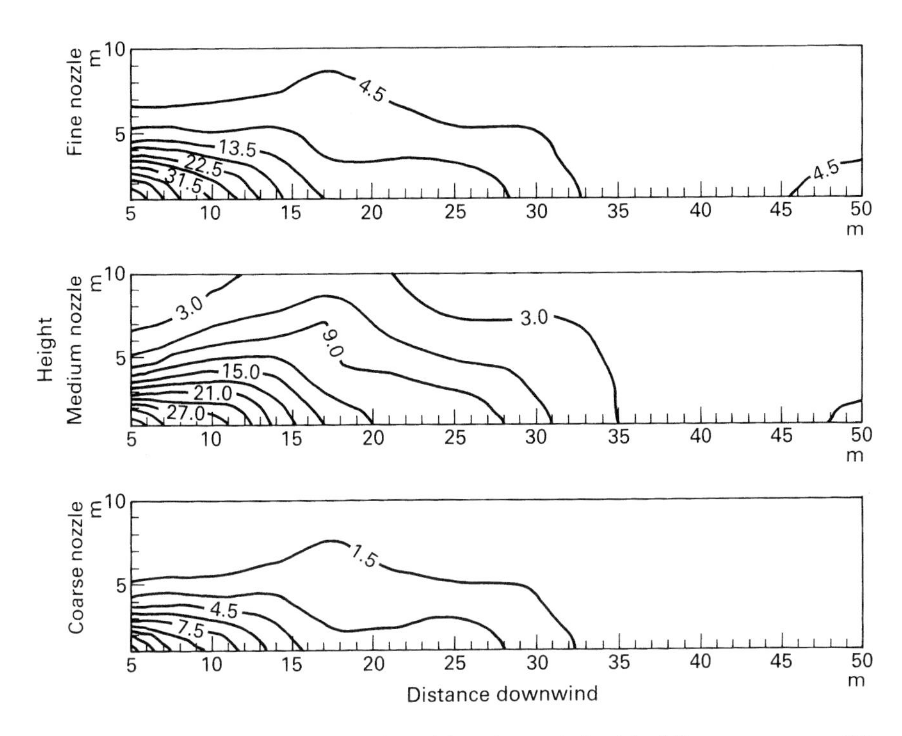

Fig. 2.2 An assessment of spray deposition downwind with different spray qualities (data supplied by C.S. Parkin).

more attention is needed to define the target and when an application is justified. Where different stages of a pest may be controlled chemically, it is important that different pesticides are used to reduce selection for resistance, and that this policy is adopted on an area-wide basis. Thus, against whiteflies with a wide range of host plants, control on horticultural crops needs to be integrated with control on other field crops such as cotton, and different insecticides used in a planned sequence. Similarly, against mosquitoes different pesticides should be applied as larvicides and adulticides.

Insect control

The concept of crop protection has aimed to reduce the population of the developmental stage of the pest directly responsible for damage within individual fields. Crop protection is most efficient when the pesticide is applied economically on a scale dictated by the area occupied by the pest and the urgency with which the pest population has to be controlled (Joyce, 1977). Control has been directed principally at the larval stage of many insect pests. This policy has been highly successful when treatments have been early to reduce the amount of larval feeding. If treatment is too late, not only is a

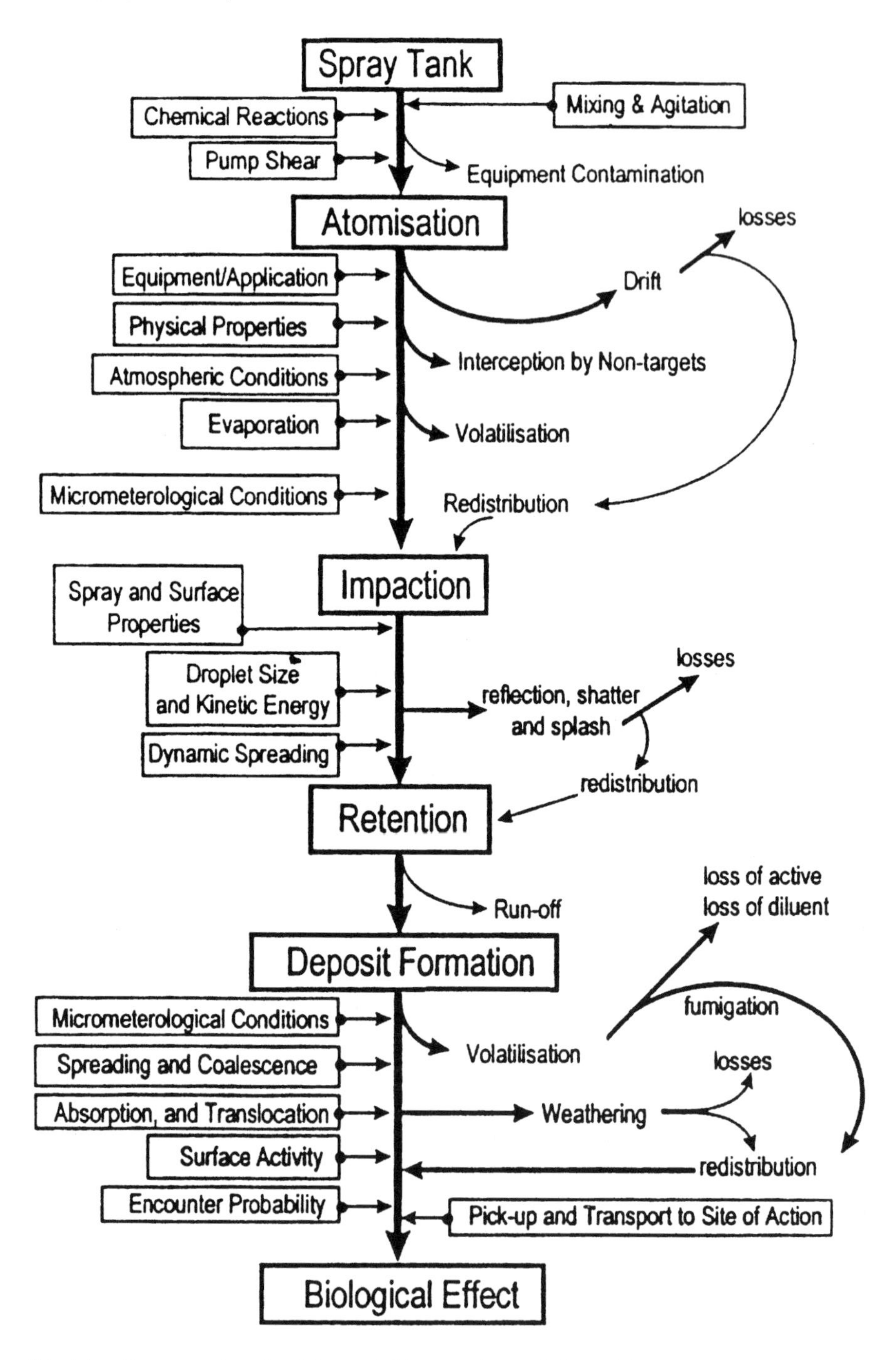

Fig. 2.3 Processes involved in pesticide transfer and deposition.

higher dose required to kill larger larvae, but also much of the damage may have already been done. Unfortunately, treatments directed at the larval stage may have little or no effect on the eggs, pupae and adults, and repeat treatments are often necessary as more larvae develop. Similarly, control of adults by spraying may result in 100 per cent mortality within a crop, but subsequent development of the immature stages provides more adults which can also have been derived by immigration. This has been well illustrated by attempts to control whiteflies (*Bemisia tabaci*), the nymphs of which are well protected from insecticide sprays because they are on the undersurface of leaves.

In a pest management programme, biological information must extend beyond a simple description of the life cycle to include data on the ecology of the pest. In particular, insect control requires an understanding of the movement of pests, between different host plants and within ecological areas. For a given pest species the target may vary according to

(1) the control strategy being adopted
(2) the type of pesticide being used
(3) the habitat of the pest
(4) the behaviour of the pest.

These factors are interrelated, but some examples of insect pests illustrate how particular targets can be defined.

Control strategy

Ideally locusts and other grasshoppers need to be controlled to prevent their immigration from breeding sites, but in many situations this has not proved to be possible, so protection of farmers' crops is also essential. In each case the target is the vegetation on which locusts are feeding. Ideally control is at the wingless immature hopper stage, but often adults also require control. The recommended technique is to apply droplets of 70–90 µm volume median diameter (VMD) which travel downwind and collect on the vertical surfaces of sparse desert vegetation (Figs 2.4 and 2.5) (Courshee, 1959; Symmons *et al.*, 1989). Johnstone (1991) selected the optimum droplet size for aerial applications in relation to wind speed and emission height. The aim is to minimise the amount of insecticide that is deposited on the ground. Courshee (1959) measured the efficiency of application on the biological target in relation to the amount emitted, and referred to the deposit per unit emission (DUE).

This technique is currently used with a range of insecticides that have been shown to be effective against locusts (FAO, 1998). Droplets with an optimum diameter of 75 µm have a volume of 221 picolitres, so the toxic dosage needs to be conveyed in the mean number of droplets likely to impact on locusts. Calculations of this type are needed to determine the volume and concentration of spray required. For logistic reasons, the recommended volume of application is 1 litre/ha. Sometimes a lower volume is effective, and if vegetation is more dense an increase in volume may be required to achieve sufficient droplets on the target. Concern about using insecticides over large areas has required environmental impact studies such as that reported by Tingle (1996) and Peveling *et al.* (1999a). The same principles apply in relation to the application of the mycoinsecticide *Metarhizium anisopliae* var *acridum* (Bateman, 1997; Hunter *et al.*, 1999) (see Chapter 17). This mycoinsecticide can be as effective as organophosphate sprays without threatening non-target arthropods (Peveling *et al.*, 1999b), an important factor when locusts are present in or near ecologically sensitive areas. To achieve the optimum droplet size, a rotary atomiser is recommended (see Chapters 8 and 13).

Hopper control is preferred as they are less mobile than swarms of adults and the infested area is relatively stationary for weeks at a time. An area of 11 000 km^2 infested with hopper bands could be treated with 35 000 litres of insecticide, whereas over 200 000 litres were needed to destroy about two-thirds of a swarm of *Schistocerca gregaria* covering 600 km^2. Forecasting and detection of locust outbreaks therefore remains essential to minimise the area over which control operations are needed.

Control of hoppers with persistent insecticides is possible by barrier spraying. This consists of using a series of parallel treated strips at right angles

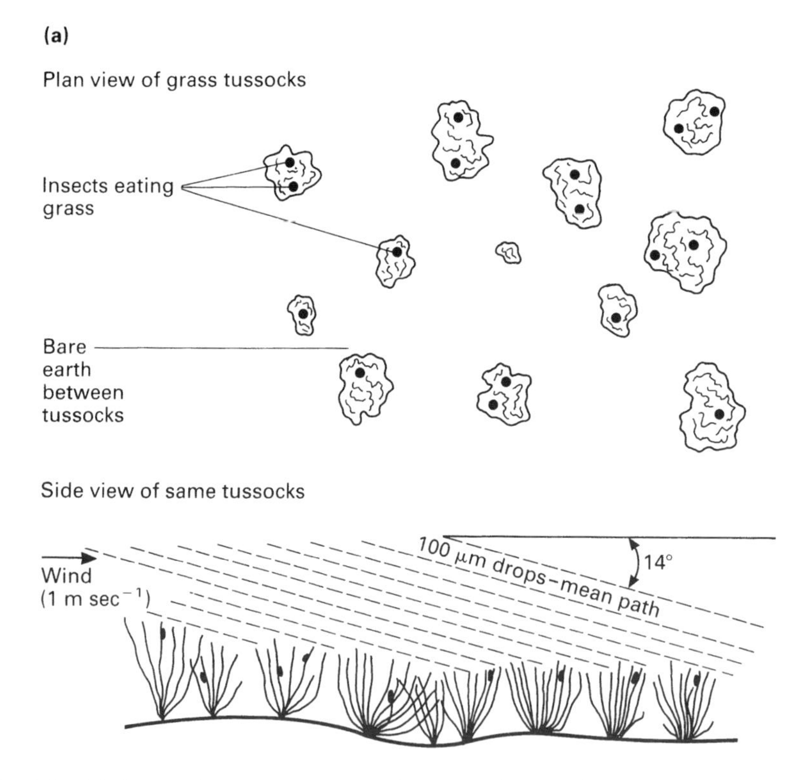

Fig. 2.4 (a) Downwind movement of droplets. (b) Track spacing used for locust hopper control with vehicle mounted sprayer. (c) Spinning disc sprayer ('ULVAmast') being used to control locusts.

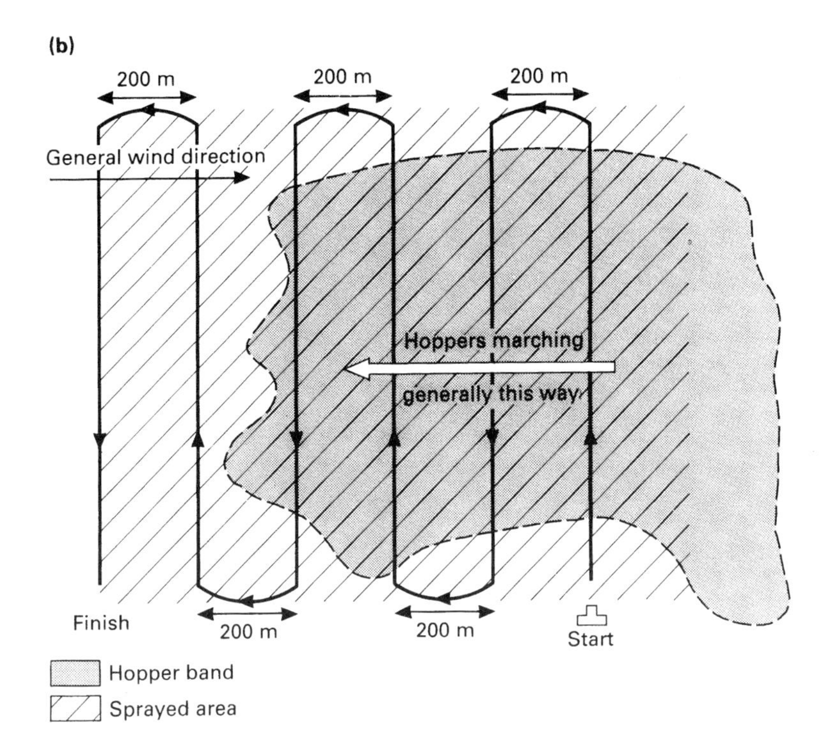
(b)
200 m
200 m
200 m
General wind direction
Hoppers marching
generally this way
Finish
200 m
200 m
Start
Hopper band
Sprayed area

(c)

Fig. 2.5 Aerial spraying of locust swarm (photo: Dick Brown).

to the wind and separated by an untreated area. The width of the treated strip and separation between strips depends on the mobility of the locust and the speed of action of the insecticide. When initially introduced against the desert locust, the accumulation of dieldrin allowed wide separation of treated barriers (Bennett and Symmons, 1972), but acylurea chitin inhibitor insecticides, such as diflubenzuron, have now been used (Cooper *et al.*, 1995). Coppen (1999) has used a simple model to optimise the use of sprayed barriers with these insecticides.

Type of pesticide

The mode of action of a pesticide can influence the selection of an application technique and timing of application An insecticide may be effective by contact, by ingestion (stomach poison) or by inhalation (fumigant effect). Similarly, fungicides and herbicides may have contact activity, or be effective within a plant by systemic activity upwards, or be translocated across leaves and in some cases, e.g. glyphosate, downwards into the rhizomes of grasses. Some pesticides have sufficient persistence that timing is less critical compared with other chemicals which break down very rapidly; however, the latter characteristic allows a pesticide to be applied closer to the time of harvesting a crop.

In the control of mosquitoes, persistent and non-persistent insecticides require different application techniques. One of the principal methods used to control domiciliary mosquitoes and interrupt malaria transmission has been the application of a persistent insecticide as a residual deposit on walls inside houses. Treatment of the latrines is particularly important to control certain species. Manually pumped compression sprayers are used for this. Persistent insecticides, especially pyrethroids, have also been applied to bednets by soaking (Rozendaall, 1989; WHO, 1989).

These techniques do not affect populations of mosquitoes away from the treated houses, so where large populations occur and transmission of a disease such as dengue needs to be interrupted, space sprays may be required. Less persistent insecticides are used in fogs (droplet size < 25 μm) at low dosages as the aim is to treat an area with droplets that remain airborne as long as possible. The optimum droplet size collected by mosquitoes is 5–15 μm (Mount, 1970) (Table 2.1). Equipment used to produce fogs is described in Chapter 11. Some chemicals, such as the natural pyrethrins, have an irritant effect which disturbs insects and causes them to fly. This is an advantage as flying insects collect more droplets than those at rest (Kennedy *et al.*, 1948).

Table 2.1 Percentage mortality of caged female *Aedes taeniorhynchus* with ULV non-thermal aerosol of technical malathion 92 m downwind (based on Mount, 1970)

Droplet diameter (μm)	Dosage[a] (kg a.i./ha)		
	0.005	0.01	0.02
6–8.0	38	100	100
8–11.0	38	100	100
11–14.0	38	98	100
13–23.0	18	52	84

[a] Based on 184 m swath.

Space treatments against mosquitoes are only effective if they are actively flying, so the best time is in the evening, especially when inversion conditions (see pp. 81–2) exist and wind velocity is low, so the spray cloud is not dispersed too rapidly. An insecticide of low mammalian toxicity is obviously needed when applications are to be close to human dwellings. An insecticide of low persistence applied with a low dosage is required as the aim is to have a short-term effect only on the population flying at the time of treatment. When this is done, it is unlikely that there would be any substantial effects on the aquatic insects or fish in seasonal wetlands (Jensen *et al.*, 1999)

Another approach to be integrated with the adulticiding is the application of a low toxicity larvicide. These may be applied to a water surface as a spray (large droplets) or as dry particulate granules or brickettes which disperse in the aquatic environment. The insecticide used for larviciding must be different from that used for adulticiding to reduce selection of insecticide resistant populations.

The contrast between residual deposits and space sprays is also evident in other situations. One example is the treatment of warehouses to prevent pests infesting stored produce. Residual treatments can be combined with fumigation of produce, but often populations of flying insects also need to be checked with a space treatment. Repetitive applications of a fog of a non-persistent chemical such as 0.4 per cent pyrethrin plus 2 per cent piperonyl butoxide at 50 ml/100 m^3 have been used.

Systemic chemicals and 'lure and kill' strategies

Systemic chemicals are redistributed in plants by upward movement, so ideally they are applied as granules in the soil or as a seed treatment. Sucking pests on leaves are controlled, provided there is sufficient soil moisture to facilitate uptake by the plant. A major advantage of a seed treatment is that very little of the pesticide is applied, and being localised it is less disruptive of non-target organisms. Treatments at planting will often protect young seedlings for up to six weeks, depending on the insecticide used and dosage applied. Crops prone to early season infestations of aphids, for example, may be treated prophylactically, especially if the insect is a virus vector. Applying such a treatment before knowing whether the pest will infest a crop is economic when there is a risk that subsequent weather conditions are liable to delay spray treatments after the pest has arrived and allow spread of the virus in the crop. One example is sugarbeet, where seed treated with imidacloprid provides good aphid control on the crop whereas spraying the undersurface of leaves for aphid control is very difficult.

In IPM programmes, instead of application of conventional insecticides, there is an increasingly important role for pheromones which can be used in mass disruption programmes or in combination with insecticides as a 'lure and kill' strategy. Various techniques are used to deploy the pheromone or other form of attractant, but it is often incorporated with the insecticide inside a trap or on a surface on which the attracted insects will walk. Examples of are the cockroach traps and 'weevil stick' treated with grandlure and an insecticide to attract the boll weevil *Anthonomus grandis*. A pheromone can be sprayed as a microencapsulated formulation, but is often used in a plastic tube or matrix that is attached to plants so that the odour permeates through the crop canopy over a period of several weeks. These techniques are unlikely to have any adverse impact on non-target species.

In view of the increasing concern about environmental pollution with chemical pesticides, biopesticides are of increasing importance. Relatively few are available but special consideration of their application is given in Chapter 17.

Pest habitat

Tsetse flies (*Glossina* spp.), vectors of pathogenic trypanosomes, are important as pests of cattle and also transmit human sleeping sickness. Different species of tsetse flies live in riverine, forest and savannah areas, in each of which control with insecticides is directed at the adult flies. Tsetse flies are unusual because the female does not lay eggs but gives birth at intervals depending on the temperature, to a single third-instar larva. The larva, which at birth is heavier than the female fly, burrows down into the soil, usually to a depth of of 1–3 cm. The larval skin hardens to form a puparium in which the tsetse becomes the fourth instar, pupa and eventually a pharate adult which emerges into the open air. Control measures have changed quite significantly. In many early control campaigns a residual insecticide was applied to selective resting sites in shaded woodland during the dry season when the area suitable for tsetse flies was restricted. Compression sprayers with a cone nozzle were normally used. Larger scale operations against savannah species used aircraft to apply sequential aerosol (droplets size around 30–40 μm VMD) treatments (Allsop, 1990; Johnstone *et al.*, 1990). Aerial spraying has been criticised because large areas have to be treated, and inevitably some ecologically sensitive areas become exposed to the spray. Thus in the Okavambo swamps in Botswana, some aquatic areas around the islands were sprayed.

More recently, more emphasis has been given to treating screens made of a cotton fabric, coloured blue, with a pyrethroid insecticide. Screens can be treated *in situ* with a compression or knapsack sprayer, but rather than take the insecticide to remote areas, it is now possible to treat screens centrally by dipping them in a drum of insecticide (Fig. 2.6). Octanol in a small plastic vial attached to the screen attracts flies which pick up a lethal dose of insecticide when they land and walk on the treated fabric. This method allows retrieval of screens that may have been vandalised or pushed over by wild animals and their cleaning before re-deployment. The technique allows villagers and those involved in tourism in game parks to be responsible for checking and treating screens.

In some situations a combination of all three methods may be needed, depending on the population of tsetse flies at a particular time.

Behaviour of the pest

Whiteflies (*Bemisia tabaci*) have increased in importance. This is partly due to the increased production of horticultural crops throughout the year assisted by irrigation, and more use of plastic greenhouses. Increasing world trade in cuttings of ornamental plants has also spread this insect. Apart from their importance on these crops, whiteflies spread to larger areas of field crops such as cotton. The adult whitefly is quite mobile and is easily disturbed, so insecticide treatments can reduce their population quite rapidly. However, the immature stages are on the undersurface of the host plant leaves where they are protected from most insecticide applications. In consequence, more adults emerge soon after a spray treatment and soon oviposit before a further spray is applied. Subsequent generations build up rapidly as the poor spray against the adults is also effective against the mobile natural enemies such as *Encarsia*

(a)

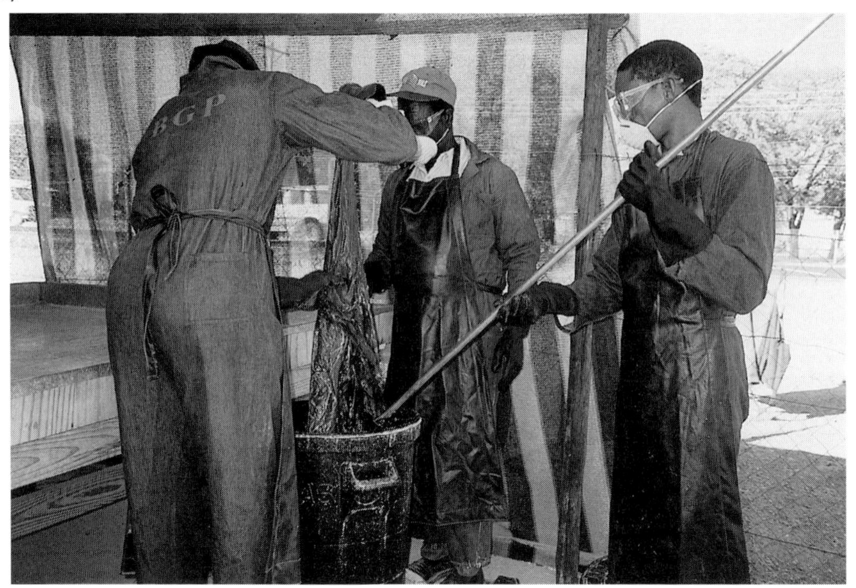

(b)

Fig. 2.6 Treatment of tsetse control screens with insecticide: (a) dipping; (b) drying screens

formosa. The problem has been exacerbated where farmers have continued to apply broad spectrum insecticide sprays over the top of plants.

The leaves of the host plants, such as cotton, aubergines, melons and others, act as umbrellas, so very little of the insecticide even reaches the upper surface of the lower leaves. Thus the behaviour of this pest on a wide range of host plants necessitates any insecticide to be directed at the undersurface of the leaves inside the crop canopy. Laboratory assessments have indicated that to kill immature stages of whitefly with a translaminar insecticide requires 20 times more chemical if it is only on the upper surface in comparison with an underleaf deposit. This is particularly important with the more selective insecticides. If soap emulsions are applied, a high volume of spray must reach the undersurface. Achieving underleaf deposits with hydraulic nozzles requires a vertical boom or drop-leg positioned in the interrow so that nozzles can be directed laterally and upwards (Lee *et al.*, 2000). Some machines employ air assistance to cause turbulence and increase deposition (see Chapter 10). Similar arguments apply to many other insect pests and pathogens (see next section). Matthews (1966) reported the need to control the first instar larvae of the red bollworm (*Diparopsis castanea*) before penetration of a flower bud or boll occurred (Fig. 2.7). This led to the use of a tailboom (see Chapter 6) to direct spray between the layers of leaves and increase deposition on the stem and petioles along which the larvae were walking.

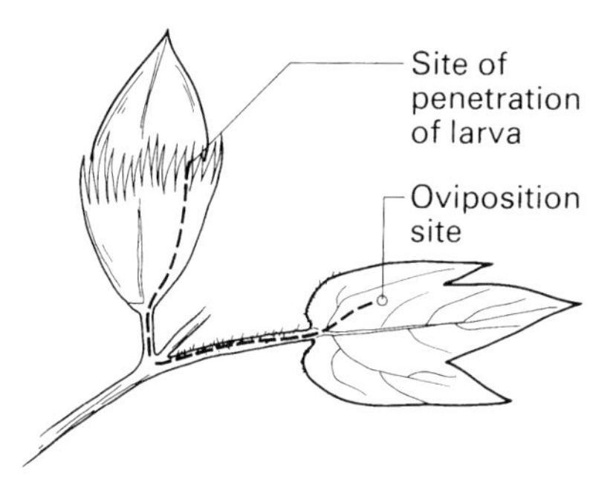

Fig. 2.7 Route of bollworm larva on cotton plant before eating its way inside boll.

Determination of the most appropriate target for deposition of a pesticide requires careful examination of the behaviour of the pest in the field throughout its life cycle. Observations should not be confined to daylight hours as many insects are more active at night. Even bollworms normally protected inside a boll may emerge onto the bracts at night, and insects that shun sunlight, such as the jassids found on the undersurface in the day, will be also on the upper leaf surfaces at night. As they are more active, an irregular coverage of insecticide will usually be adequate for jassid control in contrast to the immobile immature stages of the whitefly.

Disease control

In an IPM programme, plant diseases are suppressed preferably by choosing resistant cultivars and adopting cultural practices to minimise infection. However, certain plant pathogens are sufficiently serious to justify the application of a fungicide. A typical plant pathogen basically has four phases – prepenetration, penetration into the plant, post-invasion and finally sporulation to disperse the pathogen.

Control needs to be before the pathogen has penetrated the host plant. In practice, resistant cultivars usually exert some chemical defence which prevents an infection getting established. However, spores arriving on a susceptible host when conditions are favourable, will infect plants fairly rapidly, thus limiting the period when preventative control measures can be taken. A protectant fungicide may have to be applied several times to limit the spread of a disease. Variation in the onset of disease between seasons and areas makes it difficult to time the application of a fungicide, although minimeteorological stations can provide local weather data to provide a better analysis of whether a treatment is required. Addition of an adjuvant to increase rainfastness of a spray deposit may be beneficial, but increases in leaf or fruit area can expose plant surfaces not treated with the fungicide.

Many fungicides can be applied to curtail development of the post-invasion phase. Curative fungicides are often systemic chemicals that are moved within the plant, thus compensating for any difficulty in obtaining good coverage of the plant. For young seedlings, a systemic fungicide can be applied as a prophylactic seed treatment. Jeffs (1986) and Clayton (1993) give general accounts of seed treatment and the equipment used. Seed treatment is usually done by merchants distributing the seed rather than by individual farmers. Cereals in the UK are treated with mixtures of ethirimol + flutriafol + thiabendazole (used on barley) or by fuberidazole + triadimenol.

Most fungicide applications are directed at fruit or vegetable crops, with deciduous fruit, coffee and cocoa being the major markets for the agrochemical companies. Generally, good coverage is needed comparable to the requirements for a sessile insect pest. Systemic fungicides are easier to apply, as deficiencies in application can be compensated to some extent by the redistribution after treatment. Detailed studies can indicate which parts of a crop or plant are likely to be the initial focus of an infection; thus with crop monitoring, the area requiring initial treatments may be limited.

Weed control

Early suppression of weeds is important so that crops can get established without competition. However, in some areas of erratic rainfall, farmers may prefer to wait as long as possible before investing in chemical weed control, as insufficient rain will depress crop yields. The development of crops resistant to certain herbicides will also enable farmers to delay weed control and then use an overall over-the-crop treatment. Herbicides are increasingly used in minimum tillage farming and to avoid disturbance of crops due to mechanical cultivation. Late season herbicide application may also be beneficial, even if no increase in yield is obtained. This is because harvesting is easier in the absence of weeds and the harvested produce is cleaner. Herbicides may be applied as soil or foliar treatments before or after the emergence of the weeds, depending on the choice of herbicide (Fig. 2.8). The target may be

(1) the weed seed to prevent germination or kill the seedling immediately the seed germinates
(2) the roots, rhizomes or other underground tissues
(3) the stem, especially when applied to woody plants
(4) the foliage
(5) the apical shoots.

Choice of application technique will depend not only on the target, but also on how easily the herbicide penetrates and is translocated in plants.

A soil-acting herbicide applied before planting normally has to be effective in the upper 2–5 cm of the soil surface. Thus some of the new herbicides effective at dosages of < 1 kg are diluted in about 700 000 kg of soil. Some of these herbicides, such as triflutalin used to control *Cyperus* and annual grasses, have to be incorporated into the soil immediately to reduce losses due to volatility or photodecomposition.

Pre-emergence herbicides are applied to the soil surface at the time of sowing or immediately afterwards before the crop emerges. The former system is possible when using a seed drill, but if the crop is hand planted it is easier to complete the sowing before applying the herbicide. Pre-emergence herbicides such as simazine and the substituted ureas are applied at weed emergence as they have little effect on seedlings. Addition of a suitable surfactant, emulsifier or adjuvant oil will enhance the activity of the triazines on existing weeds. Pre-emergence herbicides are more effective if applied when the soil surface is damp, but if it is dry, shallow incorporation is possible. The residual effect will continue unless the soil surface is disturbed. Restricting an application to a band (usually 150 mm wide) along the crop row allows the inter-row soil to be cultivated mechanically. This reduces the cost of the herbicide treatment and increases infiltration of rainfall between rows.

Post-emergence herbicides can be applied after the crop has emerged. This may be before the weeds are present, in which case an overall application is possible with a selective herbicide. Selectivity is particularly important where grass weeds are in cereal fields and broad leaves are infesting broad leaved crops. Once the weeds are present it may still be possible to use a foliar-acting selective herbicide. Alternatively, it will be necessary to apply a herbicide with a directed spray, perhaps using a shield around the nozzle to prevent spray reaching the crop. Where a crop has been genetically modified so that it is resistant to a broad-spectrum herbicide such as glyphosate, then treatment can be delayed as long as the weed competition is not reducing the crop yield. Some have argued that this is ecologically beneficial, in that weed plants are

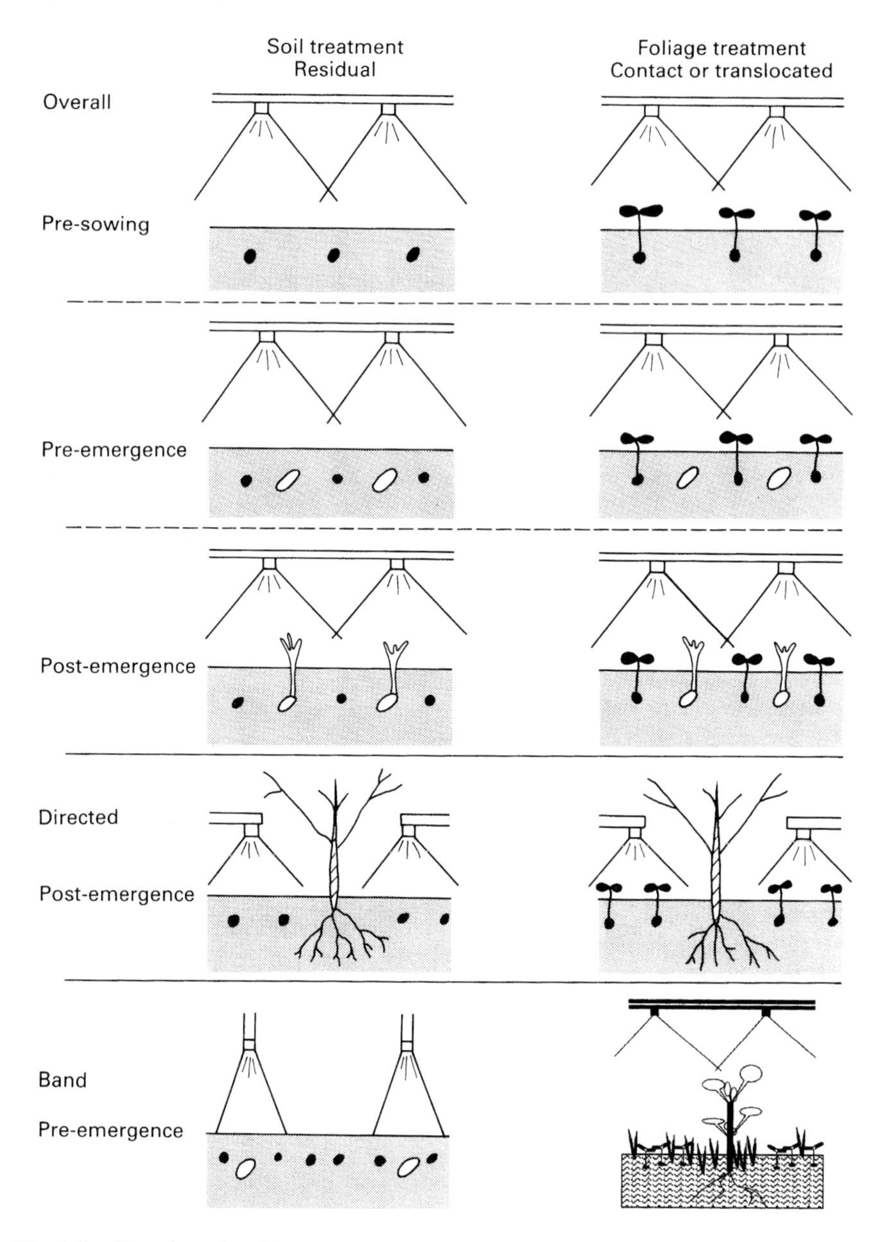

Fig. 2.8 Situations in which herbicide may be used for weed control in crops (after Fryer, 1977), with an additional diagram (bottom right) to show overall treatment of a GM crop. Weeds shown in black in contrast to crop plants.

available as food for insects and birds over a longer period than if a pre-emergence herbicide had been used. Care must be taken to ensure that the correct dose is applied as a post-emergence treatment, whether selective or non-selective on a GM crop, as an overdose can be phytotoxic and reduce yields. Whether a crop resistant to a specific herbicide could become a problem weed later in the crop rotation is one concern, while others think that weed species may become resistant to the herbicides used more extensively on GM herbicide tolerant crops.

On tree crops, the weed control treatment may be confined to an alley immediately adjacent to the tree row, leaving the inter-row sown to a grass/legume sward which can be mown. In some crops, such as oil palm, the herbicide is confined to a circle around each tree. Localised patches of weeds can be controlled by spot treatments. This may involve using a weed wiper when there are very few weeds, and a translocated herbicide can be applied. With a manually operated sprayer, patches of weeds may be treated using a full cone nozzle. More extensive patches of weeds within arable crops can now be treated by using equipment on which individual nozzles can be programmed to switch on or off by a computer using global positioning systems (GPS) (Miller *et al.*, 1997; Rew *et al.*, 1997). At present the positions of the patches of weeds need to be identified by walking the field and logging the data with a GPS. Ultimately, there is the possibility of detecting weeds from the spectral differences in crop and weed foliage (Haggar *et al.*, 1983), but at present on-line detection is limited to weeds on a bare soil background.

In applying herbicides, the narrow, often more vertical leaves of monocotyledon weeds, and the broad, mostly horizontal leaves of dicotyledons present quite different targets for spray application. There are also considerable differences in the detailed surface structure of the leaves, which affects the retention of spray droplets and the rate at which a herbicide can penetrate into the leaf. The sensitivity of different plant parts can influence the impact of a herbicide; thus the leaf axil may be the optimum target for some herbicides. Large droplets are generally advocated for herbicide application to minimise the risk of downwind drift affecting sensitive plants outside the treated field. However, droplets falling at their terminal velocity may splash off some hydrophobic leaf surfaces, resulting in poor retention. While such large droplets do fall more or less vertically and are deposited on flat horizontal leaves, smaller droplets travelling in a more horizontal plane are more likely to be deposited on the vertical leaves of grass weeds. Knoche (1994) provides a detailed review of the effect of droplet size on the performance of foliar applied herbicides. Studies with large droplets indicate the importance of the interface area, i.e. the area of leaf surface covered by droplets (Knoche and Bukovac, 2000). The addition of additional surfactant as an adjuvant may improve retention of a droplet by lowering the surface tension of the liquid, and also improve penetration. The latter may be most important as it will also reduce losses due to rain removing surface deposits. A surfactant will improve spread of a deposit, especially if the leaf is covered by dew within a day of treatment. Thus, in addition to considering the spray volume, concentration of the herbicide and droplet size (spray quality), the user may have to decide whether using an adjuvant will be economic.

Collection of droplets on targets

Droplets are collected on insects or plant surfaces by sedimentation or impaction (Johnstone, 1985). Under still conditions even small droplets will eventually fall by gravity on to a horizontal surface. For example when fogging in a glasshouse only 0.5 per cent of a *Bacillus thuringiensis* treatment was recovered on the lower surface of leaves (Burges and Jarrett, 1979). More important is the dispersal of small droplets in air currents in relation to target surfaces. There is a complex interaction between the size of the droplet, the obstacle in its path, and their relative velocity (Langmuir and Blodgett, 1946; Richardson, 1960; May and Clifford, 1967; Johnstone *et al.*, 1977; Bache and Johnstone, 1992; Bache, 1994). Parkin and Young (2000) have used computational fluid dynamics to examine collection efficiency if the adhesion of a droplet to a surface can be predicted or discounted (Fig. 2.9). Collection

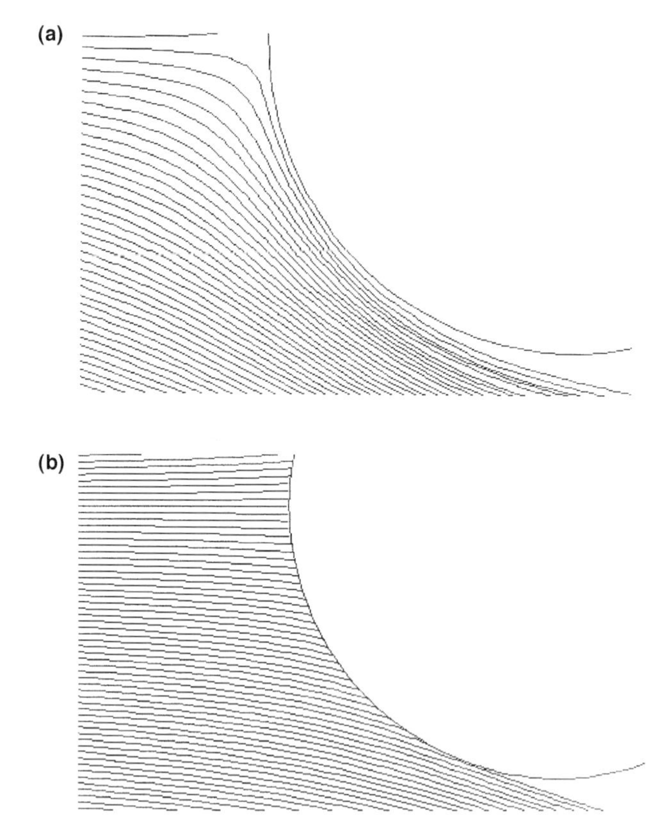

Fig. 2.9 Droplet trajectories upstream of a 10 mm diameter cylinder in a 4 m/s airstream simulated using CFD: (a) 1 μm droplets; (b) 35 μm diameter droplets.

efficiency of an obstacle in an airstream is defined as the ratio of the number of droplets striking the obstacle to the number which would strike it if the air flow had not deflected the droplet. In general, collection efficiency increases with droplet size and velocity of the droplet relative to the obstacle. It decreases as the obstacle increases in size.

The sum of the cross-sectional area of the two airstreams passing on either side of an obstacle is only about 75 per cent of the original airstream; therefore the velocity of the deflected airstream is increased. Droplets tend to flow in the airstream and miss the obstacle unless the size of the droplets and their momentum are sufficient to penetrate the boundary layer of air around the obstacle. The distance (mm) over which a droplet can penetrate still air is

$$\frac{d^2 V \rho_d}{18\eta}$$

where d is the droplet diameter (m), V is the velocity of the droplet (m/s), ρ_d is droplet density (kg/m^3) and η is the viscosity of air (N s/m^2). Even small droplets will impact if they are travelling at sufficient velocity to resist change in the direction of the airstream (Fig. 2.10). According to Spillman (1976) (Fig. 2.11), collection efficiency on flying insects is significantly less when droplets are smaller than 40 μm, but it is these small droplets that remain airborne longer and are most likely to be filtered out by insects. The effect of terminal velocity and wind speed on collection efficiency of cylinders of different diameters is illustrated in Fig. 2.12.

Most target surfaces are not smooth, and variations in the surface my cause local turbulence of the airflow. In this way, interception of a droplet or particle may occur if its path has been partially altered. Impaction of droplets on leaves depends very much on the position of the leaf in relation to the path of the

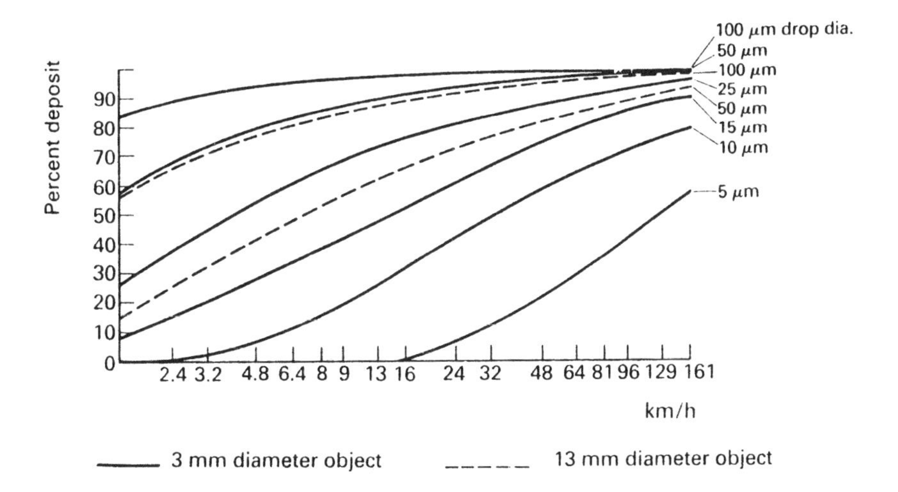

Fig. 2.10 Theoretical deposit achieved on objects of two sizes with several droplet sizes in airstreams of different velocity (after FAO, 1974).

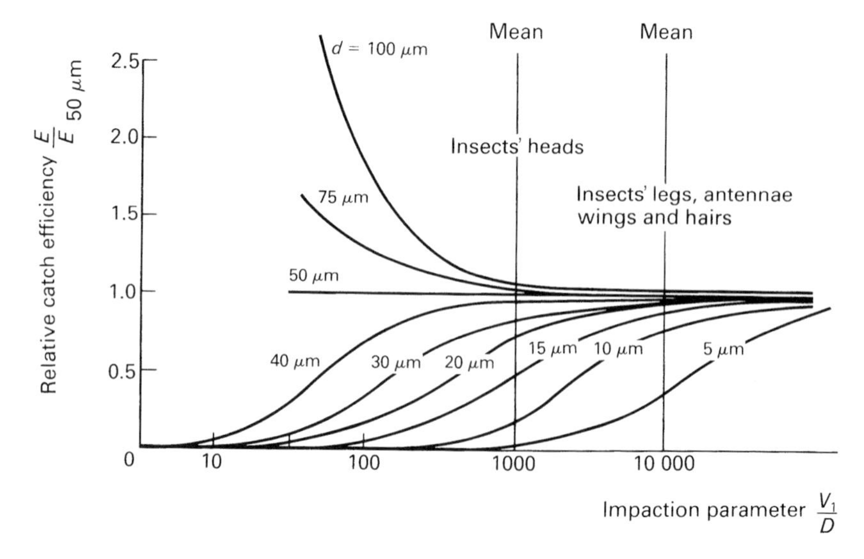

Fig. 2.11 Catch efficiency relative to 50 μm droplets over a range of impaction parameters (after Spillman, 1976).

droplet. Underleaf coverage is generally dependent on projection of droplets upwards through a crop canopy rather than downwards over the crop. More droplets are collected on leaves that are 'fluttering' in turbulent conditions and thus present a changing target pattern. If wind velocity is too great (this often happens when a high-speed air jet is used to transport droplets) the leaf may be turned to lie parallel with the airflow, so presenting the minimum area to intercept droplets. Movement of leaves during the day to optimise light interception will affect drift spraying, in that if spray droplets are released downwind from the direction of the sun, deposition will be predominantly on the upper surface of exposed leaves (Morton, 1977). Apart from the possibility of splitting application to different times of the day with different wind directions, there may be a greater need to split applications over two or three days to treat new foliage.

As mentioned previously, droplets arriving on a leaf surface may not be retained on it. Brunskill (1956) referred to an example of cabbage leaves rejecting rain falling on them in a storm. He showed that decreasing the surface tension of the spray, droplet diameter and the angle of incidence could increase retention of spray droplets on pea leaves. His studies revealed that droplets which strike a surface such as a pea leaf become flattened, but the kinetic energy is such that the droplet then retracts and bounces away. Droplets below a certain size (< 150 μm) have insufficient kinetic energy to overcome the surface energy and viscous changes, and cannot bounce. Conversely, very large droplets (> 200 μm) have so much kinetic energy that they shatter on impact. Bouncing from pea leaves is associated with the roughness of the surface. Droplets of liquid containing air bubbles are

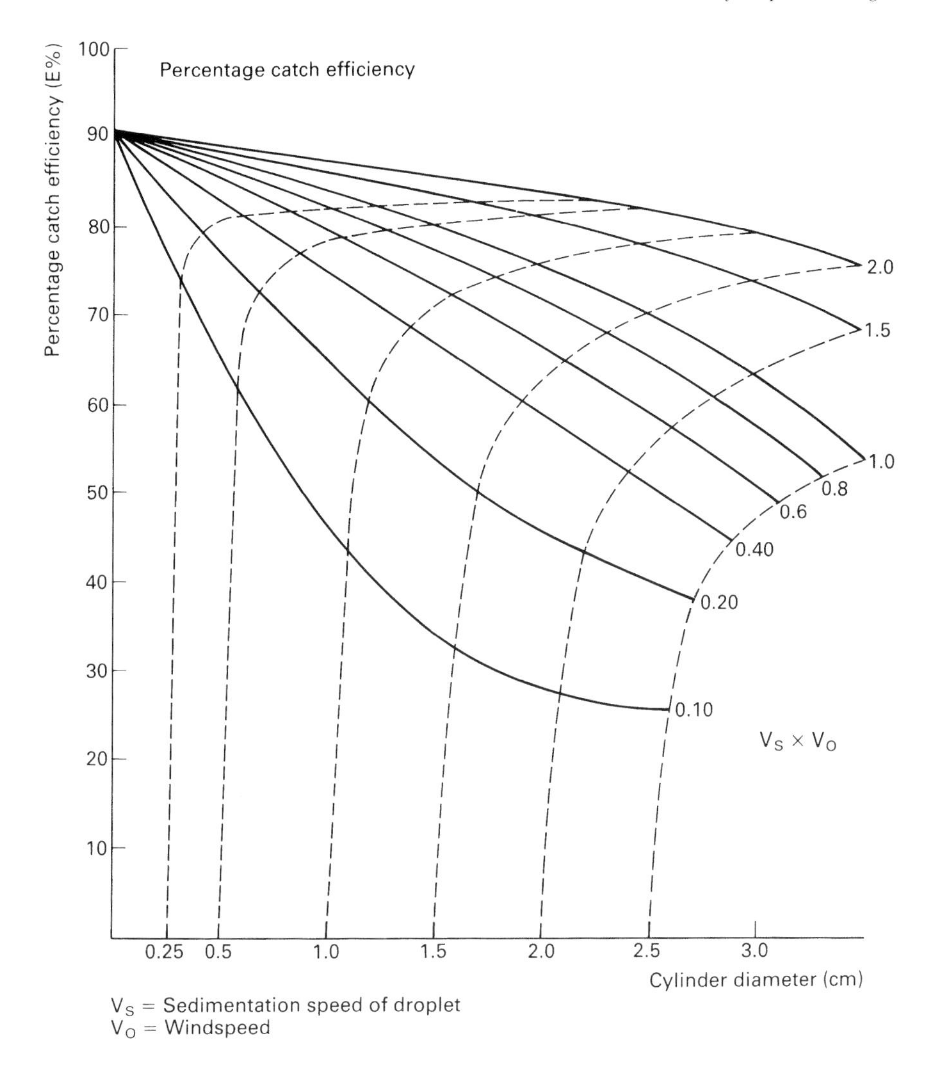

Fig. 2.12 Variation of catch efficiency with cylinder diameter, sedimentation speed and windspeed (J.J. Spillman, personal communication).

thought to behave differently, but much depends on the proportion of air within individual droplets.

Leaf roughness varies considerably between plants, and also between upper and lower surfaces (Holloway, 1970), and influences the spreading of spray droplets over leaf surfaces (Boize *et al.*, 1976). Apart from conspicuous features caused by venation, the shape and size of the epidermal cells, which may have flat, convex or hairy surfaces, influence the topography of the leaf. The cuticle itself may develop a complex surface ornamentation. Various patterns of trichomes exist on leaves, but at the extremes 'open' patterns enhance the wetting of leaves, possibly due to capillary action, while 'closed' patterns are water repellent. Holloway (1970) differentiated between the various types of superficial wax deposits on cuticle surfaces. A 'bloom' on a leaf surface occurs when these deposits are crystalline, for example when rodlets and threads are present.

When assessing wetting of leaves there are two main types of leaf surface, depending on whether the angle of contact (Fig. 2.13) is either above or below 90° (Table 2.2). With the latter group, superficial wax is not a feature, but on leaf surfaces with a contact angle above 90°, wax significantly affects wettability. Contact angles of 90–110° occur with a smooth layer of superficial wax. Above 110°, the contact angles depend on the roughness of the surface. There is a generalisation that leaf roughness is less important when the droplet size is below 150 µm, particularly as pesticides are formulated with surface active agents. Ideally, the advancing contact angle must be kept as high as possible and the receding angle as small as possible. Surface active agents (surfactants, wetters) behave differently, depending on the leaf surface, so it is not possible to formulate optimally for all uses of the pesticide. The effect of a surfactant on droplets on a leaf surface is shown in Fig. 2.14. Surfactants affect retention more on leaf surfaces, such as pea leaves that are difficult to wet (see Chapter 4). Amsden and Lewins (1966) developed a simple leaf dip test, using a 1% solution of crystal violet, otherwise known as gentian violet or methyl violet, to assess leaf wax. Sample plants held carefully using a large pair of forceps are immersed completely in the dye solution carried in a wide-necked large jar with screw top lid. On removal, the plant is shaken gently to remove surplus liquid and examined. Areas with dye show where the wax deposit is deficient or has been damaged. In the example of pea leaves, a herbicide should not be applied if plants have more than 5% of the upper surface of leaves and more than 10% of the lower surface showing dye retention. Even healthy plants will

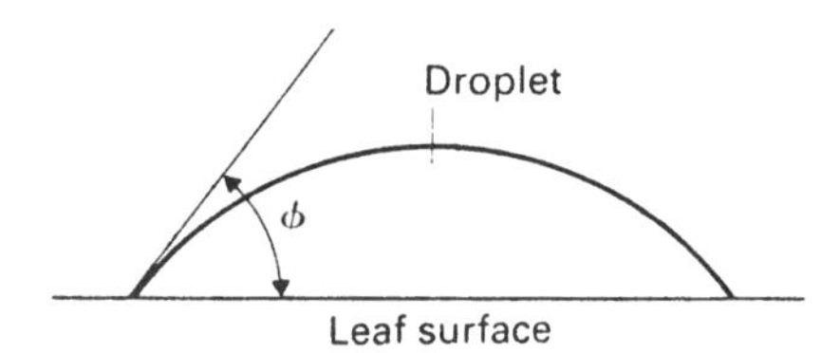

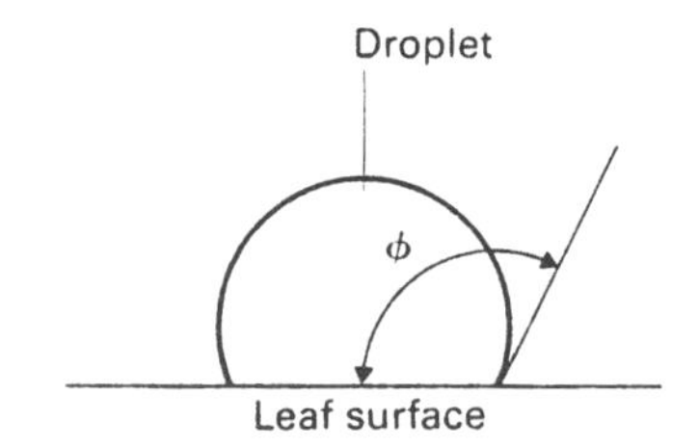

Fig. 2.13 Angles of contact.

Table 2.2 Contact angles of water on some leaf surfaces (*after* Holloway, 1970)

	Leaf surface	
	Upper	Lower
Eucalyptus globulus	170°	
Narcissus pseudonarcissus	142° 54′	
Clarkia elegans	124° 8′	159° 15′
Saponaria officinalis	100° 6′	106° 26′
Prunus laurocerasus	90° 50′	93° 32′
Rhododendron ponticum	70° 22′	43° 21′
Senecio squalidus	90° 10′	90° 15′
Rumex obtusifolius	39°	40° 5′
Plantago lanceolata	74° 23′	39° 32′

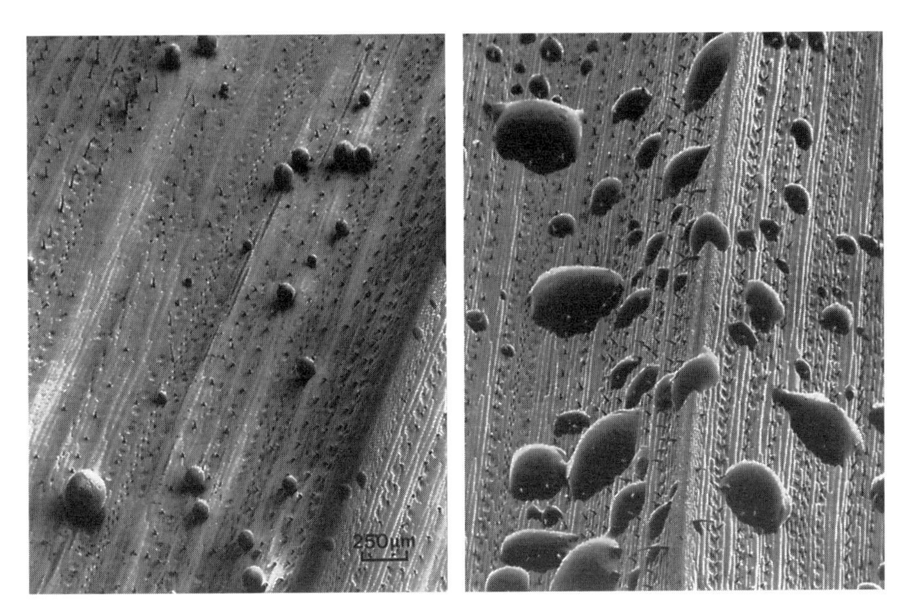

Fig. 2.14 Cryo-scanning electron micrographs of the abaxial surface of glasshouse grown wheat showing spray droplets of the plant growth regulator paclobutrazol (0.5 g/l) as 'Cultar' to show effect of adding surfactant (1 g/l 'Synperonic NP8') (from Hart and Young, 1987).

show some dye retention on the stems and tendrils. Anderson *et al.* (1987) pointed out that retention was also determined by the dynamic surface tension of the spray rather than the equilibrium surface tension. Improved retention related to dynamic surface tension was confirmed by Holloway *et al.* (2000), except for an organosilicone adjuvant with high surface activity that gave complete coverage of leaves.

Spray coverage

The philosophy used to be that all the plant surfaces had to be wetted, so high volume sprays were applied until liquid dripped from the leaves. This system of spraying to 'run-off' seldom achieves complete wetting of all parts of a dense crop canopy. Furthermore, most of the chemical is wasted as it does not remain on the plants, and because the total area is exposed to the pesticide, it undoubtedly has a major adverse impact on non-target organisms.

The trend has been to reduce the volume of liquid applied, and this has necessitated the application of discrete droplets. When discrete droplets are applied, the pesticide applicator needs to know the number of droplets required on the target area together with their distribution. Variations in distribution have less effect on control of pests when a systemic pesticide is applied, or the deposit is redistributed to the site of action. Systemic insecticides applied to the seed or as large droplets to avoid drift will be redistributed up through plants. Distribution of contact pesticides is much more important.

Mobile pests, such as jassids, are readily controlled without complete coverage, but sessile pests such as the nymphal stages of whiteflies on the undersurface of leaves require a more uniform spray distribution. Johnstone *et al.* (1972a) used 1 droplet/mm^2 so that the 100 μm diameter droplets were sufficiently close to give a high probability of a direct hit on small insects such as scale insects. When larger droplets at low density are applied, there is the chance of an insect avoiding an individual droplet. Polles and Vinson (1969) reported higher mortality of tobacco budworm larvae with 100 μm droplets of ULV malathion than with larger, more widely spaced 300–700 μm droplets which the larvae were able to detect and avoid. However, inclusion of a pheromone or food bait can attract insects to few large droplets. This is exploited with the use of protein hydrolysate + insecticide for fruit fly control.

Early attempts to assess the effect of droplet size on efficacy showed little difference when sprays were applied at volumes greater than about 200 litres/ha, but at reduced volumes, control was not always adequate (Wilson *et al.*, 1963). However, such studies were affected by the wide range of droplet sizes produced by a hydraulic nozzle. Detailed studies by Munthali and Scopes (1982) investigated the effect of uniform sized droplets of the acaricide dicofol on the egg stage of the red spider mite (*Tetranychus urticae*), using a fluorescent tracer to show the position of individual droplets of an oil-based formulation. This and later studies by Munthali (1984) and Munthali and Wyatt (1986) indicated that there was a 'biocidal area' associated with the spread of pesticide from an individual droplet (Fig. 2.15). This term had been used much earlier by Courshee *et al.* (1954) in relation to fungicide deposits. Ford and Salt (1987) discussed Munthali's results and defined biocidal efficacy as the inverse of the LD_{50}, i.e. cm^2/μg (Fig. 2.16). They suggested that effective spreading of the active ingredient from the initial deposit may involve a diffusion-controlled process. Thus the concentration of active ingredient on the leaf will decrease radially from the centre of the initial deposit. Gradually more of the active ingredient will spread over an increasing area, but the rate of diffusion

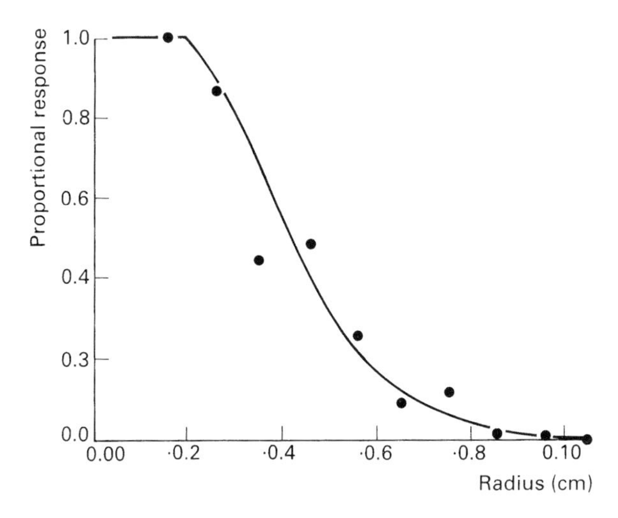

Fig. 2.15 Mortality of whitefly as a function of the radial distance from the centre of a ULV deposit containing 10 per cent w/v experimental formulation of permethrin applied to the surface of an infested tobacco leaf surface. Time after application 4 days; in-flight droplet diameter 114 μm; diameter spread factor 1.7.

will progressively decrease. A simulation model was used to examine the response to discrete droplets (Sharkey *et al.*, 1987). While modelling will indicate a maximum concentration needed to achieve control, in the field a higher concentration may be required to compensate for degradation and provide sufficient persistence of the deposit to obtain practical control. Similar experimental data were obtained with application of permethrin against the glasshouse whitefly (*Trialeurodes vaporarium*) (Abdalla, 1984; Wyatt *et al.*, 1985; Adams *et al.*, 1987).

In contrast to the sessile insects, experiments with lepidopteran larvae (*Spodoptera littoralis*) suggested the need for a critical mass of insecticide on a leaf; otherwise there was inadequate transfer of the active ingredient as it walked over the leaf surface (Ford *et al.*, 1977). Efficiency of transfer to *Plutella* larvae was increased, with better coverage obtained using small droplets (Omar and Matthews, 1987). Similar data have been reported by Hall and Thacker (1994) who showed that the LD_{50} for 100 μm droplets was ten times less than for droplets larger than 500 μm when assessing the topical toxicity to cabbage looper. Crease *et al.* (1987) showed the importance of a high viscosity oil to enhance the effect of small droplets applied at ultralow volume. Small droplets of permethrin in vegetable oil were more effective than larger droplets against *Heliothis virescens*, but droplet size was not important with water-based sprays; however, the latter were applied with a cone nozzle which would produce a wide range of droplet sizes.

Using a pesticide dose simulator (PDS) model Ebert *et al.* (1999) concluded that deposit structure plays a major role in the efficacy of a pesticide, but small

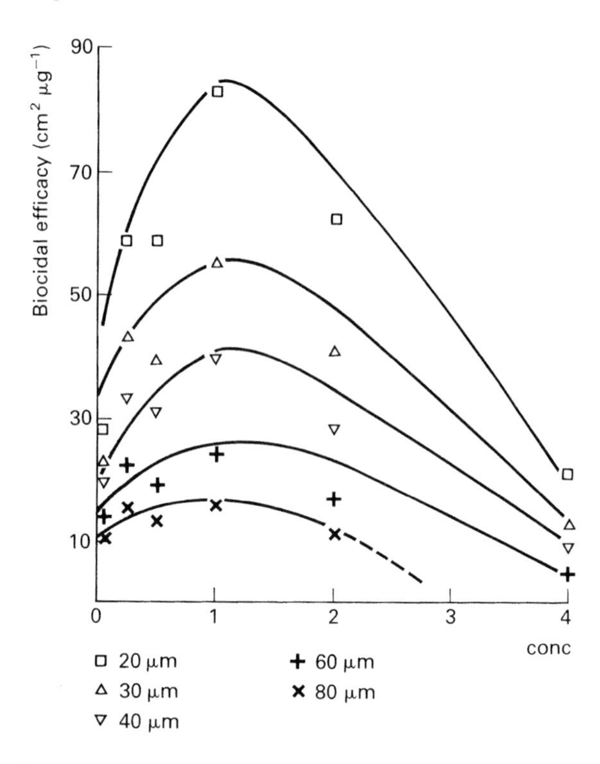

Fig. 2.16 The effect of spray concentration (% w/v) and in-flight droplet diameter on the biocidal efficacy (cm^2 μg^{-1}) of ULV formulations of dicofol applied to tobacco leaves for the control of red spider mites (*Tetranychus urticae*) (from Ford and Salt, 1987).

droplets are not always the most efficacious. Their studies with diamond back moth larvae moving and feeding on leaves treated with *Bacillus thuringiensis* showed a strong cubic interaction between droplet size*spray concentration*number of droplets, whether insect mortality or extent of protecting the leaf was measured. Clearly, there is a minimum amount of toxicant needed in the deposits transferred to an individual insect. If the insect has to encounter more small deposits with too low a dosage, it will incur more damage and may not die; conversely, if too much is deposited in each droplet, there will be considerable wastage of pesticide. Further studies are needed to investigate how bioassays should be conducted in view of the impact of deposit structure (Ebert and Hall, 1999). Studies with *B. thuringiensis kustaki* against gypsy moth larvae showed that the time to mortality increased as droplet density and droplet size decreased and larval size increased (Maczuga and Mierzejewski, 1995).

There remains a conflict between the laboratory data indicating that improved efficacy of small, but not too small droplets, can occur with the appropriate dosage, and their application in the field where small droplets are most vulnerable to downwind drift. An indication of the relative size of a droplet and an aphid tarsus is shown in Fig. 2.17. The trend to use coarser sprays could lead to more pesticide being applied within a treated field than theoretically necessary; the objective must be to see whether equipment can be designed to resolve this.

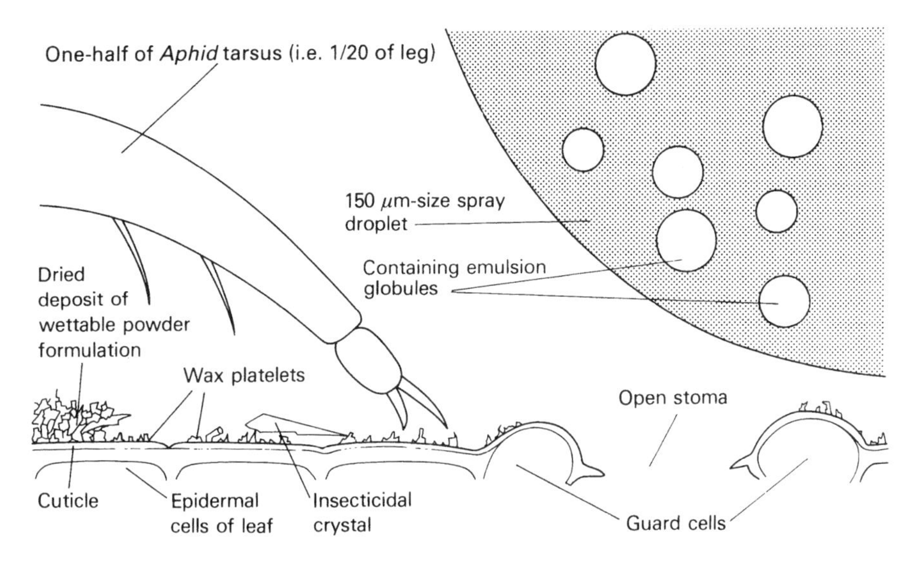

Fig. 2.17 Relative size of an aphid tarsus, spray droplet and leaf surface (from Hartley and Graham-Bryce, 1980).

Efforts of using an electrostatic spray as discussed in Chapter 9 were not very successful due to the preferential deposition on the nearest earthed surfaces. Charged sprays with an airstream were better, but there is now greater interest in using air flows of uncharged droplets to improve distribution and deposition (Matthews, 2000).

As far as fungicide application is concerned, it might also seem impossible to achieve control of a disease unless there is complete coverage, since hyphae can penetrate plants at the site of spore deposition, when suitable conditions occur. However, each particle of fungicide from a droplet has a zone of fungicidal influence, as noted earlier by Courshee *et al.* (1954). They postulated that the maximum ratio between the effective fungicidal cover and the actual cover by the deposit of the fungicidal residue of a droplet on drying is when the droplet is minimal. Initial infection of a disease such as potato blight (*Phytophthora infestans*) occurs usually in wet weather when most spores are collected on the upper surfaces of leaves (Beaumont, 1947). The spores follow the movement of raindrops to the edges of leaves where the symptoms of blight first occur. Also, a high proportion of pesticide spray deposit is redistributed over the leaf surface by rain, so however uniformly the initial deposit is applied, control can be maintained by the small proportion of deposit retained at the same sites as the spores (Courshee, 1967).

Redistribution of fungicide over the surface of plants is very important with other diseases. Coffee berry disease control has been achieved by spraying over the top of the trees with either ground (Pereira and Mapother, 1972) or aerial equipment (Pereira, 1970). Although the disease is controlled, a high proportion of the chemical applied is wasted; thus the aim should be to improve distribution of smaller quantities of pesticide in a suitable formulation, so that more of it is retained and biologically active where control is needed.

What volume of spray liquid is required?

The pesticide manufacturer should recommend the volume of spray to be applied and suggest the type of nozzle to be used. However, the application method is usually left almost entirely to the farmer's discretion. Terms such as high, medium, low, very-low and ultralow volume application have been used, but the actual volumes used for these has varied, especially between arable and orchard crops (Table 2.3). Ultralow volume (ULV) is defined as the minimum volume per unit area required to achieve economic control (Anon, 1971), and is generally associated with the use of oil-based formulations of low volatility. In the USA, it is also defined as use of <5 litres/ha, but in practice the cost of UL formulations really requires application of 0.5–1.0 litre/ha. As the cost of UL formulations has increased, the trend has been to revert to water-based formulations applied at very-low volume, where water supplies are poor. In some cases an adjuvant has been added to reduce the effect of evaporation of water from droplets in flight.

Table 2.3 Volume rates of different crops (litres/ha)

	Field crops	Trees and bushes
High volume	>600	>1000
Medium volume	200–600	500–1000
Low volume	50–200	200–500
Very-low volume	5–50	50–200
Ultralow volume	<5	<50

With the major concern about the release of pesticides in the environment, it is increasingly important to optimise the delivery system. Instead of wetting the whole target, the optimum droplet size range is selected to increase the proportion of spray which adheres to the target. Generalised indicators of optimum droplet size shown in Table 2.4 in terms of collection efficiency on insects and foliage, conflict with the adoption of coarse sprays to minimise drift. However, if a suitable droplet size is selected and an estimate made of the coverage (droplets/unit area), then the volume of spray required can be calculated (Fig. 2.18) (Johnstone, 1973b). For example, if a spray with 100 µm

Table 2.4 Optimum droplet size ranges for selected targets

Target	Droplet size	
	Diameter (μm)	Volume (picolitres)
Flying insects	10–50	0.5–65
Insects of foliage	30–50	14–65
Foliage	40–100	33–524
Soil (and avoidance of drift)	>200	>4189

diameter droplets is applied and 50 droplets/cm^2 is required, then the minimum volume is 2.5 litres/ha treated.

The target requiring treatment may be much greater than the ground area, although most recommendations refer only to the ground area occupied by a crop. Few attempts have been made to relate the dosage to the area of plant surface (μg/cm^2) as emphasised many years ago by Martin (1958) and Way *et al.* (1958). Morgan (1964) advocated selecting spray volume with tree size and Tunstall and Matthews (1961) increased spray volume in relation to the height of cotton plants (see pp. 138–9). Similarly in relation to ULV spraying,

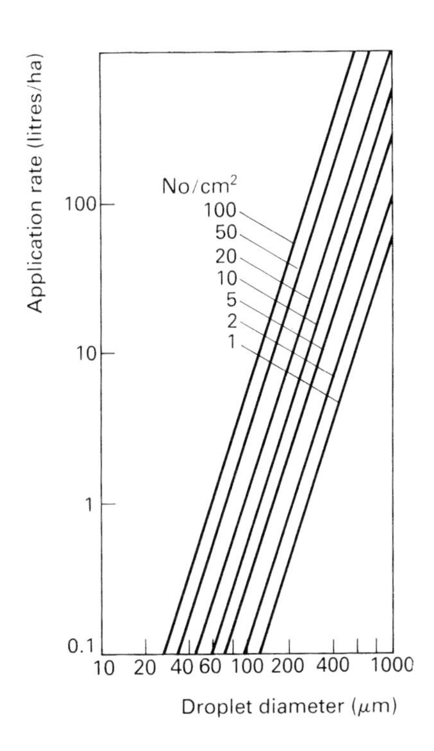

Fig. 2.18 Relation between droplet numbers, diameter and volume application rate (after Johnstone, 1973b).

Matthews (1971) changed the volume in relation to track spacing: this was decreased as cotton plants grew. Where foliage is the target, it is important to assess the leaf area index (LAI), defined as the ratio of leaf area to ground area. This will vary between crops and increase as plants are growing, but seldom exceeds about 6–7 because leaves without adequate light are usually shed. Thus if the LAI is 3 and 2.5 litres per treated hectare of foliage is needed, then the total volume should be increased to 7.5 litres/ha.

If even-sized droplets could be produced, the minimum volume that should be applied to achieve a droplet pattern of 1/mm^2 is shown in Table 2.5 (Bals, 1975b). Theoretically, very small volumes of spray per hectare are needed when it is possible to use droplets of less than 100 μm diameter, i.e. <524 picolitres per droplet. If the LD_{50} contained in a single droplet can be determined, the concentration of spray required in controlled droplet application (CDA) can also be calculated (Fig. 2.19). The application of more uniform sized droplets, i.e. CDA, is considered further in Chapter 8.

Table 2.5 Minimum spray volumes

Droplet (diameter (μm)	Spray liquid required (litres/ha) for density of 1 droplet/mm^2 applied evenly to a flat surface
10	0.005
20	0.042
30	0.141
40	0.335
50	0.655
60	1.131
70	1.797
80	2.682
90	3.818
100	5.238
200	41.905
500	654.687

Increasing spray volume does not necessarily improve coverage. With a set of nozzles, changing the flow rate merely deposits more pesticide in the most exposed target areas. Thus, there will be little or no improvement in the amount deposited on concealed sites, such as the undersurface of leaves within a crop canopy. Courshee (1967) illustrated this by plotting the cumulative percentage of targets (leaves) with different deposit densities (Fig. 2.20, line A). Doubling the the spray volume or mass application rate does not double the deposit density of each leaf (line B), but only on some of the leaves (line C). Trials on cotton with good distribution of insecticide with a multiple nozzle tailboom failed to achieve a significant increase in yield by increasing the spray concentration or volume applied (Table 2.6), but decreased yields if less than the recommended dosage was applied (Matthews

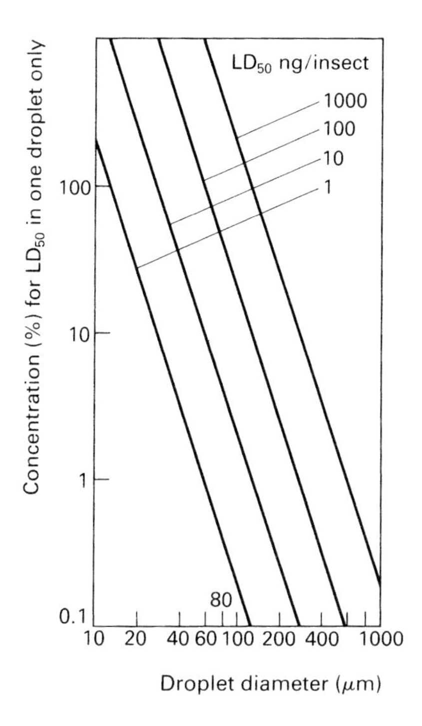

Fig. 2.19 Relation between toxicity, droplet diameter and concentration of active ingredient for one droplet to contain the LD_{50} (from Johnstone, 1973b).

and Tunstall, 1966). The lower dosage was inadequate due to the effects of weathering and dilution of the spray deposit by plant growth (Matthews, 1966).

An increase in the number of points of emission to the target by using more nozzles can achieve more uniform distribution. Furthermore, careful deployment of an air flow of suitable volume and velocity to increase turbulence within a crop canopy can also improve deposition on the more concealed surfaces within it.

Table 2.6 Effect of varying spray volume and concentration on yields of seed cotton

	Concentration (%)	Volume (litres/ha)	Yield (kg/ha)
Recommended spray	0.5	56–227	3123
Doubling concentration	1.0	56–227	3221
Increasing volume by 50%	0.5	84–340	3138
Reducing concentration by 80%	0.1	56–227	2228
Reducing volume by 33%	0.5	37–150	2819

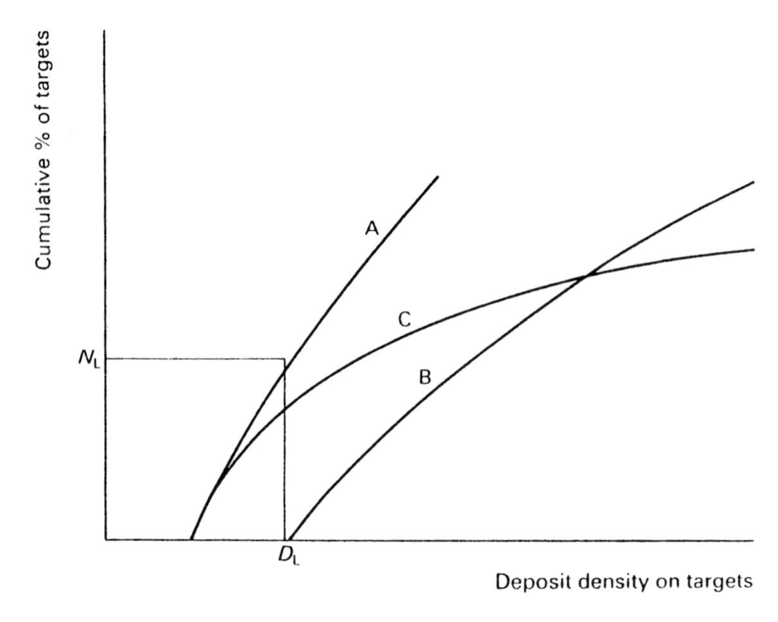

Fig. 2.20 Hypothetical deposit distribution curves on foliage. A typical distribution of doses on targets is shown by curve A. If deposit on each target were doubled by doubling the application rate, curve B would be obtained, but in practice the heavy deposits are increased while many leaves continue to receive an inadequate deposit – curve C. The minimum deposit needed may be that indicated at point D (from Courshee, 1967).

3

Formulation of pesticides

Pesticides are biologically active in extremely small quantities, and this has been accentuated by the development of the pyrethroid insecticides and other more active pesticides, so the chemical has to be prepared in a form that is convenient to use and to distribute evenly over large areas. The preparation of the active ingredient in a form suitable for use is referred to as 'formulation'. Manufacturers have their own particular skills in formulation, details of which are a closely guarded secret because of competition from rival companies. Knowles (1998) and Van Valkenburg *et al.* (1998) give general information on the principles of formulation. In the first of these books, De Raat *et al.* (1998) and Wagner (1998) describe the regulatory requirements for the European Union and the USA, respectively.

Most pesticides are formulated for dilution in water, and as the measuring of the product and transfer to the sprayer brings the user into the closest contact with the active ingredient, recent changes have aimed at reducing the risk of spillage and operator contamination. This has included developments of new types of formulation as well as improvements in packaging, including the use of water-soluble plastic containers. Changes in equipment have also made it easier to load spray liquids into the tank, by using closed filling systems or providing low level mixing facilities. Some formulations can be applied directly without dilution at ultralow volumes, but toxicological, technical and economic considerations limit the number of chemicals which can be used in this way. Most pesticides were formulated as emulsifiable concentrates, but concern about exposure of the environment to organic solvents has led to greater emphasis on particulate formulations. Where pesticides were marketed as wettable powders owing to the high cost of suitable solvents for the active ingredient, concern about the risk of inhalation hazards has also led to new formulations using dispersible granules (Bell, 1998) or suspension concentrates (Mulqueen *et al.*, 1990; Seaman 1990).

Types of formulation

A range of different formulations is usually available for each active ingredient to suit individual crop-pest and regional marketing requirements. These are now designated by a two-letter code (Table 3.1). Differences between formulated products of one manufacturer may be due to the availability of solvents, emulsifiers or other ingredients at a particular formulation plant. Registration requirements also influence the availability of certain formulations.

Table 3.1 Codes for pesticide formulations[a]

DP	Dust	EC	Emulsifiable concentrate
GR	Granule	UL	Ultralow volume
MG	Microgranule	OF	Oil-miscible flowable
BR	Briquette	TK	Technical concentrate
SC	Suspension concentrate	CS	Capsule suspension
SP	Water soluble powder	EW	Oil-in-water emulsion
WP	Wettable powder	GL	Gel
WG	Water-dispersible granule		

[a] Full list is available from the Global Crop Protection Federation (GCPF, formerly GIFAP), Brussels – Technical Bulletin No. 2 (1989).

Formulations for application as sprays

A few pesticides dissolve readily in water and can be applied as solutions. Examples are the sodium, potassium or amine salts of MCPA and 2,4-D. Owing to insolubility in water, many require formulating with surface-active agents or special solvents.

Wettable powders (WP)

These formulations, sometimes called dispersible or sprayable powders, consist of finely divided pesticide particles, together with surface-active agents that enable the powder to be mixed with water to form a stable homogeneous suspension. Wettable powders frequently contain 50 per cent active ingredient, but some contain higher concentrations. The upper limit is usually determined by the amount of inert material such as synthetic silica (HiSil) required to prevent particles of the active ingredient fusing together during processing in a hammer or fluid energy mill ('micronizer'). This is influenced by the melting point of the active ingredient, but an inert filler is also needed to prevent the formulated product from caking or aggregating during storage. The amount of synthetic silica needs to be kept to a minimum as this material is very abrasive. Apart from wear on the formulating plant, the nozzle orifice on sprayers is liable to erosion, thus increasing application rates.

Wettable powders have a high proportion of particles less than 5 μm, and all the particles should pass a 44 μm screen. Ideally, the amount of surface-active agents should be sufficient to allow the spray droplets to wet and spread over the target surface, but the particles should not be easily washed off by rain.

Wettable powders should flow easily to facilitate measuring into the mixing container. Like dusts, they have some extremely small particles, so care must be taken to avoid the powder concentrate puffing up into the spray operator's face. Most wettable powders are white, so to avoid the risk of confusing powder from partly opened containers with foods such as sugar or flour, small packs containing sufficient formulation for one knapsack sprayer load have been used in some developing countries (Gower and Matthews, 1971). Water-soluble plastic sachets of some products have now superseded this type of packaging, so the whole package is added to the spray tank.

The wettable powder should disperse and wet easily when mixed with water and not form lumps. To ensure good mixing, some pesticides should be pre-mixed with about 5% of the final volume of water and creamed to a thin paste. When added to the remaining water the pre-mix should disperse easily with stirring and remain suspended for a reasonable period. The surface-active or dispersing agent should prevent the particles from aggregating and settling out in the application tank. The rate of sedimentation in the spray tank is directly proportional to the size and density of the particles (see p. 78). Suspensibility is particularly important when wettable powders are used in equipment without proper agitation; for example, many knapsack sprayers have no agitator. Suspensibility of a wettable powder suspension is checked by keeping a sample of the suspension in an undisturbed graduated cylinder at a controlled temperature (WHO, 1973). After 30 min a sample is withdrawn halfway down the cylinder and analysed. The sample should contain at least 50 per cent of the pesticide.

Polon (1973) has described methods of preparing wettable powder formulations. Some wettable powders contain too much surface-active agent and foam when air is mixed in the spray liquid. Foam within the spray rig may cause intermittent application, and is prevented by keeping air out of the spray system. No more than 10 ml of foam should remain in a 100 ml graduated cylinder 5 min after mixing a sample of spray at field strength. Foam can be dispersed by silicones such as Silcolapse.

Wettable powders should retain their fluidity, dispersibility and suspensibility, even after prolonged storage. Containers should be designed to that even if wettable powder is stored in stacks, the particles are not affected by pressure and excessive heat, which may cause agglomeration. The World Health Organization requires tests for dispersibility and suspensibility after the wettable powder concentrate has been exposed to tropical storage conditions. Poor quality wettable powders are difficult to mix and readily clog filters in the spray equipment.

Normally, wettable powder formulations are not compatible with other types of formulation, although some have been specially formulated to mix with emulsions. Mixing wettable powders with an emulsion frequently causes flocculation or sedimentation, owing to a reaction with the surface-active agents in the emulsifiable concentrate formulation. Sometimes a small quantity of an emulsifiable concentrate can be added to a wettable powder already diluted to field strength, but compatibility should always be checked before mixing in the field.

Water dispersible granules (WG)

To overcome the problems associated with wettable powders, the powder can be granulated with a highly water soluble or water absorbing material and binding agent to form dispersible granules (Fig. 3.1) (Wright and Ibrahim, 1984) (Table 3.2). These granules disintegrate rapidly when mixed with water and essentially form a particulate suspension similar to that of a wettable powder. Bell (1989) and Woodford (1998) describe various techniques for producing water-dispersible granules including pan granulation, fluid bed granulation and spray drying.

Fig. 3.1 Dispersible granule formulation on mixing with water (courtesy Giba-Geigy, now Novartis).

Suspension concentrates (SC)

Farmers generally prefer to use a liquid formulation, as it is easier to measure out small quantities for use in closed systems. Furthermore, some environmental authorities have restricted the use of certain solvents and surfactants. These factors have led to more interest in suspension concentrates in which a particulate is pre-mixed with a liquid. With an aqueous base they are less hazardous to use. Initially these colloidal suspensions had a short shelf life, as

Table 3.2 Example of a wettable powder formulation

75% Wettable powder	(wt %)
Technical	76.5
Inert filler: HiSil 233 (hydrated silicon dioxide)	21.0
Wetting agent, e.g. Igepon T.77	1.5
Dispersing agent, e.g. Marasperse N	1.0
	100.0

the pesticides sedimented to form a clay deposit which was not easily resuspended. Advances in milling of particles (Dombrowski and Schieritz, 1984) and improved dispersing agents (Heath *et al.*, 1984; Tadros, 1989) have significantly enhanced the shelf life of aqueous based suspension concentrates, which are often referred to as 'flowables' (Fraley, 1984). Mixtures of two pesticides, which are not easily co-formulated in an EC, can be formulated as an SC. Following rain, deposits of an SC formulation were retained better than an EC or WG formulation on cotton leaves (Pedibhotla *et al.*, 1999).

Where possible, the agrochemical industry is replacing emulsifiable concentrates with suspension concentrates. Tadros (1998) gives a detailed account of suspension concentrates.

Emulsifiable concentrates (EC)

An important component in these formulations is the surfactant emulsifier, a surface-active agent, which is partly hydrophilic and partly lipophilic. There are four types of surfactants, namely anionic (negatively charged hydrophilic group), cationic (positively charged hydrophilic group), non-ionic (uncharged) and amphoteric (with both positive and negatively charged hydrophilic groups). A pesticide dissolved in a suitable organic solvent such as xylene or cyclohexanone cannot be mixed with water, because the two liquids form separate layers. The addition of an emulsifier enables the formation of a homogeneous and stable dispersion of small globules, usually less than 10 μm in size, of the solvent in water. The small globules of suspended liquid are referred to as the disperse phase, and the liquid in which they are suspended is the continuous phase. The concentration of many emulsifiable concentrate formulations is usually 25% w/v active ingredient. One of the lowest concentrations available commercially is 8% w/v tetradifon, but manufacturers prefer to use the highest concentration possible, depending on the solubility of the pesticide in a particular solvent. Some pesticides, such as carbaryl, cannot be formulated economically as an emulsifiable concentrate because the solvents in which the active ingredient is soluble are too expensive for field use.

Van Valkenburg (1973) discussed the factors which affect the stability of an emulsion which involves a complex dynamic equilibrium in the disperse phase–interface–continuous phase system. The stability of an emulsion is improved by a mixture of surfactants as the anionics increase in solubility at higher temperatures, whereas the reverse is true of non-ionic surfactants (Van Valkenburg, 1973). Becher (1973) lists a number of emulsifiers, together with a numerical value for the hydrophile–lipophile balance (HLB). An unstable emulsion 'breaks' if the disperse phase separates and forms a 'cream' on the surface, or the globules coalesce to form a separate layer. Creaming is due to differences in specific gravity between the two phases, and can cause uneven application.

Agitation of the spray mix normally prevents creaming. Breaking of an emulsion after the spray droplets reach a target is partly due to evaporation of the continuous phase (usually water) and leaves the pesticide in a film which may readily penetrate the surface of the target. The stability of emulsions is affected by the hardness and pH of water used when mixing for spraying and also conditions under which the concentrate is stored. High temperatures and frost can adversely affect a formulation.

Choice of solvent may also be influenced by its flashpoint so as to reduce possible risks of fire during transportation and use, especially with aerial application. For example, naphthenes are too inflammable for use as insecticide solvents. Emulsifiable concentrates have been applied without mixing in water, but their use as a ULV formulation is not advisable owing to the high volatility of the solvent.

Emulsions pre-mixed with a small quantity of water to form a mayonnaise-type formulation deteriorate in storage, so are not used. Miscible oil formulations are similar to emulsifiable concentrates, but contain oil in place of, or in addition to, the organic solvent. These products are less volatile and more suitable for applications in hot, dry climates.

Invert emulsions

Use of a viscous invert (water-in-oil) emulsion has been considered for aerial application of herbicides to minimize spray drift (Pearson and Masheder, 1969) but has not been accepted because of the need for specially designed equipment. Invert suspensions have been used as drift control agents (Hall *et al.*, 1998).

Encapsulated pesticides

Microencapsulated formulations have been developed primarily for volatile chemicals, e.g. pheromones (Fig. 3.2) (Hall and Marr, 1989) and for controlled release of the pesticide. There are three basic processes:

(1) a physical method of covering a core with a wall material
(2) a phase separation in which microcapsules are formed by emulsifying or dispersing the core material in an immiscible continuous phase, or
(3) by using the second process followed by an interfacial polymerization reaction at the surface of the core.

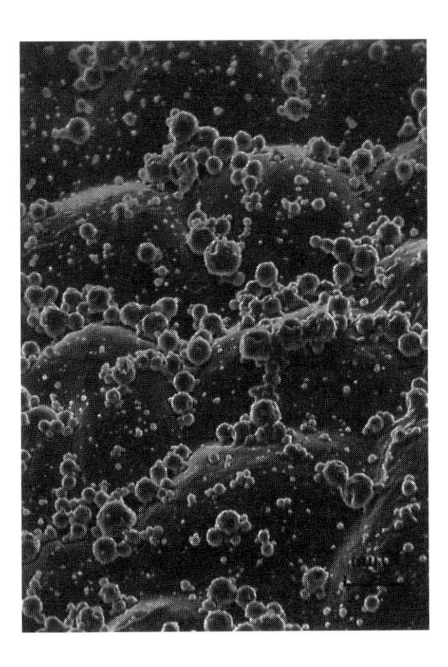

Fig. 3.2 A microencapsulated formulation of a pheromone on the adaxial surface of a cotton leaf (photo: ICA Agrochemicals, now Zeneca).

These processes are discussed in detail by Marrs and Scher (1990) and Tsuji (1990, 1993), Scher *et al.* (1998) and Perrin *et al.* (1998). The persistence of a deposit can be controlled by varying the wall thickness and type of polymer as well as the size of the microcapsule. Special materials to screen the effect of UV light can be incorporated in the capsules. Microcapsules of less than 10 μm diameter have been sprayed very effectively, but the wall thickness relative to the actual capsule size needs to be optimized for specific pesticides and their intended use. Beneficial insects are less exposed (Dahl and Lowell, 1984), although it has been argued that bees can collect capsules because their size is similar to pollen grains. An insecticide can be targeted at foliar feeding lepidoptera when the capsule wall is ruptured only when it reaches the alkaline gut (Perrin, 2000). Specificity can be increased, especially if a suitable attractant is used with a stomach poison, for example in leaf-cutting ant control (Markin *et al.*, 1972). Retention of microcapsules applied in suspension in water with stickers such as Acronal 4D is good on foliage, even after rain (Phillips and Gillham, 1973). In practice slow-release characteristics of microcapsules are particularly useful for application of chemicals which affect the behaviour of insects (Campion, 1976). Application of the pheromone disparlure was reported by Beroza *et al.* (1974), dicastalure by Marks (1976) and gossyplure by Campion *et al.* (1989). Evans (1984) has described a soluble acrylic system for applying the pheromone gossyplure.

Some surfactants solubilised in the aqueous phase of an insecticide microcapsule dispersion can provide a second barrier to the microcapsule wall. This enables the dermal toxicity of a product to be reduced without affecting the release rate of insecticide to such an extent that its activity is reduced to an unacceptable level. A micro-encapsulated formulation of fenitrothion was as effective as the UL formulation against grasshoppers, with some aggregation of capsules in large droplets (Holland and Jepson, 1996).

Ultralow volume formulations

When small spray droplets are used to achieve effective coverage, the evaporation of the droplets needs to be minimized. The decrease in size of droplets between the nozzle and the target as a result of evaporation is discussed in relation to meteorological factors in Chapter 4. A few insecticides such as malathion can be applied as the technical material without any formulation, although there is no need for such high concentrations of active ingredients. The number of such pesticides is very restricted, but it has been possible to formulate a biopesticide with lipophilic spores in a UL formulation. In many cases the viscosity has to be adjusted with a suitable solvent which is normally used to dissolve chemical pesticides. Barlow and Hadaway (1974) investigated a number of solvents to determine which were sufficiently non-volatile for a spray deposit to remain liquid for days or weeks rather than minutes or hours (Table 3.3). Although meteorological factors considerably influence rates of evaporation, they concluded that a suitable solvent should have a boiling point of at least 300°C at atmospheric pressure.

Table 3.3 Volatility of single compounds from cellulose papers at 25°C (from Barlow and Hadaway, 1974)

Compound	Boiling point at 760 mmHg (°C)	Volatility (g/m^2 per day)
n-Decane	174	2030
Isophorone	215	290
n-Hexadecane	287	2.7
Dibutyl phthalate	340	0.05

In addition to low volatility, a solvent suitable for ULV application should have a low viscosity index, i.e. the same viscosity at different temperatures, should be compatible with a range of chemicals and not be phytotoxic. The specific gravity should be high to increase the terminal velocity of small droplets, and pesticides should readily dissolve in it. Viscosity is particularly important in relation to flow rate of liquid to the nozzle. The risk of phytotoxicity is reduced with small droplets. Solvents with all these characteristics are not available (Table 3.4), so a mixture of solvents may be used which overcomes the problem to some extent and is a compromise between persistence and the need for the spray droplet to spread and penetrate an insect

Table 3.4 Physical properties of solvents (from Maas, 1971) (italic type signifies undesirable characteristics)

		Dissolving power	Volatility	Viscosity	Phytotoxicity
I	Low boiling aromatic hydrocarbons, e.g. xylene and solvent naphtha	Good	*High*	Low	Low
II	High boiling aromatic hydrocarbons, e.g. Iranolin, KEB	Good	Low	Low	*High*
III	Aliphatic hydrocarbons, e.g. white spirit kerosene	*Poor*	*Medium*	Low	Low
IV	High boiling alcohols, e.g. nonanol	*Medium*	Low	Low	*High*
V	Ketones, e.g. cyclohexanone	Good	*High*	Low	*Medium*
VI	Special solvents, e.g. pine oil and tetralin	Good	Low	Low	*High*
VII	Vegetable oils, e.g. cottonseed oil and castor oil	*Poor*	Low	*High*	Low
VIII	Glycolethers and glycols	*Medium*	Low	Low	Low
Ideal	ULV solvent	Good	Low	Low	Low

cuticle or plant surface. If droplet size is increased to allow for the volatility of one component of a mixture, fewer droplets can be sprayed from a given volume, reducing the coverage of the target, for example because of the cube relationship between diameter and volume of a droplet, doubling the diameter from 75 to 150 μm reduces the number of droplets to one-eighth. Data on the volatility of certain commercial formulations used in aerial spraying of cotton in Swaziland are given by Johnstone and Johnstone (1977). Solvents used in ULV formulations should have no detrimental effects on the application equipment and fabric of aircraft.

Several different mixtures of a low volatility oil and a more volatile solvent have been tried (Johnstone and Watts, 1966; Coutts and Parish, 1967). Vegetable oils such as cotton seed and soyabean oil have been in some ULV formulations (Scher, 1984). Special solution formulations of carbaryl, DDT and dimethoate (Maas, 1971) were applied successfully on cotton at 2.5 litres/ha, using a sprayer with a spinning disc nozzle (Matthews, 1973) and also with aerial application (Mowlam, 1974), but phytotoxicity was evident on foliage if droplets were too large. The high cost of these special formulations has limited their use, so products diluted in water have been used at 10–15 litres/ha, i.e. very low volume (VLV) on narrow swaths with the spinning disc held close to the crop (Mowlam *et al.*, 1975; Nyirenda, 1991; Cauquil and Vaissayre, 1995). In some areas, molasses has been added as an anti-evaporant and the volume of water reduced to 5 litres/ha (Gledhill, 1975).

Farmers using these VLV techniques still have to prepare the spray as with conventional hydraulic spraying, but are using a more concentrated spray. While this is suitable for some pesticides, the original concept of using ULV treatments was to eliminate mixing on the farm. In some situations pre-packaged products are available to fit directly on the sprayer without any further mixing. Progress in this direction requires closer collaboration between the equipment and chemical manufacturers to match the pesticide product with the sprayer. Products suitable for electrostatic spraying are referred to in Chapter 9. Other products, including several herbicides, are marketed for spinning disc sprayers. These products usually contain either a vegetable oil or refined mineral oil. The latter are selected with a minimum of unsulfonated residue (UR) of 92% to reduce phytotoxicity. A light mineral oil alone or with a fungicide is used at very low volume (VLV) 10–30 litres/ha to arrest development of banana leaf spot 'Sigatoka' (*Mycosphaerella musicola*) (Klein, 1961). Copper fungicide in a heavy alkylate oil was more resistant to field weathering when applied to control angular leaf spot of cucumber (Mabbett and Phelps, 1976). Superior deposition was achieved with ULV copper sprays for control of early blight of tomato (Mabbett and Phelps, 1974). Carbendazin fungicide mixed with a high-grade paraffinic oil was successfully applied to groundnuts for *Cercospora* control (Mercer, 1976). In general, ULV formulations should be checked for phytotoxicity at the proposed field application rate and also at double the rate, using the correct droplet size. Multiple applications may be required to detect any undesirable symptoms which do not show after a single application. Ideally, studies on phytotoxicity should include measurements of photosynthesis and respiration in the crop. Excessive rates of application or too large a droplet are likely to have an adverse effect on plants. The lower surface tension of oils allows greater penetration through stomata of certain leaves, and also through lipoidal leaf cuticles. As these oils are such poor solvents, a suitable solvent and cosolvent to dissolve sufficient active ingredient may have to be added. An advantage of mineral oils is that spraying equipment is not corroded or otherwise affected.

Specially formulated oils such as rapeseed oil plus emulsifier may be added to other formulations mixed with water, specially for controlled droplet application (Barnett, 1990). The proportion of oil in the final spray will depend on the volume applied; usually only 1–2 litres of oil per hectare can be used economically. Further evaluation of this type of carrier is needed with a range of chemicals on different crops to establish optimum concentrations and application rates. Less active ingredient may be required against some pests or weeds when formulations based on oils are used, because the chemical is spread more effectively on the target and is less likely to be washed off plant surfaces.

Uptake of the active ingredient may be reduced if high concentrations of active ingredient are applied in minimal volumes when users attempt to apply the same dose per unit areas as used in HV sprays. This may be caused by localized toxic effects preventing further absorption of the active ingredient.

Fog formulations

In thermal fogging machines, an oil solution of insecticide is normally used. Kerosene or diesel oil is a suitable solvent, provided the solution is clear and no sludge is formed. If a sludge is present a cosolvent, such as heavy aromatic naphtha (HAN) or other aromatic solvent, with a flash point in excess of 65°C should be used. Consideration of flash point is particularly important to avoid the hot gases igniting the fog. Wettable powder formulations have been used, but are normally mixed with a suitable carrier. Certain carriers are based on methylene chloride and a mixture containing methanol. Pre-mixing the powder with some water is advisable, especially with certain wettable powder formulations, so that a clod-free suspension is added to the carrier. Care must be taken to ensure that the viscosity of the fogging solution allows an even flow, and that powder formulations remain in suspension, as the spray tank on fogging machines is not equipped with an agitator. Pesticides such as pirimiphos methyl which have a fumigant effect are ideally applied as an aerosol spray or fog, provided the appropriate concentration is retained for a sufficient time.

Smokes

The pesticide is mixed with an oxidant and combustible material which generates a large amount of hot gas. Water vapour with carbon dioxide and a small quantity of carbon monoxide is produced when a mixture of sodium chlorate and a solid carbohydrate (e.g. sucrose) is used with a retarding agent such as ammonium chloride. The pesticide is not oxidized, as sugars are very reactive with chlorate. Care has to be taken in the design and filling of smoke generators to avoid an explosion and to control the rate of burning. The high velocity of the hot gas emitted from the generator causes the pesticide to be mixed with air, before condensation produces a fine smoke. The period of high temperature is so short that breakdown of the active ingredient is minimal. Smokes have been used in glasshouses and in warehouses and ships' holds. Care must be taken to avoid the smoke diffusing into nearby offices or living quarters, which should be evacuated during treatment.

A special form of smoke generator is the mosquito coil. The coils are made from an extruded ribbon of wood dust, starch and various other additives and colouring matter, often green, together with natural pyrethrins or allethrin. MacIver (1963, 1964a,b) gives a general description of the coils and their biological activity. Each coil is usually at least 12 g in weight and should burn continuously in a room without draughts for not less than 7.5 hours. Chadwick (1975) suggested that the sequence of effects of smoke from a coil on a mosquito entering a room is deterrence, expulsion, interference with host finding, bite inhibition, knockdown and eventually death. The coils provide a relatively cheap way of alleviating the nuisance from mosquitoes during the night.

Dry formulations

Dust

Dust is a general term applied to fine dry particles usually less than 30 μm diameter. Most dust formulations contain between 0.5 and 10 per cent of active ingredient. Transport of large quantities of inert filler is expensive, so a manufacturer may ship more concentrated dusts that are diluted before use in the country importing them. Sulphur dust is applied against some pathogens without dilution.

The concentrate is prepared by impregnating or coating highly sorptive particles with a solution of the pesticide. Alternatively, it may be made by mixing and grinding together the pesticide and a diluent in a suitable mill. The concentrate is then mixed, usually with the same diluent, to the strength required in the field (Table 3.5). Diluent fillers with high surface acidity or alkalinity, or a high oil absorption index, need to be avoided as the formulation would be unstable. Suitable materials for the diluent or carrier are various clay minerals such as attapulgite, often referred to as fuller's earth, montmorillonite or kaolinite (Watkins and Norton, 1955). Forms of silica or almost pure silica such as diatomite, perlite, pumice or talc are also used. Diatomite is composed of the skeletons of diatoms and, like all the other materials mentioned above except talc, it is highly abrasive to the insect cuticle, and can have an insecticidal effect (David and Gardiner, 1950). In the tropics, road dust drifting into hedges and fields is often very noticeable and can upset the balance of insect pests and their natural enemies. Dusts have been used to protect stored grain without an insecticide, but mortality is less as the moisture content is increased (Le Patourel, 1986).

Apart from sulphur dust used mainly as a fungicide, dusts are no longer used as much as the extremely small particles are prone to drift downwind, and winnowing of independent particles of active ingredient and diluent can also

Table 3.5 Examples of dust formulations

Component	Content (%)
(*a*) Dust concentrate	
50% Sevin (carbaryl) dust concentrate	
Technical carbaryl	50.5
a Montmorillonite clay, e.g. Peak clay	49.5
	100.0
(*b*) Field strength dust (by dillution)	
5% Sevin dust	
50% dust concentrate	10.2
a Montmorillonite clay, e.g. Pikes Peak clay	9.8
a Kaolinite clay, e.g. Barden A.G. clay	80.0
	100.0

occur (Ripper, 1955; Eaton, 1959). Often only 10–20% or less of a dust is deposited on the target (Courshee, 1960), so most dusts are now used to treat seeds, to protect horticultural crops grown in long narrow polythene tunnels where the water in sprays can exacerbate fungal diseases, and in farm stores. Seed can be treated centrally by seed merchants, but the product used in the treatment should contain a warning colour and bitter ingredient to prevent such seeds being eaten by humans, birds or farm animals.

Most dust is removed from foliage by rain, although the very small particles can adhere very effectively to plant surfaces. In some cases redistribution by rain can be advantageous; thus dusts applied to the 'funnel' of maize plants may be washed to where stalk-borer larvae penetrate the stem. However, most farmers now prefer to use larger microgranules to control this pest. Dusts with a low content of active ingredient (0.5% a.i.) with a short persistence such as malathion and pirimiphos methyl, have been mixed with grain (Hall, 1970), but there is concern about residues, even though surface deposits are removed by washing before the grain is ground into flour and cooked.

Granules

Large, discrete dry particles or granules are used to overcome the problem of drift, although care is essential during application to avoid fracture or grinding of the granules to a fine dust, which could be dangerous if inhaled or touched. This is particularly important, because pesticides, which are too hazardous to apply as sprays, such as aldicarb, are formulated as granules. These granules may be coated with a polymer and graphite to improve the flow characteristics and reduce the risk of operator contamination. Granules are prepared by dissolving the pesticide in a suitable solvent and impregnating this onto a carrier which is similar to those used in dust formulations, namely attapulgite or kaolin. Other materials that have been used include vermiculite, coal dust, coarse sand and lignin (Allen *et al.*, 1973; Wilkins, 1990; Humphrey, 1998). Sometimes a powder is made and the granule formed by aggregation (Whitehead, 1976). Goss *et al.* (1996) provide a recent review of granule formulation.

The choice of the inert carrier will depend on the sorptivity of the material, its hardness and bulk density (Table 3.6) (Elvy, 1976). Bulk density is especially important in relation to the volume of the product that has to be transported. Like dusts, the concentration of active ingredient is usually less than 15%, so transport costs per unit of active ingredient are high. The rate of release of a pesticide from the granules will depend on the properties of the pesticide, solvent and carrier, but the period of effectiveness is often longer than that obtained with a single spray application. The coating of a granule and the thickness of it can be selected to control the rate of release to increase persistence.

When an infestation can be predicted, a prophylactic application of granules may be more effective than a spray, especially if weather conditions prevent sprays being applied at the most appropriate time. Uptake of a pesticide by a

Table 3.6 Properties of dust diluents and granule carriers

	Bulk density (g/dm^3)	Specific gravity	pH
Oxides	144–176	2.0–2.3	5–8
Silicon			
Diatomite			
Graded sands			
Calcium	448–512	2.1–2.2	12–13
Hydrated lime			
Sulphates	784–913	2.3	7–8
Gypsum			
Carbonates	769–1073	2.7	8–9
Calcite			
Silicates	480–833	2.7–2.8	6–10
Talc	448	2.7–2.9	6–7
Pyrophyllite	608–705	2.2–2.8	6–10
Clays	480–561	2.6	5–6
Montmorillonite	432–496	2.6	7
Kaolinite			
Attapulgite			

plant may be negligible if the soil is dry and movement of chemical to the roots is limited, so granules of certain pesticides are more suitable on irrigated land where soil moisture can be guaranteed. Conversely, there may be phytotoxicity under very wet conditions. Granules have been used extensively in rice cultivation where they are broadcast, but the main advantage of granules is that they can be placed very precisely, so less active ingredient may be required and there is less hazard to beneficial insects. They are often placed alongside seeds or seedlings at planting, but spot treatment of individual plants is possible later in the season. In Africa, control of the stalk-borer of maize has been achieved with a 'pinch' of granules dropped down each maize 'funnel' (Walker, 1976). Banana plants may be treated with granules to control burrowing nematodes. Granules are often applied by hand, but this should be discouraged even if the person wears gloves. Simple equipment with an accurate metering device is available for both placement and broadcast treatments (see Chapter 12). With more precise placement, there is also less hazard to beneficial insects.

Despite the advantage of not mixing the pesticide on the farm, there has been a rather slow acceptance of granule application. One main drawback is that equipment required for granule application is more specialised than a sprayer and, with a smaller range of products available in granule form, farmers are reluctant to purchase a machine with a limited use. Granules are often applied at sowing, in which case the applicator has to be designed to operate in conjunction with a seed-drill or planter. Secondly, development of suitable equipment has been hampered by a lack of research to determine the best means of distributing granules to maximise their effectiveness, especially with herbicides where uniform distribution is essential. Variation in the quality of granules has also caused difficulties in calibrating equipment. Granules have been categorised by mesh size, the numbers indicating the coarsest sieve through which all the granules pass, and on which the granules are retained (Table 3.7), but similar samples may have quite different particle size spectra (Table 3.8) (Whitehead, 1976). The Agriculture (Poisonous Substance) Regulations in the UK require that not more than 4% by weight shall pass a 250 μm sieve, and 1% by weight through a 150 μm sieve when the more toxic pesticides are formulated as granules (Crozier, 1976). The size range affects the number of particles per unit area of target (Table 3.8).

Table 3.7 Sieve analysis of two samples described as 8/22 mesh granules

Pass mesh no.	Retained by mesh no.	Percentage of granules in sample	
8	12	2	10
12	16	36	60
16	22	42	30

Table 3.8 Estimated number of attapulgite granules per unit area

Mesh size	Particle size (μm)	Calculated no. of particles/m[a] applying 1 kg/ha
8/15	2360–1080	32
15/30	1080–540	253
20/40	830–400	817
30/60	540–246	2712
80/120 (microgranule)	200–80	78125

[a] The number of granules per kg will depend on whether dried or calcined granules of attapulgite are used; number of granules per plant can be calculated knowing the plant density.

Larger granules (8/15 mesh) which fall easily, even through foliage, are used principally for application in the soil or to water surfaces, for example to control mosquito larvae or aquatic weeds. Granules, including soft ones such as bentonite which release the toxicant quickly, have been widely used to control various pests in paddy fields where they can be broadcast by hand. Movement of insecticide is partly by systemic action, but also some chemical is carried by capillary action between stems and the leaf bases. With some pesticides, there may also be a localised fumigant effect. Chemical is lost if granules are carried out of the fields in irrigation water, so smaller microgranules (80–250 μm particle size) which adhere to foliage are used for application to rice plants. Size 30/60 mesh granules are normally used for stalk-borer control on maize.

Impregnating fertiliser granules with pesticides has been considered to eliminate the cost of an inert carrier and save time and labour during application. Apart from the possible breakdown of the pesticides when combined with fertiliser, there may be different requirements for timing and placement of toxicants and nutrients, although a broadcast application of certain herbicides plus fertiliser ('herbiliser') has been used with success (Ogborn, 1977). Walker (1971) reviewed the subject of residues following the application of granules.

Dry baits

Pesticides are sometimes mixed with edible products or sometimes with inert materials, usually to form dry pellets, or briquettes, which are attractive to pests. Using bran as a bait has controlled cutworms and locust hoppers, and banana bait has been used in cockroach control. Baits have also been used to control leaf-cutting ants (Lewis, 1972; Phillips and Lewis, 1973) and slugs. Maize and rodenticides have been mixed in wax blocks for rat control in palm plantations. Peregrine (1973) reviewed the use of toxic baits. A major problem with pelleted baits is that domesticated animals can eat them, and they disintegrate readily in wet weather and are then ignored by the pests. Non-pelleted baits go mouldy very rapidly, but a silicone waterproofing agent can be added to delay mould development. For invertebrate pests such as ants, the bait can be dispersed in the infested area, but mammalian pests may develop 'bait shyness' especially if dead animals are left near a bait station. Pre-baiting or a mixture of poisoned and unpoisoned baits reduces this.

Dry fumigants

Aluminium phosphide compressed in small, hard tablets with ammonium carbonate, on exposure to moisture releases the fumigant phosphine, together with aluminium hydroxide, ammonia and carbon dioxide. The tablets can be distributed evenly throughout a mass of grain in stores. Normally, no appreciable evolution of the fumigant occurs immediately, and respirators are not required if application is completed in less than one hour. The exposure period for treatment is usually three days or longer, so precautions must be taken to avoid personnel becoming affected. Other fumigants such as methyl bromide are supplied as liquefied gases under pressure in special containers, and their use is described in Chapter 14.

Other formulations

Pressure packs

The pesticide active ingredient is dissolved in a suitable solvent and propellent, and packaged under pressure in a 'pressure pack', commonly referred to as an aerosol can. Apart from pesticides, these are used to dispense a wide range of products such as hair lacquers, paints and deodorants. The pressure pack is a convenient but expensive means of producing aerosol droplets, and is used as a replacement for the 'Flit gun' (Fig. 3.3), which is less expensive, but requires manual effort to force air through a nozzle to which the insecticide is sucked by a venturi action. Propellants such as butane, carbon dioxide and nitrogen are now more commonly used, as fluorinated hydrocarbon use has been discontinued.. Because the propellant is confined at a temperature above its boiling point, opening of the valve on the top of the container (Fig. 3.4)

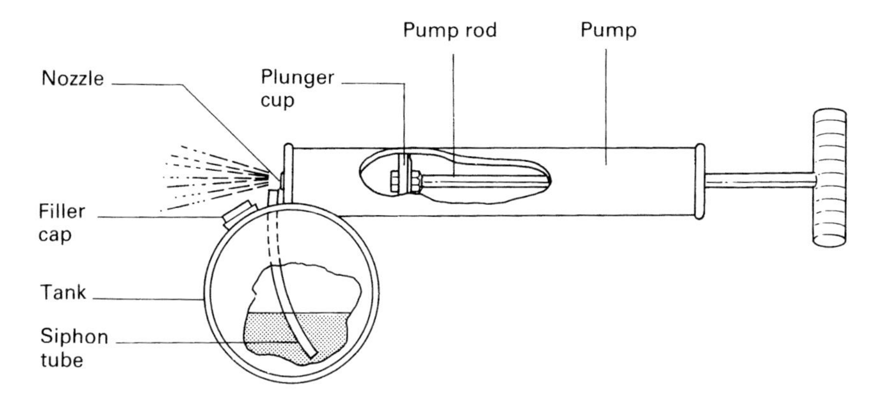

Fig. 3.3 Flit gun.

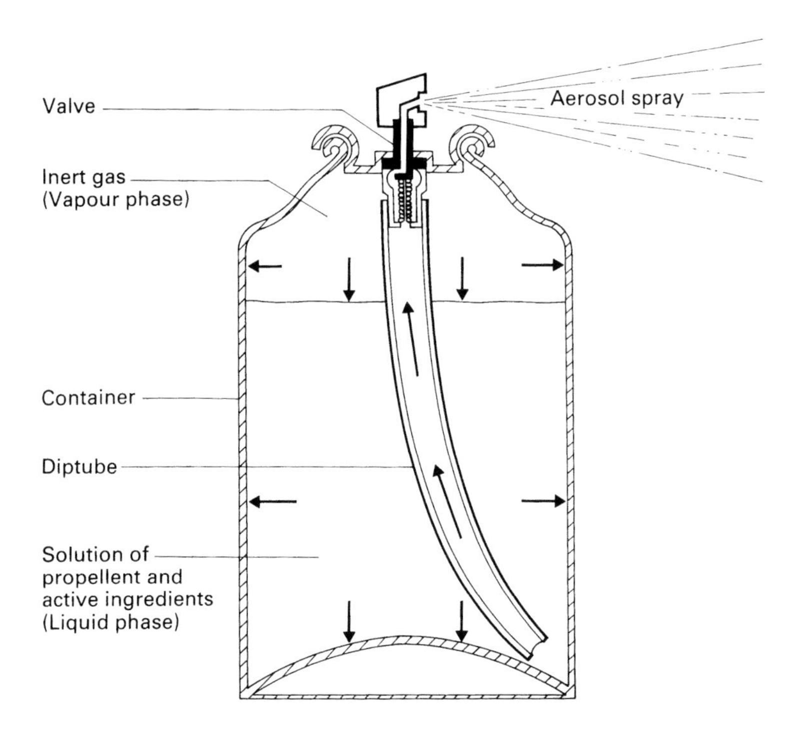

Fig. 3.4 Cross section of typical pressure pack.

allows the pressure inside the container to force the contents up a dip tube and through the valve, which is essentially a pressure nozzle. However, as the propellant reaches the atmosphere, some of it flashes from a liquid to a gas and causes the solution of active ingredient to break up into droplets. Further evaporation of the propellant and solvent causes a reduction in droplet size between the nozzle and target; hence the pressure pack should not be held too close to the target otherwise an uneven deposit will be obtained.

Typical valves continue to operate while the valve is depressed, but others incorporate a metering chamber, so that the quantity of product discharged can be controlled. The standard orifice is usually about 0.43 mm. When a coarse spray is required, the amount of propellant is reduced, and the valve may incorporate a swirl chamber as on cone nozzles. Finer sprays in a wider cone are obtained when the orifice in the valve has a reverse taper. The problem for some of the alternative propellants such as compressed carbon dioxide is that the pressure decreases as the pressure pack empties. The pressure increases with temperature when a liquefied compressed gas is used. Droplet size decreases rapidly after formation of droplets at the nozzle when a very volatile solvent such as xylene is used.

Tar distillates

These contain a mixture of aromatic compounds such as benzene, naphthalene and anthracene, tar bases such as pyridine and tar acids including phenol and cresol. This mixture starts to boil at 210°C, but less than 65 per cent of the volume has boiled at 350°C. Tar distillates have been used as dormant sprays.

Solubilised formulation

Solubilisation refers to the mixing of a water-soluble pesticide with oil by using suitable surfactants as cosolvents. The aim is to produce a formulation which increases penetration of the bark or leaf cuticle, but permits translocation of the active ingredient in the plant. The technique has been tried with some herbicides (Turner and Loader, 1974). The effect of leaf-applied glyphosate was improved compared with an aqueous spray, but the cost of oil and solvents was too great.

Banding materials

Localized application of pesticide to the trunk of trees can be achieved by banding. Grease bands have been used to trap insects climbing trees.

Plastic strips

Some volatile insecticides such as dichlorvos have been impregnated onto plastic strips. When hung in a closed room or room with minimal ventilation, the concentration of vapour released from the strip is sufficient to control some insects. This type of formulation is also used to dispense pheromones.

Paints and gels

Some insecticides have been incorporated into paints applied to surfaces where insects such as cockroaches may walk. Experiments have also included a systemic insecticide with a latex paint applied to the inside of pots to protect young seedlings from aphids and other sucking pests (Pasian *et al.*, 2000). Dispersion of herbicides makes it difficult to control aquatic weeds, so slow release formulations (e.g. gels) have been used (Barrett, 1978; Barrett and Logan, 1982; Barrett and Murphy, 1982).

Adjuvants

A wide range of non-pesticidal products has now been marketed as adjuvants (Fig. 3.5) (Hall *et al.*, 1993; Thomson, 1998). Manufacturers of these products claim that their use will enhance the performance of a pesticide and in some cases reduce the amount of active ingredient that needs to be applied. Many agrochemical companies believed that mixing an adjuvant with their pesticide was unnecessary due to their precise formulation, but with the extensive range of targets for a pesticide, it has become increasingly recognised that an adjuvant may be required in certain situations. Indeed, some agrochemical

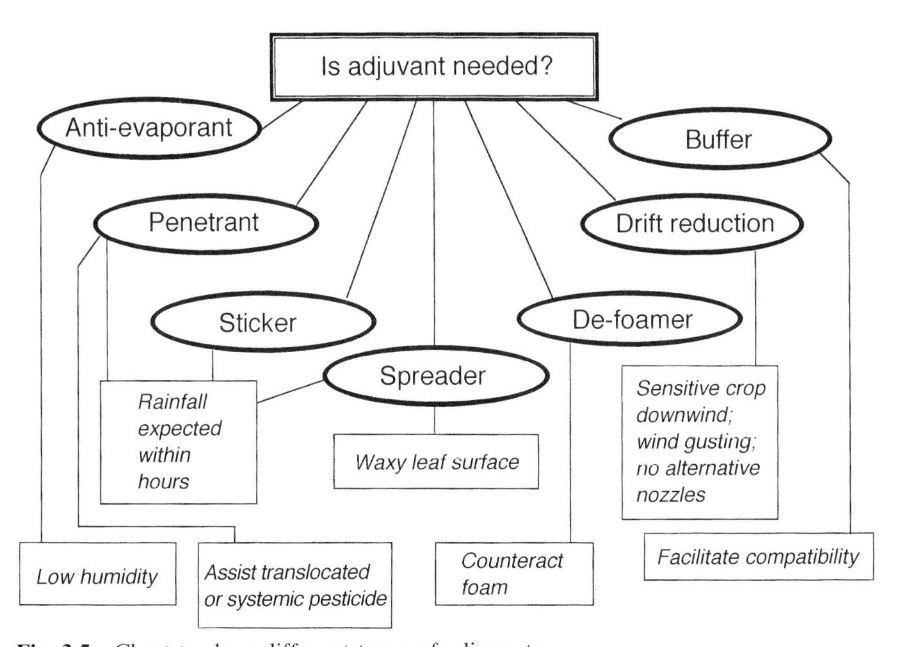

Fig. 3.5 Chart to show different types of adjuvants.

companies have marketed specific adjuvants for tank mixing with their pesticide. Although not pesticides, adjuvants do require registration as they affect the performance of pesticides (Chapman and Mason, 1993; Chapman *et al.*, 1998).

Adjuvants can be divided into several distinctly different types, although some will act in more than one way (Fig. 3.5). Adding a surfactant to a spray may improve the spread of a droplet across a hydrophobic surface, such as a waxy leaf. A surfactant can also improve penetration of the spray deposit through the leaf cuticle and thus reduce losses if rain occurs soon after application. Addition of an oil with emulsifier can reduce the proportion of small droplets in a spray and also decrease the effect of evaporation on droplet size, thus enhancing deposition with less drift (Western *et al.*, 1999).

Refined white oils have also been added to emulsifiable concentrate sprays applied at HV to improve penetration of the toxicant where the cuticle is particularly resistant to uptake of water-based sprays. Control of scale insects on citrus and other crops is a good example of this, where the addition of a suitable oil improves the effectiveness of an insecticide.

Reduction but not elimination of drift, particularly with aerial application, is assisted by adjuvants containing thickening agents, such as polysaccharide gum with thixotropic properties, alginate derivatives, hydroxethyl cellulose and various polymers. Using bioassays, Thacker *et al.* (1994) demonstrated that a polymeric adjuvant significantly reduced the amount of drift of an organophosphate insecticide. In the USA, Hewitt *et al.* (1999a) provide a review of drift control agents as a background to future guidelines on assessing their impact in minimising drift. These adjuvants can affect spray distribution, although such effects are less important with translocated herbicides than with a contact chemical (Downer *et al.*, 1998a). Sanderson *et al.* (1994) reported that agitation affected the polymer structure, and the addition of some of these has created mixing problems and increased costs. Some of the polymers used in drift retardant adjuvants lose their effectiveness after being re-circulated through the pump (Zhu *et al.*, 1997). In many cases changing the spray nozzles to provide a coarser spray, preferably with a narrow range of droplet sizes, or applying granules has more effectively reduced drift.

The persistence of a formulation can be improved by adding ‘stickers’, but care must be taken to avoid protecting the deposit so much that the availability of it to a pest is reduced. Persistence to rain-washing was improved by formulating wettable powders with amine stearate (Phillips and Gillham, 1971). Such formulations were effective on foliage that was difficult to wet, for example cabbage leaves (Amsden, 1962), and were also useful where new growth was insufficient to justify repeat applications. The addition of a surfactant to increase penetration of glyphosate has been used in tropical areas (Turner, 1985), but rain within one hour can remove 75 per cent of a herbicide deposit (Reddy and Locke, 1996). Apart from the use of various additives (including oils), improved rain-fastness can also be achieved with fine particles, which are not readily washed off by rain. Even a tropical thunderstorm fails to remove all the dust from surfaces, as electrostatic forces hold the particles to a surface. Advantages of small size and slow release of a pesticide are combined with microencapsulated formulations.

The addition of an adjuvant may affect the droplet spectrum of a nozzle due to changes in dynamic surface tension and viscosity. Butler Ellis and Tuck (1999, 2000) give data on the effects of adjuvants on fan nozzles. Changes in temperature can also influence spray droplet formation; thus the VMD decreased with increasing temperature (10–50°C) when the drift retardant Nalcatrol II was added, but droplet size increased with an organo-silicone surfactant (Downer *et al.*, 1998b). Adjuvants may also affect the distribution across the spray swath (Chapple *et al.*, 1993a; Hall *et al.*, 1993). The addition of a surfactant should also be checked to ensure that no phytotoxicity occurs at the concentration used. Some surfactants can interact with the epicuticular wax on leaves depending on the oxyethylene chain length (Knoche *et al.*, 1992). Regular conferences report effects of adjuvant use with different pesticides (e.g. Foy, 1992; Foy and Pritchard, 1996).

Choice of formulation

Formulations have usually been selected on the basis of convenience to the user. Farmers who have large tractor-mounted sprayers fitted with hydraulic agitation prefer liquid concentrates which can be poured into the tank or transferred straight from the container, as a volume of concentrate is much easier to measure than to weigh out a powder. Nevertheless, in many parts of the world, the less expensive wettable powder or granule is used extensively. Pre-packaging selected weights of dry particulate formulations facilitates having the correct dosage for knapsack or tractor equipment, by eliminating the need to weigh them on the farm. When these dry formulations are packaged in a sachet made from a water soluble polymer, the whole sachet can be placed in the spray tank or induction hopper (Fig. 3.1). Particular care is needed to avoid touching the surface of these sachets with wet hands/gloves and keeping them protected in a dry container until needed. The water soluble polymer may be slow to dissolve at low temperatures. Wettable powders have a rather better shelf life than emulsifiable concentrates, an important factor when it is difficult to forecast requirements accurately.

Barlow and Hadaway (1947) showed that a particulate deposit was more readily available to larvae walking on sprayed leaves than an emulsifiable concentrate, perhaps because the leaf absorbed the emulsion more readily. Large-scale trials on cotton confirmed that by using a wettable powder instead of an emulsifiable concentrate formulation against bollworm, farmers obtained a higher yield at less cost (Matthews and Tunstall, 1966). Similarly, DDT wettable powder was recommended for residual deposits on walls of dwellings for mosquito control. With powder formulations, particle size is important, especially with some pesticides such as insect growth regulators. In general, micronization of a formulation provides finer particles which are more effective for contact pesticides than when coarse particles are present.

When stomach poisons are applied, surface deposits are effective against leaf-chewing insects, but less so against borers which often do not ingest their first few bites of plant tissue. The effectiveness of stomach poison can be improved by the addition of a feeding stimulant, such as molasses, to these sprays. Carbaryl wettable powder is relatively ineffective against the noctuid *Helicoverpa armigera*, but up to 20 per cent molasses added to the spray gave improved larval mortality and also considerable mortality of moths feeding on the first three nights following application.

Choice of formulation has often been dictated by the availability of equipment in developing countries. Low percentage concentration dusts and granules can often be applied by hand or shaken from a tin with a few holes punched in it, when a sprayer is not available. On the other hand, farmers may be reluctant to use granules where neither labour nor specialised equipment is readily available. Shortage of water in many areas has dictated the use of dusts or granules, but high transport costs have favoured the use of highly concentrated formulations as these are less bulky.

Phytotoxicity is another factor in determining the choice of formulation, as some plants, or indeed individual varieties, are susceptible to certain solvents and other ingredients, such as impurities due to the use of cheap solvents. Phytotoxic effects may be caused by chemical burning, physically by droplets on the plant surface acting as lenses which focus the sun's rays on the plant tissue, or by subsequent effects on plant growth.

The use of formulations of low concentration reduces the toxic hazard when measuring and mixing the concentrate material with diluent or applying it undiluted. This hazard can be reduced by properly designed and standardised containers, which allow the concentrate to be either poured easily or pumped directly into the spray tank. The greatest danger occurs when the spray operator does not wear protective clothing, especially when only a small quantity is required from a large container. Spillage may occur over the operator's hands or feet, or a splash may contaminate the eyes or skin. Water-soluble plastic sachets eliminate handling of concentrate formulations, but strong outer packaging is needed to keep the sachets dry until required. Some liquid products are now packed in containers which incorporate a measure (Fig. 3.6).

Usually, what is readily available and the price decide the choice of formulation. In general, the cheapest in terms of active ingredient are wettable powder formulations and those with the highest amount of active ingredient per unit weight of formulation. When assessing costs, the whole application technique needs to be considered, since the use of a particular formulation may affect the labour required, the equipment and spraying time. Whichever formulation is chosen, users must read the instructions with great care before opening the container. Manufacturers attempt to provide clear instructions on the label of each container, but limitations of pack size may restrict information on the label, in which case an information leaflet is usually attached to the container. Care must be taken to avoid premature loss of labels, and important information can also be easily obliterated by damage under field

Fig. 3.6 Litre container incorporating measure to dispense small volumes of liquid pesticide formulations.

conditions. If in doubt about the correct dosage rate and method of use, alternative sources of information, such as the appropriate pesticide manual or crop pest handbook, should always be checked before starting a pesticide application programme.

4

Spray droplets

Importance of droplet size in pest management

The aim must be to optimise the amount of pesticide deposited on the intended target with minimal losses elsewhere. Unfortunately, most sprays contain a large number of droplets that vary in size, so some contain too much pesticide, while others may be too small and are particularly prone to movement outside a treated area – the spray drift that needs to be avoided wherever possible. Pesticide sprays are generally classified according to droplet size with particular reference to volume median diameter (VMD) in micrometres (Table 4.1). In the UK, spray quality is based on the assessment of the droplet size spectrum measured using a laser system (see below) rather than just the VMD shown in Table 4.1 (Doble *et al.*, 1985). This system was updated by Southcombe *et al.* (1997), as its use had been taken up by other countries. Reference nozzles were used to demarcate the separation between the

Table 4.1 Classification of sprays[a] according to droplet size

Volume median diameter (μm)	Size classification		
< 25	Fine aerosol	fog[b]	very fine spray
26–50	Coarse aerosol		
51–100	Mist[c]		
101–200	Fine spray		
201–300	Medium spray		
< 300	Coarse spray[d]		

[a] Standards for spray quality classification use specified reference nozzles to demarcate the boundary between different spray qualities. This table is a guide for those unable to do a full assessment of spray quality.
[b] The term 'fog' is used in the UK for treatments with a VMD < 50 μm and with more than 10% by volume below 30 μm
[c] Mist treatments must have less than 10% by volume < 30 μm.
[d] In the USA, an additional extra coarse spray is used.

different spray qualities because the various measuring instruments can give different numerical results. Womac *et al.* (1999) report on the differences in droplet sizes obtained from nozzles produced by different manufacturers and indicated that a dedicated set of reference nozzles would improve the overall uniformity in classification thresholds. Womac (2000) evaluated over a hundred nozzles for use as reference nozzles to support an ASAE standard (X-572) which is the American equivalent of the BCPC nozzle classification system.

As indicated earlier, the original studies were made with standard fan nozzles, so now data from operating the nozzles in a wind tunnel are also considered as a measure of the drift potential of the nozzle (Miller *et al.*, 1993; Herbst and Ganzelmeier, 2000). This was necessary to accommodate rotary atomisers as well as air inclusion and twin fluid nozzles that produce droplets containing air bubbles. Instead of quoting a droplet size, the terms 'very fine', 'fine', 'medium', 'coarse' and 'very coarse' are used to indicate the spectrum produced by a nozzle at a given operating pressure. Such information will increasingly be included on pesticide labels (Hewitt and Valcore, 1999).

Space treatments require very small droplets that remain airborne, so fogs (see Chapter 11), also referred to as aerosols, are used. Mists are considered ideal for treating foliage with very-low or ultralow volumes of spray liquid. The droplets in the 50–100 μm range can move downwind, but are small enough to be transported within a crop canopy by air turbulence and be deposited on leaves. When drift must be minimised a 'medium' or 'coarse' spray is required, irrespective of the volume applied. However, even when a coarse spray is applied, with most standard hydraulic nozzles there will be a proportion of the spray volume emitted as very small droplets that can drift. The proportion of small droplets has been decreased with certain new nozzle designs described in Chapter 5. Although more prone to downwind drift, especially in hot weather with thermal upcurrents, a fine spray may be required where good coverage of foliage is required, especially on the more vertical leaves of cereal crops. An electrostatic charge on the droplets can improve deposition on exposed leaves, but deposition will be poor on the lower leaves unless there is sufficient air turbulence to improve penetration of the crop canopy. Where a fine spray is required drift can be reduced by also using a downwardly directed air flow.

The most widely used parameter of droplet size is the volume median diameter (VMD) measured in micrometres (μm). A representative sample of droplets of a spray is divided into two equal parts by volume, so that one half of the volume contains droplets smaller than a droplet whose diameter is the VMD and the other half of the volume contains larger droplets (Fig. 4.1). A few large droplets can account for a large proportion of the spray and so can increase the value of the VMD. The number median diameter (NMD) is when the droplets are divided into two equal parts by number without reference to their volume, thus emphasising the small droplets. The ratio between the VMD and NMD will give an indication of the range in sizes of droplets within a spray. If a spray was produced with a VMD/NMD ratio of unity, then all the

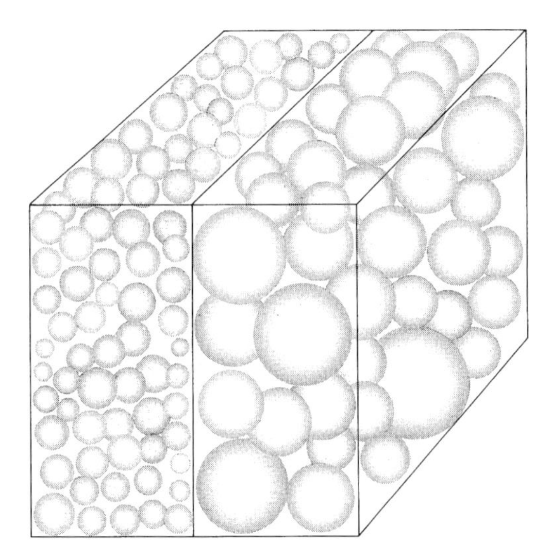

Fig. 4.1 Diagrammatic representation of the VMD – half of the volume of spray contains droplets larger than the VMD, while the other half has smaller droplets.

droplets would be of the same size. Table 4.2 indicates droplet parameters for a range of different nozzles. NMD is more difficult to measure, so the range of droplet size is more often referred to by the 'span'. This is the difference in diameter for 90 and 10 per cent of the spray by volume divided by the VMD.

$$\text{Span} = \frac{D_{0.9} - D_{0.1}}{D_{0.5}}$$

Bateman (1993) and Maas (1971) favour the volume average diameter (VAD), which is the diameter of the droplets representing the total volume of spray divided by the number of droplets. The VAD can also be expressed as the average droplet volume (ADV) in picolitres. Dividing 10^{12} by the ADV will give the estimated number of droplets of uniform size that can be obtained from a litre of liquid. Lefebvre (1989) describes other measures of droplet size used mainly in relation to studies of combustion and other industrial uses of spray nozzles. Butler Ellis and Tuck (2000) preferred to use Sauter mean diameter for aerated droplets produced by air induction nozzles, due to the variation in amount of air within droplets.

When choosing a given droplet size for a particular target, consideration must be given to the movement of spray droplets or particles from the application equipment towards the target. The magnitude of the effects of gravitational, meteorological and electrostatic forces on the movement of droplets is influenced by their size. The size of individual droplets has not always been considered in the past, as most nozzles produce a range of droplet sizes. When a high volume of spray is applied, droplets coalesce to provide a continuous film of liquid on the surfaces which are wetted. As indicated

Table 4.2 Some examples of spray droplet size data for different nozzles[a]

Nozzle	Flow rate	Spray liquids used	r.p.m. or pressure	VMD	NMD
Vortical	35 ml/min	Deodorised kerosene	0.2 bar	12	
Electrodynamic	6 ml/min	ED blank	25 kV	48	46
Spinning cup 52 mm dia.	30 ml/min	ULV	15 000 r.p.m.	70	42
Standard fan 80°	200 ml/min	Water + wetting agent	300 kPa	99	22
Standard fan 80°	800 ml/min	Water + wetting agent	300 kPa	145	13
Cone D4/25	1.47 l/min	Water + wetting agent	500 kPa	228	42
Air inclusion[b]	0.96 l/min	Water + wetting agent	200 kPa	500	
Spinning disc 90 mm dia.	60 ml/min	Water + wetting agent	2000 r.p.m.	260	
Airshear	50 ml/min	Risella oil	85 m/s[c]	132	25
Airshear	400 ml/min	Water + wetting agent	85 m/s	282	35
Airshear	480 ml/min	Water + wetting agent	100 m/s	90	24

[a] See subsequent chapters for description of different nozzles.
[b] Performance of air inclusion fan nozzles varies between manufacturers.
[c] Air velocity at nozzle.

earlier, there is greater concern about effects of pesticides reaching non-target organisms, so it is imperative where possible to select a droplet size, or at least as narrow a range of sizes as possible, to increase the proportion of spray that is deposited on its intended target.

The theoretical droplet density obtained if uniform droplets were distributed evenly over a flat surface is given Table 4.3. The number of droplets available from a specified volume of liquid is inversely related to the cube of the diameter; thus the mean number falling on a square centimetre n of a flat surface is calculated from

$$n = \frac{60}{\pi}\left(\frac{100}{d}\right)^3 Q$$

where d is the droplet diameter in micrometres and Q is the volume is spray (litres applied) per hectare.

Volumes of 50–100 litres/ha will become more accepted as growers increasingly relate application to the amount of foliage that needs protection.

Table 4.3 Theoretical droplet density when spraying 1 litre evenly over 1 ha

Droplet diameter (mm)	Number of droplets/cm^2
10	19099
20	2387
50	153
100	19
200	2.4
400	0.298
1000	0.019

In the UK, growers can increase the concentration of a.i. in a spray by up to ten times that specified on the label if not specifically prohibited to do so and provided the maximum dose rate is not exceeded, but the grower must accept responsibility for using any variation of the label recommendations such as a reduced dosage.

Movement of droplets

Effect of gravity

A droplet released into still air will accelerate downwards under the force of gravity until the gravitational force is counterbalanced by aerodynamic drag forces, when the fall will continue as a constant terminal velocity. Terminal velocity is normally reached in less than 25 mm by droplets smaller than 100 μm diameter, and in 70 cm for 500 μm droplets. The size, density of the contents of the droplet, and the shape of the droplet, together with the density and viscosity of the air, all affect terminal velocity. Thus,

$$V_t = \frac{gd^2Q_d}{18\eta}$$

where V_t is the terminal velocity (m/s), d is the diameter of the droplet (m), Q_d is the droplet of density (kg/m^2), g is gravitational acceleration (m/s^2), η is the viscosity of air in newton seconds per square metre (1 N s/m^2 = 10 P (poise)) equal to 181 μP at 20°C. This equation is usually referred to as Stokes' law.

The most important factor affecting terminal velocity is droplet size. The terminal velocity for a range of sizes of spheres is given in Table 4.4, and is approximately the same for liquid droplets within this range, but droplets may be deformed due to aerodynamic forces so the diameter is reduced and terminal velocity is less than calculated for a sphere. Owing to their low terminal velocities, droplets of less than diameter 20 μm will take several minutes or longer to fall in still air. Examples of the time to fall to ground level when released from a height of 3 m are shown in Table 4.4. Small droplets are thus exposed to the influence of air movement over a longer period. In a light

Table 4.4 Terminal velocity (m/s) of spheres and fall time in still air

Droplet diameter (μm)	Specific gravity		Fall time from 3 m (sp.gr = 1)
	1.0	2.5	
1	0.00003	0.000085	28.1 h
10	0.003	0.0076	16.9 min
20	0.012	0.031	4.2 min
50	0.075	0.192	40.5 s
100	0.279	0.549	10.9 s
200	0.721	1.40	4.2 s
500	2.139	3.81	1.65 s

breeze, for example a constant wind velocity of 1.3 m/s parallel to the ground, a 1 μm droplet released from 3 m can theoretically travel over 150 km downwind before settling out. In contrast a 200 μm droplet can settle less than 6 m downwind if the droplet remains the same size.

If air moved smoothly over a flat surface (laminar flow), the distance S that droplets travel downwind could be predicted from the equation

$$S = \frac{HU}{V_t}$$

where H is the height of release U is the wind speed and V_t is the terminal velocity of droplets.

The Porton method of spraying described by Gunn *et al.* (1948) utilised this principle to spray by aircraft, by adjusting spray height within practical limits inversely with wind speed to deposit spray with droplets of a given size at a fixed distance downwind of the source. This technique is still the basis for drift spraying against locusts where droplets of 70–90 μm are released so that they move downwind and are collected on the vegetation on which locusts are feeding (Courshee, 1959).

In practice, airflow is not laminar. Surface friction affects the flow of air, even over a flat surface, so that wind speed is zero at ground level. The topography of the land will also influence air movement. However, the presence of a crop will cause crop friction and significantly affect the flow of air. On large fields with a crop of uniform height, e.g. wheat, there will be less crop friction than in an intercrop with a tall and low crop, such as grounduts with maize.

Large droplets (>200 μm) will be deposited rapidly by sedimentation, so spray drift will not be a problem. Thus, in areas immediately adjacent to an ecologically sensitive area such as a water course, a buffer zone is needed unless a coarse spray is applied to reduce drift. However, large droplets in a coarse spray will follow a vertical path and will be collected on predominently on horizontal surfaces. If not collected on foliage, such droplets will fall to the soil surface. Smaller droplets will give better coverage of foliage as their trajectory will be increasingly affected by air flows, and thus their path from the nozzle will change direction. Droplets moving in a more horizontal plane can be impacted on the more vertical parts of crops, i.e. the stems and petioles as well as the more vertical leaves of monocotyledon crops. Small droplets released above a flat field can travel long distances, but foliage of a crop can filter out most of the droplets (Payne and Shaefer, 1986). Few droplets will be deposited on the undersides of leaves unless the nozzle is positioned to spray upwards and/or there is air turbulence moving the leaves or an upwardly directed airflow (see Chapter 10).

Effect of meteorological factors

The proportion of spray which reaches the target is greatly influenced by local climatic conditions, so an understanding of the meteorological factors affecting the movement of droplets necessitates information on the climate close to the ground. The basic factors are temperature, wind velocity, wind direction and relative humidity.

Air temperature is affected by atmospheric pressure which decreases with height above the ground, so that if a mass of air rises without adding or removing heat, it expands and cools. A decrease in temperature of approximately 1°C for every 100 m in dry air is referred to as the adiabatic lapse rate. If the temperature decreases more rapidly, a super-adiabatic lapse rate exists. Under these conditions, a mass of air which is close to the ground, warmed by radiation from the sun, will start to rise and continue to do so while it remains hotter and lighter than surrounding air. These convective movements of air result in an unstable atmosphere, and thus turbulent conditions, such as those associated with the formation of thunderstorms when large changes in wind speed and direction often occur (Fig. 4.2). Turbulence can occur at night under monsoonal conditions.

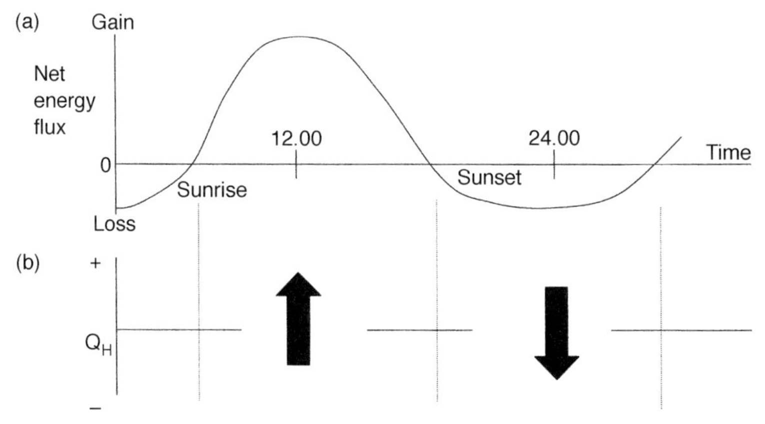

Fig. 4.2 Schematic representation of diurnal variations in (a) net energy flux; (b) sensible heat flow

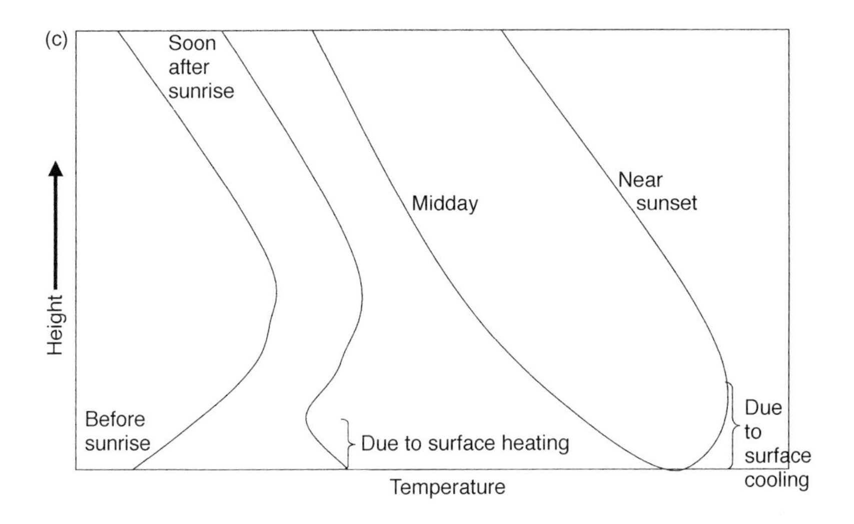

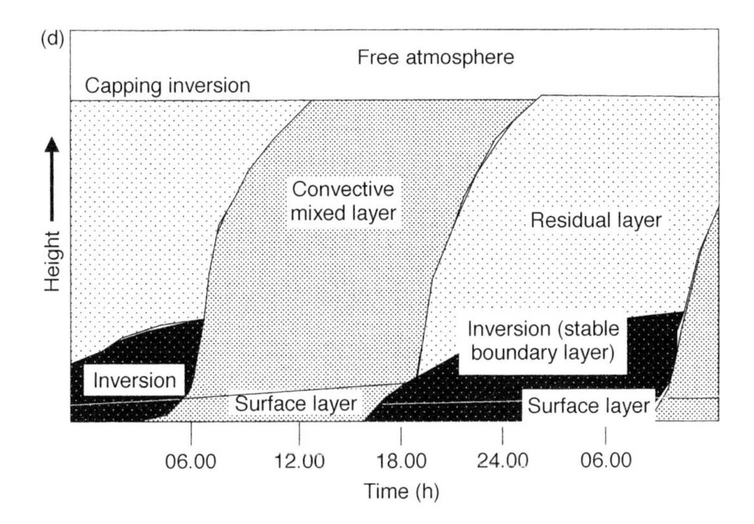

Fig. 4.2 (c) air temperature profiles and (d) boundary layer structure for a period of fine weather with clear skies (from Bache & Johnstone, 1992).

A temperature decrease less than the adiabatic lapse rate inhibits upward movement of air so the atmosphere is stable. When the ground loses heat by radiation and cools more rapidly than the air above it, air temperature increases with height and an inversion condition exists (Fig. 4.2d). Inversions typically occur in the evening when there is a clear sky following a hot day, and may persist until after dawn and until the sun heats up the ground.

Fog or early morning mist occurs during inversion conditons, when wind velocity is low and air flow approaches a smooth laminar state, so there is little turbulence (Fig. 4.3a). Irregularities in the ground surface cause masses of air to be mixed by friction, so eddies develop. These can cause rapidly fluctuating gusts, lulls and changes in wind direction to occur. This mixing of air may destroy an inversion or it may persist at a higher level. Therefore the stability of the atmosphere is affected by the movement of masses of air from convection caused by thermal gradients and surface friction determined by local topography. Roughness of vegetation, causing resistance to air flow, is one of the factors influencing surface friction (Fig. 4.4). Bache and Johnstone (1992) discuss in greater detail the dispersion of spray in relation to the microclimate associated with crops. Earlier accounts are given by Sutton (1953), Pasquill (1974) and Oke (1978).

Measurements of air turbulence include the dimensionless Richardson number (Richardson, 1920) and stability ratio SR (Coutts and Yates, 1968).

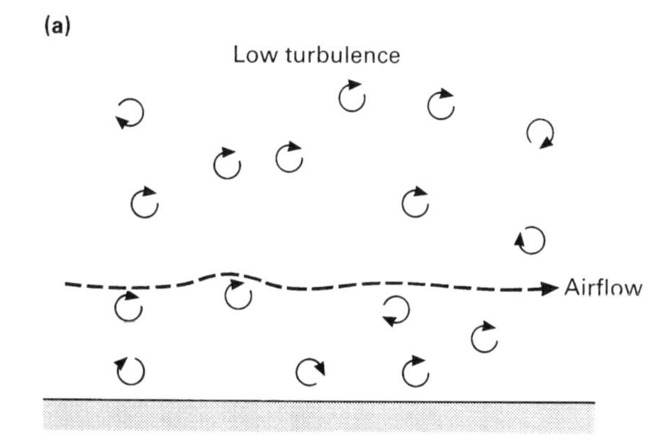

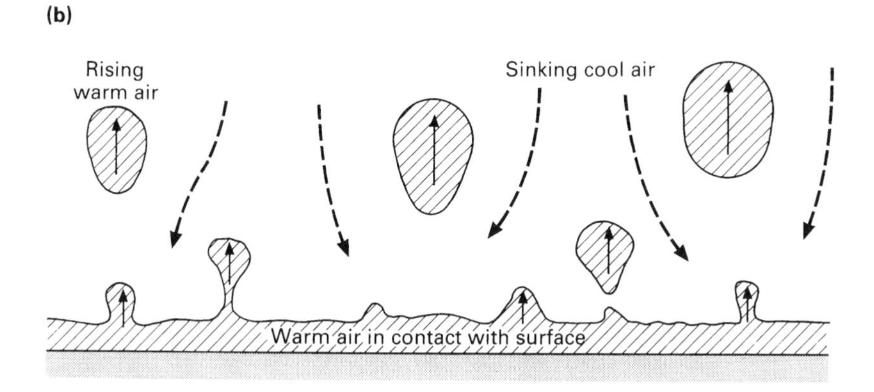

Fig. 4.3 (a) Stable inversion conditions. (b) Air turbulence caused by surface heating – super adiabatic lapse rate conditions.

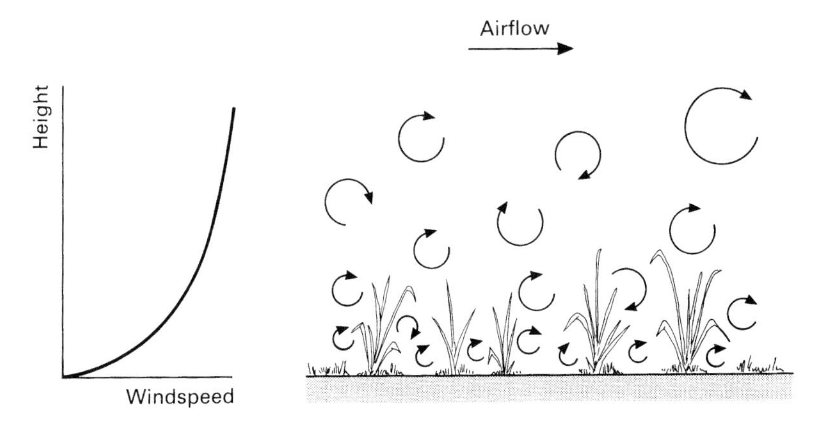

Fig. 4.4 Air turbulence caused by surface friction.

$$SR = \frac{T_2 - T_1}{U^2} \times 10^5$$

where T_2 and T_1 are the temperatures (°C) at 10 and 2.5 m above ground level and U is the wind velocity (cm/s) at 5 m. The gustiness of a wind is not taken into account.

A positive SR value indicates temperature inversion conditions, which are ideal for applying a cold or thermal fog to control mosquitoes. This fortunately often coincides with mosquito activity in the evening or early morning. A negative stability ratio occurs when there is turbulent mixing. Normal lapse rate and mild mixing conditions prevail if the SR is at or near zero. Convection is usually less on cloudy days.

A multidirectional anemometer may be used to record variations in wind speed within crop canopies, especially in orchards, to assess the impact of an air assisted sprayer on spray distribution. Mini-meteorological stations are also available for growers to use and can be used in conjunction with computer models to optimise timing of an application in relation to a pest or disease (Fig. 4.5) (Leonard *et al.*, 2000).

Effect of evaporation

The surface area of the spray liquid is increased very significantly when dispersed as small droplets, especially when the diameter of the droplet is less than 50 μm (Fig. 4.6). A droplet will lose any volatile liquid over this surface area. The rate of evaporation decreases as the evaporation from a droplet saturates the surrounding air, but as droplets move further apart, this effect is diluted. Changes in concentration of the components of the spray liquid due to non-volatile components may depress the vapour pressure of any solvent within the formulation. Many of the older emulsifiable concentrate formulations contained highly volatile organic solvents. The main concern is that the

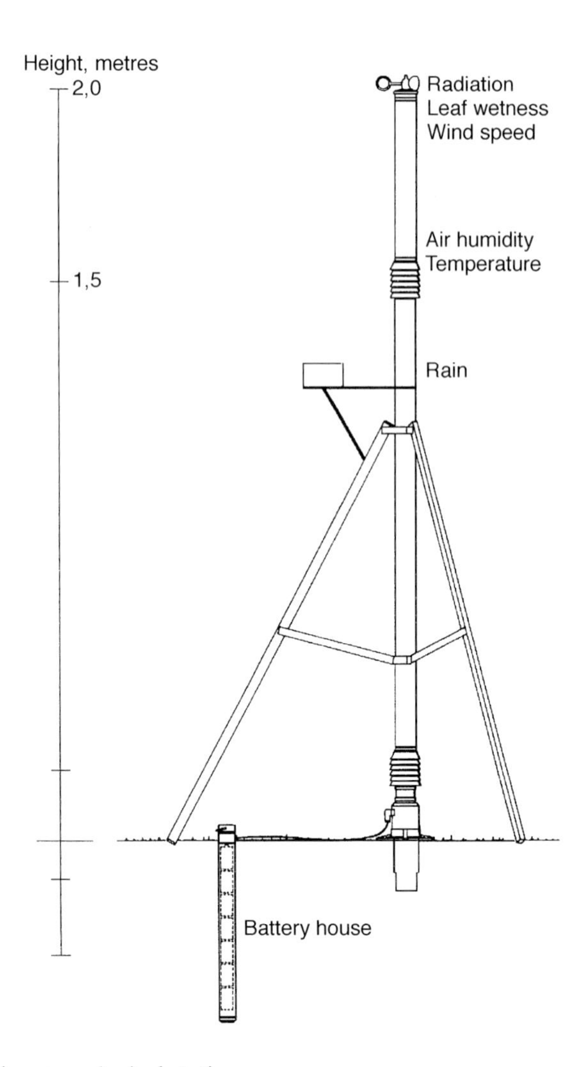

Fig. 4.5 Mini-meteorological station.

most widely used diluent of pesticidal sprays is water, which is volatile. Thus evaporation of diluent during the flight of droplets will cause droplets to shrink in size and become more vulnerable to movement by air flows.

Spillman (1984) indicated that the diameter of freely falling water droplets ($>150\,\mu m$) decreased linearly with time, but below $150\,\mu m$ the rate at which the diameter decreased increased by about 27 per cent. This change seems to be associated with the fall in Reynolds number, such that at values greater than four a toroidal vortex of trapped air becomes saturated and reduces the rate of evaporation from part of the surface (Fig. 4.7). Batchelor (1967) had shown that the liquid within a falling droplet would follow certain streamlines (Fig. 4.8) from which Spillman postulated that because volatile liquid evapo-

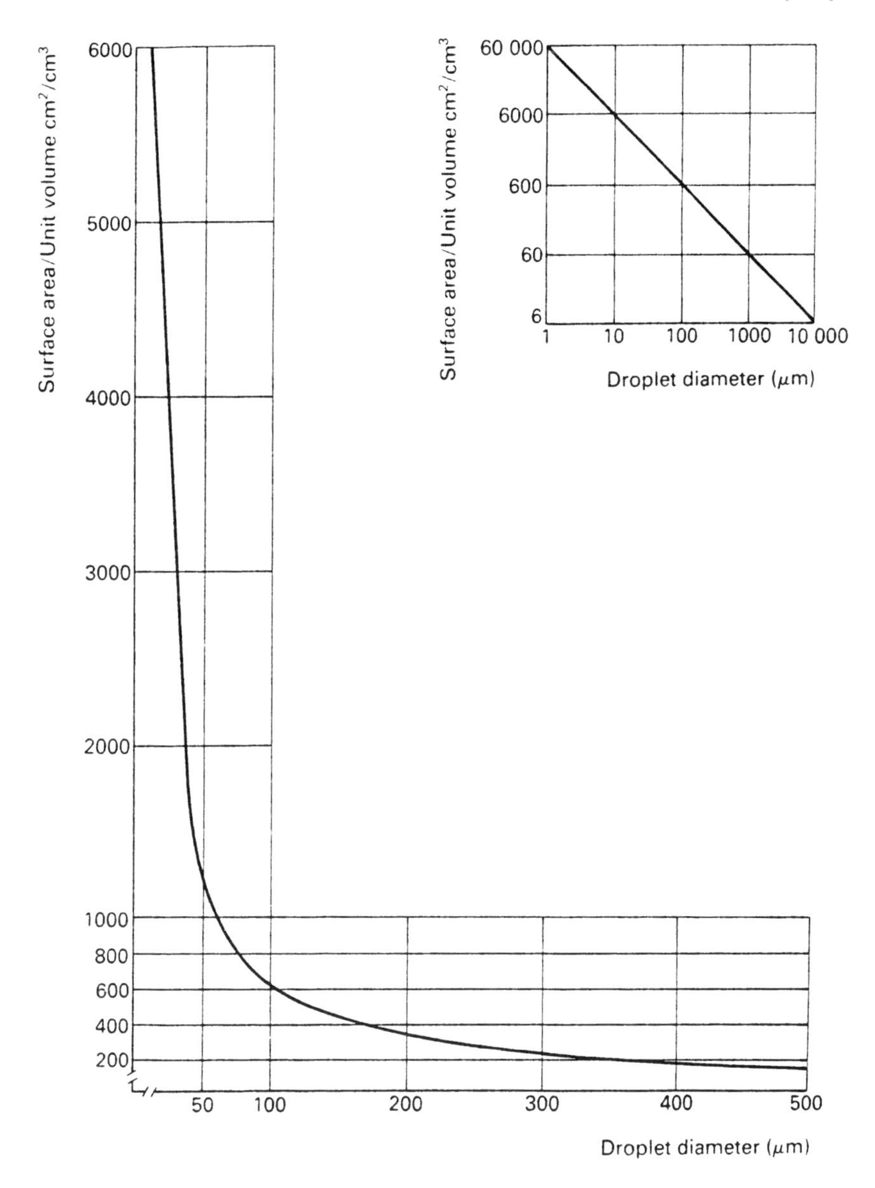

Fig. 4.6 Rate of increase of specific surface or reduction of droplet diameter (after Fraser, 1958).

rated from the surface, the concentration of any involatile component will increase. If the involatile component has a higher viscosity, the surface velocity will decrease, and this can result in a more rigid skin of involatile material over the surface. Studies suggested that if 20–30 per cent molasses is added to a spray, the thickness of the skin was 1.5–3 μm for 70–100 μm droplets. Similar effects have been noted when an oil adjuvant is mixed with the spray.

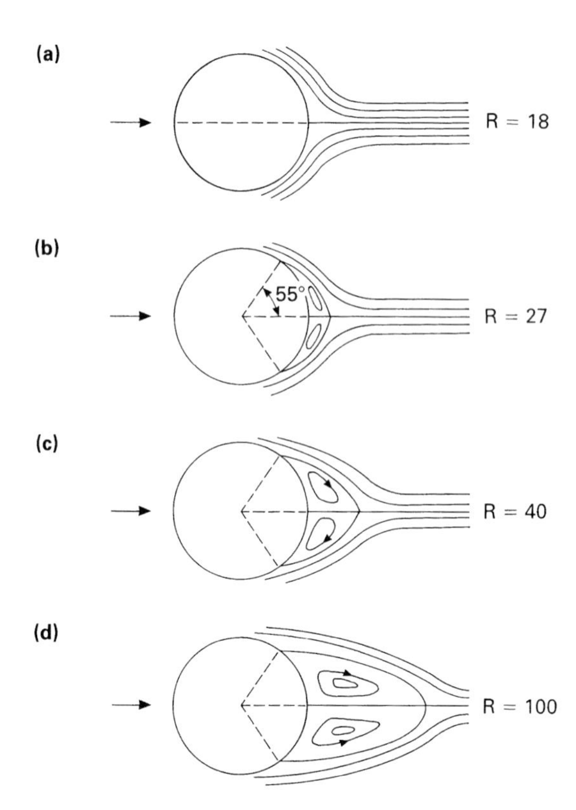

Fig. 4.7 Development of the rear toroidal vortex behind a sphere as Reynolds number increases (Spillman, 1984).

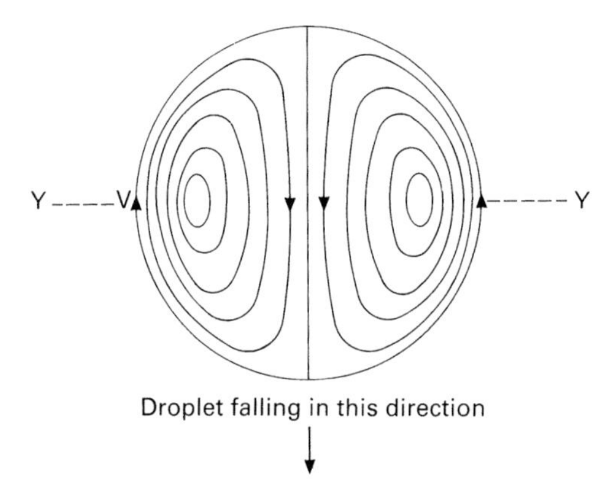

Fig. 4.8 Streamlines of the flow induced by surface friction on a falling droplet (from Batchelor, 1967).

The interaction between the water within a droplet and moisture in the surrounding air is very complex, but a simple equation indicates the lifetime t of a water droplet measured in seconds (Amsden, 1962)

$$t = \frac{d^2}{80\Delta T}$$

where d is the droplet diameter (μm); ΔT is the difference in temperature (°C) between wet and dry thermometers (i.e. a measurement of relative humidity).

Data (Table 4.5) show that the small droplets of water have a very short lifetime, so if a pesticide spray loses all the diluent, it creates a very small particle of concentrated chemical which may then be carried over much longer distances by air flows. Thus in hot, dry conditions, it is important to consider the use of a non-volatile adjuvant to ensure that the droplets will not shrink below a minimum size. It is also stresses the need for an involatile carrier in ULV applications. Johnstone and Johnstone (1977) recommended that spraying water-based sprays at 20–50 litres/ha with 200–250 μm droplets should cease if ΔT exceeds 8°C or the temperature exceeds 36°C. Lower values are needed if smaller droplets are applied.

Table 4.5 Lifetime and fall distance of water droplets at different temperatures and humidities

	Conditions A[a]		Conditions B[b]	
Initial droplet size (μm)	Lifetime to extinction (s)	Fall distance (m)	Lifetime to extinction (s)	Fall distance (m)
50	14	0.5	4	0.15
100	57	8.5	16	2.4
200	227	136.4	65	39

[a] Temperature 20°C, ΔT 2.2°C and RH 80%.
[b] Temperature 30°C, ΔT 7.7°C and RH 50%.

The theoretical distance (cm) a water droplet of diameter d (μm) will fall due to gravity before all the water has evaporated is given by

$$\frac{1.5 \times 10^{-3} d^4}{80\Delta T}$$

Droplet dispersal

A droplet will follow the resultant direction (V_r) depending on the combined effects of gravity (V_f) mean wind velocity (V_x) and turbulence (V_z) which can be upward when convection forces prevail (Fig. 4.9). This can be clearly seen when spray is drifted over several rows during unstable conditions. A proportion of the spray may not be collected within the crop being treated.

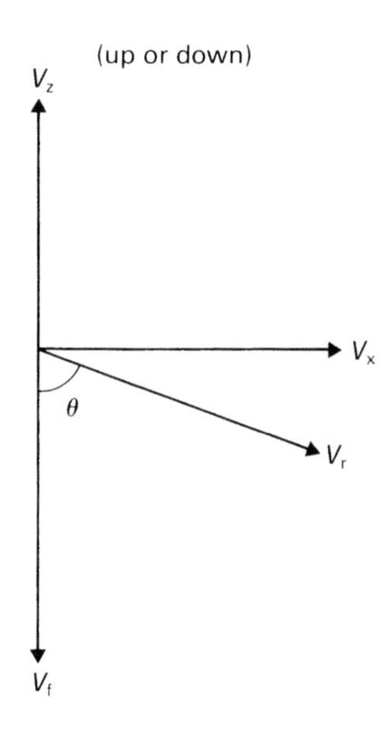

Fig. 4.9 Resultant direction of a droplet (V_r) depending on the magnitude of effects of gravity, wind and convective air movement (after Johnstone *et al*.,. 1974).

Bache and Sayer (1975) found that peak deposition of small droplets downwind was proportional to the height of the nozzle and inversely proportional to the intensity of turbulence, whereas larger droplets are relatively unaffected by turbulence and sediment according to the HU/V_t relationship. Upward movement of small droplets (< 60 μm) is counterbalanced by downdraughts which return the droplets elsewhere; thus when relatively small areas are involved (hectares rather than square kilometres) the downdraughts may deposit droplets on a totally different area, contaminating other crops or pastures.

Evidence of this has been clearly demonstrated when an untreated crop, susceptible to a particular pesticide, shows distinctive symptoms of damage. A good example of this is when cotton has 'strap leaf' due to 2,4-D herbicide, which may be detected considerable distances from the site of application. Early morning is often considered the best time to apply herbicides (Skuterud *et al*., 1998), provided there are no small droplets. If an inversion persists, such small airborne droplets could disperse in any direction.

Studies on droplet dispersal under field conditions are not easy due to variations in meteorological conditions, variations in droplet size, and the complexity of sampling. Many different techniques have been used to sample spray droplets downwind. Flat sheets have been widely used to measure fallout of the larger droplets, but smaller targets have been used for airborne spray to increase the capture efficiency. Bui *et al*. (1998) evaluated a number of different samplers and Amin *et al*. (1999) report studies with air samples for aerosol and gaseous pesticides. Hewitt *et al*. (2000) included field studies with a range of hydraulic nozzles, while similar studies have been made in several countries. However, as results in the field are quite variable due to changes in wind speed and direction, most attention to the quantification of spray drift is being given to wind tunnel studies. Using a single nozzle and 2 mm diameter polythene lines as collectors in a wind tunnel, Phillips and Miller (1999) and Walklate *et al.* (2000a) then employed a model to relate laboratory measurements to field data. This system has been used to classify drift from boom sprayers in relation to determining buffer zones.

Some comparisons in the field have been made by spraying simultaneously with different machines, each applying a different tracer (Johnstone and Huntington, 1977). Similarly Sanderson *et al.* (1997) reported using an aircraft with a separate spray system for each wing so the two different dyes simulataneously traced the distribution of different formulations of a herbicide. Parkin *et al.* (1985) used two food dyes – red erythrosine and water blue, while Cayley *et al.* (1987) suggested using a series of chlorinated esters as tracers. Babcock *et al.* (1990) used the ninhydrin reaction to quantify deposits of an amino acid. Cross and Berrie (1995) used fluorescent tracers Tinopal CBS-X or Uvitex OB (with 10% Helios per litre) to assess orchard sprayers and photographed deposits on leaves under UV light for analysis with a computer image analyser. Payne (1994) used a fluorescent tracer suspended in tripropylene glycol monomethyl ether to aerial sprays applied in different wind conditions. Background fluorescence on some leaves can be a problem in assessing deposits. An alternative is the use of chelated metal salts; thus Cross *et al.* (2000) used manganese, zinc, copper and cobalt salts to compare different spray treatments on strawberries. Choice of tracer is particularly important when assessing spray distribution on food crops.

A more complex approach was made in Canada to examine the dispersion of a spray cloud from an aircraft. Mickle (1990) used a light detection and range (LIDAR) laser beam to scan the spray cloud and sample 1 m^3 of cloud at distances of 1 km. Similar studies showed that movement of airborne droplets was primarily dependent on the stability of the atmosphere, and that widespread dispersal of a small amount of pesticide is inevitable (Miller and Stoughton, 2000). A LIDAR technique has also been used to assess the structure of orchard crops to assist development of improved spraying techniques (Fig. 10.18) (Walklate *et al.*, 2000b)

The biological effect of downwind drift of insecticides was assessed by bioassays using two day old *Pieris brassicae* larvae on potted plants (Davis *et al.*, 1994). Similarly, herbicide drift was examined on young seedlings of ragged-robin (*Lychnis flos-cuculi*) in conjunction with deposition of fluorescein on various sampling receptors (Davis *et al.*, 1993)

Spray distribution

Fluorescent tracers are also used more generally to demonstrate differences in the distribution of deposits on plants (Patterson, 1963). Insoluble micronised powders such as Lunar Yellow or Lumogen have been widely used (Fig. 4.10), but require careful mixing with a suitable surfactant before being diluted with water to the correct volume. A pre-formulated fluorescent dye suspension is available in some countries. A fluoresence spectrophotometer can be used to assess leaf surface deposits (Furness and Newton, 1988). Uvitex 2B has also been widely used, as it retains sensitivity at low concentration (Hunt and Baker, 1987). Downer *et al.* (1997) examined the effect of several water soluble fluorescent tracers on the droplet spectra produced by a fan nozzle. They showed that addition of a tracer can increase the proportion of small droplets produced; thus care needs to be taken in matching the spray quality with a pesticide spray when using a tracer to study downwind drift.

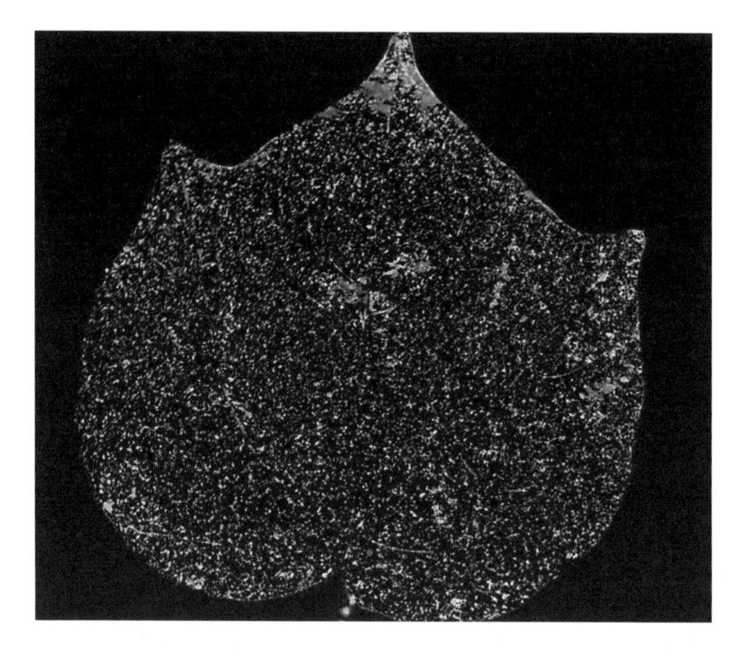

Fig. 4.10 Fluorescent spray deposit on cotton leaf (photo: ICI Agrochemicals, now Zeneca).

Samples of the target, often leaves, are collected and examined under UV light. Care must be taken to avoid looking directly at the UV light. Samples are usually sorted into arbitrary categories depending on the amount of cover obtained (Staniland, 1959). Courshee and Ireson (1961) and Matthews and Johnstone (1968) did chemical analysis of a sub-sample of leaves from each of the arbitrary categories to relate actual deposits with coverage. This allowed examination of a very large sample of leaves but limited the need for chemical analyses. Quantitative analysis of deposits is possible with light stable soluble tracers that are easily removed from the surface and measured with a fluoresence spectrometer. Murray *et al.* (2000) have also used ranked set sampling in spray deposit assessment and introduced an image analysis system that enables images to be stored. Carlton (1992) developed a method of washing deposits from both the upper and lower leaf surfaces.

The behaviour of individual droplets landing on leaf surfaces was studied using a videographic system (Reichard *et al.*, 1998; Brazee *et al.*, 1999). Similar studies have been reported by Webb et al. (2000) who investigated the impact of droplets on pea leaves. The aim of such studies has been to develop a model to predict retention or loss of pesticide droplets from leaf surfaces. With a single droplet generator the reflection height of droplets rebounding from a leaf surface could be determined. The effect of different surfactants on aqueous spray droplets landing on an adaxial leaf surface was also quantified. Another model examining deposition within a cereal crop canopy has analysed the crop architecture, aiming to use crop height as the main parameter in the model (Jagers op Akkerhuis *et al.*, 1998).

Studies of droplets on leaf surfaces are also possible using scanning electron microscopy and cathodoluminescence (Fig. 4.11) (Hart, 1979; Hart and Young, 1987). Detailed spatial distribution can be provided by elemental mapping with a scanning electron microscope with energy dispersive X-ray (EDX) analysis. Fluorescence microscopy and autoradiography have also been used to trace deposits (Hunt and Baker, 1987), while Dobson *et al.* (1983) used neutron activation analysis to determine the amount of dysprosium in spray deposits in a crop and up to 100 m downwind. Salyani and Serdynski (1989) reported experimenting with a sensor to provide an electrical signal proportional to the amount of spray deposit.

Variations in spray coverage in farmers' crops are now usually demonstrated by clipping pieces of water sensitive paper to various parts of the crop. Stapling a folded paper to a leaf will show differences between upper and lower surface deposits. The papers can be made by treating glossy paper with a water sensitive dye such as bromophenol blue. The paper is yellow when dry, but aqueous droplets produce blue stains of the ionised dye (Turner and Huntington, 1970). Suitable papers are commercially available, but need to used carefully as the whole surface can turn blue in humid conditions and unless held on the edge, the surface readily shows blue fingerprints. Plain Kromekote cards have also been used when spraying a dye (e.g. Sundaram *et al.*, 1991).

Determination of spray droplet size

Measurement of spray droplets in flight is now done using one of several instruments that have a laser light beam through which droplets are projected. Sampling is either spatial (measuring droplets simultaneously within a defined space, within a section of the laser beam, i.e. number/m^3) or temporal (measuring droplet flux passing through a defined sampling volume over a set time, i.e. number per m^3 per second) (Fig. 4.12). In one type of instrument, the sampling volume is defined by the intersection of two laser beams. If all the droplets travel at the same speed both methods would be equivalent, but in

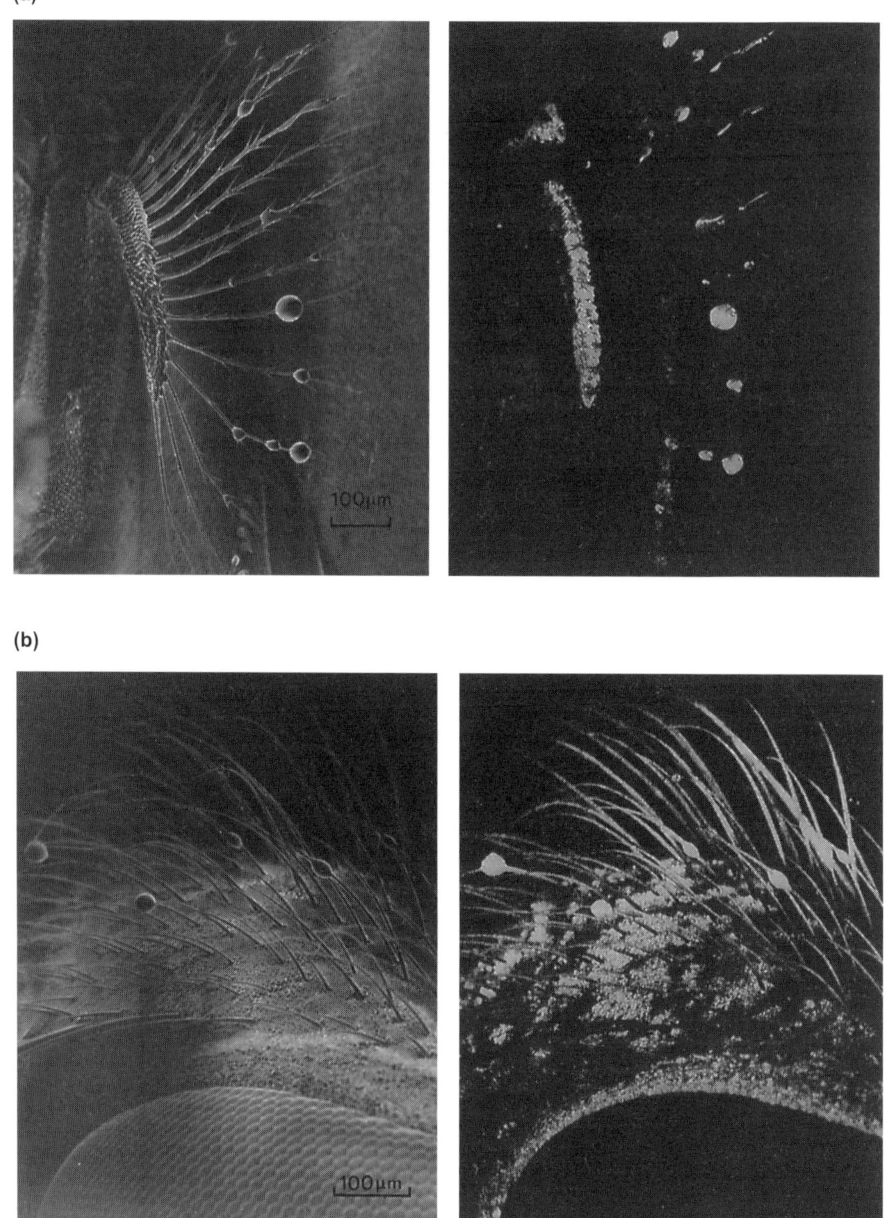

Fig. 4.11 Cathodoluminescence image of 'Ulvapron' oil containing 'Uvitex OB' and Brilliant Yellow R (photo: ICI Agrochemicals, now Zeneca). Spray droplets on tsetse fly: (a) arista; (b) near compound eye.

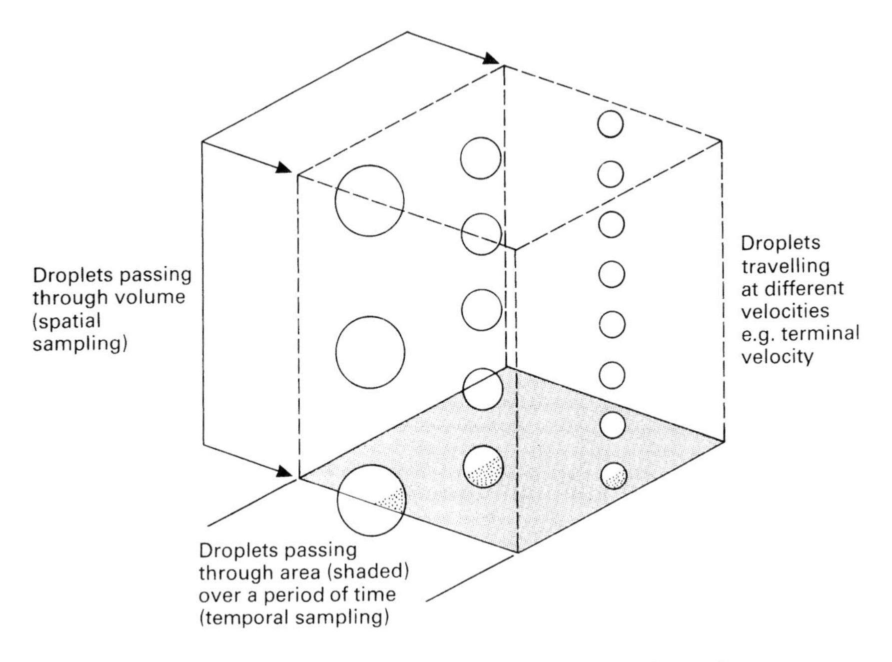

Fig. 4.12 Diagrammatic representation of spatial and temporal sampling.

practice droplets will differ in their velocity through the laser beam (Frost and Lake, 1981). This is particularly true with nozzles that produce a wide range of droplet sizes. There is the additional problem with droplets that contain air bubbles, as some instruments will see the bubbles as extra droplets.

Light scattering

The laser light diffraction technique developed by Swithenbank *et al.* (1977) has been extensively used for pesticide spray droplet analysis (Combellack and Matthews, 1981a; Arnold, 1983a; Hewitt, 1993). The Malvern Particle Size Analyser and a similar Sympatec instrument was also used by the US Spray Drift Task Force to generate generic data for the US Environmental Protection Agency (EPA) (Barry *et al.*, 1999; Hewitt *et al.*, 1996, 1999a,b, 2000). Spray is directed through the laser beam within one focal length of the lens so that light diffracted by the droplets is focused on a special photodetector in the focal plane of the lens (Fig. 4.13). The detector consists of 31 concentric, semicircular photosensitive rings which convert the light into an electrical energy signal processed by the computer. Data is normally analysed using a Model Independent program to obtain the best fit of the measurements. The volume of spray in different size classes is calculated, the size range being dependent on the focal length of the lens.

Data for hydraulic fan nozzles can be obtained with either the major axis of

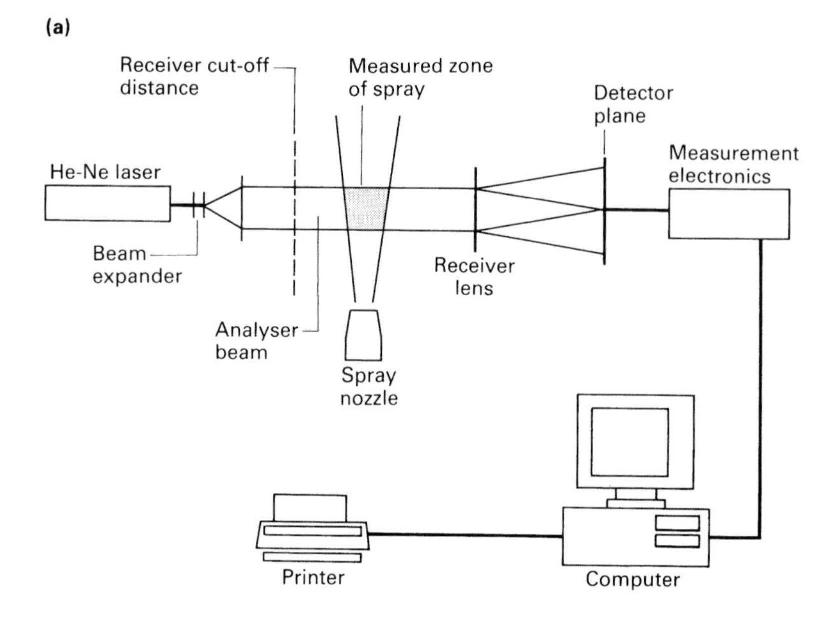

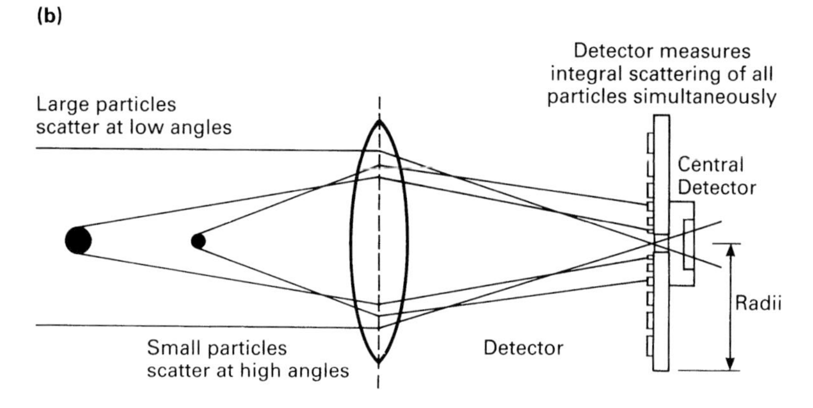

Fig. 4.13 (a) Layout of a Malvern light diffraction particle size analyser. (b) Diffraction of two different sized droplets. (c) Photo detector. (d) Transform property of receiver lens.

the fan in line with the laser beam, or alternatively individual parts of the spray can be examined with the fan perpendicular to the beam (Arnold, A.C., 1983). Cone nozzles need to be assessed more carefully as the droplet sizes will be affected by the part of the cone sampled (Combellack and Matthews, 1981a). In the USA, a Malvern mounted in a high speed wind tunnel measured droplets from nozzles used on aircraft (Hewitt *et al.*, 1994a,b). Apart from comparisons between different types of nozzles, flow rates and operating pressures, effects of formulation on droplet spectra have been measured

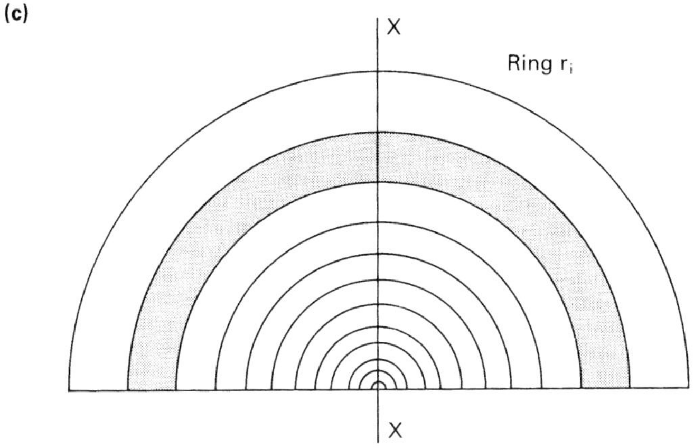
(c)
X
Ring r_i
X
Each detector element is an annular ring collecting light scattered between two solid angles of scatter w_1 and w_u

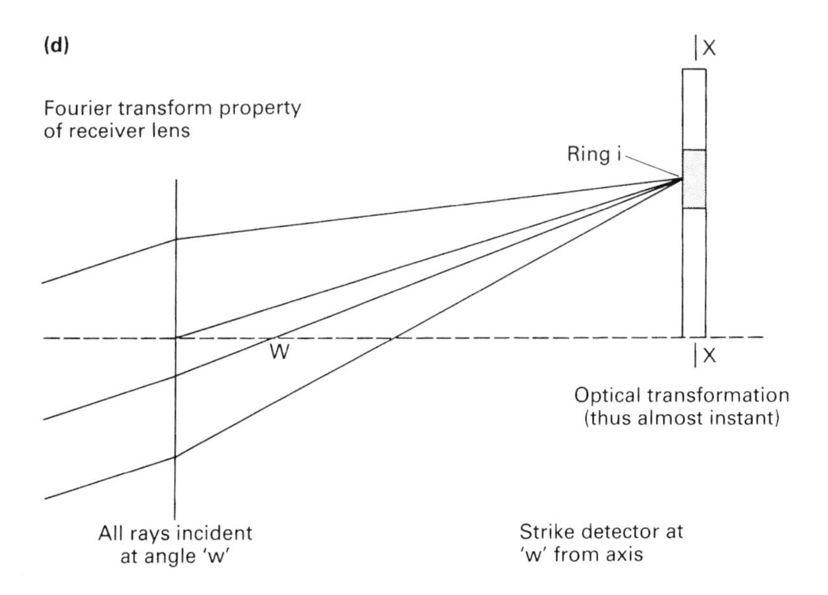
(d)
X
Fourier transform property of receiver lens
Ring i
W
X
Optical transformation (thus almost instant)
All rays incident at angle 'w'
Strike detector at 'w' from axis

(Combellack and Matthews, 1981b). The instrument can also be used to measure small particles in suspension in a glass cell mounted in the laser beam to check the formulation of a biopesticide (Bateman and Chapple, 2000).

Arnold (1987) and Teske *et al.* (2000) have compared the Malvern with another light scattering instrument, the Particle Measuring System (PMS) (Knollenberg, (1976) developed to measure particles in clouds. In this system of temporal sampling, droplets passing through a focused laser beam form a shadow on a photodiode array (Fig. 4.14). Droplet size is a function of the number of elements obscured by the passage of a droplet. Any droplet which is

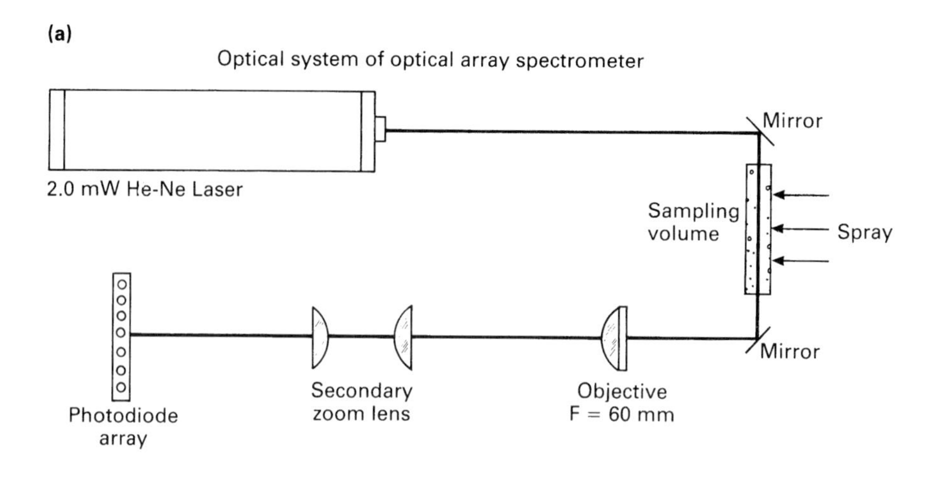

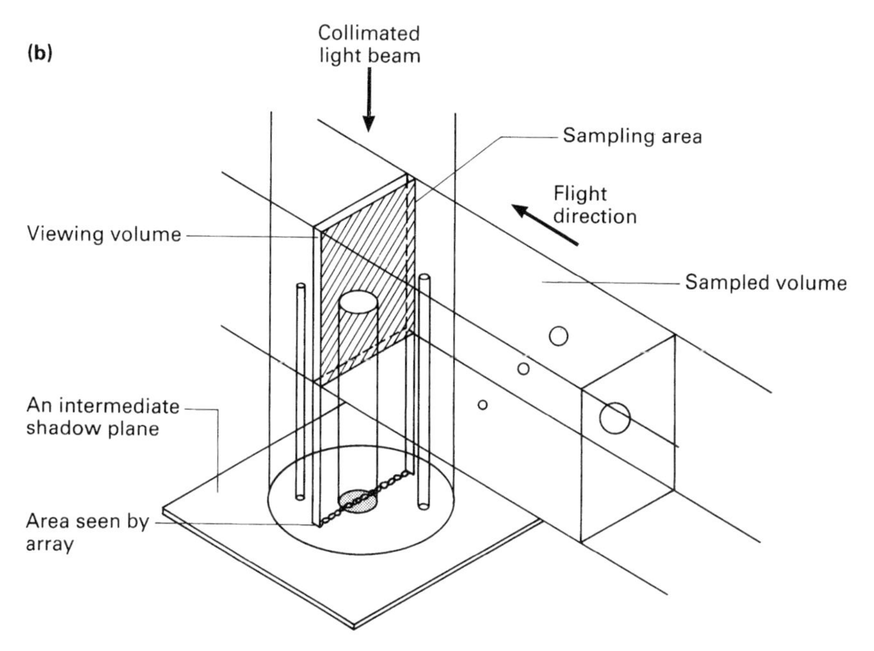

Fig. 4.14 (a) Optical system of optical array spectrometer. (b) Three dimensional diagram to show shadow effect on detector.

not in the correct plane produces an out-of-focus pattern, so the computer is programmed to accept or reject data depending on the shadow produced. As the sampling volume is small, most nozzles have to be moved on an x–y grid to obtain a representative sample of a spray (Lake and Dix, 1985). The PMS has been used in wind tunnels (Parkin *et al.*, 1980) to assess orchard sprayers (Reichard *et al.*, 1977) and on aircraft (Yates *et al.*, 1982).

Laser Doppler droplet sampling

In this system (Aerometrics and Dantec equipment) a beamsplitter and lenses are used with a continuous laser to provide two intersecting beams, where interference fringes are produced (Bachalo *et al.*, 1987; Lading and Andersen, 1989). A droplet passing through the intersection of the two beams produces modulated scattered light (a Doppler burst signal), the spatial frequency of which has to be measured to size the droplets (Fig. 4.15). Also, droplet velocity is measured, as it is proportional to the temporal frequency of the modulation. A forward light scattering angle of 30° is usually used. Three detectors are used to detect the phase shifts and avoid ambiguity in measurements. As there is a greater probability of larger droplets crossing the edge of the sample volume and producing an inadequate signal, the computer program adjusts for variation of sample volume. The droplet size range is dependent on the optics, but is usually with 50 class sizes. The total range can cover droplet diameters from 1 to 8000 μm. An on-line computer system provides real-time displays of droplet spectra. Using this type of equipment with an *x*–*y* grid, Western *et al.* (1989) compared hydraulic nozzles with a twin fluid nozzle.

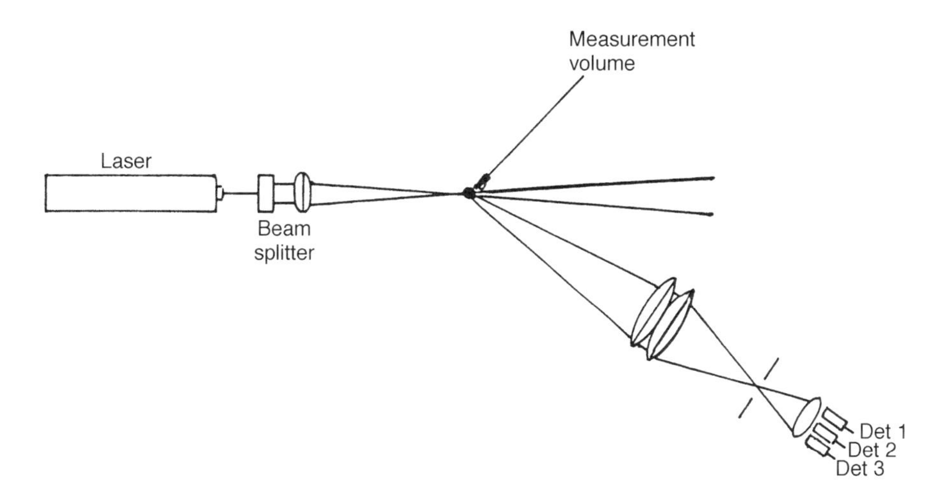

Fig. 4.15 Principle of phase Doppler droplet sizing.

When Tuck *et al.* (1997) compared the Doppler particle analyser with a two-dimensional imaging probe (PMS), they showed that each instrument produced different droplet size and velocity distributions, but both instruments were useful provided their limitations were recognised.

Other techniques

A new system involves a pulsed laser to capture an image of the spray which can be subsequently analysed. By using a double flash of the laser, each droplet is recorded twice so that droplet velocity is also calculated. The system also enables the behaviour of droplets to be observed as they impinge on leaf surfaces (Fig. 4.16).

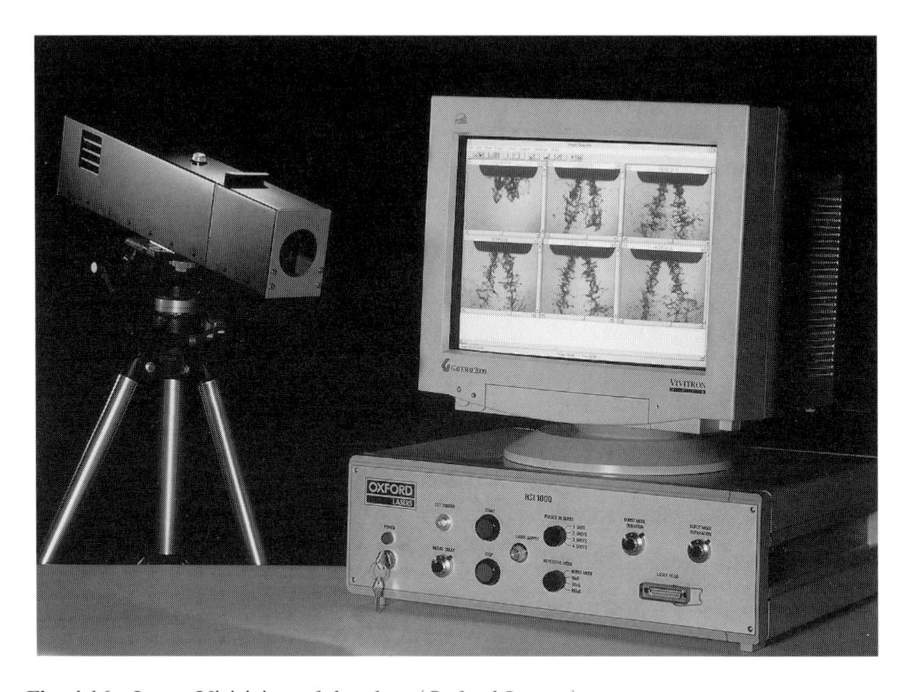

Fig. 4.16 Laser Visisizing of droplets (Oxford Lasers).

In-line holography was used to measure droplets in flight (Dunn and Walls, 1978), but the reconstruction of the holograms and their analysis is very slow compared with the newer laser techniques. Another technique has used hot-wire anemometry to assess the size of droplets, particularly small aerosol droplets (Mahler and Magnus, 1986) (Fig. 4.17). Using a portable instrument, this technique was used to measure droplets in a crop canopy (Adams *et al.*, 1989).

Despite the capability of measuring droplets in flight, it is also important to measure droplets that have landed on a surface. Himel (1969a) pointed out that, ideally, the actual leaf surface should be used, and by spraying a known concentration of fluorescent particles (FPs) he estimated droplet size based on the number of individual particles deposited as discrete droplets on leaves. The method is most suitable for droplets in the range 20–70 μm if the spray contains a uniform suspension of 2×10^8 FPs/ml, but counting individual particles is very tedious as a doubling of droplet diameter increases the number of particles per droplet eightfold (Fig. 4.18). Some sampling methods indicate the presence of spray but are not suitable for droplet sizing. One example is the use of a yarn of acrylic and nylon fibres with many fine hairs

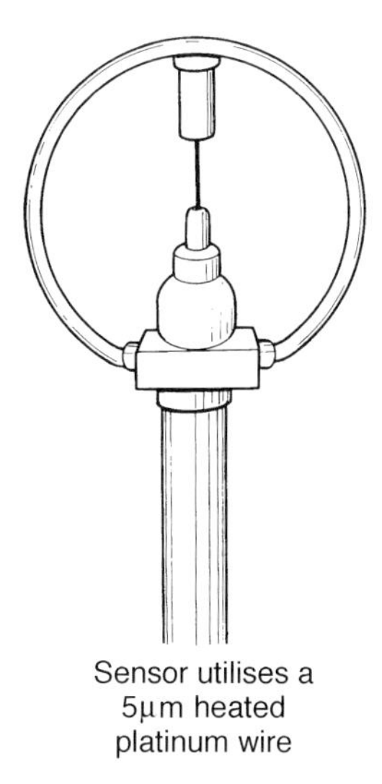

Fig. 4.17 Hot-wire droplet sensor.

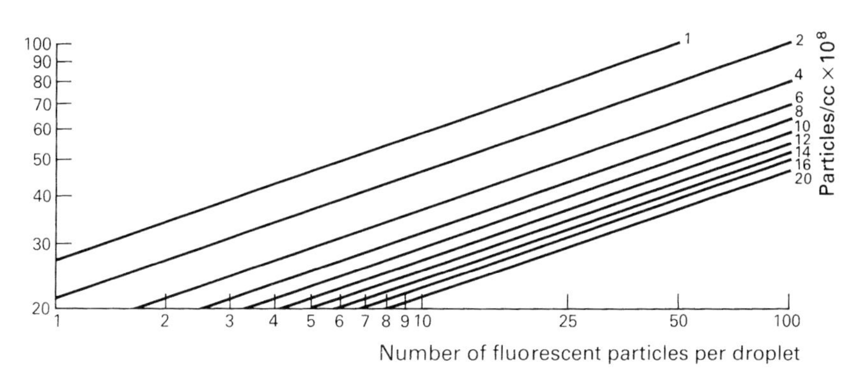

Fig. 4.18 Number of fluorescent particles (FPs) in droplets of different size according to the concentration of FPs in the spray (after Himel, 1969a).

which will collect small droplets very efficiently for chemical or fluorometric analysis (Cooper *et al.*, 1996). The 'harp' with extemely fine tungsten wires was proposed for measuring small droplets, but is only suitable for an involatile spray liquid (McDaniel and Himel, 1977).

Sampling surface

Sampling droplets in the field requires their collection on a suitable surface on which a mark, crater or stain is left by their impact. A standard surface is magnesium oxide, obtained by burning two strips of magnesium ribbon, each 10 cm in length, below a glass slide so that only the central area is coated uniformly. The slide should be in contact with a metal stand to prevent unequal heating of the glass. On impact with the magnesium oxide a droplet (20–200 μm diameter) forms a crater which is 1.15 times larger than the true droplet size (May, 1950). The difference in size between the crater and the true size is the spread factor. The reciprocal of the spread factor is used to convert the measurements of craters (or stains) to the true size; thus for magnesium oxide the factor is is 0.86. The factor is reduced to 0.8 and 0.75 for measuring droplets between 15 and 20 μm and 10 and 15 μm, respectively. The magnesium oxide surface is less satisfactory for smaller droplets, and those above 200 μm may shatter on impact. Droplets below below 100 μm may bounce unless they impinge at greater than terminal velocity. If using water, the addition of a colour dye will facilitate seeing the droplets on the white surface. Glass slides coated with Teflon have been used to assess droplets of relatively involatile insecticides applied in mosquito control (Carroll and Bourg, 1979). The water sensitive paper mentioned previously has also been used to collect droplets for sizing, but as the stains can increase in diameter with time, its use to indicate percentage area covered is preferred (Salyani and Fox, 1999). However, treatment with ethyl acetate can be used to make the stains more permanent. In addition, the spread factor will vary according to the formulation and droplet size (Thacker and Hall, 1991). Plain glossy white card, such as Kromekote card, can be used if a water soluble colour dye (e.g. lissamine scarlet or nigrosine) or oil-soluble (e.g. waxoline red) dye is added to the spray, depending on the type of liquid being used. Paper sensitive to oils and especially certain solvents can also be used. Salyani (1999) used acetone vapour to stabilise the stains caused by the droplets.

Water droplets can be collected on a grease matrix, but the droplets must be covered with oil to prevent evaporation reducing their size. A suitable matrix has one part of petroleum jelly and two parts of a light oil (risella oil or medicinal paraffin). No spread factor is needed as the droplets resume their original shape on the surface of the matrix.

Sampling technique

Although widely used, the technique of waving a slide through a spray cloud is not a very efficient way of sampling. Droplets less than 40 mm in diameter are not collected as efficiently as larger droplets. Sampling of airborne aerosol droplets is either with a cascade impactor which requires a vacuum pump (May, 1945) or by using a battery operated electric motor to rotate the slides (Cooper *et al.*, 1996) or preferably narrower rods (Lee, 1974). Alternatively, sampling surfaces can be placed in a horizontal position within a settling chamber and sufficient time allowed for the smallest droplets to sediment on them.

Measurement of droplets

One method is to view the sample of droplets with a microscope fitted with a graticule such as a Porton G12 graticule (Fig. 4.19). The microscope must have a mechanical stage to line up the stains or craters on the sampling surface with a series of lines on the graticule. The distance between these lines from the baseline *Z* increases by a $\sqrt{2}$ progression. A stage micrometer is needed to calibrate the graticule. Use of the graticule is laborious if large numbers of samples require measurement, and alternative methods have been devised to increase speed and accuracy. Automatic scanning of samples using an image-analysing computer is much more rapid, provided the image of droplets is sharply contrasted against the background (Jepson *et al.*, 1987; Last *et al.*, 1987). Wolf *et al.* (1999) used DropletScan software to measure droplets collected on water sensitive cards.

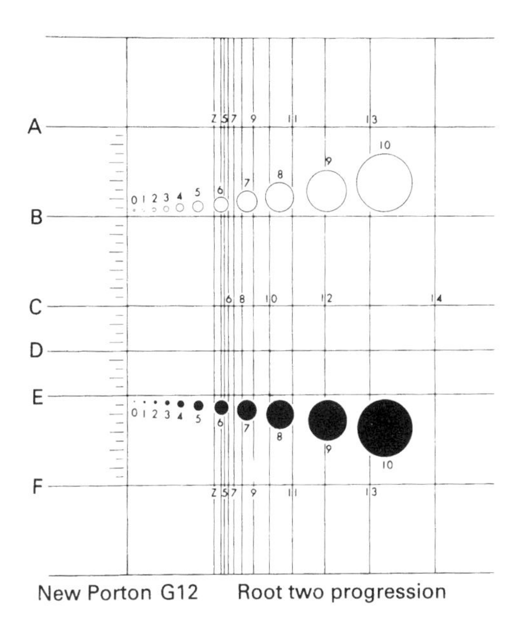

Fig. 4.19 Porton G12 Graticule (courtesy: Graticules Ltd.).

Calculation of number and volume median diameter

A computer program (Cooper, 1991) can be used to calculate these parameters, but the following notes provide an example of the calculation shown in Table 4.6. The graticule is calibrated by measuring the distance between the *Z* line and one of the outer lines, e.g. 13; the true size is then calculated on the basis of the spread factor, and then the distance between *Z* and each of the other lines calculated on the $\sqrt{2}$ progression. The mean size is the average size of the limits of each class size. The number of droplets measured in each class size is recorded in column *N*. The percentage of droplets is then calculated, and the cumulative percentage plotted on log probability paper against the mean diameter (Fig. 4.20). As the volume of a sphere is $\pi d^3/6$ and $\pi/6$ is a common factor, the cube of the mean diameter is calculated and multiplied by

Table 4.6 An example of the calculations required to determine the NMD and VMD (from Matthews, 1975)

Graticule no.	Upper class size (D)	True size upper limit (d)	Mean size (d_m)	Number in class (N)	N (%)	ΣN (%)	d_m^3	Nd_m^3	Nd_m^3 (%)	ΣNd_m^3 (%)
4		13.2								
5		18.8	16	33	6.5	6.5	4096	135168	0.3	0.3
6		26.5	22.6	97	19.1	25.6	11543	1119671	2.3	2.6
7		37.5	32	150	29.6	55.2	32768	4915200	10.0	12.6
8		53	45.25	143	28.3	83.5	92652	13249236	26.9	39.5
9		75	64	66	13.0	96.5	262144	17301504	35.1	74.6
10		106	90.5	17	3.3	99.8	741217	12600689	25.5	100.1
11		150	128							
12		212	181							
13	780	300	256							
			Total:	506			Total:	49321468		

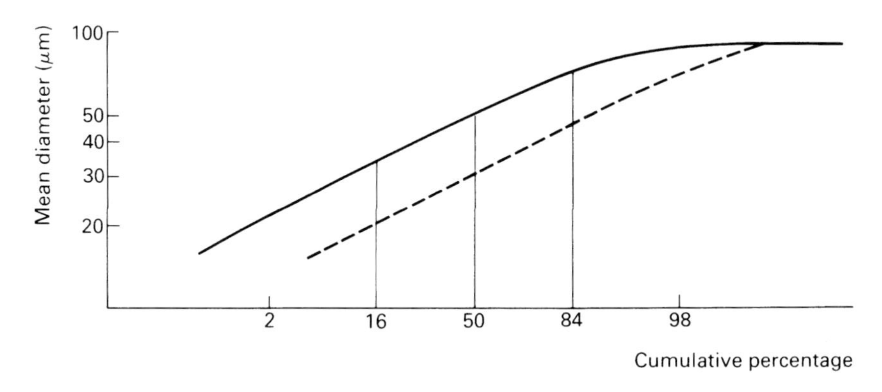

Fig. 4.20 Graph of cumulative percentage against mean droplet size to calculate the volume (solid curve) and number (broken curve) median diameter of the spray.

the number of droplets in that class ($Nd\text{m}^3$). These figures are then expressed as percentages of the total volume of the sample, and the cumulative percentages plotted on the same graph (Fig. 4.20). The NMD and VMD are then read at the 50% intersect.

5

Hydraulic nozzles

All sprayers have three features in common. Spray liquid is held in a container (spray tank) from which it is moved by pumps, pressure or a gravity-feed system to one or more outlets called nozzles. A nozzle is strictly the end of pipe through which liquid can emerge as a jet. In this book, the term 'nozzle' is used in the wider sense of any device through which spray liquid is emitted, broken up into droplets and dispersed at least over a short distance. Principally natural air movements influence further distribution of spray droplets, although on certain sprayers an airstream is used to direct droplets towards the appropriate target as described in Chapter 10.

In addition to hydraulic nozzles, sprayers may be fitted with other types using gaseous, centrifugal, kinetic, thermal and electrical energy (Table 5.1) to produce the spray droplets. A detailed description of atomisation and sprays is given by Lefebvre (1989), with special reference to the requirements in combustion technology. There is no universal nozzle, different designs being used to achieve the appropriate droplet spectrum. In this chapter the most common types of hydraulic nozzle are described, while alternative atomisers are included in Chapters 8–11. Major manufacturers of nozzles now have their own web sites that provide the latest information on the availability of different nozzles, which can also be purchased via the internet.

Most of the pesticide formulations described in Chapter 3 are diluted in water and applied through hydraulic nozzles. These nozzles meter the amount of liquid sprayed and form the pattern of the spray distribution in which the liquid breaks up into droplets. The droplet spectrum will depend on the output, spray angle of the nozzle and operating pressure, and this determines the spray quality. Correct choice of nozzle is therefore essential to ensure that expensive pesticides are applied effectively at the correct rate.

Table 5.1 Different types of nozzle and their main uses

Energy	Type	Uses
Hydraulic	Deflector	Coarse spray mainly for herbicide application
	Standard fan	Spraying flat surfaces, e.g. soil and walls[a]
	Pre-orifice fan	Fan pattern with reduced drift potential
	Air induction	Low drift potential, droplets contain air bubbles
	Boundary	Edge of boom to minimise deposit in buffer zone
	Offset	Lateral projection of spray, e.g. roadside
	Even-spray	Band sprays
	Cone	Foliar sprays, especially dicotyledon plants
	Solid stream	Spot treatment
Gaseous (see Chapters 10 and 11)	Twin fluid	Various, provide greater flexibility with control of both air and liquid flow
	Air shear	High velocity air stream to project droplets into trees and bushes
	Vortical	Aerosol (cold fog) space sprays
Centrifugal (see Chapter 8)	Spinning disc, Cage	Application of minimal volumes with controlled droplet size. Slow rotational speed: large droplets for placement sprays. Fast rotational speed: mist/aerosols for drift and space sprays
Thermal (see Chapter 11)	Fog	
Electrostatic (see Chapter 9)	Annular	ULV electrostatically charged spray

[a] Volume of spray depends on surface, i.e. runoff occurs at approximately 25 ml/m^2.

Types of hydraulic nozzle

Production of droplets

A large range of hydraulic nozzles have been designed in which liquid under pressure is forced through a small opening or orifice so that there is sufficient velocity energy to spread out the liquid, usually in a thin sheet which becomes unstable and disintegrates into droplets of different sizes. The pressure of liquid through the nozzle, surface tension, density and viscosity of the spray liquid and ambient air condition influence the development of the sheet. A minimum pressure is essential to provide sufficient velocity to overcome the contracting force of surface tension and to obtain full development of the spray pattern. For most nozzles the minimum pressure is at least 1 bar (14 p.s.i.), but higher pressures are often recommended when a finer spray is required, especially for fungicide and insecticide applications. An increase in pressure will increase the angle of the spray as it emerges through the orifice and also increase the flow rate in proportion to the square root of the pressure. Flow rate divided by the square root of the pressure differential is equal to a constant, commonly termed the flow number (FN).

Fraser (1958) noted three distinct modes of sheet – these are perforated, rim, and wavy-sheet disintegration (Fig. 5.1) – but only one mechanism of disintegration in which separate filaments of liquid break up into droplets. Perforated sheet disintegration occurs when holes develop in the sheet and, as

(a)

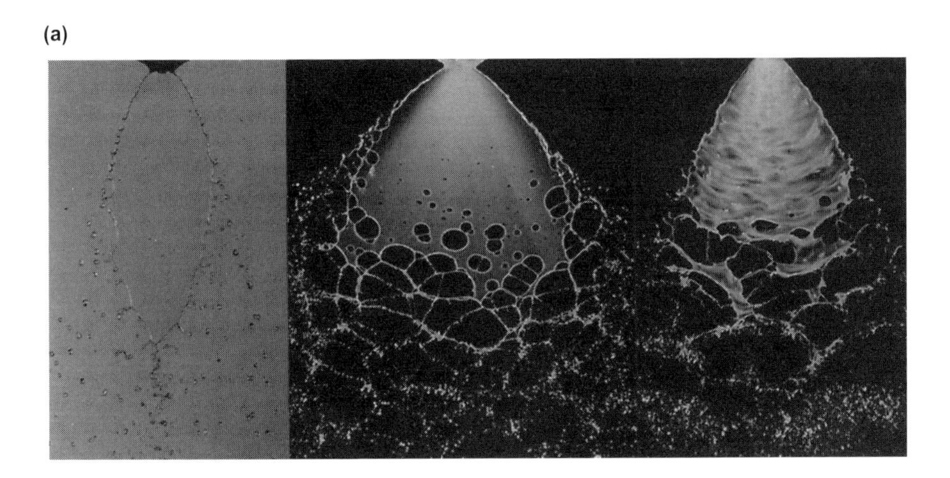

(b)

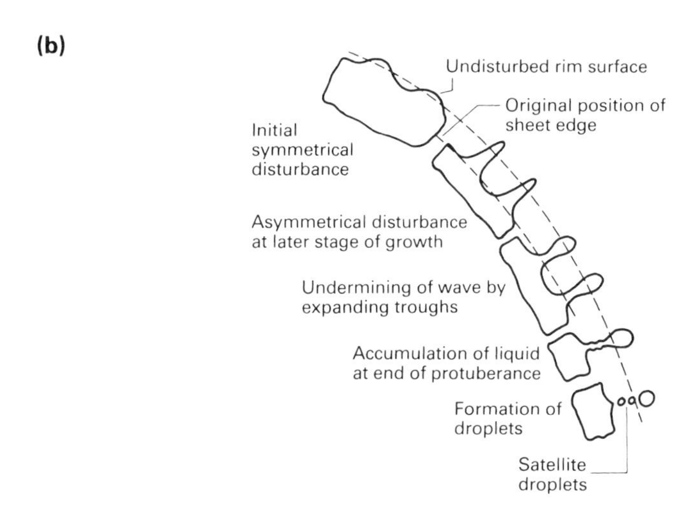

Fig. 5.1 (a) Rim, perforated sheet and wavy-sheet disintegration (photos: N. Dombrowski). (b) Diagram showing rim disintegration.

they expand, their boundaries form unstable filaments which eventually break into droplets. In rim disintegration, surface tension contracts the edge of the sheet to form rims from which large droplets are produced at low pressure, but at higher pressures threads of liquid are thrown from the edge of the sheet. Rim disintegration is similar to droplet formation from ligaments thrown from a centrifugal energy nozzle. Whereas in perforated sheet and rim disintegration, droplets are formed at the free edge of the sheet, wavy-sheet disintegration occurs when whole sections of the sheet are torn away before reaching the free edge (Clark and Dombrowski, 1972).

Recent studies using laser systems to measure droplet spectra combined with high-speed photography have examined the effects of emulsions on droplet production (Butler Ellis *et al.*, 1997a). They showed that emulsions resulted in perforated sheet formation, giving larger spray droplets than when spraying water alone. The influence of different adjuvants on the break-up is complex, with some such as Ethokem reducing the VMD compared to water, with break-up occurring further from the nozzle (Butler Ellis *et al.*, 1997b). When the sheet breaks up closer to the nozzle orifice, the VMD is generally larger, for example when the viscosity is increased by the addition of an oil plus emulsifier. Conversely, where the sheet remains stable and is stretched before breaking up into droplets, the thinner sheet forms a spray with a smaller VMD. The droplets vary considerably in size (Fig. 5.2) in the range 10–1000 μm, owing to the irregular break-up, so the volume of the largest droplets

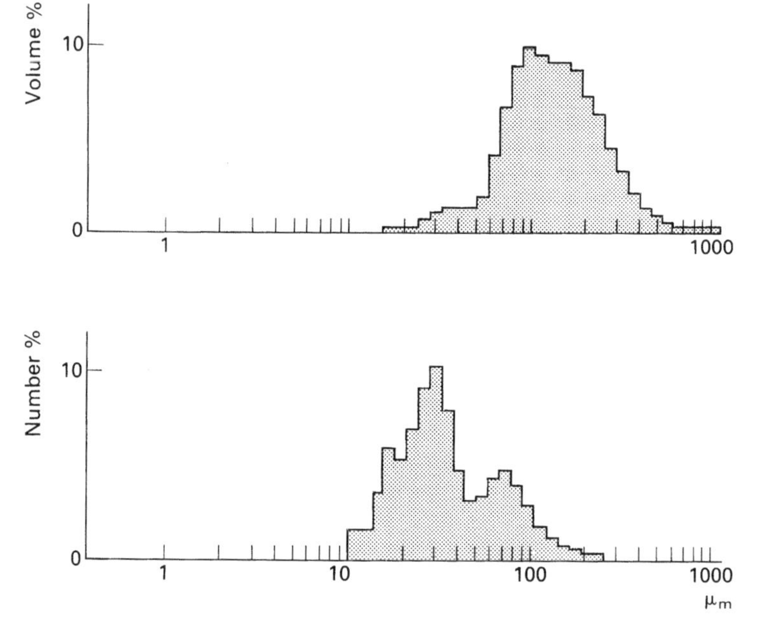

Fig. 5.2 Example of droplet distribution from a fan nozzle from Malvern particle size analyser.

is a million times that of the smallest. Their average size decreases with an increase in pressure and increases with a larger orifice. The range of sizes is less at the higher pressures, especially in excess of 15 bar. During forward movement of the sprayer, inwardly curling vortices are formed on either side of a flat-fan nozzle, so that small droplets are carried in a low energy trailing plume and are subsequently more vulnerable to drift away from the intended target (Young, 1991). Mokeba *et al.* (1998) have modelled the meteorological and spraying parameters that affect dispersion from the nozzle.

Particular interest has been directed at producing coarser sprays with fewer small droplets vulnerable to downwind movement by the wind. Thus in addition to standard types of hydraulic nozzles, several variations in design have been developed (see section below). With all the different types of hydraulic nozzle, the spray formation mechanism is similar, although changes in droplet size with some types makes it difficult to develop a model to predict the effects of adjuvants and spray drift potential (Butler Ellis and Tuck, 1999). An example of this is that the proportion of small droplets in a spray tends to increase with higher temperatures. However, when a polymer (e.g. Nalcotrol) was added to reduce drift the proportion of small droplets was not significantly influenced by temperature, although the VMD decreased with increasing temperature (Downer *et al.*, 1998b). Womac *et al.* (1997) published a set of droplet data for selected nozzles applying water, water + surfactant and a water–crop oil mix to assist users select nozzles for applying herbicides.

Components of hydraulic nozzle

Hydraulic nozzles consist of a body, cap, filter and tip. Various types of nozzle body are available with either male or female threads, or special clamps, sometimes with hose shanks, for connecting to booms (Fig. 5.3), and some nozzle tips are designed to screw directly into a boom without a special body or cap. On most large sprayers, the cap is attached to the body with a bayonet fitting. The body and cap of some nozzles have a hexagonal or milled surface or wings to facilitate tightening and eliminate leaks. The cap should be tightened by hand and where a seal is used, care should be taken to avoid damaging it. These components are more frequently moulded in plastic such as Kematal. Some nozzles are not provided with a filter, but as spray liquid is readily contaminated with dust or other foreign matter that can block the nozzle tip, a suitable filter should be used in the nozzle body, although on many sprayers a large capacity filter is in line with the boom. A 50-mesh filter is usually adequate, except for very small orifice tips when an 80-, 100- or 200-mesh filter may be needed. A coarse strainer, normally equivalent to 25-mesh, may be used with large orifice nozzles (Fig. 5.4). A filter fitted with a small spring and bolt valve as an antidrip device is not recommended because the spray operator is easily exposed to spray liquid when changing the nozzle tip. A diaphragm check valve is preferred as an antidrip device. It consists of a synthetic rubber diaphragm held by a low pressure spring held in place by a

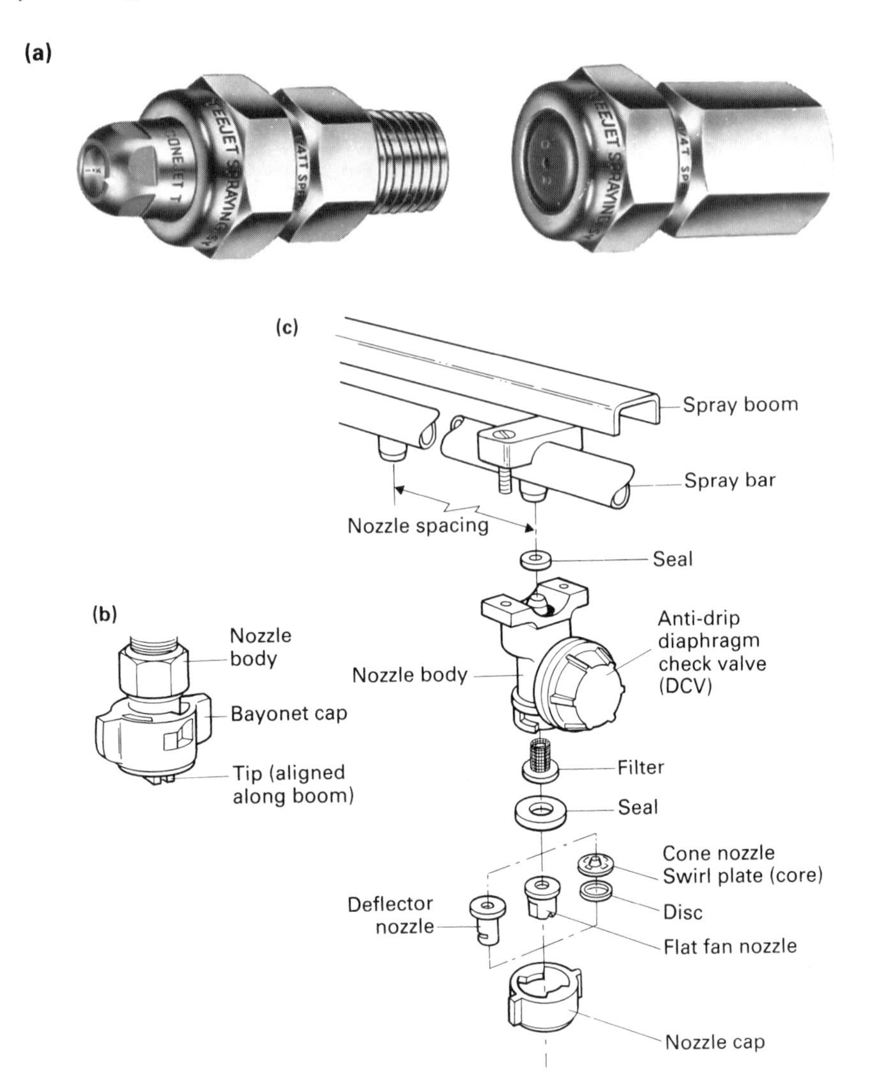

Fig. 5.3 (a) Hydraulic nozzles, male and female nozzle body (Spraying Systems Co.). (b) Plastic nozzle with bayonet cap and diaphragm check valve. (c) Exploded view of nozzle (Lurmark Ltd.).

Fig. 5.4 Strainer, 50-mesh and 100-mesh filters (Spraying Systems Co.).

separate cap (Fig. 5.5). This valve can be replaced, especially on manually operated equipment by a 'constant flow valve' that in addition to being an antidrip device, ensures that liquid flows to the nozzle at a constant rate and/or pressure (see pp. 137–8).

Most nozzles are now manufactured from engineering plastics, rather than the traditional brass. The important aspect is that the components should not be affected by a wide range of chemicals. However, the orifice

Fig. 5.5 (a) Diaphragm check valve. (b) Diaphragm check valve incorporated into nozzle body (Spraying Systems Co.).

can be easily abraded by particles, so many users prefer nozzle tips made in ceramics, although the plastic Kematal nozzles are inexpensive and easily replaced when worn. Plastic tips are sometimes more resistant to abrasion than metal tips because moulded tips have a smoother finish. The surface of metal tips has microscopic grooves as a result of machining and drilling the orifice; the rough finish presumably causes turbulence and enhances the abrasive action of particles suspended in a spray liquid. The threads of some nozzle bodies and caps manufactured in plastic are easily damaged by constant use, especially if they are over-tightened with a spanner. Various hydraulic nozzle tips are manufactured to provide differences in throughput, spray angle and pattern. The tip and cap of some nozzles are integrated. Ceramic and stainless steel tips are now often mounted in a plastic outer section.

Each manufacturer has its own system of identifying different nozzles, including colour coding, so an independent code has been introduced to be recommended without referring to an individual manufacturer. The code uses four parameters to describe a nozzle: the nozzle type, spray angle at a standard pressure, flow rate and the rated pressure (Table 5.2). As an example, F110/1.6/3 refers to 110 degree fan nozzle, 1.6 litres/min output at 3 bar. Information relating to standard fan nozzles is available for farmers in the UK on a chart (Fig 5.6) obtainable from the British Crop Protection Council. There is now a colour code (ISO 10625: 1996) which indicates the flow rate of fan type nozzles. Nozzles from the major manufacturers now conform to this standard (Table 5.3), but as different colour schemes were used, older nozzles need to be checked. The main advantage of a colour is that the user can readily see if all the nozzles on a spray boom have the same output. Choice of nozzle will also depend on the spray spectrum produced, so a system of spray categories is used to indicate the 'quality' of the droplet spectrum (Doble *et al.*, 1985) (Table 5.4).

Table 5.1 indicates a range of different types of hydraulic nozzle, details of which are given in the following section.

Table 5.2 Code for describing nozzles

Code	Nozzle type	Spray angle	Nozzle output	Rated pressure
F	Standard fan	Given in degrees (if known)	Given in litres per minute	Normally output is rated at 3 bar pressure, but some LP nozzles are rated at 1 bar
FE	Fan with even spray			
RD	Reduce drift, pre-orifice fan			
LP	Low pressure fan			
AI	Air inclusion			
D	Deflector			
HC	Hollow cone			
FC	Full cone			
OC	Offset fan			

Table 5.3 Colour code for fan nozzles based on nozzle output

BCPC nozzle code	Colour	Example of nozzle
F110/0.4/3	Orange	11001
F110/0.6/3	Green	110015
F110/0.8/3	Yellow	11002
F110/1.2/3	Blue	11003
F110/1.6/3	Red	11004
F110/2.0/3	Brown	11005
F110/2.4/3	Grey	11006
F110/3.2/3	White	11008

Nozzle tips

Deflector nozzle

A fan-shaped spray pattern is produced when a cylindrical jet of liquid passes through a relatively large orifice and impinges at high velocity on a smooth surface at a high angle of incidence (Fig. 5.7). Within most deflector nozzles, spray is projected at an angle away from the plane of the nozzle body. A relatively new design of deflector nozzle suitable for use on a tractor boom, ensures that they can be used instead of conventional fan nozzles without having to adjust the orientation of the boom (Figs 5.9, 5.10b). This is due to the internal design of the nozzle tip. The angle of the fan depends upon the angle of inclination of the surface to the jet of liquid. Droplets produced by this nozzle are large (>250 µm VMD) and there can be more spray at the edges of the fan (spray 'horns'). The deflector nozzle is normally operated at low pressures and has been widely used for herbicide application to reduce the number of small droplets liable to drift. When applying herbicides, the spray is normally directed downwards, but when used on a lance, the nozzle can be inverted to direct spray sideways under low branches. The effect of nozzle orientation on the spray pattern has been reported by Krishnan *et al.* (1989).

Deflector nozzles have been widely used where blockages could occur if a smaller elliptical fan nozzle orifice were used, and also where a wide swath is required with the minimum number of nozzles. They are sometimes referred to as flooding, anvil or impact nozzles. One type, known as the CP nozzle, is used on aircraft (see Chapter 13). This type of nozzle has been produced in plastic, colour-coded according to the size of orifice. A full circular pattern can be obtained if the side of the nozzle is not shrouded. Deflector nozzles have been used on fixed pipes in citrus orchards to apply nematicides, herbicides and systemic insecticides, metered into the irrigation water, around the base of individual trees. A deflector nozzle has also been used as part of a twin-fluid nozzle in which droplet formation and dispersal are affected by combinations of liquid and air pressure (see below).

(a)

BCPC Nozzle Card - For 110° Flat Fan Nozzles

Use the following tables and notes to help you choose the best nozzle for your application.

- Follow the pesticide label recommendations for spray quality wherever possible;
- Check reduced drift nozzles are suitable for the pesticide product and target;
- Renew all nozzles at least annually or when damaged;
- Set 110° nozzles at 35 to 50cm above the target or the crop;
- Nozzle fans are usually offset by at least 5° on the spray boom;
- Use multi-head nozzle bodies to simplify changing nozzles size and type.

Spray Quality and Nozzle Outputs

Typical of 110° conventional flat fan nozzles (not reduced drift fan nozzles).
Note: Check with your nozzle supplier for the actual spray quality for their nozzles.

Nozzle code		11001	110015	11002	11003	11004	11005	11006	11008
ISO colour		Orange	Green	Yellow	Blue	Red	Brown	Grey	White
Pressure - Bar	1.5	0.29	0.42	0.56	0.85	1.13	1.41	1.70	2.26
	2.0	0.33	0.49	0.65	0.98	1.31	1.63	1.96	2.61
	2.5	0.37	0.55	0.73	1.10	1.46	1.82	2.19	2.92
	3.0	0.40	0.60	0.80	1.20	1.60	2.00	2.40	3.20
	3.5	0.43	0.65	0.86	1.30	1.73	2.16	2.59	3.45
	4.0	0.46	0.69	0.92	1.39	1.85	2.31	2.77	3.69
		Nozzle output = litres/minute							

Spray Quality	Fine	Fine/Medium	Medium	Medium/Coarse	Coarse

Spray Volume, Speed, Nozzle Output & Calibration Equations

Spray volume litres/ha	Speed		
	5 km/h	8 km/h	10 km/h
80	0.33	0.53	0.66
100	0.42	0.67	0.83
150	0.62	1.00	1.25
200	0.83	1.33	1.67
250	1.04	1.66	2.09
300	1.25	2.00	2.50
400	1.67	2.67	3.33
500	2.08	3.33	4.17
	Nozzle output - litres/minute		

Speed km/h = **360** ÷ **seconds** per 100 metres

Volume = **Output x 600** ÷ **Speed** ÷ **Nozzle Space**
litres/ha litres/min km/h metres

Calculating nozzle outputs and pressures:

P1 = First pressure P2 = Second pressure
Q1 = First output Q2 = Second output

To calculate new pressure:

$$P2 = (Q2 \div Q1)^2 \times P1$$

To calculate new output:

$$Q2 = \sqrt{(P2 \div P1)} \times Q1$$

Fig. 5.6 British Crop Protection Council Guide Chart for flat fan nozzles with reference to spray quality, and suitability for different spraying operations.

(b)

Typical Uses in Cereals

Adapted from '*Guide to Selecting Nozzles*' and reproduced by permission of the Home Grown Cereals Authority (HGCA). Always check with the pesticide suppliers before using reduced drift nozzles.

Nozzle types	Conventional Flat Fan		Pre-Orifice Reduced Drift		Air-Induction Reduced Drift
Spray quality	FINE	MEDIUM	MEDIUM	COARSE	Air-inclusions
Likely drift potential	High	Medium	Low	Very low	Very low
Soil herbicides		OK		OK	Best
Grass weed herbicides	OK	Best			
Other herbicides		Best	OK	OK	OK
Fungicides - foliar	OK	Best	OK		OK
Fungicides - late	Best	OK			OK
Insecticides - autumn	OK	Best			OK
Insecticides - ear	Best	OK			OK

Nozzle Suppliers, Codes and Materials

'11003' / Blue size given as a common example.

Supplier	Flat Fan	Pre-Orifice	Air-Induction	Materials*
Lurmark	03 F110 UB FanTip	LD 03 F110 UB LO-Drift	DB 03 F120 DriftBETA	P, S
Spraying Systems	XR 11003 XR TeeJet	DG 11003 Drift Guard	AI 11003 Air Induction	S/P, P, C/P
Hardi	S F-03-110 ISO F110 Flat Fan	S LD-03-110 ISO LD Low Drift	S INJET 03 INJET, B-JET	P, C/P
Tecnoma Berthoud	RFX - AFX 110-03 Flat Fan	RLX - ALX 110-03 Low Drift	RRX - ARX 110-03 Air-Injection	P - C/P
Lechler	LU 120-03 Multirange	AD 120-03 Anti Drift	ID 120-03 Air-Injektor	P, S, C/P
Albuz	API Blue - 11003 Fan	ADI Blue - 11003 Drift Reduction	AVI 11003 AVI Anti Drift	C/P
Sprays International	110-SF-03 Standard	110-LD-03 Enviroguard	03 Pneu'Jet	P
Billericay	110-03, TC 110-03 Flat Fan, TipCap	110-03 Multi Drop	03 Air Bubble Jet	P
Agrotop			TDO3 TurboDrop	C/P

* Nozzle tip materials: P - Plastic, S – Stainless steel, C – Ceramic, C/P – Ceramic tip in plastic body, S/P – Stainless steel tip in plastic body.

Standard fan nozzle

If two jets of liquid strike each other at an angle greater than 90°, a thin sheet is produced in a plane perpendicular to the plane of the jets. The internal shape of a fan nozzle is made to cause liquid from a single direction to curve inwards so that two streams of liquid meet at a lenticular or elliptical orifice. The shape of the orifice is very important in determining not only the amount of liquid emitted but also the shape of the sheet emerging through it, particularly the

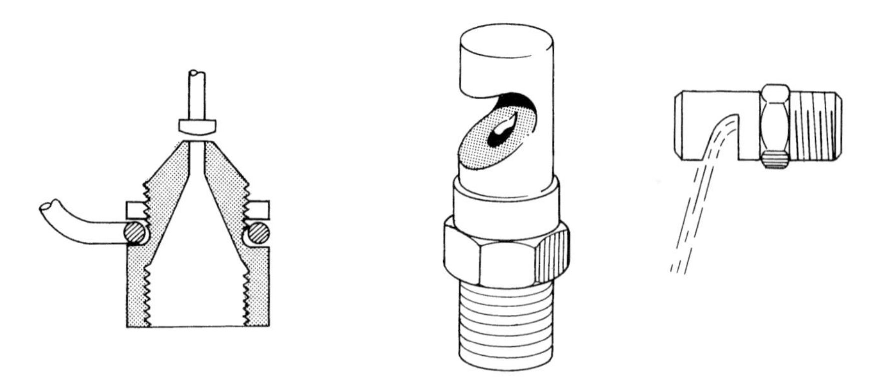

Fig. 5.7 Deflector nozzles (after WHO, 1974).

spray angle. The angle and throughput of fan nozzles used for applying pesticides are normally measured at a pressure of 3 bar. Snyder *et al.* (1989) give data for fan nozzles used in industrial applications and show the effect of viscosity, surface tension and nozzle size on the Sauter mean diameter over a wide range of pressures. An example of a range of fan nozzles is shown in Fig. 5.9.

Many farmers prefer to use 110° rather than 80° or 65° nozzles to reduce the number required on a boom or to lower the boom to reduce the effect of drift, although droplets are on average smaller with the wider angle. Boom height is very important and computer simulations have predicted more drift from 80° angle nozzles 50 cm above the crop compared to 110° nozzles at 35 cm height (Hobson *et al.*, 1990). Boom height can also be reduced by directing the spray forwards, instead of directly down into the crop. Some manufacturers produce a twin fan, so that one is directed forwards at an angle while the other fan is directed at a similar angle backwards. Other manufacturers provide a twin cap that will hold two fan nozzles. Angling the spray will improve coverage on the more vertical leaves.

The spray pattern usually has a tapered edge with the lenticular shape of orifice (Fig. 5.8), and these nozzles may be offset at 5° to the boom to separate overlapping spray patterns and avoid droplets coalescing between the nozzle and target. Great care must be taken to ensure that all the nozzles along a boom are the same and that they are spaced to provide the correct overlap according to the boom height and the crop which is being sprayed. Details of the position of nozzles and boom height on tractor sprayers are given in Chapter 7. Fan nozzles are ideal for spraying 'flat' surfaces such as the soil surface and walls. They have been widely used on conventional tractor and aerial spray booms and on compression sprayers for spraying huts to control mosquitoes (Gratz and Dawson, 1963).

Standard fan nozzles produce a relatively high proportion of droplets smaller than 100 μm diameter, especially at low flow rates and high pressures. Sarker *et al.* (1997), using a F110/0.8/3 nozzle at 300 kPa in a wind tunnel,

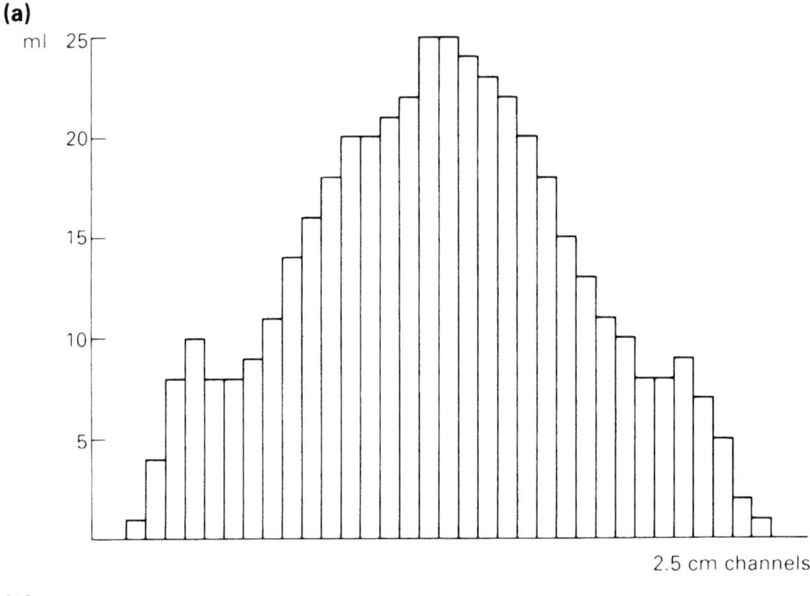

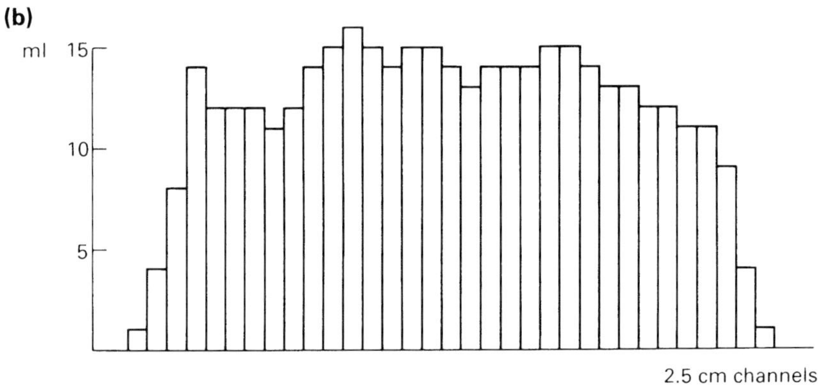

Fig. 5.8 (a) Spray pattern with a fan nozzles. (b) Spray pattern with an even-spray nozzle.

showed that drift potential increased as dynamic surface tension of the spray liquid decreased. Spray drift in these tests also increased marginally with an increase in viscosity.

A number of other fan nozzles are now available as alternative to a standard fan nozzle (Fig 5.9).

Low pressure fan nozzle

Low-pressure fan nozzles provide the same throughput and angle of a conventional fan tip, but at a pressure of 1 bar instead of 3 bar (Bouse *et al.*, 1976). Other low drift nozzles have to a large extent superseded these.

Table 5.4 Spray quality for agricultural nozzles

Spray quality	Retention on difficult leaf surfaces	Used for	Drift hazard
Very fine	Good	Exceptional circumstances	High
Fine	Good	Good coverage	Medium
Medium	Good	Most products	Low
Coarse	Variable	Soil applied herbicides, but with aerated droplets is also suitable for foliar application of systemic or translocated pesticides	
Very coarse	Poor	Liquid fertiliser	Very low

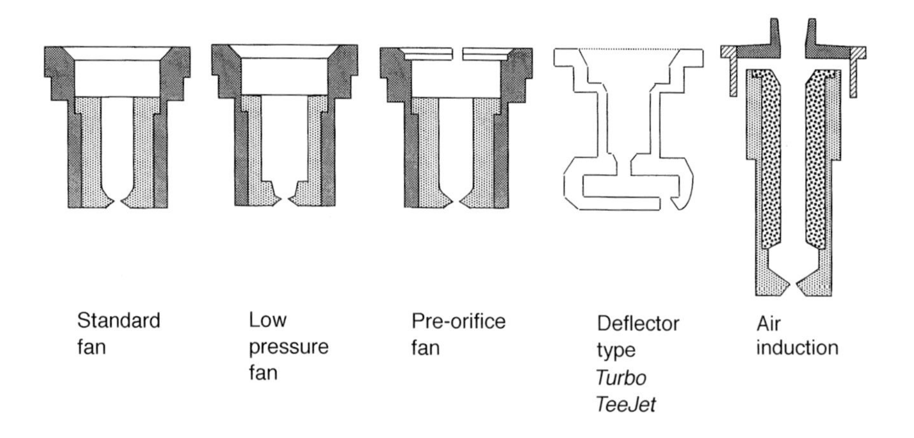

Fig. 5.9 Alternative designs of flat-fan nozzle, including one design of deflector nozzle.

Pre-orifice fan nozzles

Another modification of a fan nozzle is to incorporate a second orifice upstream of the tip. This is referred to as a 'pre-orifice' nozzle. The aim is to decrease the pressure through the nozzle and thus reduce the proportion of spray in droplets smaller than 100 μm (Barnett and Matthews, 1992).

Air induction nozzles

Another design based on the foam type nozzle, is the 'air-induction' nozzle which has an air inlet so that a venturi action of liquid through the nozzle sucks in air (Cecil, 1997; Piggott and Matthews, 1999). The nozzles produce larger droplets, many of which contain one or more bubbles of air. The presence of the air bubbles makes analysis of the droplet spectra more difficult with some laser equipment, when individual air bubbles are measured as droplets. Generally these nozzles produce a coarse spray (Fig. 5.10c) with less risk of spray-drift, but there is a wide variation in the spray quality produced with these nozzles due to the design of the venturi system (Piggott and Matthews, 1999). Etheridge *et al.* (1999) and Wolf *et al.* (1999) report similar droplet size data for herbicide sprays and compare an air induction nozzle with several other fan nozzles. The significant effect on spray droplet spectra from these nozzles by the addition of an adjuvant was demonstrated by Butler Ellis and Tuck (2000), who confirmed the variation between air induction nozzles of the same output from different manufacturers (Fig. 5.11).

When an extremely coarse spray is applied, the number of droplets deposited per unit area is reduced unless the spray volume is increased. It has been suggested that the presence of air bubbles in the large droplet reduces the risk of a droplet bouncing off a leaf surface. Deposition on horizontal targets was better with air induction nozzles and similar to standard fan nozzles on vertical surfaces in wind speeds up to 4 m/s (Cooper and Taylor, 1999). Biological results, especially with systemic pesticides have been very acceptable, but Jensen (1999) reported that efficacy of some herbicides can be significantly reduced with low volume air induction nozzles. Wolf (2000) reporting on trials with 19 different herbicides, showed that in some cases the low drift nozzle performed better than conventional nozzles, and suggested that the coarsest spray should be avoided with contact herbicides and when targeting grassy weeds. Deposits on oats were poor with low drift nozzles (Nordbo *et al.*, 1995). When used on an air-assisted orchard sprayer, Heinkel *et al.* (2000) obtained as good control of scab and powdery mildew with air induction nozzles as hollow cone nozzles with certain fungicides, presumably due to redistribution of the active ingredient from spray deposits.

Boundary nozzles

A variation of the air induction nozzle provides a half spray angle, so that when fitted as the end nozzle of a boom, spray is directed down at the edge of the crop and not beyond into the field margin (Taylor *et al.*, 1999).

Even-spray fan nozzle

A narrow band of spray requires a rectangular spray pattern when herbicides are applied to avoid under-dosing the edges of the band, so a fan nozzle with an ‘even-spray’ pattern is required (Fig. 5.8b), especially with pre-emergence herbicides.

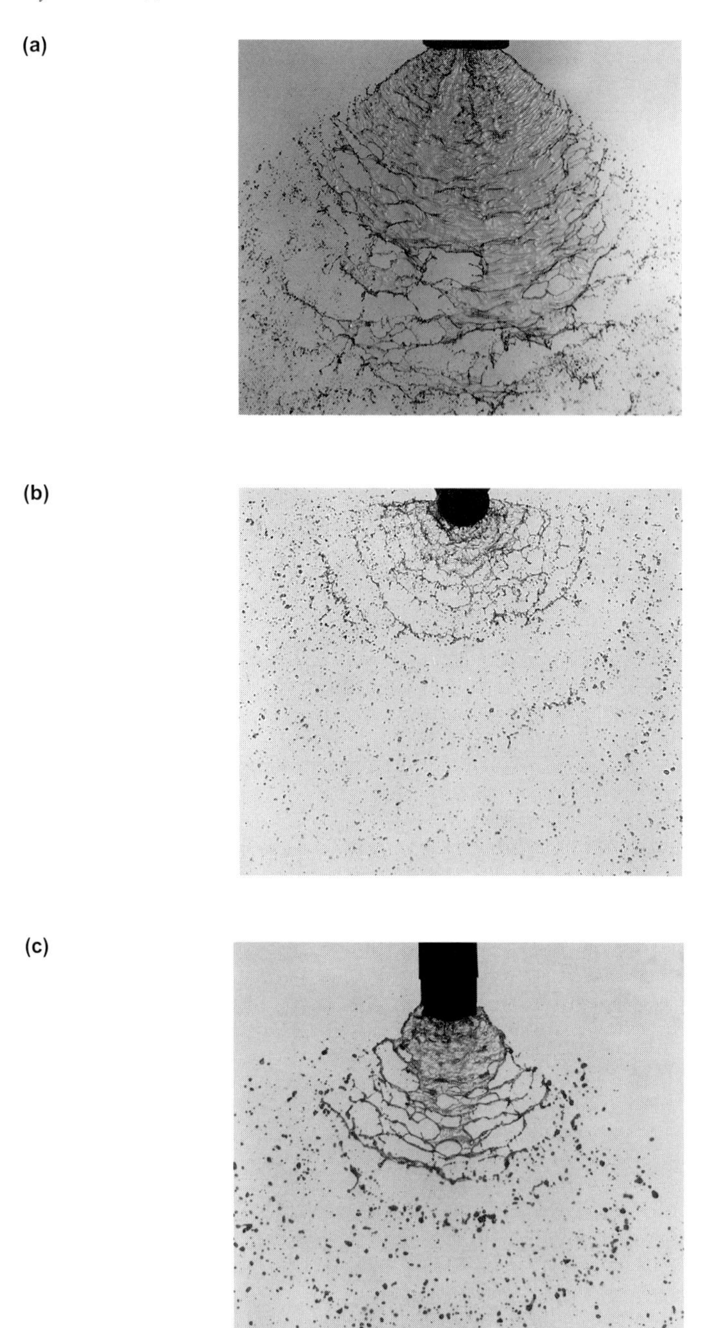

Fig. 5.10 (a) Standard fan nozzle spray pattern; (b) deflector nozzle (Turbo TeeJet); (c) air induction nozzle (Photos: SRI).

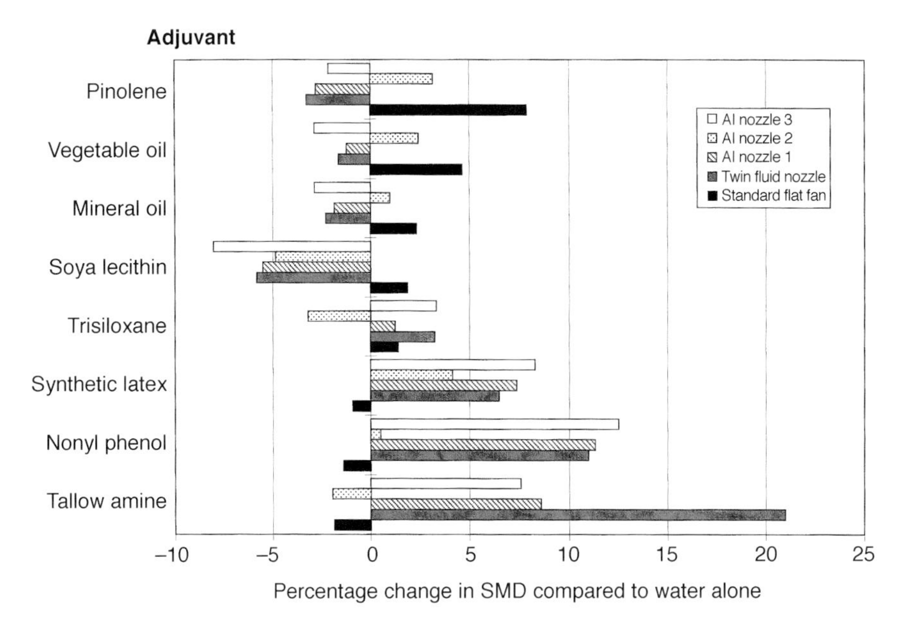

Fig. 5.11 Percentage change in Sauter mean diameter for eight adjuvants compared to water with five nozzles (SRI).

Cone nozzle

Liquid is forced through a swirl plate, having one or more tangential of helical slots or holes, into a swirl chamber (Fig. 5.12). An air core is formed as the liquid passes with a high rotational velocity from the swirl chamber through a circular orifice. The thin sheet of liquid emerging from the orifice forms a hollow cone (Fig. 5.13) as it moves away from the orifice, owing to the tangential and axial components of velocity. A solid cone pattern can be achieved by passing liquid centrally through the nozzle to fill the air core; this gives a narrower angle of spray and larger droplets. Some authors (e.g. Yates and Akesson, 1973) referred to the cone nozzles as centrifugal nozzles because the liquid is swirled through the orifice, but droplets are formed from the sheet of liquid in the same manner as with other hydraulic nozzles, so the term 'centrifugal' should be reserved for those nozzles with a rotating surface (spinning disc).

A wide range of throughputs, spray angles and droplet sizes can be obtained with various combinations of orifice size, number of slots or holes in the swirl plate, depth of the swirl chamber and the pressure of liquid. Some manufacturers designate orifice sizes in sixty-fourths of an inch; thus D2 and D3 discs have orifice diameters of 2/64 in (0.8 mm) and 3/64 in (1.2 mm), respectively.

Reducing the orifice diameter, with the same swirl plate and pressure, diminishes the spray angle and throughput (Table 5.5). The smaller the

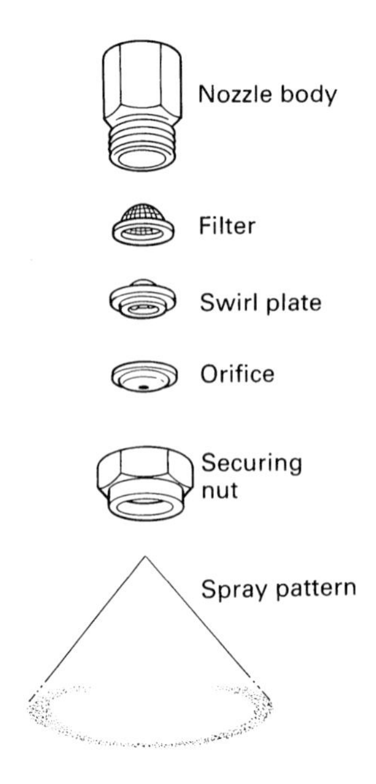

Fig. 5.12 Diagram of cone nozzle.

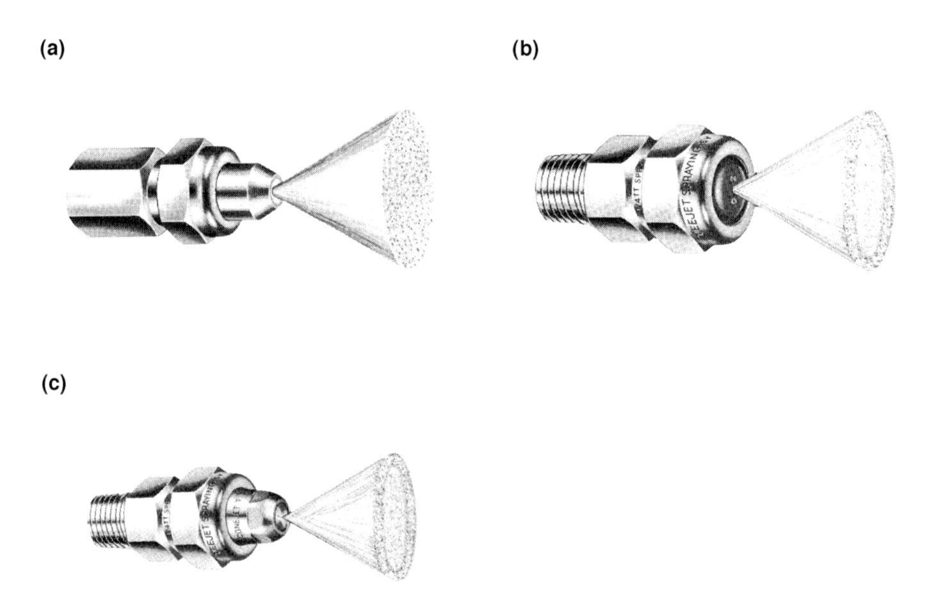

Fig. 5.13 (a) Solid cone. (b) Hollow-cone nozzle – disc type. (c) Hollow cone nozzle – 'Cone-Jet' type (Spraying Systems Co.).

Table 5.5 Effect on throughput and spray angle of certain combinations of disc and swirl plate of hollow–cone nozzles

Orifice	Orifice diameter (mm)	Swirl plate	Pressure (bar) 1.03		Pressure (bar) 2.8	
			Throughput (litres/min)	Angle (°)	Throughput (litres/min)	Angle (°)
D2	1.04	13	0.22	41	0.30	67
		25	0.38	32	0.61	51
		45	0.49	26	0.76	46
D4	1.60	13	0.31	64	0.45	79
		25	0.68	63	1.10	74
		45	0.83	59	1.36	69
D6	2.39	25	1.06	77	1.67	85
		45	1.32	70	2.20	79

openings are on the swirl plate, the greater the spin given to the spray. Also a wider cone and finer spray are produced with a smaller swirl opening. An increase in pressure for a given combination of nozzle and swirl plate increases the spray angle and throughput. On most cone nozzles, the orifice disc and swirl plates are separate parts. The depth of the swirl chamber between swirl plate and orifice disc can be increased with a washer to decrease the angle of the cone and increase droplet size. Where cone nozzles with a low flow rate are used, the swirl slots are cut on the back of the disc, and closed by a standard insert. On some nozzles, the flow rate can be adjusted if some of the liquid in the swirl chamber is allowed to return to the spray tank. Bode *et al.* (1979) and Ahmad *et al.* (1980, 1981) have investigated the use of these by-pass nozzles.

Variable-cone nozzles are available in which the depth of the swirl chamber can be adjusted during spraying, but this type of nozzle is suitable only when a straight jet or wide cone is needed at fairly short intervals, as intermediate positions cannot be easily duplicated. However, these nozzles are no longer generally recommended as the user is exposed to pesticide when adjusting the angle of spray, unless a special spray gun is used where a trigger mechanism adjusts the nozzles.

Cone nozzles have been used widely for spraying foliage because droplets approach leaves from more directions than in a single plane produced by a flat fan, although the latter can penetrate further between leaves of some crop canopies.

When a second chamber is positioned immediately after the orifice (Fig. 5.14), the proportion of small droplets is decreased. Air is drawn into this second chamber and mixes with swirling liquid, the net result of which is the production of larger, aerated droplets. This additional chamber on a nozzle operated at 2.8 bar can reduce the proportion of droplets of less than 100 μm diameter from over 15% to less than 1% (Brandenburg, 1974; Ware *et al.*, 1975). This type of nozzle is used for application of herbicides. An air induction cone nozzle is also available.

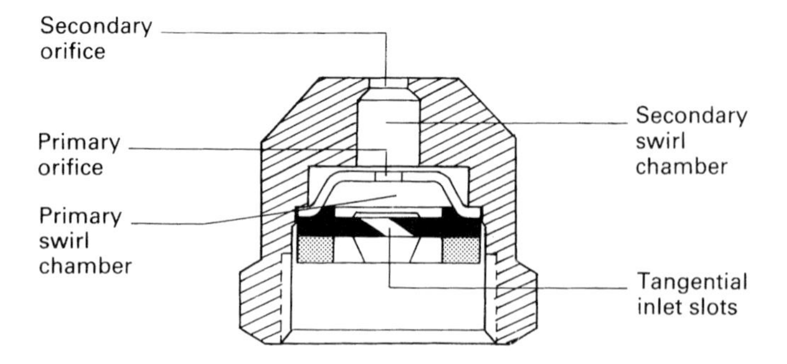

Fig. 5.14 Diagram of 'Raindrop' nozzle.

Plain jet or solid stream nozzle

This nozzle is similar to a cone nozzle but without a swirl chamber, and sometimes may have more than one orifice. It is used for various purposes including the spot treatment of weeds, young shrubs or trees with herbicide, and has been used to project spray to pods high in the canopy of cacao trees. This type of nozzle is used to apply molluscicides to control vectors of schistosomiasis to ponds and at intervals along canals where there is insufficient flow of water to redistribute chemical from a point source at the head of the canal. A long thin plastic tube attached to a solid stream nozzle has been used to inject pesticides into cracks and crevices for cockroach control.

Foam or air-aspirating nozzle

These nozzles were used primarily with additional surfactant to produce blobs of foam to indicate the end of the spray boom (Fig. 5.15). The use of 'tramlines' (p. 174) has reduced this need. Studies on the application of herbicide

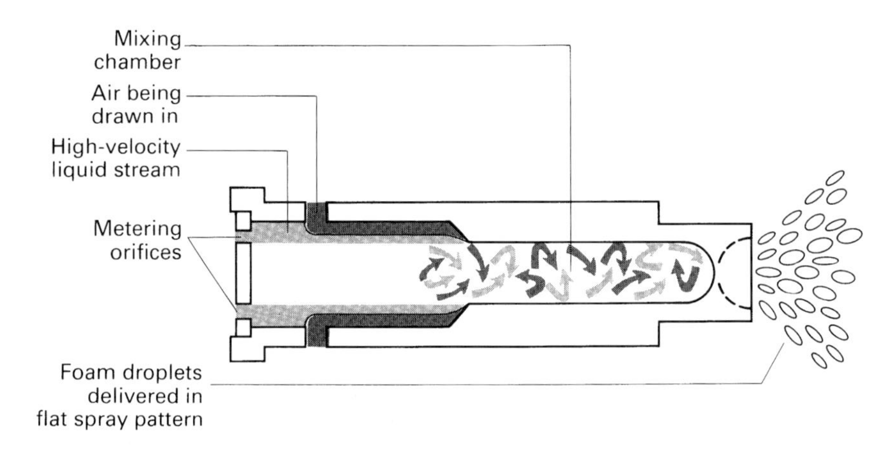

Fig. 5.15 Diagram of foam nozzle.

(Bouse *et al.*, 1976) were not followed by large scale use, but recently the need to have 'no spray' or buffer zones has led to greater use of air induction nozzles (see p. 14 and p. 116).

Intermittent operation of hydraulic nozzles

The idea of reducing the volume of spray applied from hydraulic nozzles by using a solenoid valve to provide an intermittent flow to the nozzle has been investigated previously, but has come into prominence in relation to precision agriculture. Giles and Comino (1989) described the control of liquid flow rate by positioning the nozzle directly downstream of the valve. A 10:1 flow turndown ratio (TDR) can be achieved by interrupting the flow while independently controlling the droplet spectrum by adjusting the pressure (Giles, 1997), but droplet size spectra were slightly affected over a 4:1 range in flow (Giles *et al.*, 1995a). The predominant effect of reduced flow was to produce slightly larger droplets, but the effect was so slight that the VMD was not significantly changed. Droplet velocity and energy were slightly reduced, as intermittency was increased (Giles and Ben-Salem, 1992). Changes in flow rate with pressure and duty cycle of the valve for an XR8004 (F80/1.6/3) nozzle is shown in Fig. 5.16, while the VMD is indicated for different flow rates in Fig. 5.17. The fitting of the solenoid allows a farmer to use one nozzle, e.g. F80/2.4/3, and apply flow rates down to the equivalent of a nozzle with half the flow rate at the same pressure. By adjusting the pressure from the cab, the user can change to a coarser spray while spraying near sensitive areas. Droplet spectra

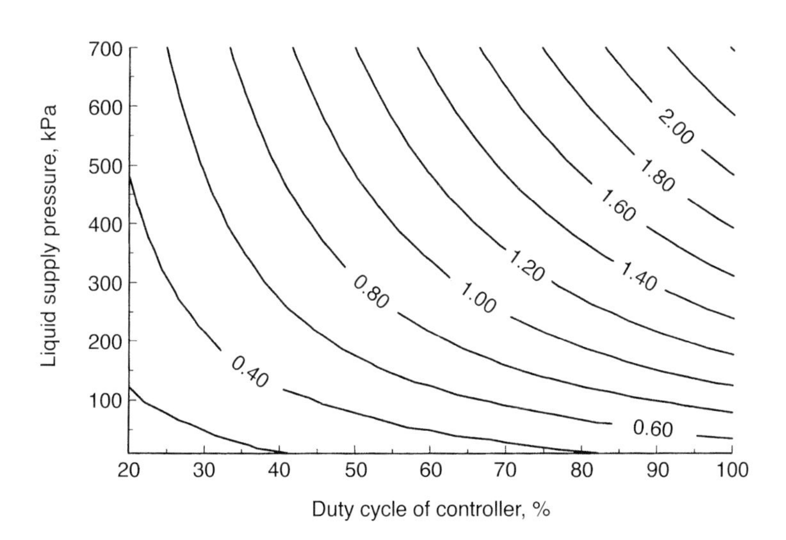

Fig. 5.16 Flow control envelope for XR8004 flat-fan nozzle from 70 kPa to 700 kPa liquid supply pressure and 20% to 100% duty cycle of valve. Isoquants are flow rates in litres per minute.

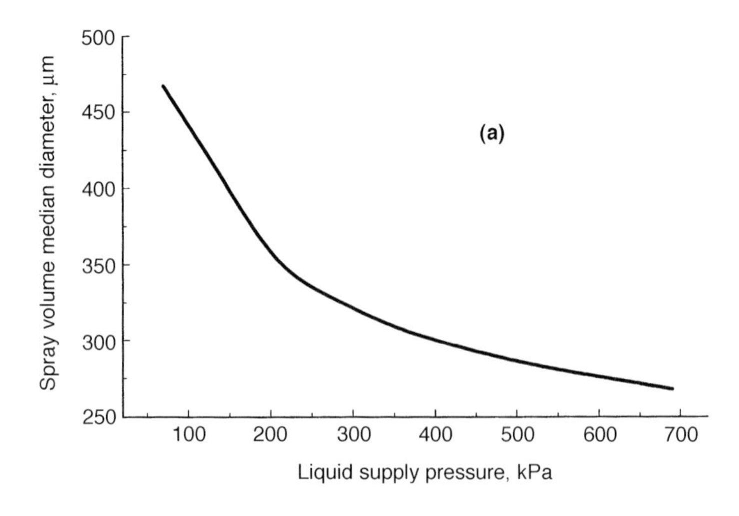

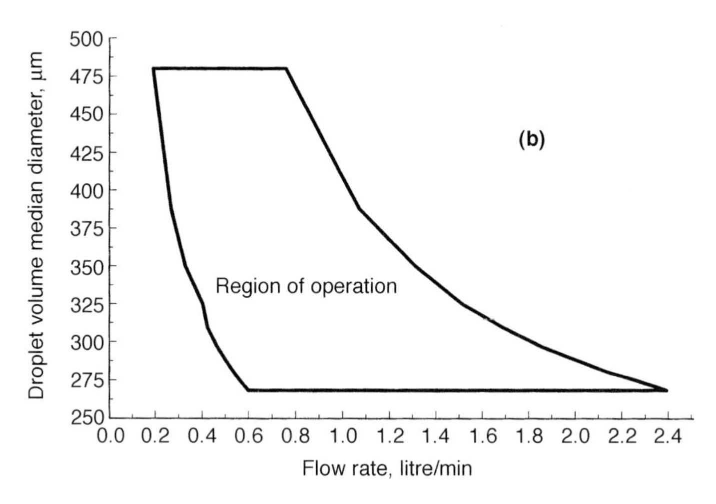

Fig. 5.17 (a) Volume median diameter of spray emitted from an XR8004 flat-fan nozzle over a range of liquid supply pressures from 70 to 700 kPa (Spraying Systems Co.). (b) Flow rate – droplet size control envelope for an XR8004 flat-fan spray nozzle over a liquid supply range 70–700 kPa and a valve duty cycle range of 20–100% (from Giles, 1997).

were affected if flow rate was decreased to 10%, and were generally more consistent with low flow rate nozzles than at higher flow rates (Ledson *et al.*, 1996).

Gaseous energy nozzle ('twin-fluid')

Some of the hydraulic nozzles have been adapted to become twin fluid nozzles in which air is fed into the liquid before it reaches the nozzle orifice. Other nozzles involving air shear are considered in Chapters 10 and 11.

With twin fluid nozzles, sometimes referred to as pneumatic nozzles, the spray quality will be affected by nozzle design, air supply pressure and spray liquid characteristics, e.g. viscosity and flow rate. In the 'Air Tec' nozzle (Fig. 5.18), air, fed into a chamber under pressure, is mixed with the spray liquid before emission through a deflector nozzle. This produces aerated droplets. By controlling both the air and liquid pressure, spray quality can be adjusted and low volumes applied per hectare with a relatively large orifice in the nozzle. Spray drift from this nozzle was significantly lower than that obtained from flat fan nozzles operated at 100 litres/ha (Rutherford *et al.*, 1989). This only applies if the nozzle is not used at too high an air pressure (> 10 bar) or very low flow rates (< 0.5 litres/min per nozzle) (Western *et al.*, 1989), otherwise drift could be exacerbated (Cooke and Hislop, 1987). Similarly, potential drift from the aerated droplets applied at 100 litres/ha was reported to be no greater than with conventional flat fan nozzles applying 200 litres/ha (Miller *et al.*, 1991). Subsequent studies indicated that in comparison with conventional low-volume fan nozzles, one design of twin fluid nozzle in a wind tunnel test produced drift intermediate between a standard fan and pre-orifice low drift nozzles (Combellack *et al.*, 1996). Womac *et al.* (1998) report similar assessments for the 'Air Tec', 'Air Jet' and another design 'LoAir' using water, a vegetable oil and mineral oils. They found that increasing the liquid flow rate

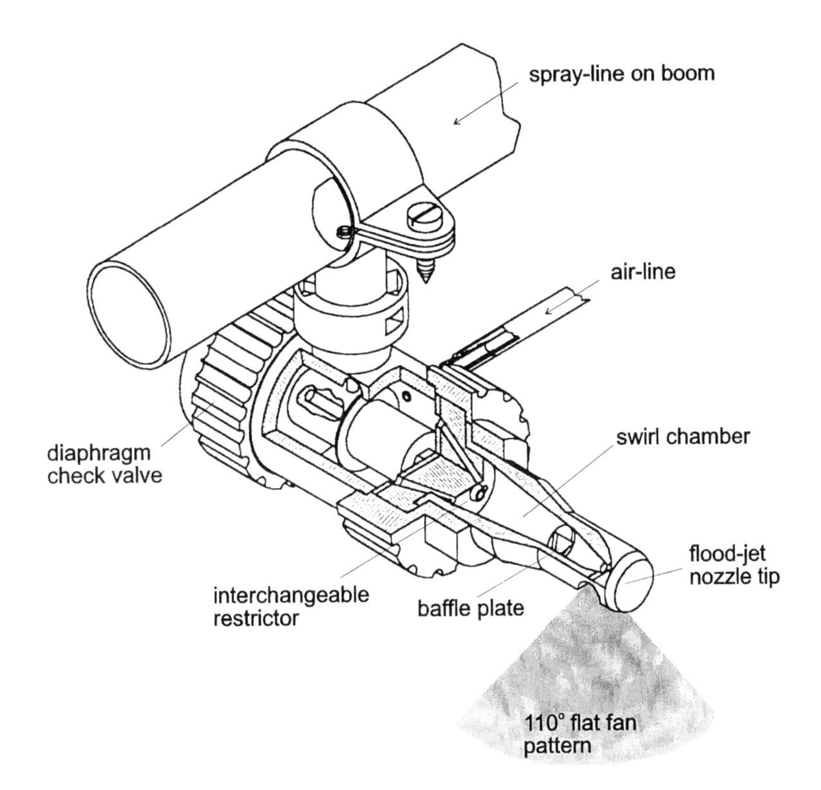

Fig. 5.18 'Airtec' twin-fluid nozzle.

increased the VMD and decreased the airflow, and that the proportion of small droplets (< 105 μm) was inversely and non-linearly proportional to the VMD. Atomisation of oils tended to produce small droplets and increased air pressure and flow rate also reduced droplet size, but in a way unique to each nozzle design.

Where sprays are applied at fast tractor speeds, there is a need to be able to adjust flow rate while maintaining a similar droplet size range. Combellack and Miller (1999) refer to nozzles needing a TDR of up to four. The TDR is defined as the difference between the lowest and highest flow rate divided by the lowest flow rate. Miller and Combellack (1997) also considered that a nozzle using less air volume and pressure was needed. The 'Air Tec' can require 25 litres of air per minute while a similar 'Air Jet' nozzle required up to 50 litres of air per minute. In practice the 'Air Tec' normally uses less than 10 litres of air per nozzle per minute, exceeding 10 litres of air per minute to achieve a fine or very fine spray at certain liquid flow rates. Subsequent work has shown that air consumption is reduced if air is delivered to a venturi nozzle insert where the greatest vacuum is produced, thus 5–8 litres of air per minute is sufficient (Combellack and Miller, 1999).

Kinetic-energy nozzle

A filament of liquid is formed when liquid is fed by gravity through a small hole, for example in the rose attachment fitted to a watering can, or the simple dribble bar, which can be used for herbicide application. The liquid filament when shaken breaks into large droplets.

Checking the performance of hydraulic nozzles

Calibration of flow rate

Flow rate or throughput of a hydraulic nozzle can be checked in the field by collecting spray in a measuring cylinder for a period measured with a stop-watch. Constant pressure is needed during the test period, so a reliable pressure gauge should be used. Output of nozzles mounted on a tractor sprayer boom can be measured by hanging a suitable jar over the boom to collect spray, but direct reading flowmeters are also available. Those fitted with electronic devices rely on battery power and need to be checked and calibrated. Throughput of nozzles at several positions should be checked to determine the effect of any pressure drop along the boom. The pressure gauge readings may require checking, as gauges seldom remain accurate after a period of field use. More accurate results can be obtained by setting up a laboratory test rig, with a compressed-air supply to pressurise a spray tank and a balanced diaphragm pressure regulator to adjust pressure at the nozzle. An electric timer operating a solenoid valve can be used to control the flow. The test rig should have a large pressure gauge frequently checked against standards and positioned as close to the nozzle as possible. The throughput of liquid, usually water, sprayed through the system can be measured in three ways:

(1) in a measuring cylinder
(2) in a beaker which is covered to prevent any losses due to splashing and the weight of liquid measured (Anon., 1971)
(3) a suitable flowmeter can be incorporated in the spray line.

For gaseous, centrifugal-energy and other nozzles, flow rate can be determined by placing a known volume of liquid into the spray tank and recording the time taken for all the liquid to be emitted while the sprayer is in operation.

Spray pattern

Various patternators have been designed to measure the distribution of liquid by individual or groups of nozzles. An early design was described by Thornton and Kibble-White (1974). Liquid monitored through a flowmeter is sprayed from one, two or three nozzles on to a channeled table and collected in a sloping section which drains into calibrated collecting tubes at the end of the channels. Separation of the channels is by means of brass knife-edge strips, below which are a series of baffles to prevent droplets bouncing from one channel to another. Whether droplet bounce need be prevented is debatable, as nozzles are often directed at walls, the soil or other solid surfaces where bouncing occurs naturally. Spray distribution has been measured satisfactorily

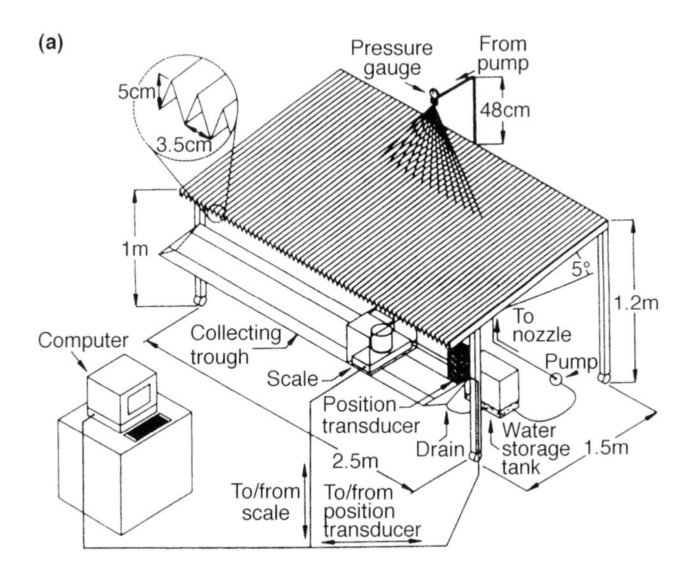

Fig. 5.19 (a) Automated spray nozzle patternator.

Fig. 5.19 (b) 'Spray scanner'.

with a simple patternator consisting of a corrugated tray. The nozzle is usually mounted 45 cm above the tray and connected to a similar spray line as that described for calibration of throughput. The standard width of each channel is 5 cm, although on some patternators each channel is 2.5 cm wide.

The main development has been in the way in which the volume of liquid in each collecting tube is measured. Patternators can now have a weighing system or an ultrasonic sensor that is moved across the top of each collecting tube and transfers data directly to a computer (Ozkan and Ackerman, 1992) (Fig. 5.19a). According to Richardson *et al.* (1986), the pattern can vary with successive runs with individual nozzles. Patternators can be positioned under a tractor boom to investigate variation in spray distribution along its length (Rice, 1967; Ganzelmeier *et al.*, 1994), and are now used to check the

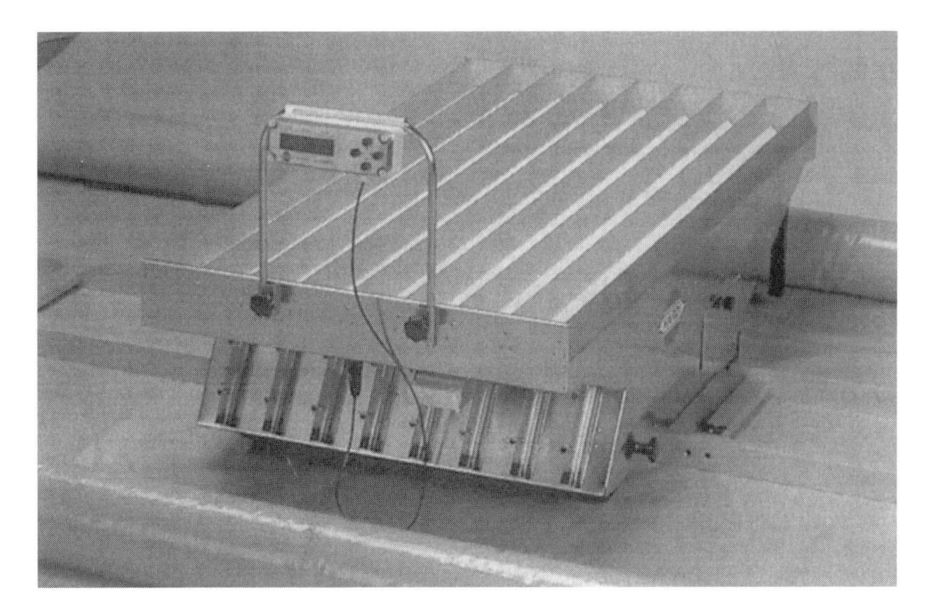

Fig. 5.19 (c) Close up of scanner.

performance of sprayers on farms. Another approach is to have a small number of channels on a trolley that moves along rails positioned under the spray. Data from this scanner (Fig. 5.19b) is downloaded to a PC.

Young (1991) used a two-dimensional patternator to assess the magnitude of a trailing plume from a stationary nozzle in a headwind, and thus assess the drift potential. Subsequently Chapple *et al.* (1993b) endeavoured to relate the pattern of a single static nozzle with the pattern obtained with a moving boom. Data from a patternator is mainly of relevance where spray is directed at a flat surface, but is less satisfactory for a complex crop canopy. In Germany, a vertical patternator was designed to assess the spray pattern from air-assisted orchard sprayers (Fig. 5.20) (Kummel *et al.*, 1991).

Nozzle erosion

The orifice of the nozzle tip is enlarged during use by the combined effects of the spray liquid's chemical action and the abrasive effect of particles, which may be the inert filler in wettable powder formulations or, more frequently, foreign matter suspended in the spray. This is referred to as nozzle-tip erosion and results in an increase in liquid flow rate, an increase in droplet size and an alteration in spray pattern. Increase in flow rate can result in over-use of pesticides and increased costs. This will occur especially where large areas are involved, so throughput should be checked regularly and the tip replaced when the cost of the cumulative quantity of pesticide wasted equals the cost of a replacement nozzle tip, if the rate of erosion is fairly regular (Kao *et al.*, 1972).

Fig. 5.20 Vertical patternator for orchard sprayer.

Rice (1970) reported increases in throughput of 49–63% with brass nozzle tips after 300 h wear with a 1% copper oxychloride suspension, whereas with stainless steel, ceramic and plastic tips, throughput increased only by 0–9% over the same period.

Over 70% of sprayers examined in a survey in the UK had at least one nozzle with an output that varied more than 10% from the sprayer mean. In extreme cases the maximum output was three times that of the minimum throughput (Rutherford, 1976). Beeden and Matthews (1975) and Menzies *et al.* (1976) have reported effects of erosion on cone nozzles. In Malawi, throughput increased by an average of 10% after 35 ha sprays, so a farmer should replace nozzle tips after three seasons. When water with a large amount of foreign matter in suspension is collected from streams or other sources, a farmer is advised to collect it on the day before spraying and allow it to settle overnight in a large drum. Nozzle tips should be removed after each spray application and carefully washed to reduce any detrimental effect of chemical residues. When cleaning a nozzle, a hard object such as a pin or knife should not be used, otherwise the orifice will be damaged (see Chapter 15).

Assessment of the effect of abrasion on a nozzle tip can be made in the laboratory by measuring the throughput before and after spraying a suspension of a suitable abrasive material. A suitable test is to spray 50 litres of a suspension containing 20 g of synthetic silica (HiSil 233) per litre (Jensen *et al.*, 1969; WHO, 1990), but other materials which have been used include white corundum powder which abrades nozzles similarly to HiSil but in one-third of the time. A test procedure for nozzle wear is also given by Reichard *et al*. (1990). Langenakens *et al*. (2000) showed by static and dynamic patternation that the quality of the distribution of worn nozzles was significantly worse.

6

Manually-operated hydraulic sprayers

Hydraulic-energy nozzles continue to be more widely used than the other types described in Chapters 8–11. Several new designs are now commercially available and add to the considerable flexibility which can be achieved by interchanging the tips in a standard nozzle body to provide a wide range of outputs and spray patterns at low cost. Hydraulic nozzles are used on a wide range of sprayers from a simple hand-syringe type to equipment mounted on aircraft. Hand-operated equipment is described in this chapter, power-operated equipment in Chapters 7 and 10, while aerial equipment is discussed in Chapter 13.

Sprayers with hydraulic pumps

Syringes

There are various types of simple syringe-type sprayers in which liquid is drawn from a reservoir into a pump cylinder by pulling out the plunger; the liquid is then forced out through a nozzle on the compression stroke. This type of sprayer has been replaced either by the double-acting slide pump or hand compression sprayer. A small syringe-type sprayer is useful for spot treatment, for example individual *Striga* plants can be killed in maize if there is a low infestation; a volume of 1 ml can be applied as a coarse spray to an area of 25 cm diameter. Some of these syringes have a simple means of adjusting the volume dispensed (Fig. 6.1). A syringe-type sprayer is also used to inject systemic insecticides into holes previously bored into trees.

The double-acting slide pump consists of a piston pump in which one valve is mounted at the inlet end of the cylinder, and a second valve is in a tube which is used as a piston. A handle grip is positioned on both the pump cylinder and piston. Holding the piston handle-grip firmly and moving the cylinder in and out continuously, while directing the nozzle, operates the pump. On the first

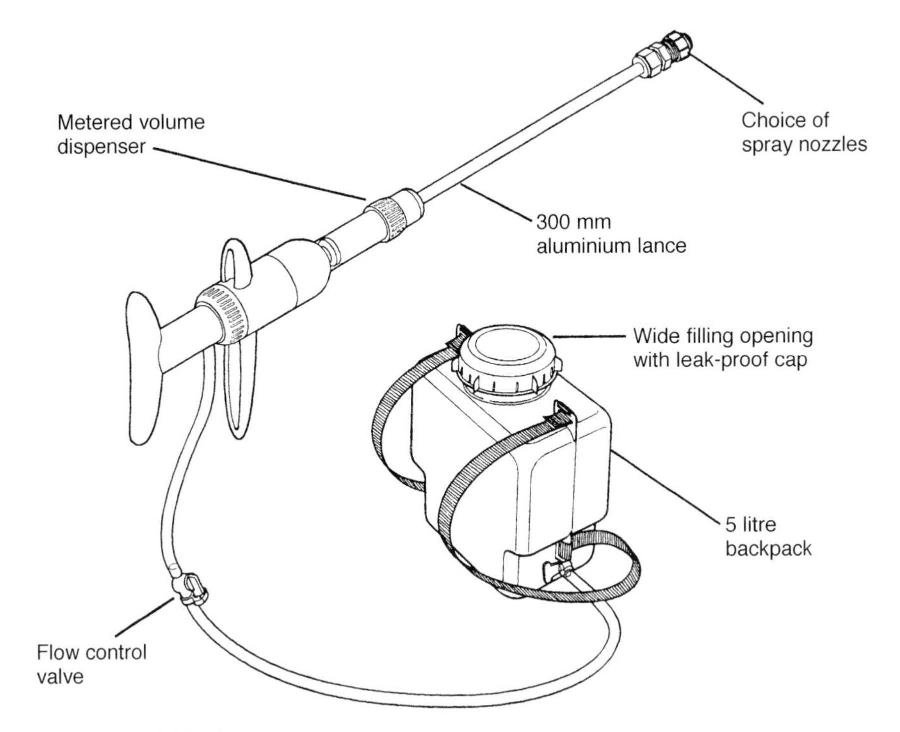

Fig. 6.1 Variable dose spot sprayer.

stroke, liquid from a separate container is drawn into the cylinder past the inlet valve, and on the return stroke, the inlet valve is closed and the liquid is forced past the second valve into the piston tube. The piston seal is usually a thick washer. If this is too tight in the cylinder, pumping effort will be excessive, and conversely, if too loose, pumping will be inefficient. An effective seal is also needed between the piston rod and end of the cylinder, otherwise spray liquid will leak over the operator's hands. A rubber stop between the cylinder and the piston handle-grip cushions the end of the pump stroke. The piston handle grip is usually enlarged to contain a small pressure chamber to even out the variation in pressure between strokes of the pump. A knapsack container or ordinary bucket can be used to contain the liquid, which must be kept stirred, especially if wettable powder suspensions are being applied. The delivery tube to the pump usually has a filter fitted at the inlet end.

Because of the narrow diameter of the pump cylinder (15–20 mm) and long stroke (25–40 cm), quite high pressures can be obtained with small nozzles. Output varies with the length of pump stroke, so syringes are not suitable for precise application of pesticide. Various types of hydraulic nozzles can be used, but these sprayers are often fitted with an adjustable cone nozzle. Both hands are required for operation, so the pump is very tiring to use, even when spraying is over a short period.

For small-scale home garden use, a 1 or 0.5 litre container with a trigger-operated pump is often suitable for intermittent operation. Continuous use is tiring.

Stirrup pump

This version of the double-acting pump requires two operators – one to work the pump while the other directs the nozzles. The lower end of the pump is immersed in the spray liquid in a bucket. The user needs to keep one foot on a footrest or stirrup on the ground next to the bucket while pumping. Ideally, the position of the stirrup can be adjusted to allow buckets of different depths to be used. Agitation is seldom provided, but on some sprayers there is a paddle agitator in the bucket which may cause splashing when the bucket is nearly empty. Agitation is also possible by recycling some of the liquid through a nozzle mounted at the lower end of the pump. The outlet of the pump is near the top and is connected to a hose, usually 6 m in length, with a lance, at the end of which any type of hydraulic nozzle may be fitted. Longer hoses are difficult to handle. Spraying is continuous because an air chamber is incorporated in the spray line, and on some models pressures up to 10 bar can be obtained fairly easily.

With stirrup pumps, great care has to be taken to avoid spillage of toxic chemicals as the liquid is in an open container. These pumps are strongly constructed and should withstand considerable wear in the field if properly cleaned after use. The pump may be a simple plunger type or a piston with a cup washer or gasket gland, which should be smeared with grease, especially if it is made of leather. Stirrup pumps have been used to treat apple orchards in India, but preference is now given to motorised portable line systems. In vector control, larvicides can be applied to the surface of water and to spray a residual insecticide on the wall surface of dwellings, but compression sprayers are now preferred. Specifications for a stirrup pump have been published by the World Health Organisation (WHO/EQP/3.R3). They have been used to apply molluscicides to water and, by removing the nozzle, can also be used to transfer liquids from a container to a sprayer.

Knapsack sprayers – lever-operated

The lever-operated knapsack sprayer (LOK) continues to be the most widely used small sprayer (Fig. 6.2). They were initially developed to treat vines with fungicides in the late nineteenth century (Lodeman, 1896; Galloway, 1891). A lever-operated sprayer consists of a tank which will stand erect on the ground and fit comfortably on the operator's back like a knapsack when in use, a hand-operated pump, a pressure chamber, and a lance with an on/off tap or trigger valve and one or more nozzles. FAO and some national Standards organisations have published specifications for this type of sprayer (FAO, 1998). Some aspects of these specifications are aimed at improving operator safety.

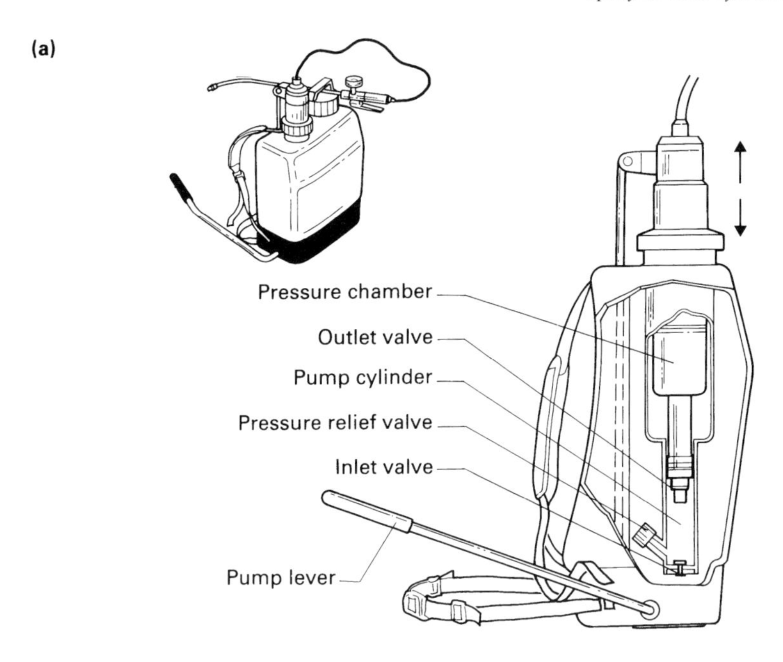

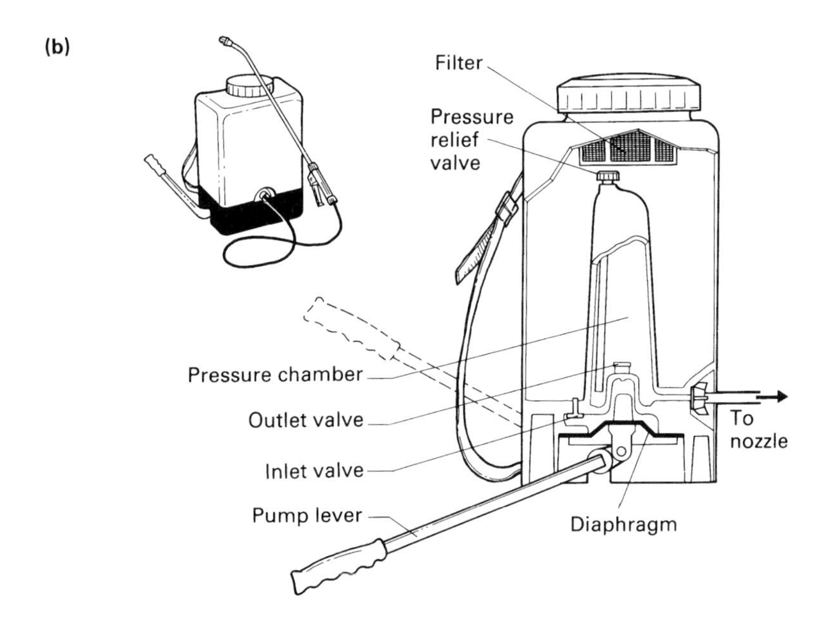

Fig. 6.2 Lever-operated knapsack sprayers: (a) piston type; (b) diaphragm pump type (courtesy: British Crop Protection Council).

The tank is now usually moulded from polypropylene or an alternative plastic that is extremely resistant to most of the agrochemicals used. To reduce the effect of sunlight on the plastic, a UV-light inhibitor is incorporated into the plastic. A few manufacturers can supply a tank made from brass or stainless steel. In Europe, regulations concerning the weight that can be lifted by a person limit the capacity of the tank; thus most tanks carry about 15 litres. Smaller 10 litre sprayers are also available. Plastic tanks can be moulded to fit the operator's back more comfortably than was possible with metal tanks. The design of the tank must avoid any outer surfaces that might collect spray liquid to which the user will be exposed. The volume of spray in the tank is usually indicated by graduated marks, moulded into the tank. To facilitate filling, the tank should have a large opening not less than 95 mm in diameter at the top. This large opening permits operators to put their gloved hands inside the tank if necessary for cleaning. The lid covering this opening must fit tightly. An air vent in the lid must have a valve to prevent any spray liquid splashing out and down the operator's back. When filling the tank, there should be a filter in the opening to remove particles that might damage the pump or block nozzles. The filter should be positioned deep inside the tank, so that liquid will not splash back on the spray operator. As some pesticides are now available in a water-soluble sachet or are formulated in a tablet, the filter should be designed to hold the pesticide while water is poured into the tank. The filter can have a mark to indicate the capacity of the tank to avoid the risk of over-filling. Some filters have a 50 mesh strainer, but most have a coarser mesh at this stage to allow rapid filling.

Lever-operated knapsack sprayers usually have an underarm lever, but some farmers prefer sprayers with an over-arm lever as this keeps the hand away from crop foliage; however, operating this lever for any length of time causes blood to drain from the arm and fatigue occurs very easily. There is either a piston or diaphragm pump (Fig. 6.2), which has to be continually operated at a steady rate. The piston pump is more common as higher pressures at the nozzle can be obtained, but some users prefer the diaphragm pump, especially when applying suspensions that may cause erosion of the piston chamber. The pump is operated by movement of a lever, which is pivoted at some point on the side of the tank. Many sprayers have the facility to change the lever from left- to right-arm operation.

To use the lever efficiently, the sprayer must fit comfortably on the operator's back so that the straps can be adequately tightened. Easily adjustable straps made of suitable rot-proof, non-absorbent material should be wide enough (40–50 mm) to fit comfortably over the shoulder without cutting into the neck. A waist strap is essential to reduce movement of the tank on the operator's back while pumping, and enable the load to be taken on the hips. Straps fitted with a hook to clip under the edge of the protective skirt of the tank tend to slide out of position easily, especially when the sprayer is not full, and are not recommended.

When using the sprayer, liquid is drawn through a valve into the pump chamber with the first stroke. With the return of the lever to the original position, liquid in the pump chamber is forced past another valve into a pressure chamber. The first valve between the pump and the tank is closed during this operation to prevent the return of liquid to the tank. A good seal between the pump piston and cylinder is obtained by a cup washer or 'O' ring. Abrasive materials suspended in the spray will cause excessive wear of the pump, also the chemicals in some formulations cause the seal to swell and prevent efficient operation of the pump. Air is trapped in part of the pressure chamber and compressed as liquid is forced into the chamber. This compressed air forces liquid from the pressure chamber through a hose to the nozzle. The size of the pressure chamber varies considerably on different types of knapsack sprayers (160–1300 ml), but should be as large as possible and at least ten times the pump capacity. Considerable variations in pressure will occur with each stroke if the capacity of the pressure chamber is inadequate, but even with a pressure chamber strongly constructed to withstand these fluctuations in pressure, a small variation in pressure occurs while spraying unless a pressure regulating valve is fitted to the lance.

The valves on each side of the pump can be either of a diaphragm type or a ball valve. Some operators prefer the ball valve, which is usually made of polypropylene. Pitting of the side of the ball valve or collection of debris in the ball-valve chamber may cause the liquid to leak past the valve. Also, the ball valve is easily lost when repairs are carried out in the field. The alternative is a diaphragm valve, made of various materials such as synthetic rubber (e.g. Viton) or certain plastics. The chemicals, or more often the solvents, used in some formulations can affect the material and cause the valve to swell up. This causes the valve to stick and block the passage of liquid through the pump unless there is adequate space for the diaphragm valve to move. With many knapsack sprayers an agitator or paddle is fitted to the lever mechanism, or directly to the pressure chamber, to agitate the spray liquid in the tank. On a few sprayers, part of the pump's output is recirculated into the tank to provide agitation. Agitation has been essential when spraying certain pesticides to reduce settling of particles inside the tank, which can occur rapidly with some wettable powder formulations. The pressure chamber and pump are fitted outside the tank of some sprayers to facilitate maintenance, but they are more vulnerable to damage if the sprayer is dropped. The pressure chamber may be fitted with a relief valve so that the operator cannot over-pressurise it. This should not be used as a pressure control valve and should be touched only when the tank is empty and clean.

To start spraying with a lever-operated knapsack, the lever is moved up and down several times with the trigger valve closed, so that pressure is built up in the pressure chamber. The trigger valve is opened and the operator continues to pump steadily with one hand while spraying. Inevitably there are variations in pressure at the nozzle unless a regulating valve is fitted adjacent to the nozzle or trigger valve. Several designs of regulator are available and of these the constant flow valve is the lightest (Fig. 6.3) (McAuliffe, 1999; Eng *et al.*, 1999). It operates at a set pressure so that the user cannot adjust the output in the field. Different versions are available to provide 1, 1.5, 2 and 3 bar

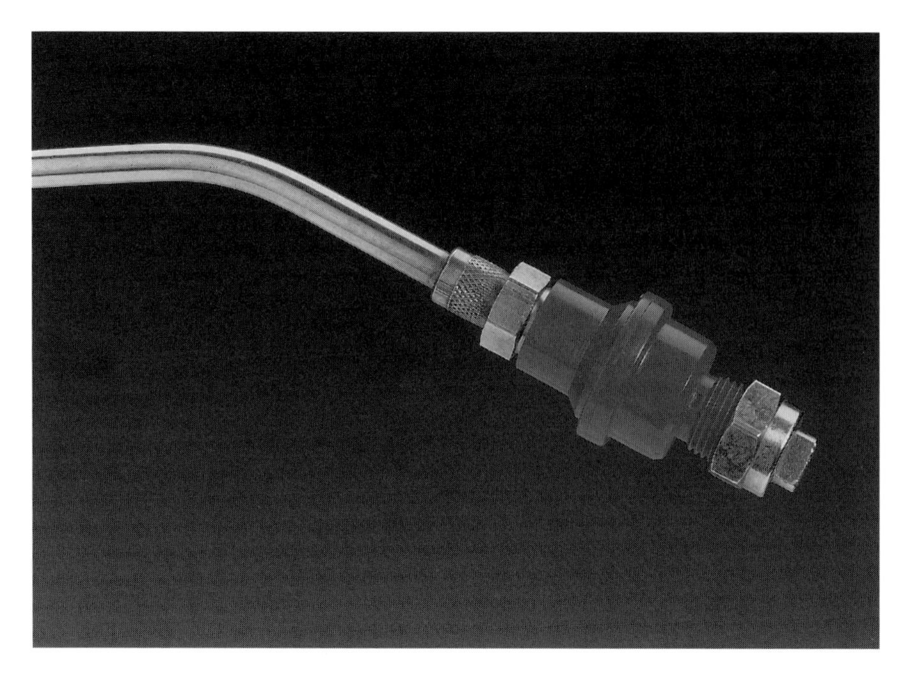

Fig. 6.3 Control flow valve (Global Agricultural Technology Engineering).

operating pressures. The lowest pressure is required where spray drift must be avoided, whereas the 3 bar version is intended for applications where a higher pressure is recommended. The 1.5 and 2 bar control flow valves provide a compromise suitable in many circumstances for either herbicide, fungicide or insecticide application. On some vertical spray booms it may be useful to fit a low pressure control flow valve near the top of the crop to minimise drift, yet have a higher pressure valve close to the bottom of a crop canopy to obtain better coverage.

Most lever-operated knapsack sprayers are fitted with a simple lance with usually one, but sometimes two, nozzles at the end. Continuous operation of the lever makes it difficult to direct the lance precisely at a target, so in certain circumstances the compression sprayer is preferred. A major problem is that the operator tends to walk towards where he is directing his spray and then through foliage which has been treated, thus becoming contaminated with pesticide, particularly on the legs (Tunstall and Matthews, 1965; Sutherland *et al.*, 1990; Thornhill *et al.*, 1996; Machado-Neto *et al.*, 1998). On various occasions, therefore, adaptations on the knapsack sprayer have been developed either to improve safety, obtain a better distribution of spray droplets or to increase the speed of spraying.

An example of this is the fitting of wide-angle nozzles onto the back of the spray tank for treating rice crops so that the operator walks away from the spray (Fernando, 1956). Pairs of nozzles are used on the tailboom (Fig. 6.4) as the plants increase in size, so that good distribution is achieved throughout the

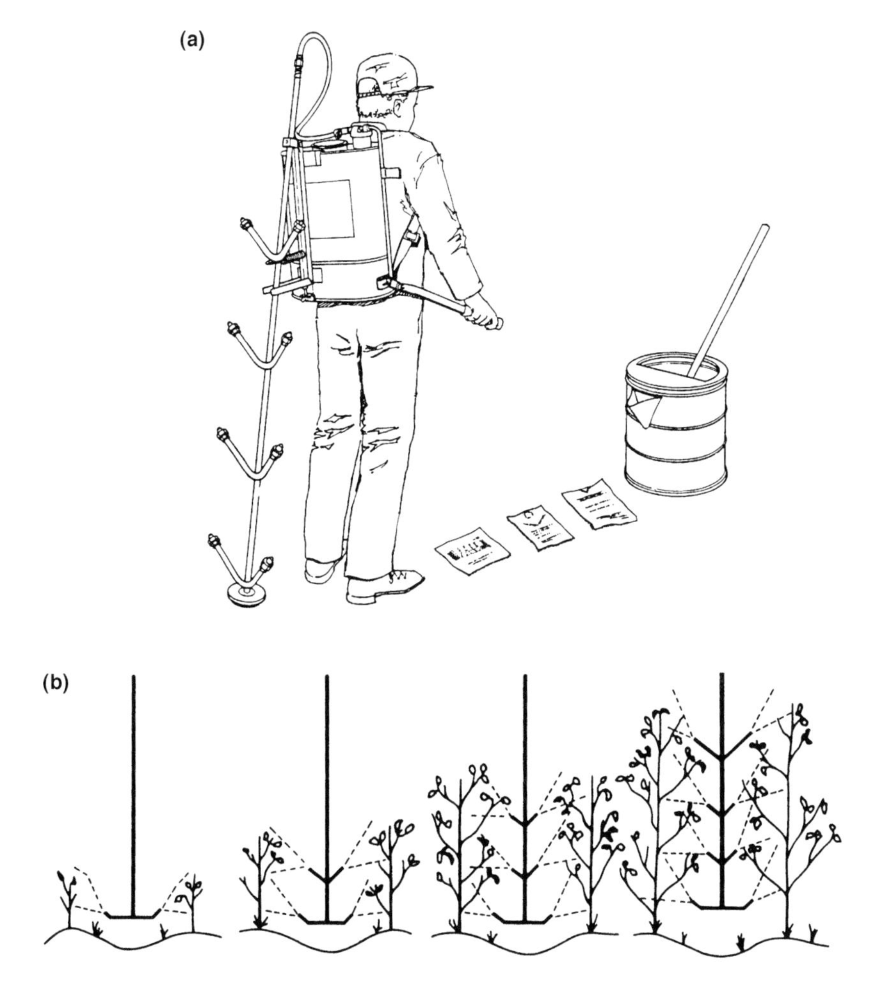

Fig. 6.4 (a) Knapsack sprayer with tailboom for spraying cotton. (b) Variation in the number of nozzles with plant height.

crop canopy (Tunstall *et al.*, 1961, 1965). By angling the nozzles upwards, underleaf coverage is increased, thus improving control of insects and pathogens located there, including whiteflies. A tailboom has also been used to spray coffee. To improve the speed of spraying, a horizontal boom has been developed for spraying more than one row of cotton at a time (Cadou, 1959), and for applying fungicides to groundnuts (Johnstone *et al.*, 1975).

In addition to the conventional lance, some sprayers have a short boom with two or three nozzles fitted to the end of the lance. Extendable lances made of bamboo, glass-reinforced plastic (GRP), carbon fibre or aluminium may be used to spray trees up to about 6 m in height. A gooseneck at the end of a lance is useful for spraying some inaccessible sites; similarly, other specialised nozzle arrangements have been used to spray special targets such as pods resting on stems of cacao trees. The nozzles may be shielded so that herbicide sprays can be applied close to a susceptible plant or tree.

The design and efficiency of operation of trigger valves on lances vary considerably. The handle should fit comfortably in the operator's hand, so that the valve is easy to operate. A clip mechanism to hold the valve open for prolonged spraying is useful, provided it can be released easily. Ideally there should be a clip to hold the valve closed when not in use. Unfortunately, many valves leak, particularly after abrasive particles have been sprayed, so that regular maintenance of the valve seating is needed with replacement springs. Hall (1955) has described a test procedure for trigger valves (WHO, 1990).

Some commercially available lever-operated sprayers are strong enough only if used for short periods and frequently leak. In an assessment of sprayers in Malaysia, nearly half the knapsack sprayers leaked (Cornwall *et al.*, 1995). When crops require several treatments, a farmer requires a robust sprayer. Mechanised durability tests can be carried out to assess whether the pump mechanism will operate without any problem for at least 250 hours (Matthews *et al.*, 1969; Thornhill, 1982). The main faults have been poor linkage, inadequate strength of the lever, poor design or strength of certain components such as strap hangers, and the poor capacity and design of the pumps.

The performance of lever-operated sprayers has been recorded in the field by using a small portable recording pressure gauge. Comparison of number of strokes required to maintain various outputs and pressures can be a useful guide to the efficiency of the different sprayers commercially available (Matthews, 1969; Thornhill, 1985). To avoid the drudgery of manual pumping, there are now motorised versions, and some have an electrically operated pump powered by a rechargeable battery. Field workers preferred the latter in trials in plantations in Malaysia (Fee *et al.*, 1999).

Stretcher and other sprayers

Stretcher sprayers are seldom used now. They have rather heavy piston pumps, which are operated by hand by a rocking motion of a long handle or by a foot pedal. A very strong pressure chamber with the pump allows pressures up to 10 bar or more to be obtained. They are normally operated in one position with one or more hoses up to about 14 m in length to supply lances to which various hydraulic nozzles are fitted. A separate spray container is needed and has to be carried separately from the pump. Two people are needed to carry the pump and operate it, while two or more others are needed to handle the lances. Stretcher sprayers have been most useful for spraying large or tall trees or widely spaced bush crops. Similar sprayers are sometimes mounted on a wheelbarrow frame, together with a spray tank.

Compression sprayers

Compression sprayers have an air pump to pressurise the spray tank, which is never completely filled with liquid. Space is needed above the liquid so that air can be pumped in to create pressure to maintain the flow of liquid to the nozzle. Usually, a mark on the side of the tank indicates the maximum capacity of liquid at about two-thirds of total capacity. These sprayers vary in size from the small hand sprayers, suitable for limited use by gardeners, to large shoulder mounted sprayers usually of 10 litre capacity. Some may be carried as a knapsack. As no agitation is provided, these sprayers need to be shaken occasionally if using wettable powder formulations to prevent the suspension settling out.

Hand sprayers

A small tank usually made of plastic and of 0.5–3 litres capacity is pressurised by a plunger-type pump to a pressure of up to 1 bar (Fig. 6.5). Often a cone nozzle, the pattern of which can sometimes be adjusted, is fitted to a short delivery tube. The on/off valve is sometimes a trigger incorporated into the handle. Hand sprayers are useful for spraying very small areas where it is inconvenient to pump continuously.

Fig. 6.5 Hand-carried compression sprayer.

Shoulder-slung and knapsack compression sprayers

There were two types – the ordinary and pressure-retaining types – but use of the latter has been discontinued due to their weight and necessity for routine testing of the strength of the tank. There are a large number of non-pressure retaining compression sprayers with which the air pressure is released before refilling the tank with liquid.

The compression sprayers (Fig. 6.6) are pressurised by pumping before spraying commences, in contrast to the continuous pumping needed with

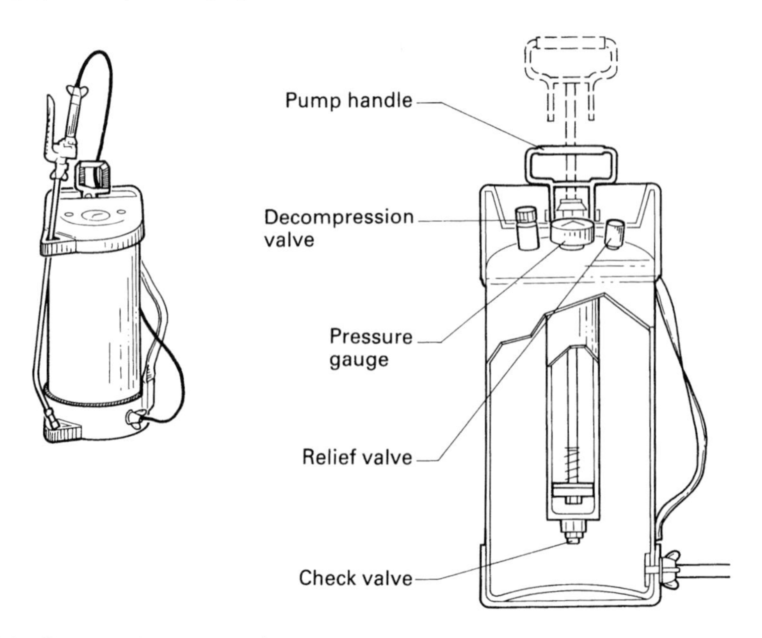

Fig. 6.6 Compression sprayer (courtesy: British Crop Protection Council).

lever-operated sprayers. This allows the operator to give more attention to directing the nozzle at the correct target. The pump is screwed in as part of the lid of the tank on the simpler and cheaper compression sprayers. The action of screwing the pump into the tank before each pressurisation can damage the threads, so limiting the life of the sprayer. Another problem with this design is that when the pump is removed to refill the tank, it can contaminate the surface on which it is placed and may transfer dirt into the tank when it is replaced. The tank lid and pump are separate on the more durable designs of compression sprayer. Ideally, this type of sprayer should be fitted with a pressure gauge so that the operator knows what pressure is in the tank. A pressure gauge may not be provided, in which case the operator is instructed to pump a given number of strokes to achieve the working pressure. Some sprayers have a safety valve which releases excess pressure if the operator pumps too much.

When using a compression sprayer, the tank pressure decreases very rapidly as soon as the operator starts spraying. Operators may have to stop and repressurise the tank before they can discharge the total contents from it. With the decrease in pressure at the nozzle while spraying, the output will decrease (Fig. 6.7) and droplet size will increase. To apply a pesticide more uniformly, it is essential to fit a pressure-regulator or constant flow valve to the tank outlet or lance. After use, the whole sprayer must be cleaned with water and the pump used to flush liquid through the valve and nozzle, so that any particulate formulation does not dry out inside and cause a potential blockage when the sprayer is used again.

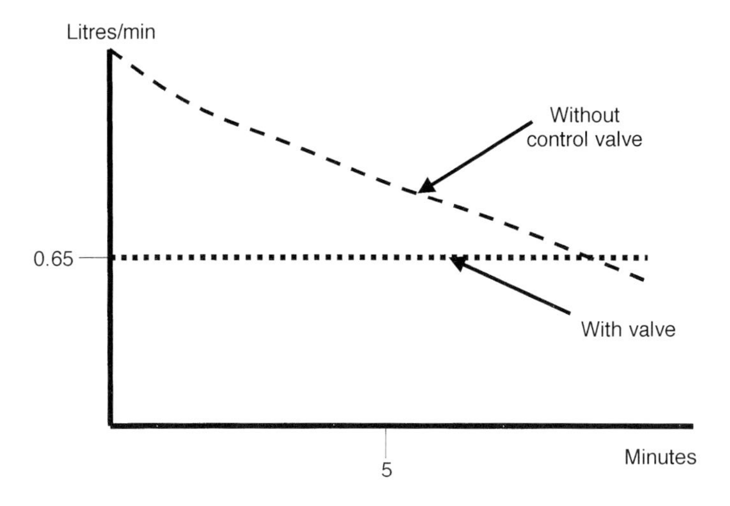

Fig. 6.7 Compression of nozzle flow rate from a compression sprayer with a control flow vale and without a valve.

On some occasions, some pressure may still be inside the tank when the operator has discharged the spray liquid and needs to refill the tank. This is released on the first quarter turn of the lid or pump, when a hissing sound indicates the escape of air. On some sprayers, the lid cannot be moved until a valve is activated to release the pressure. The valve is either in the lid or on the top of the tank. The tank expands and contracts slightly during normal operation. To assess the durability of the tank, this is simulated by pressurising the tank to 4.5 bar for 11 seconds, releasing the pressure and then repeating this for 12 000 cycles. The sprayer must be completely filled with water during this test. Further tests at 7 bar are usually carried out after dropping the sprayer in set positions to detect any weakness caused by the drop tests (Hall, 1955). Manufacturers should not use rivets in metal tanks used as compression sprayers, although one particular version has a strengthening bar that is held in place by two rivets. Some people may feel that plastic tanks will not be as strong as metal tanks but, in general, blow-moulded plastic tanks so far tested can stand pressures in excess of 7 bar, which is usually far above that obtained with the hand pumps provided with the equipment. Degradation of the plastic in sunlight (or UV light) has occurred, possibly by interaction with pesticides impregnated on the tank wall. The strength of these tanks is thus impaired. The base of the tank is usually provided with a skirt for protection against wear and also to enable the sprayer to stand firmly on the ground. On some sprayers, a footrest is attached to the skirt to assist pumping. The skirt serves as a backrest on some sprayers and is the lower fixing point for straps.

As with the lever-operated sprayers, the sprayer should have as large a tank opening as possible to facilitate filling. This has become more important where a pesticide is provided in a water-soluble sachet, so that the sachet can be put easily into the tank. A wide lid also allows operators to put their gloved hands through the opening and clean the inside of the tank. Unfortunately, very few of the sprayers of this type have an adequately wide tank opening.

The hose outlet is often at the base of the tank to avoid leaving any liquid inside and to eliminate a dip tube inside the tank, but this hose nipple is often broken when the sprayer is accidentally dropped. The better types of compression sprayers have the hose opening at the top of the tank and a clamp is also provided to hold the lance when not in use. When the lance is left to trail in the mud while the sprayer is being refilled, the possibility of nozzle blockages is increased. Thornhill (1974a) has described the adaptation of a container used for dispensing soft drinks as a compression sprayer.

Compression sprayers (Fig. 6.8) have been widely used on farms and also in vector-control programmes. WHO specification WHO/EQP/l.R4 was developed to ensure that reliable equipment was used to spray a residual deposit of insecticide on walls to control mosquitoes (Fig. 6.9). The standard technique recommended by WHO for indoor residual spraying was to apply 757 ml/min through an 8002 fan nozzle when operated at 2.8 bar. Unless a pressure regulator is fitted, the working pressure decreases from 3.8 to 1.7 bar (Brown *et al.*, 1997). Normally, the sprayer is charged initially to at least 3.8 bar with about fourteen pump strokes per bar (one stroke for each p.s.i.) and usually needs repressurisation once during a 10 min period to discharge 7.5 litres. The lance is held 45 cm from the wall and moved at a steady speed of 0.64 m/s up and down the walls, covering a 75 cm swath (with 5 cm overlap) each time. The same technique has been used for a number of different insecticides. If a constant flow valve at 2 bar is used, apart from more uniform spraying, the decrease in output from 757 to 650 ml/min allows a longer time for spraying per sprayer load (Fig. 6.7).

The same type of sprayer has been used to apply larvicides, but a solid stream or cone nozzle is used instead of a fan nozzle. An experienced sprayman using a solid-stream nozzle with a 'swinging wand' pattern can treat an 8–10 metre swath when walking at a steady 2 m/s. The nozzle is pointed above the horizontal so that the liquid trajectory reaches a maximum distance. If it is pointed down at the water surface there will be localised overdosing. The jet from the nozzle breaks up into a band of droplets, which overlap with each swing of the lance. The solid-stream nozzle is also useful when directing spray into cracks and crevices in houses so that an insecticide is deposited in the resting sites of cockroaches and other household pests. A cone nozzle is used if a wider band of spray is needed on irregularly shaped objects in areas such as at the backs of sinks and boilers.

Pressure-regulating valve

Apart from the control flow valves referred to above, there are adjustable pressure-regulating valves suitable for use on knapsack compression sprayers; they differ from that used on powered equipment in that there is no by-pass

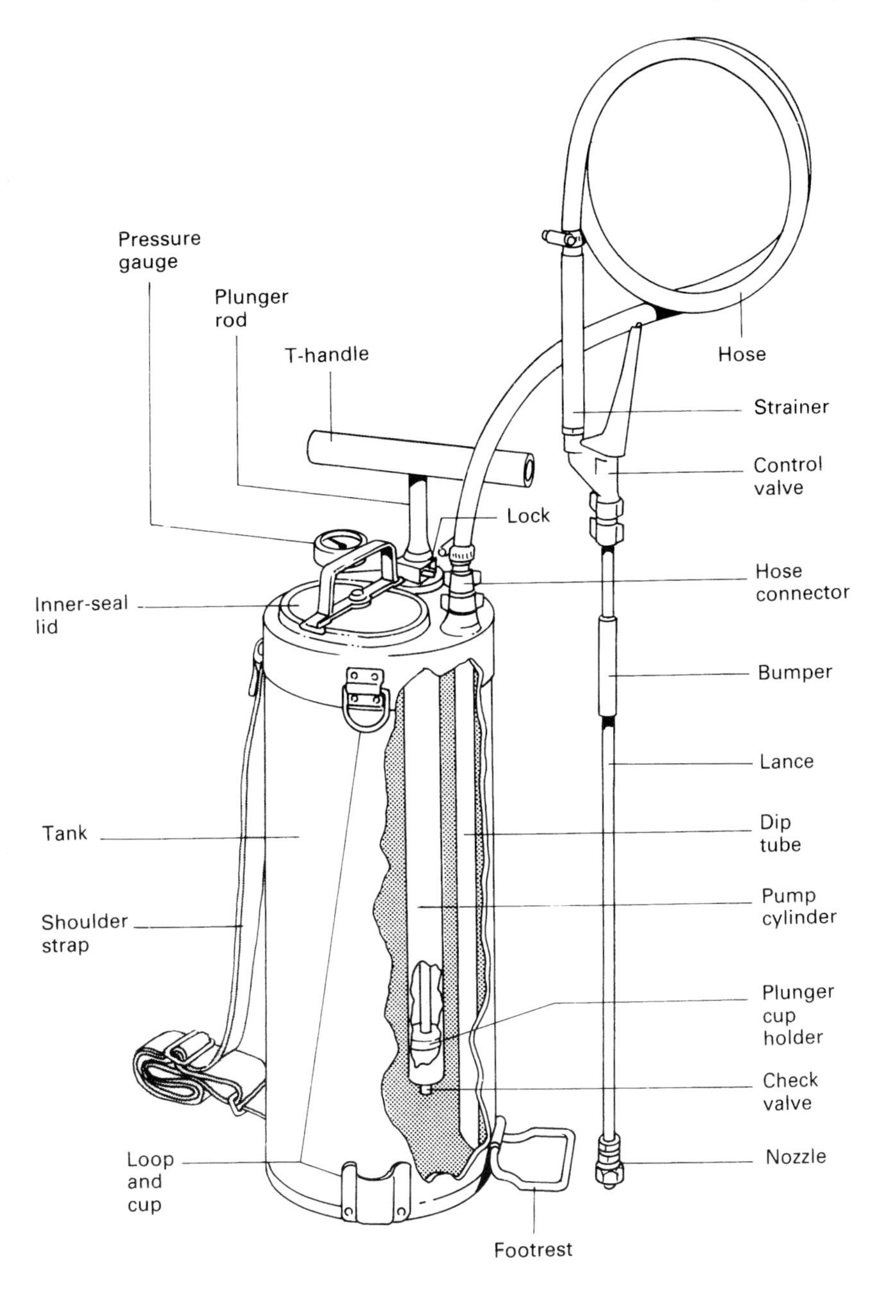

Fig. 6.8 Compression sprayer to meet WHO specification (courtesy: H.D. Hudson Manufacturing Co.).

Fig. 6.9 Residual spraying against mosquitoes in dwellings (photos: R. DaSilva, WHO).

return to the spray tank. The valve consists of a diaphragm, the movement of which is effected by a spring. The tension of the spring can be adjusted by means of a screw so that output pressure can be varied from 0.71 to 4 bar, irrespective of the input pressure, which can be up to 18 bar. The main problem with an adjustable valve is that the user is never sure what the output pressure is unless there is also a reliable accurate pressure gauge on the lance.

Calibration of knapsack sprayers

The label of the pesticide should be examined to see if a volume application rate, spray quality, nozzle or spray concentration is recommended. The desired nozzle is selected and its output measured during 1 minute. When using lever-operated sprayers, a control flow valve, or alternatively a pressure gauge, should be fitted as close to the nozzle as possible, and the lever operated evenly with a full stroke to maintain as uniform a pressure as possible. The operator will need to practise before achieving an even pumping rate. Having determined the output from the nozzle in litres/min, the rate per unit area treated can be calculated, knowing the swath width and walking speed.

$$\frac{\text{Output (litres/min)}}{\text{Swath (m)} \times \text{Speed (m/min)}} = \text{Volume application rate (litres/m}^2\text{)}$$

Thus with a swath of 1 m, walking at 60 m/min and a flow rate of 0.6 litres/min, the volume of spray per square metre is

$$\frac{0.6\ \text{litres/min}}{1\ \text{m} \times 60\ \text{m/min}} = 0.01\ \text{litres/m}^2 \quad \text{or} \quad \times 10\,000 = 100\ \text{litres/ha}$$

Alternatively, if you measure speed in km/h, then

$$\frac{600 \times \text{Output (l/min)}}{\text{Swath (m)} \times \text{Speed (km/h)}} = \text{Volume application rate (litres/ha)}$$

Thus if your flow rate is 2.2 litres/min over a 1.7 m swath and your speed is 3.8 km/h, then your application rate is 204 litres/ha.

If the application rate is incorrect, other nozzles should be tried. When the most suitable nozzle has been selected, the volume applied can be rechecked by measuring the distance walked and time taken to spray a known quantity. For example if a full tank load of 15 litres is applied in 25 minutes, the output is 0.6 litre/min which checks against the earlier calibration, the volume per hectare being given by

$$\frac{15 \times 10\,000\ \text{m}^2\ \text{(i.e. 1 ha)}}{\text{Distance travelled (m)} \times \text{Swath width (m)}} \times \text{Application rate (litres/ha)}$$

If the distance travelled was 1.5 km with a swath of 1 m, the application rate was 100 litres/ha. When the output is low, the sprayer can be calibrated more quickly by using a smaller volume in the tank.

Some manufacturers supply a calibrated container (Fig. 6.10) which can be fitted to the nozzle so that the spray is collected while treating a known area

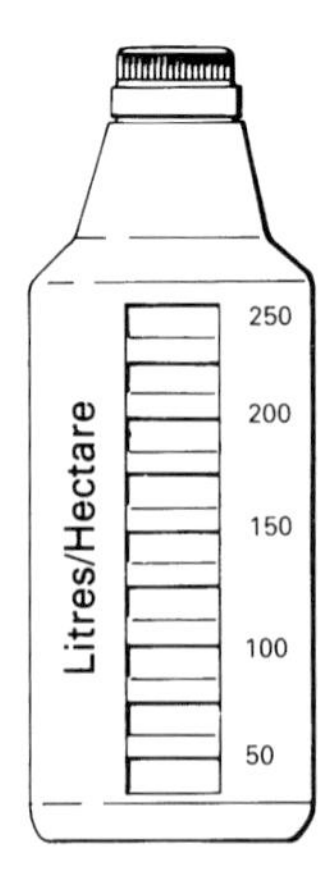

Fig. 6.10 A calibrated bottle to assist calibration of knapsack sprayers.

($25\,m^2$). This method is particularly useful when training teams of spray operators, because individuals can see their own output and adjust their speed of walking or rate of pumping to get the required output.

Another method is to measure the time to walk 100 metres and the swath width, then measure the output of the sprayer for the same time that it took to walk 100 metres. Then the volume per hectare is the output in millilitres divided by the swath width (m); then divide answer by ten.

Handymist sprayer

In China, Tu (1990) designed a lever-operated sprayer with an air pump. The sprayer is essentially a compression sprayer but, in addition to pressurising the liquid, some of the air is fed to a single twin fluid nozzle. The manually operated sprayer provides an inexpensive means of applying a very low volume mist sprayer and has been used to spray insecticides on rice and cotton. The sprayer has not been widely used.

Disposable container dispenser

A disposable container dispenser (DCD) was designed to fit manually operated sprayers, so that only water is put in the lever-operated knapsack or compression sprayer container (Craig *et al.*, 1993). When the water passes through a specially designed trigger valve that incorporates a flow control, pesticide is metered into it at a set dilution rate. The aim was to reduce exposure of the operator to the pesticide as there was no longer a need to measure out small quantities of pesticide product to put in the sprayer. However, the design was not suitable for all pesticide formulations and the container could not be rinsed after use. The intention was to return the container for refilling, but this has not been adopted by the chemical industry.

Peristaltic pump

Liquid is forced through a piece of rubber of plastic tubing by progressive squeezing along the wall of the tube. A peristaltic pump, operated by rotating cams, attached to the wheel of a small sprayer can be used to deliver small volumes to individual nozzles.

7

Power-operated hydraulic sprayers

Various power-operated sprayers have been designed and range in size from small, hand-carried engine-driven pump units to large self-propelled sprayers (Fig. 7.1). Usually a series of nozzles are mounted along a boom. Small units have a two- or four-stroke internal combustion engine or an electrically operated pump, and are mounted on a knapsack, a wheelbarrow, or a frame for fitting on a vehicle. These small sprayers are mostly used where treatment is not required over extensive areas. Otherwise the tractor-mounted boom sprayers with a pump driven from the power take-off (p.t.o.) is used, especially for field crops, to apply 50–500 litres/ha. Larger-capacity tanks may be mounted on trailers, or as saddle tanks alongside the tractor engine to spread the load more evenly. In Germany, a survey indicated that 42% of large sprayers were mounted on the three-point linkage, while 33% were trailed, and 15% were mounted on a vehicle. Some large sprayers are self-propelled, instead of using the normal farm tractor, but these sprayers are used only on farms with sufficient flat land to allow the use of booms up to 36 m in width and where the capital outlay is justified by their usage (see Chapter 19). Animal drawn sprayers have also been used in some countries. Arable crop sprayers may now be fitted with air-assistance to improve the distribution of spray within a crop canopy. These sprayers are described in Chapter 10.

Tractor-mounted sprayers

Tank design

A typical layout for a modern tractor-mounted sprayer is shown in Fig. 7.2. Most tractors have a standard three-point linkage on which the sprayer is mounted. The capacity of the tank is restricted by the maximum permitted weight specified for the tractor; half the sprayers in the UK have a tank of

Fig. 7.1 Various types of power-operated hydraulic sprayers (photos: Allman & Co.).

less than 750 litres capacity. Weights may be needed on the front end of the tractor, particularly the small tractors, to maintain stability. The farmer may prefer to use a smaller tank to reduce compaction of soil under the tractor paths, but if tank capacity is too low, frequent refilling may be required. The choice of spray tank size is also discussed in Chapter 19 in relation to other variables.

Most modern sprayers have tanks constructed with a corrosion-resistant material such as multilayer plastic. The tank should have a large opening (>300 mm) so that the inside can be scrubbed out if necessary. A large opening

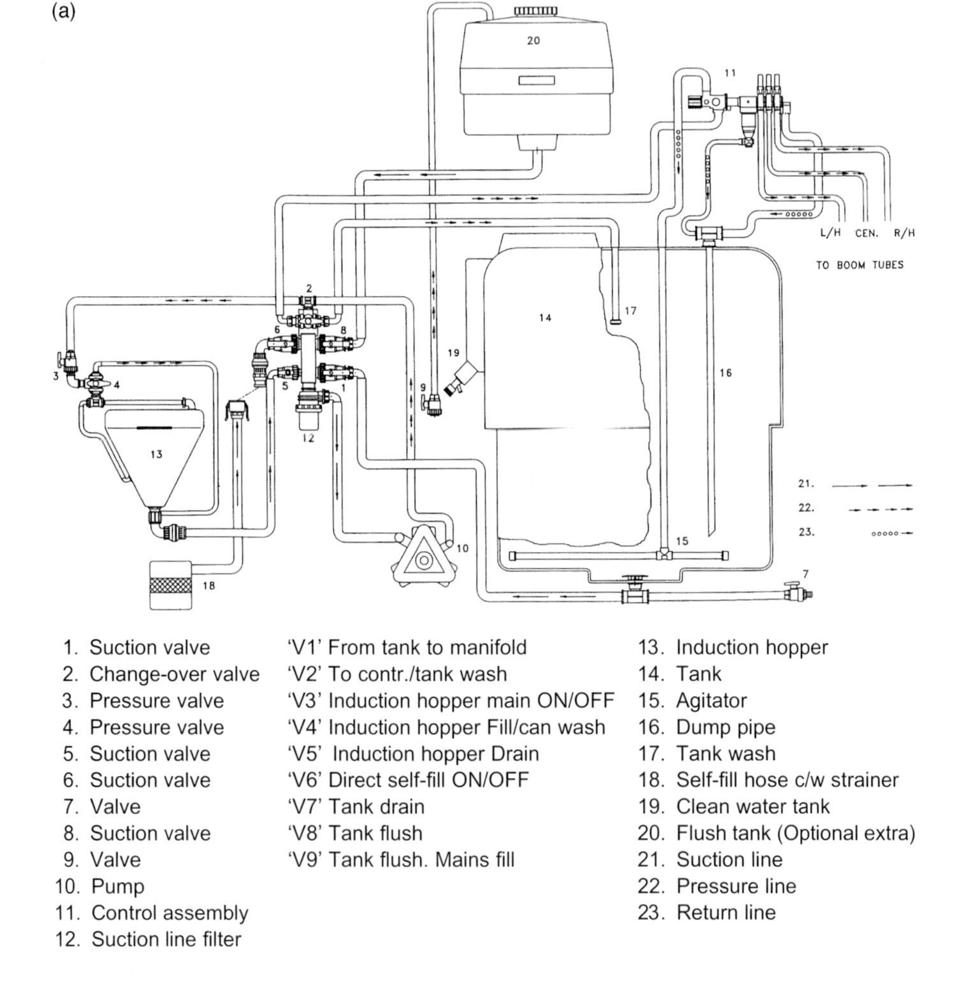

Fig. 7.2 (a) Layout of tractor-mounted sprayer (Allman & Co.). (b) Lower level induction bowl. (c) Nozzle to wash containers in induction bowl. (d) Locker for PPE. (e) and (f) Closed transfer systems.

(b)

(c)

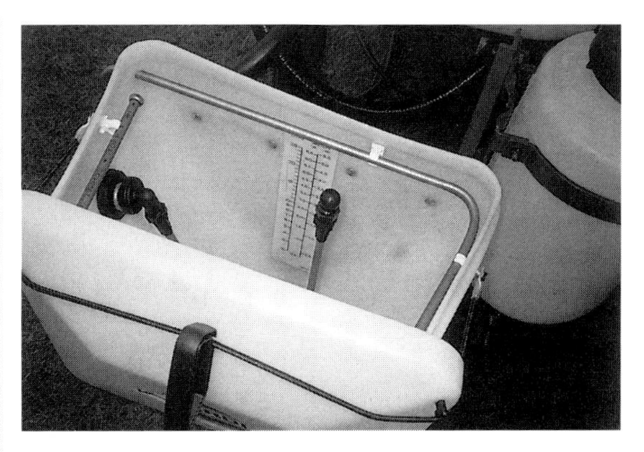

(d)

(e)

(f)

also facilitates filling the sprayer, but to improve safety for the operator, new sprayers are fitted with a self-fill system and in some cases a separate low-level mixing chamber. A large basket-type filter should fit into the opening, which is closed by a tight-fitting lid. The tank should have a drainage hole at its lowest point and a sight gauge visible to the tractor driver. The bottom of the tank should be fitted with a sparge-pipe agitator, that is a pipe with a line of holes along its length to give a series of jets of liquid to scour the tank bottom. Instead of a sparge pipe, a nozzle may be used to swirl the liquid over the tank bottom. Mechanical agitation is not recommended as, when the tank is nearly empty, the paddles may be only partly immersed and mix in air to cause foaming. A 200 litre drum has been used as a cheap tank on a sprayer, but this is not recommended, as rust is liable to occur quite rapidly inside metal drums.

Pumps

A number of different types of pump are used on tractor-mounted sprayers. Selection of the appropriate pump will depend on the total volume of liquid and pressure required for supplying all the nozzles and agitating liquid in the tank. The type of spray liquid will also influence the choice of pump, particularly the materials used in its construction. A comparison of pumps is given in Table 7.1.

Diaphragm pump

The basic part of the diaphragm pump is a chamber completely sealed at one end by a diaphragm (Fig. 7.3). The other end has an inlet and outlet valve. Liquid is drawn through the inlet valve by movement of the diaphragm enlarging the chamber, thus creating suction, and on the return of the diaphragm, it is forced out through the outlet valve. Some pumps have only one diaphragm, but usually two, three or more diaphragms are arranged radially around a rotating cam. This actuates the short movement of each diaphragm in turn to provide a more even flow of liquid instead of an intermittent flow or 'pulse' with an individual diaphragm. In any case a compression chamber, sometimes referred to as a surge tank, is required in the spray line if not incorporated in the pump to even out the pulses in pressure with each 'pulse' of the pump. These pumps are rather more complex as several inlet and outlet valves are required, but maintenance is minimal as there is less contact between the spray liquid and moving parts. Care must be taken to avoid using chemicals which may affect the diaphragms or valves. In general, diaphragm pumps are used to provide less than 10 bar pressure, but maximum pressures of 15–25 bar are attainable.

Piston pump

Liquid is positively displaced by a piston moving up and down a cylinder; thus

Table 7.1 Summary of types of pumps

	Diaphragm	Piston	Centrifugal	Turbine	Roller
Materials handled	Most; some chemicals may damage diaphragm	Any liquid	Any liquid	Most; some may be damaged by abrasives	Emulsion and non-abrasive materials
Relative cost	Medium/ high	High	Medium	Medium	Low
Durability	Long life	Long life	Long life	Long life	Pressure decreases with wear
Pressure ranges (bar)	0–60	0–70	0–5	0–4	0–20
Operating speeds (r.p.m.)	200–1200	600–1800	2000–4500	600–1200	300–1000
Flow rates (l/ min)	1–15	1–15	0–30	2–20	1–15
Advantages	Wear resistant Medium pressure	High pressures Wear resistant Handles all materials Self-priming	Handles all materials High volume Long life	Can run directly from 1000 r.p.m. p.t.o. High volume	Low cost Easy to service Operates at p.t.o. speeds Medium volume Easy to prime
Disadvantages	Low volume Needs compression chamber	High cost Needs compression chamber	Low pressure Not self-priming Requires high-speed drive	Low pressures Not self-priming Requires faster drive for 540 r.p.m. p.t.o. shafts	Short life if material is abrasive

the output is proportional to the speed of pumping and is virtually independent of pressure (Fig. 7.4). Piston pumps require a positive seal between the piston and cylinder and efficient valves to control the flow of liquid. To provide greater durability, the pump cylinder may have a ceramic sleeve. Owing to their high cost in relation to capacity, piston pumps are not used very much on tractor sprayers, but are particularly useful if high pressures up to 40 bar are required. A compression chamber is also required with these pumps. Piston pumps are less suitable for viscous liquids.

(a)

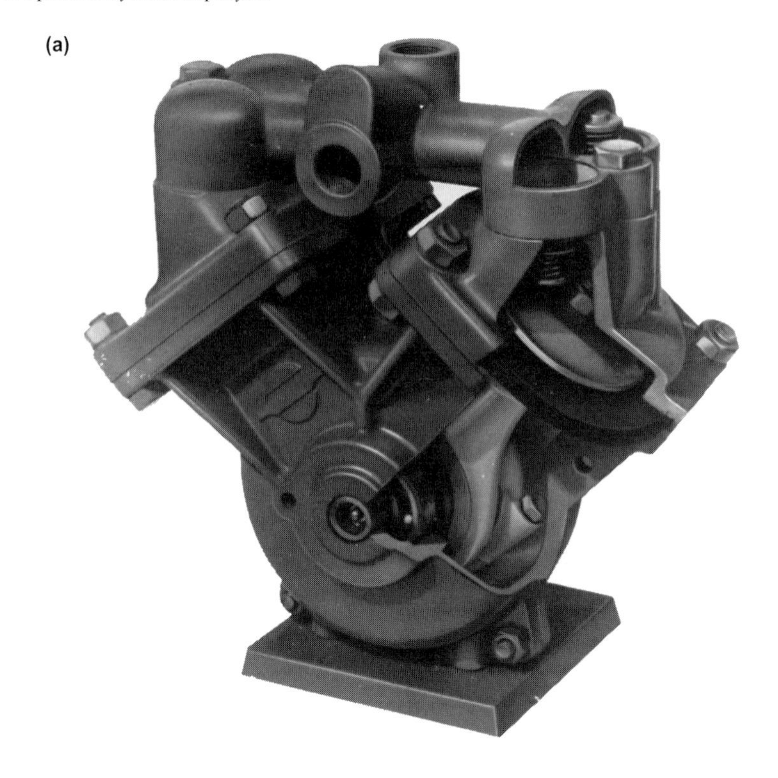

(b)

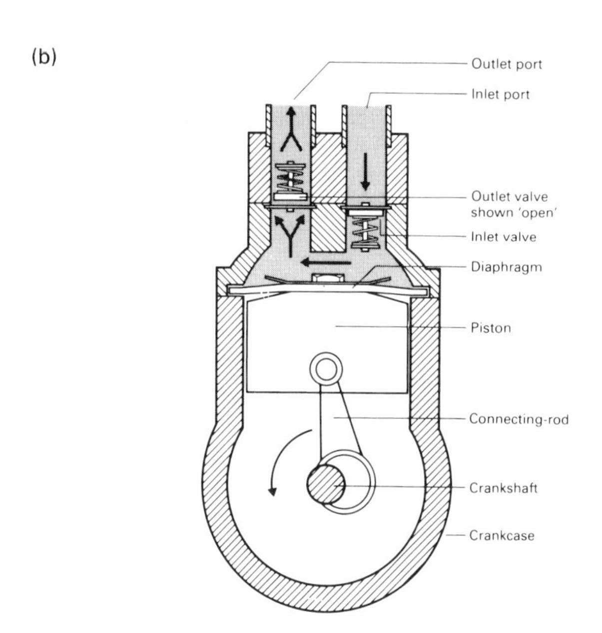

Fig. 7.3 (a) Diaphragm pump partly cut away to show diaphragm and valves (Hardi, UK). (b) Diagram to show construction.

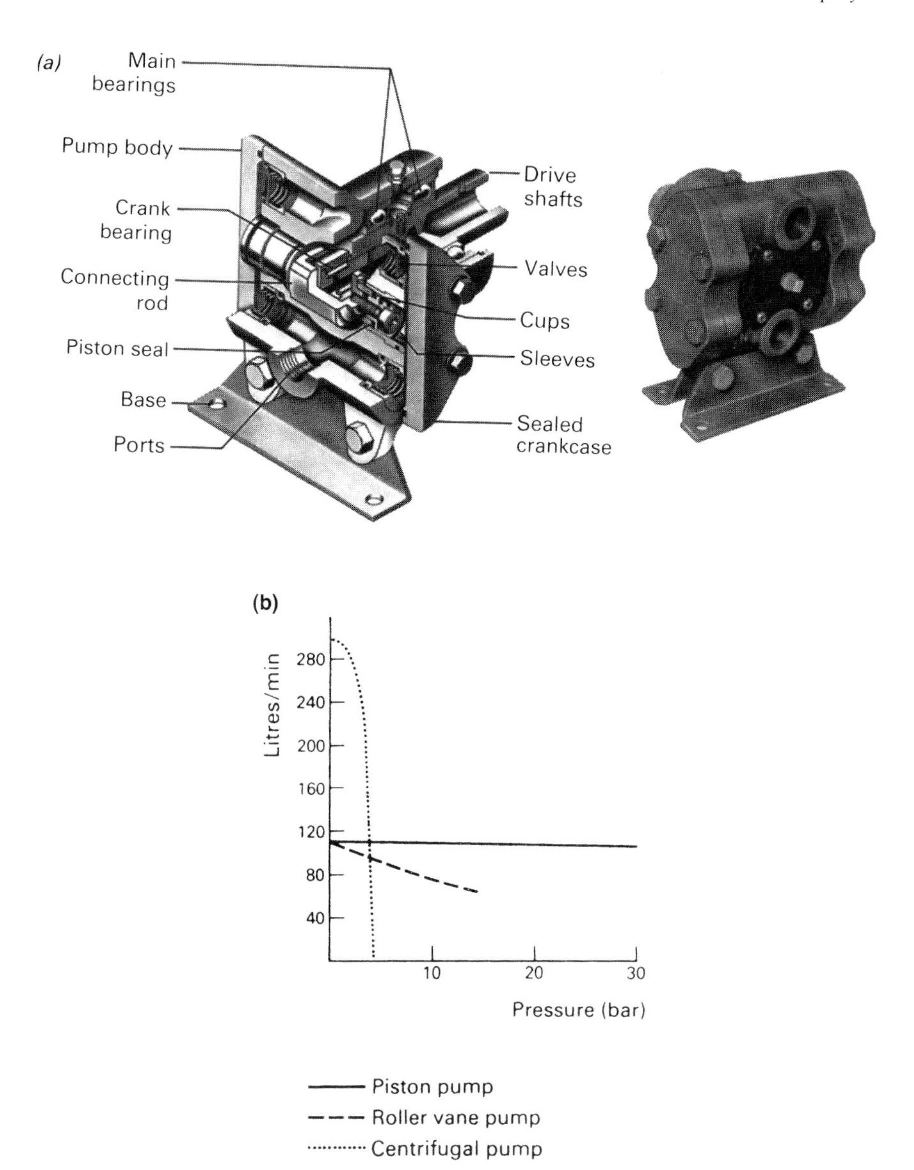

Fig. 7.4 (a) Piston pump cutaway and intact (photo: Delavan). (b) Performance of piston pump related to other types. Note: a compression chamber or surge tank must be placed with either a piston or diaphragm pump to even out pulses of pressure.

Centrifugal pump

An impeller with curved vanes is rotated at high speed inside a disc-shaped casing, and liquid drawn in at its centre is thrown centrifugally into a channel around the edge. This peripheral channel increases in volume to the outlet port on the circumference of the casing (Fig. 7.5). Centrifugal pumps are ideal for large volumes of liquid, up to 500 litres/min at low pressures. They can be used up to 5 bar, but the volume of liquid emitted by the pump decreases very rapidly when the pressure exceeds 2.5–3 bar. The pressure will increase slightly if the outlet is closed while the pump is running, and then slippage occurs

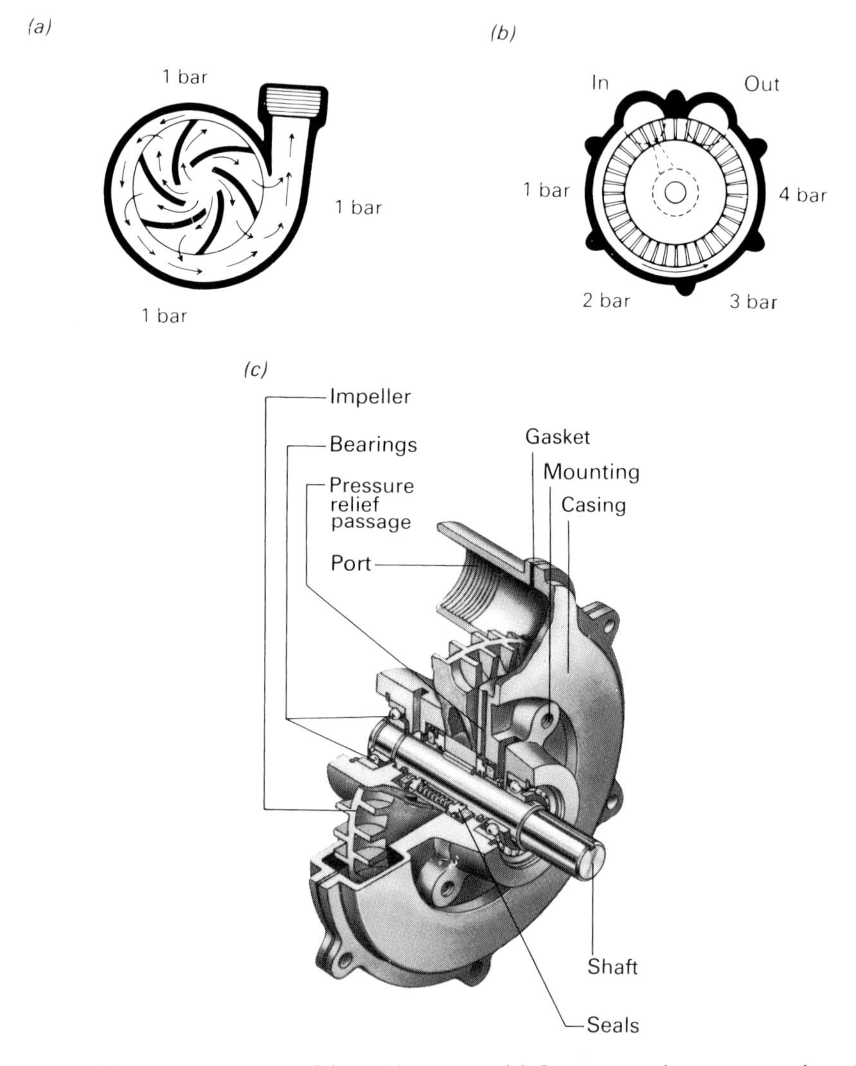

Fig. 7.5 (a) Centrifugal pump. (b) Turbine pump. (c) Cutaway to show construction of turbine pump (photo: Delavan).

without damage to the pump. Viscous liquids and suspensions of wettable powders and abrasive materials can be pumped. The seals on the shaft are liable to considerable wear as the pumps are operated at high speeds, but there is less wear on other parts as there are no close metal surface contacts. Instead of mounting a centrifugal pump directly on the p.t.o., a belt or pulley drive is required to obtain sufficient rotational speed of the pump. The pump may also be driven by a hydraulic motor from the tractor. Centrifugal pumps with a windmill drive are frequently used on aircraft spray gear. These pumps are not self-priming, and should be located below the level of liquid in the tank.

Pressure is increased in the turbine pump with a straight-bladed impeller in which liquid is circulated from vane to channel and back to the vane several times during its passage from the inlet to outlet port.

Gear pump

Gear pumps (Fig. 7.6) are seldom used, and have been superseded by either the roller-vane or diaphragm pumps. The gear pump consists of two elongated meshed gears, one of which is connected to the tractor. The gears revolve in opposite directions in a closely fitting casing, the liquid being carried between the casing and the teeth to be discharged as the teeth enmesh once more. Any damage or wear to the gears or the casing results in a loss in efficiency; therefore these pumps should not be used to spray wettable powders or where dirty water is used for spraying. A spring-loaded relief valve is usually incorporated in the pump to avoid damage caused by excess pressure. Outputs of 5–200 litres/min can be obtained with pressures up to 6 bar, although they are usually operated at lower pressures. These pumps were normally made in brass or stainless steel but engineering plastics have also been used.

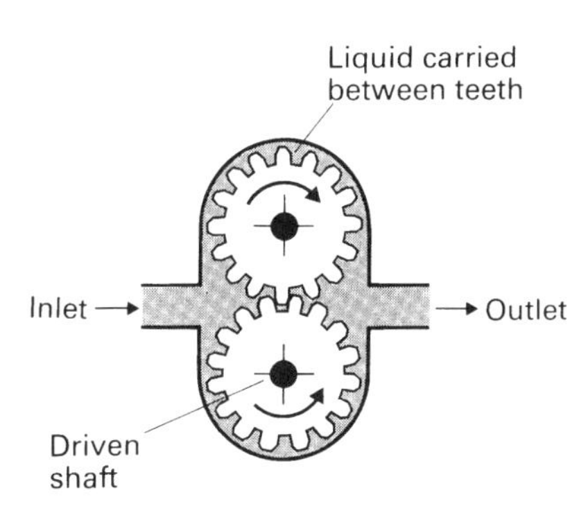

Fig. 7.6 Gear pump.

Roller-vane pump

This pump (Fig. 7.7) has an eccentric case in which a rotor with five to eight equally spaced slots revolves. A roller moves in and out of each slot radially and provides a seal against the wall of the case by centrifugal force. Liquid is

(a)

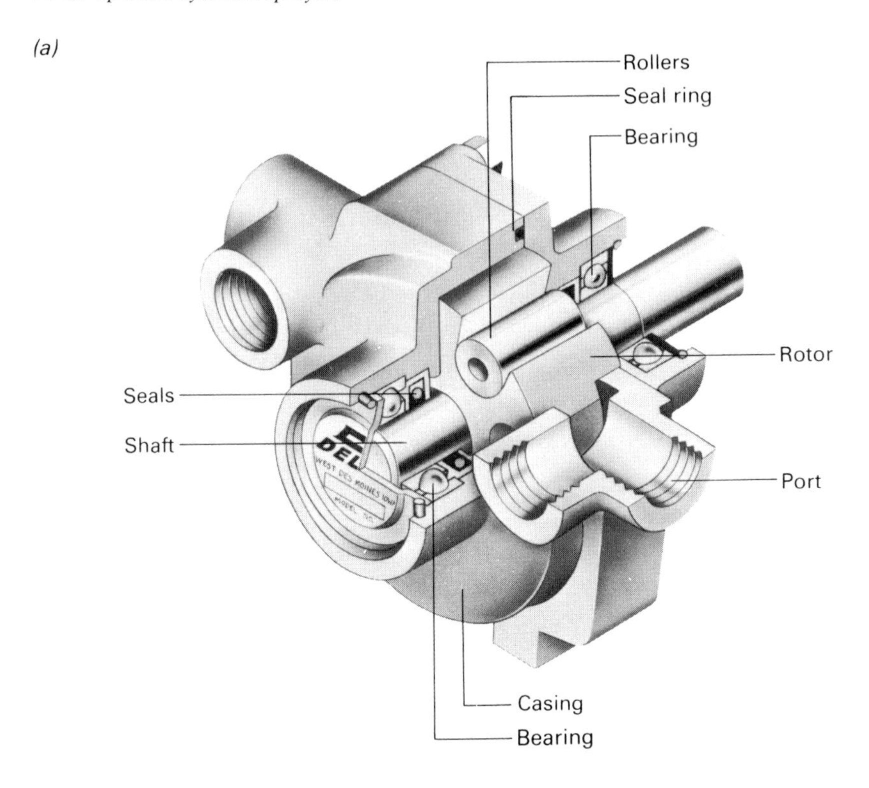

(b)

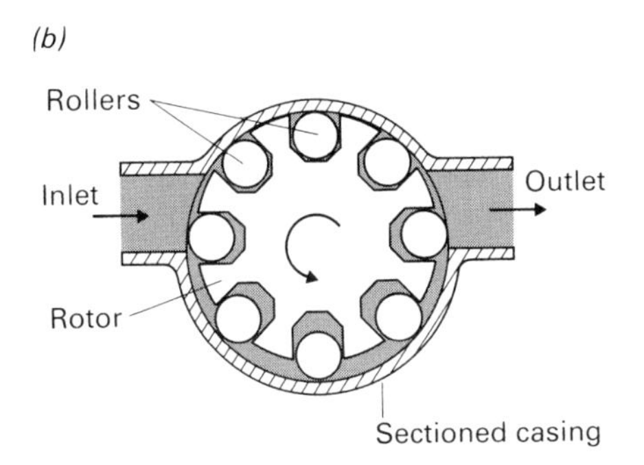

Fig. 7.7 Roller vane pump: (a) cutaway diagram to show construction (photo: Delavan); (b) diagram to show action of pump.

forced into the expanding space between the rotor by atmospheric pressure on the liquid in the tank as the rollers pass the inlet port on one side of the pump creating a low pressure area. As the space contracts again, liquid is forced through the outlet port. The pump is easily primed. Nylon or Teflon rollers are resistant to most pesticides, including wettable powder suspensions. Rubber rollers are recommended to pump water and wettable powders when the pressure does not exceed 7 bar. However, sand particles contaminating water supplies can abrade and damage the pump, so a filter between the spray tank and the pump inlet is essential to reduce the damage. The rollers can be replaced when necessary or the whole pump returned to the manufacturers for reconditioning. The case is usually made of cast iron or corrosion-resistant Ni-Resist, and has replaceable Viton, Teflon or leather, shaft seals. The pumps are usually designed to operate at p.t.o. speeds of 540–1000 r.p.m. with outputs from 20 to 140 litres/min, with pressures up to a maximum of 20 bar, although at higher pressures output and pump life are reduced. Output is approximately proportional to speed. The roller-vane pump is compact in relation to its capacity and is readily fitted to the p.t.o. and attached to a torque chain on the tractor. Before mounting, the pump shaft should be turned by hand, or with the aid of a wrench, to check that it turns easily in the proper direction.

Filtration

Careful filtration of the spray liquid is essential to prevent nozzle blockages during spraying. Apart from a filter in the tank inlet, a filter, or line strainer, must protect the pump on its input side (Fig. 7.2), and each individual nozzle, should have a filter. At the nozzle, the apertures of the filter mesh should be not more than half the size of the nozzle orifice. The line strainer should have a large area, ideally of the same mesh or slightly coarser than that used in the nozzle filter, to cope with the capacity of the pump. The line strainer should be positioned to collect debris on the outside of the mesh at the bottom of filter, so that blockage is unlikely to occur, even if debris has collected (Fig. 7.8). All filters should be regularly inspected and cleaned. Some manufacturers provide 'self clean' filters. With these it is possible to back-flush debris collected on the screen. While suitable for temporarily cleaning the filter to complete spraying in the field, it is better to ensure that the screen is cleaned each day.

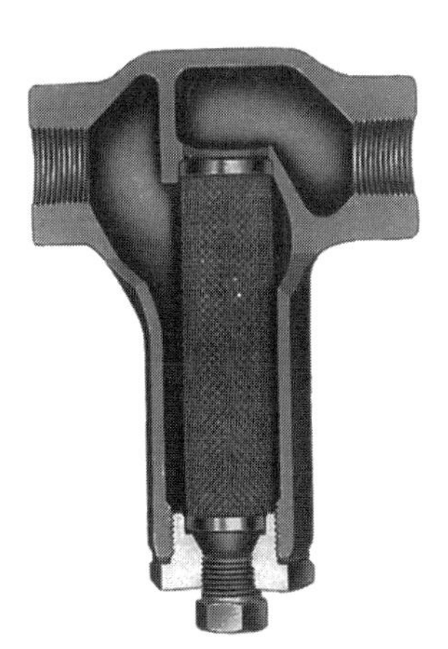

Fig. 7.8 Line strainer (photo: Spraying Systems Co.).

Pressure control

A pressure-regulating valve (PRV) (Fig. 7.9) controls flow of spray liquid from the pump to nozzles. This consists of a spring-loaded diaphragm or ball valve that can be set at a particular pressure. When this pressure is exceeded, the valve opens and the excess liquid allowed into a by-pass return to the spray tank; this causes hydraulic agitation of the spray liquid. The return flow should be through a suitable agitator at the bottom of the tank to ensure thorough circulation of the liquid. Some sprayers have a separate flow line to the agitator in addition to the by-pass line from the pressure-regulating valve. When the pressure gauge is mounted next to the valve, readings have to be checked against pressures measured at the nozzles, so that account is taken of any drop in pressure between the valve and the nozzles. The drop in pressure to the end of a boom depends on the capacity of the boom, output of the nozzles and input pressure. It is important that the bore of the boom is adequate for the nozzles being used. Ideally, the output and pressure of liquid from the pump is in excess of total requirements of the nozzles, so that hydraulic agitation in the spray tank is continuous and sufficient to keep wettable powders in suspen-

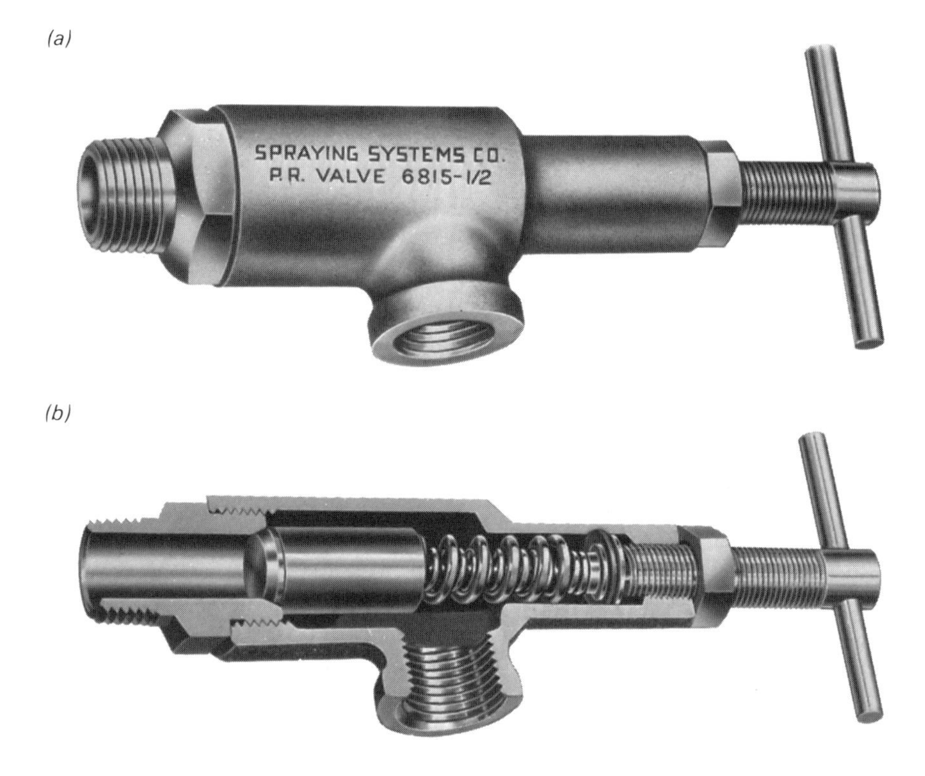

Fig. 7.9 (a) Pressure relief valve (photo: Spraying Systems Co.); (b) cutaway to show construction.

sion, even when spraying at maximum output. Unfortunately, pressure gauges do not remain reliable under field conditions and the gauge and sprayer calibration should be checked regularly. The life of a pressure gauge can be increased if a diaphragm (Fig. 7.10) protects it. Some are filled with glycerine to dampen vibration of the needle. A gauge should have a large dial to facilitate reading.

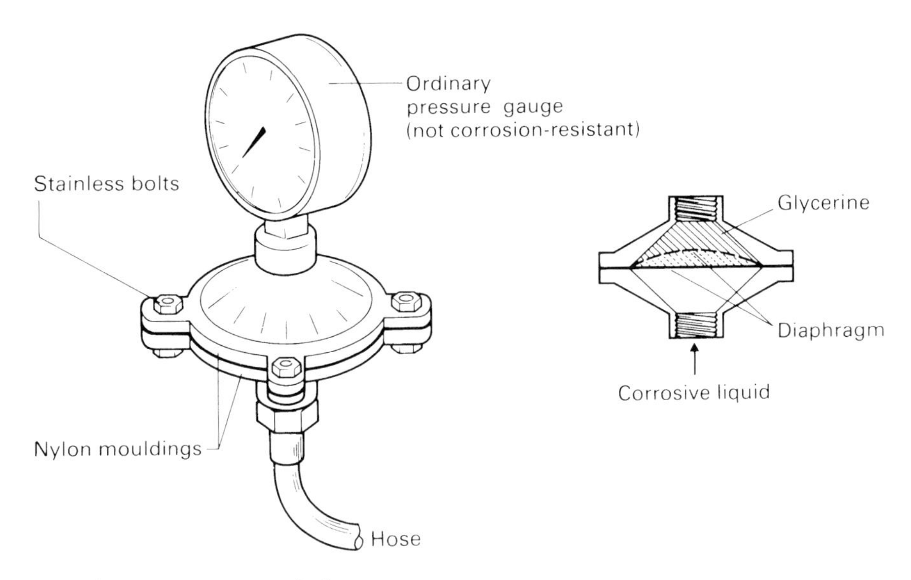

Fig. 7.10 Pressure gauge isolator.

Between the PRV and the nozzles, an on/off valve is positioned so that the tractor driver can easily operate it. Often there is a simple mechanical lever for the driver to operate, but for the totally enclosed safety cabs, electrically operated solenoid valves (Fig. 7.11) are required for remote control and to avoid pipes containing pesticides being in the cab. Closed cabs with charcoal-filtered air-intake units minimise exposure compared to half-open tractor cabs (Vercruysse *et al.*, 1999b). Some electronic devices are available to provide the tractor driver with a digital display of the area covered, output, speed and other variables (Allan, 1980). Such remote control devices are likely to be used more frequently due to safety regulations. When the spray boom is divided into three sections, left, right and central, the main valve is often a seven-way valve, so that individual sections, pairs or the whole boom can be operated. This is particularly useful when the edges of fields are being treated and part of the boom is not required. On some sprayers, liquid in the boom can be sucked back to the tank when the valve is closed. This may result in excess foaming and care must be taken to avoid damage to the pump if the sprayer is empty.

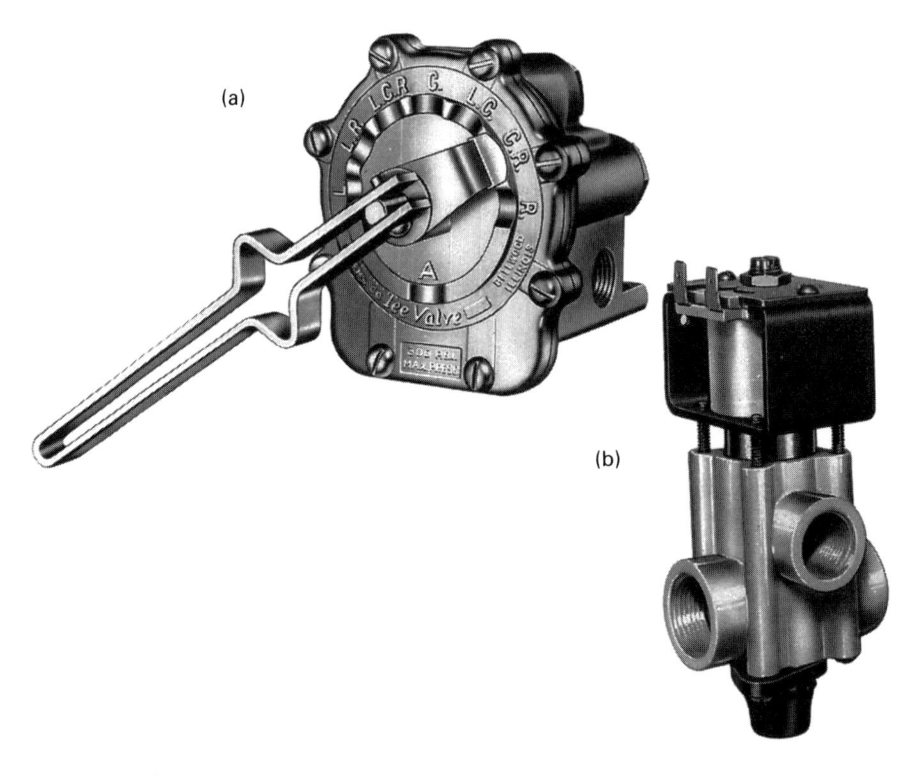

Fig. 7.11 (a) Seven-way tap for different boom sections. (b) Solenoid valve (photos: Spraying Systems Co.).

Spray booms

For most farmers the width of the boom is fixed. A suitable boom width for the fields can be calculated from

$$\text{Boom width} = \frac{\text{Area requiring treatment}}{\text{Time available} \times \text{Tractor speed}} = \frac{\text{m}^2}{\text{h} \times \text{m/h}}$$

Thus, if a farmer has a 100 ha field which needs treating in 3 days (6 h actual spraying per day) at a speed of 8 km/h, the minimum boom width required is 6.94 m (i.e. 7 m). Sprayer requirements should be based on completing the spray programme within 3 days in any one week to allow for rain, wind, equipment maintenance and other delays. On this basis, 1 m of boom is required for each 13.5 ha to be treated. Over the last two decades, farms in the UK have moved from boom widths of about 12 m to 18–24 m depending on the width of the seed drill and position of the tramlines, as this reduces the number of passes across fields. Variation in spray deposit is liable to increase with wide booms, due to greater movement of the end of the boom relative to the ground unless the land is very even. The pump output in litres/min is given by

$$\frac{\text{Swath (m)} \times \text{Application rate (litres/ha)} \times \text{Velocity (km/h)}}{600}$$

For example, with a 24 m boom travelling at 8 km/h, the pump capacity required to apply 200 litres/ha is 64 litres/min to allow for agitation. In practice, the cereal farmer also needs to choose a boom width related to the width of the seed-drill.

Boom design

Most spray booms are mounted at the rear of the spray tank, except for a few which are in front of the tractor to facilitate band applications of herbicides when the farmer needs to see the position of the nozzles in relation to the rows. The front boom position should not be used when spraying insecticides because the operator moves towards the spray. Booms are generally designed in three or more sections so that the outer sections can be folded for transport and storage. During spraying, the outer sections are often mounted so that they are moved out of the way by any obstruction which is hit. Manufacturers have used various methods to pivot and fix the boom sections for easy handling. Normally, the booms are unfolded by hand, but on some sprayers, positioning of the boom can be controlled through the hydraulic system without the operator leaving the tractor.

During field spraying, movements of the boom, including vertical bounce, horizontal whip or both, cause uneven distribution of pesticide, which is accentuated as booms increase in width. Due to the yawing, the boom may be stationary at times in relation to ground speed, so causing an overdose of pesticide. The rolling movement varies the height of nozzles relative to the crop, and thus the pattern of overlap is affected. Ideally, the boom should be as rigid as possible over its length and mounted centrally in such a way that as little as possible of the movement of the tractor is transmitted to the boom. Any breakaway mechanism should be strong and return the outer boom quickly and positively into its correct position. Booms constructed as stiff cantilevers have been shown to be better than other types (Nation, 1982). An inclined-link boom suspension was developed to allow articulation between the boom and sprayer in both rolling and yawing planes (Nation, 1985). Instead of a passive suspension, a boom can now be fitted with an active suspension in which a sensor detects the height of the boom relative to the crop and controls its position (Frost, 1984; Marchant, 1987; O'Sullivan, 1988; Frost and O'Sullivan, 1988; Marchant and Frost, 1989).

The effectiveness of shields fitted to spray booms has been investigated to assess whether drift can be reduced. Shielded booms did reduce off-target drift, but solid shields decreased on-swath uniformity in contrast to a perforated shield (Wolf *et al.*, 1993). A double-foil shield used with standard flat fan nozzles produced the best deposit and reduced drift by 59 per cent in wind tunnel tests (Ozkan *et al.*, 1997). By using a simple bluff plate attached to a boom driven at high speed (20–40 km/h) and applying low volume (11–15 litres/ha) sprays, oncoming air is deflected over the plate to create an area of stalled air where the nozzles are situated. Turbulence created assists downward movement of droplets into the crop canopy (Furness, 1991). Drag created by the plate requires a powerful tractor and requires more energy to maintain the high speeds, but in a new design the front is angled to deflect air over the top. Similarly Enfalt *et al.* (2000) used a plastic sheet mounted on a parallelogram frame to shield nozzles angled backwards behind the shield. Felber (1988) mounted a simple bar below the boom to open the cereal canopy and allow better penetration to the lower canopy for fungicide application.

Nozzles on spray boom

A wide range of hydraulic nozzles (see Chapter 5) can be used on a boom. Certain organisations have issued charts to guide farmers in the selection of nozzles (Powell *et al.*, 1999). Some farmers use twin-fluid nozzles that require the fitting of an air compressor and related pipes to deliver air to each nozzle (see also Chapter 10). The nozzle body may be screwed into openings along the boom, but often the boom incorporates special nozzle bodies clamped to the horizontal pipe (Fig. 7.12). Sometimes the liquid is carried to the nozzles in a plastic tube so that spacing between nozzles can be adjusted by sliding the nozzle body along the boom (vari-spacing). Choice of nozzle tip depends very much on the material being sprayed, the volume of liquid needed and the ultimate target, so that the output (litres/min), spray pattern quality and angle and droplet size are appropriate. Cone nozzles are preferred for application of insecticides and fungicides to foliage of broad-leaved crops, while fan nozzles are mostly used when treating wheat and similar cereal crops or the soil with any pesticide. To reduce drift, herbicides may be sprayed at low pressure,

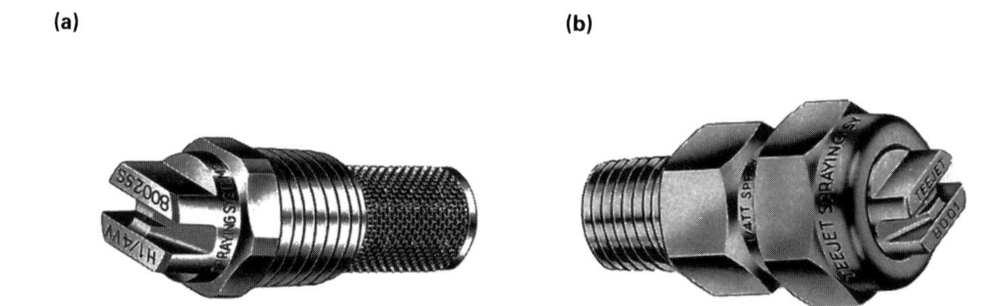

Fig. 7.12 Different systems of fitting nozzles to spray booms: (a) conventional nozzle body screws into boom; (b) nozzle tip screws directly into boom; (c) nozzle body clamps to pipe boom; (d) nozzle fixed to L-section boom with hose between nozzles; (e) vari-spacing; (f) bayonet fitting of nozzle tip; (g) double swivel nozzle on down pipe (photos: Spraying Systems Co.); (h) nozzle turret for rapid selection of different nozzles in the field (Lechler).
Note: Although metal nozzles are shown, many plastic nozzles are now used.

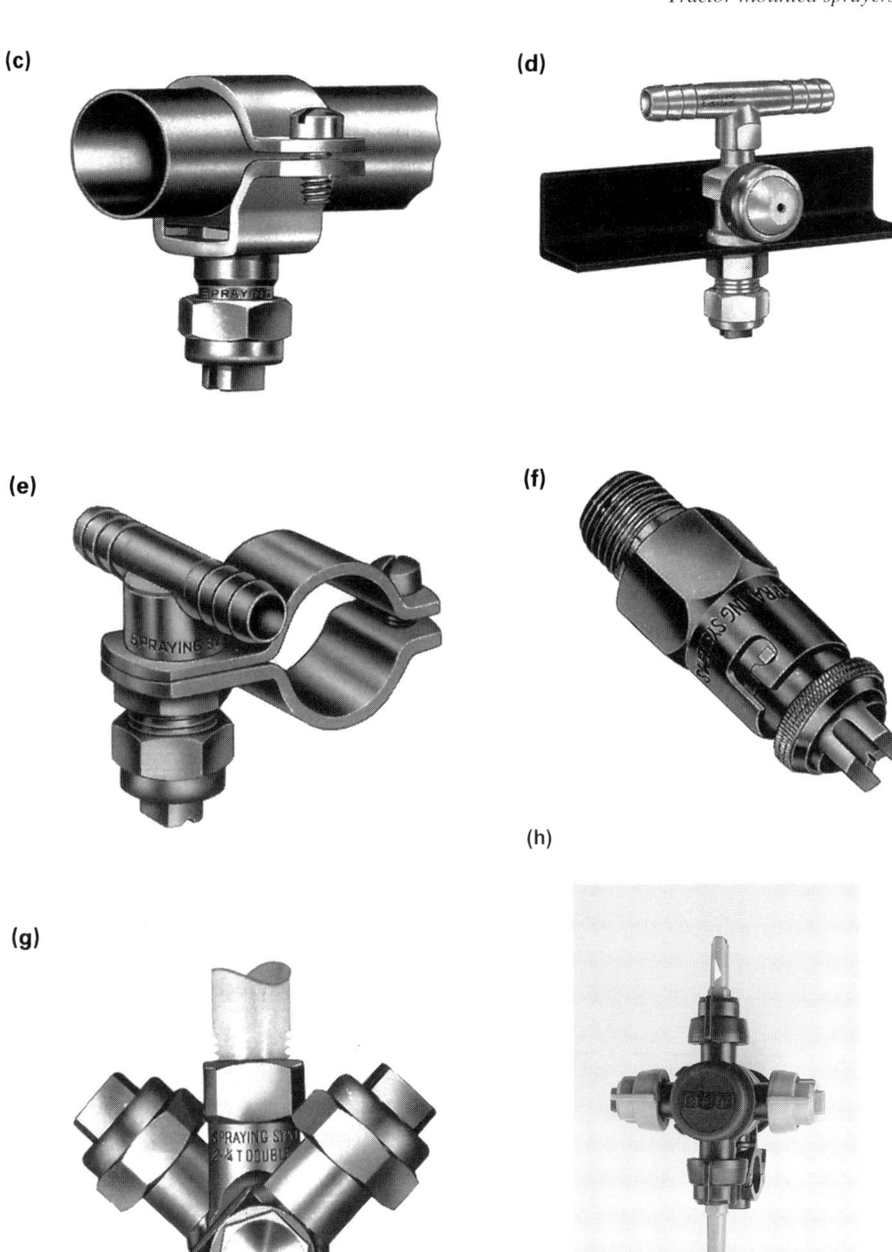
(c)
(d)
(e)
(f)
(h)
(g)

usually 1–2 bar. A check valve should be used with each nozzle to prevent dripping of liquid if the sprayer is stationary.

The throughput for each nozzle can be determined from the output of the pump and the number of nozzles on the boom, thus

$$\frac{\text{Nozzle throughput}}{\text{litres/min)}} = \frac{\text{pump output (litres/min)}}{\text{Number of nozzles}}$$

$$= \text{Pump output (litres/min)} \times \frac{\text{Nozzle spacing (m)}}{\text{Boom length (m)}}$$

For example, with a pump output of 18.6 litres/min on a 12 m boom with nozzles spaced at 0.5 m

$$\text{Nozzle throughput} = 18.0 \times \frac{0.5}{12} = 0.775 \text{ litres/min}$$

The spacing between nozzles along the boom is often fixed, and the height of the boom should be adjusted according to the type of nozzle being used. In particular, attention must be given to the spray angle and pattern, which are affected by pressure. The pattern from each nozzle has to be overlapped to achieve as uniform a distribution of spray as possible across the whole boom (Fig. 7.13); indeed some operators use a double overlap. If the boom is set too low, excessive overlap occurs and results in an uneven distribution. The 'peaks' and 'troughs' occur with both fan and hollow-cone nozzle, but are generally more pronounced with the latter. Uneven distribution is also obtained if the boom is set too high (Fig. 7.14 and Table 7.2).

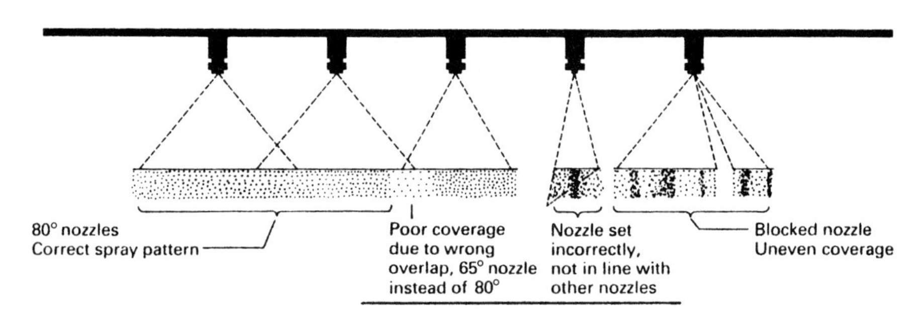

Fig. 7.13 Correct overlapping of the spray patterns is required across the boom.

Potential spray drift is increased with faster forward speeds, thus with 110° standard fan nozzles emitting 0.6 litres/min at 3 bar on a 12 m boom, drift at 5 m downwind was increased 51% when the tractor speed was increased from 4 to 8 km/h. This change in forward speed had a greater effect than increasing pressure up to 4 bar. This was because an increased pressure increased the downward velocity of droplets as well as producing smaller droplets (Miller and Smith, 1997). Nordby and Skuterud (1975) recommended that the boom should be about 40 cm high and the working pressure of fan nozzles kept below

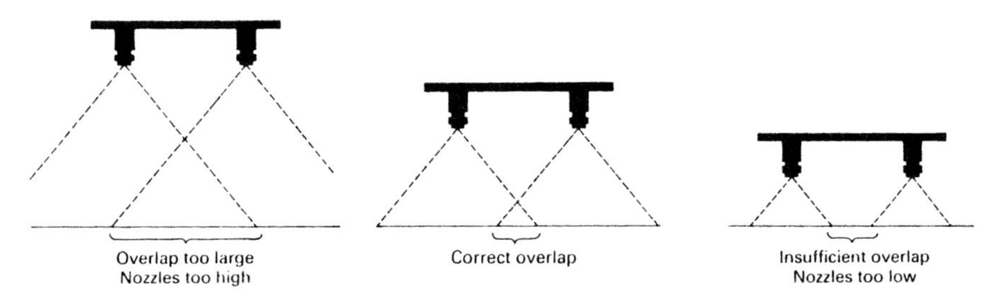

Fig. 7.14 Correct height above the crop is essential.

Table 7.2 Variation in boom height (cm) above crop or ground with different nozzle spacing along the boom and spray angles.

Nozzle angle	Nozzle spacing along boom (cm)		
(deg)	46	50	60
65	51	56	66
80	38	46	50
110	24	27	29

2.5 bar. Furthermore, they felt that herbicide sprays should be applied only when wind speeds are less than 3 m/s. Downwind drift is undoubtedly increased if the boom height is too high (Miller, 1988). Lower boom heights are possible if a fan spray is directed back at an angle instead of pointing vertically down on the crop (some nozzles are specially made to do this) or if wide-angle nozzles are used, but the wider the spray angle, the greater the risk of producing more very fine droplets. The distribution can be checked by spraying water onto a dry surface or placing strips of water-sensitive paper across the swath, or by adding a dye such as lissamine scarlet to the water, a record of the distribution being obtained by spraying across a band of white paper. If the spray pattern is uneven, the throughput of each nozzle must be checked (see p. 126). A computer model showed that for a boom set at the optimum height, the coefficient of variation increases continuously with increases in boom roll angle, due to the changes in nozzle height, rather than a change in the angle of the spray (Mawer and Miller, 1989). Electronic instruments to measure flow rate can be used to check the evenness of the output across a boom, but the actual output should be checked by collecting liquid in a calibrated container.

Most tractor-mounted booms have nozzles arranged in a horizontal line and directed downwards, but for spraying some crops to get underleaf cover, 'droplegs' are fitted so that nozzles can be directed sideways or upwards into the foliage (Fig. 7.15). When tailbooms were used on tractor equipment, they were pivoted on the horizontal boom and held by a strong spring. Also, the bottom section of the boom was mounted on a flexible coupling to avoid

Fig. 7.15 Tractor sprayer with vertical booms between row crop.

damage if the boom touched the ground. In front of the booms, a curved guard was needed to ease the passage of the boom through the crop. Movement through some crops is possible only if the sprayer is used regularly along the same rows and in the same direction.

Some chemicals are applied in a band, usually 18 cm wide, along the crop row to reduce the cost of chemical per hectare. Band spraying requires a higher standard of accuracy in the selection and positioning of the nozzles, which are often mounted on the seed-drill (Fig. 7.16). In one system in the USA where the nozzles are mounted so that they can be rotated up to 90° on a vertical axis, the user can control the bandwidth of a fan nozzle (Fig. 7.17). Guidance systems have been developed to ensure more precise positioning of the nozzle above small plants (Giles and Slaughter, 1997). A similar guidance system was developed to treat vegetation alongside roadways to avoid herbicide being applied to bare areas (Slaughter *et al.*, 1999).

Calibration of a tractor sprayer

The importance of careful calibration cannot be over-stressed. One method is to select the gear to give a p.t.o. speed of 540 r.p.m. and a forward speed which results in an acceptable level of boom movement. Next, a 100 m length is marked out and, with the tractor moving at the required speed as it passes the first mark, the length of time taken to cover the 100 m to the next marker is measured. The forward speed (km/h) = 360/measured time (s). The nozzle

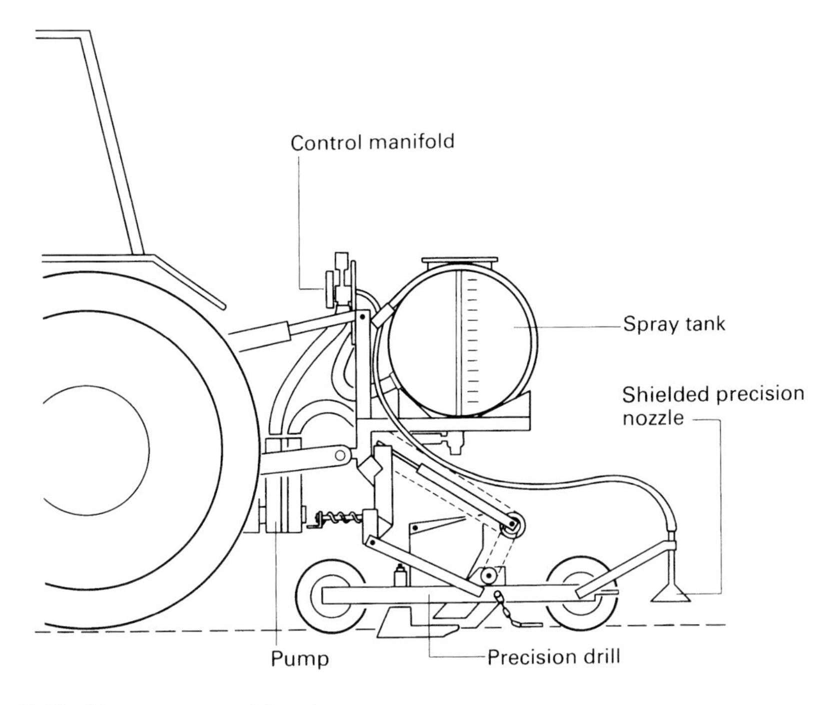

Fig. 7.16 Tractor-mounted band sprayer.

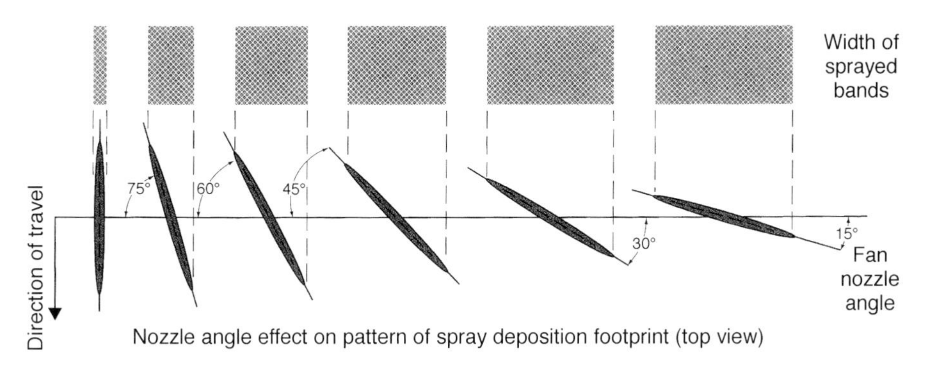

Fig. 7.17 Nozzle body with capstan to allow rotation up to 90° to vary swath width for band application.

spacing (m) is measured and the output required per nozzle is calculated as follows:

$$\text{Nozzle output (l/min)} = \frac{\text{Volume application rate (l/ha)} \times \text{Speed (km/h)} \times \text{Nozzle spacing (m)}}{600}$$

For example,

$$\text{Nozzle output} = \frac{200\,\text{l/ha} \times 6\,\text{km/h} \times 0.5\,\text{m}}{600}$$
$$= 1\,\text{l/min}$$

The nozzle is then selected from the information in the manufacturer's charts to emit the correct volume at the appropriate pressure and achieve the spray quality required. With the spray boom set up, the output of the nozzles is checked.

Another method, which can be used to check the calibration of the sprayer is to calculate the time, required to spray 1 hectare, thus:

$$\text{Time required (min)} = \frac{600}{\text{Swath (m)} \times \text{Speed (km/h)}}$$

Note the effective swath is the distance between each nozzle along the boom multiplied by the number of nozzles; for example, if 30 nozzles are spaced at 0.5 m intervals the swath is

$$30 \times 0.5\,\text{m} = 15\,\text{m}$$

The tractor speed can be checked by measuring the distance covered in metres when travelling for 36 seconds in a gear selected to give approximately the correct speed with a p.t.o. speed of 540 r.p.m. This distance divided by 10 gives the speed in kilometres per hour.

By knowing the time required to spray 1 hectare, the volume applied per hectare can be measured by filling the spray tank to a mark, operating the pump at the required pressure with the tractor stationary and the p.t.o. running at 540 r.p.m. for this period of time, and then carefully measuring the amount of water required to refill the sprayer to the mark. If the volume is within 5 per cent of that required, the pressure regulator can be adjusted slightly to raise or lower the pressure. However, adjustment of pressure must be avoided, because droplet size spectrum and spray angle are also affected and nozzle throughput is in proportion to the square root of the pressure; thus pressure needs to be doubled to increase throughput by 40 per cent. Alternatively, the speed of travel can be adjusted or, if necessary, different nozzles will be required. It is useful to keep different sets of nozzles to provide different spray qualities. Some sprayers have sets of nozzles in a rotating nozzle body to enable a change of nozzle tip to be made very easily. This may be particularly important when treating the edge of a field close to a watercourse when a LERAP rated nozzle – coarse spray – is required to avoid drift. If any adjustments are necessary, the sprayer calibration should be repeated.

The calibration can also be made by travelling over a known distance and measuring the volume (litres) applied. If the distance travelled is selected by dividing the boom width (m) into 1000, the volume measured multiplied by 10 is in litres per hectare. The pump pressure and speed of travel must be constant.

With a band spray, the application rate can be calibrated as described above, but as only a proportion of the area is actually sprayed; the rate per treated area will be higher in proportion to the ratio between the width of the treated plus untreated band and the treated band, thus:

$$\underset{\text{(l/ha)}}{\text{Volume applied to surface area}} \times \frac{\text{Treated band width} + \text{Untreated band width}}{\text{Treated band width}}$$
$$= \text{Volume applied to band (l/ha)}$$

For example, if 20 litres/ha is applied but confined to a band 20 cm wide along rows 100 cm apart, the volume applied to the band will be

$$20 \times \frac{100}{20} = 100 \text{ litres/ha}$$

Details of any calibration of the sprayer should be recorded for future reference (Table 7.3).

Table 7.3 Record of calibration

Tractor – make — registration
— tank capacity litres

Calibration	Tractor gear	Throttle setting (r.p.m.)	Ground speed (km/h)	Nozzle-tip size	Pressure (bar)	Output (l/h)	Area per load[a] (ha/tank)
1							
2							
3							

[a] $\frac{\text{Tank capacity (litres)/}}{\text{Output (litres/ha).}}$

A sprayer should be cleaned and checked regularly. The main faults reported include worn nozzles, boom defects, damaged hoses, leaks and faulty pressure gauges. In many countries a sprayer must be officially examined at intervals of usually about 3 years to ensure that it is properly maintained. This mandatory examination by mobile inspection teams has led to an improvement in the general condition of sprayers, due to the financial consequences of a sprayer failing the test (Langenakens and Pieters, 1997). A stock of spare parts should be readily available. In particular, it is wise to keep spare nozzle-tips and take some to the field during spraying. If a nozzle is blocked, a replacement can be quickly fitted to avoid the need to clean a blocked nozzle in the field. The output of each nozzle should be checked periodically to ensure that it has not increased (see p. 129). The cost of a replacement nozzle is negligible in comparison with costs of the pesticides sprayed. The interval between checks will depend on the volume and type of liquid sprayed.

Swath matching

Matching the end of one swath with the next is not easy, especially in a closely spaced cereal crop. As the passage of the tractor wheels through the crop for fertilizer and pesticide application can reduce yield, especially at advanced stages of crop growth, farmers now leave gaps for the wheels of the tractor. These are referred to as 'tramlines' (Fig. 7.18a). Tillering, and more grains per ear on the plants adjacent to the gaps, almost compensates for the reduced plant population. With tramlines there is a small saving in seed, operations subsequent to drilling are quicker, and late applications, if needed, are more likely to be applied at the correct time.

The tramline system requires the width of the seed drill, fertilizer spreader and sprayer to match (Fig. 7.18b). Tramlines are established by blocking appropriate drill coulters at the required intervals across the field. The seed cut-off mechanism can be operated automatically on certain drills. Tractor tyre widths may also necessitate a slight displacement of the coulters on either side of the tramlines. Other systems are available, for example when the seed drill cannot be altered, a herbicide such as paraquat is sprayed behind the tractor wheels. The disadvantages of the technique are that it is more expensive, accurate marking is essential and greater care is needed to avoid spread of the herbicide to other areas. The headland operations and weed control must be carefully planned to ensure continuity of clean tramlines.

Fig. 7.18 (a) Field with tramlines (photo: J.W. Chafer).

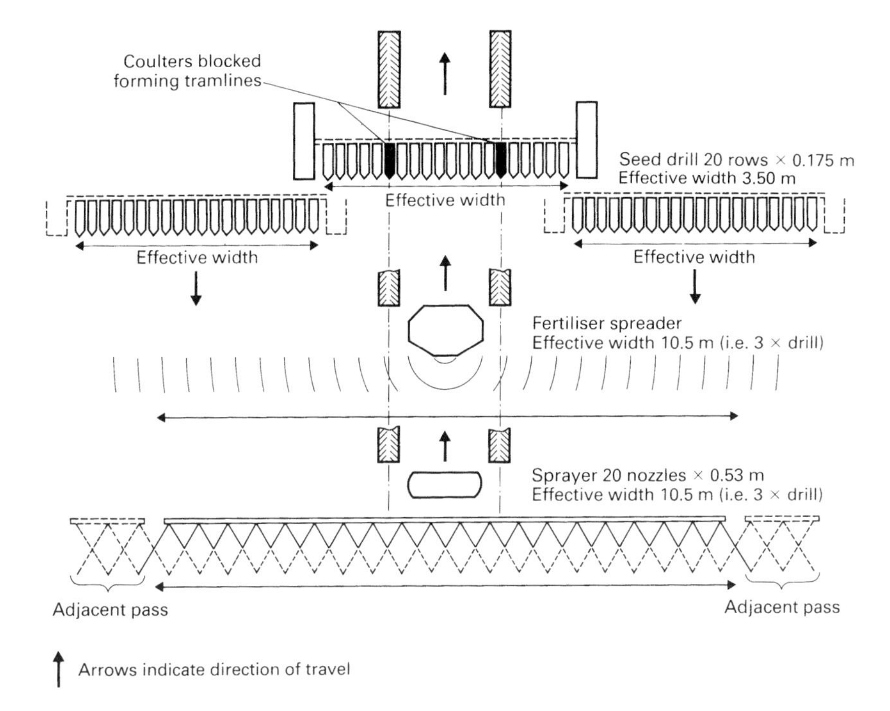

Fig. 7.18 (b) Diagram to show formation of tramlines by matching seed drill, fertilizer spreader and sprayer.

Increased attention to rabbit and hare control may be required, since these vertebrate pests may use tramlines as 'runs' into the fields. It is well worth spending some time measuring out each swath and having fixed marks to indicate the centre of each swath, even on row crops. Damage to bushy plants, like cotton, caused by the passage of the tractor is less than expected owing to plant compensation. Even when a tractor with a clearance of only 48 cm at the front axle was driven over two rows of cotton, it was more profitable than growing an alternative low crop, such as groundnuts, along the 'pathways' (Tunstall *et al.*, 1965).

The tramline system is now so widely used that alternative methods, such as the foam marker at the end of a boom, are seldom used. Often the crop is sown right up to the edge of the field and no headland is available for turning. When the turn is made inside the crop, the crop will be overdosed if spray is applied during the turns. It is preferable to spray two swath widths around the field, and then treat the remainder of the field by spraying swaths parallel to the longest side of the field. The p.t.o. is kept running during the turn to keep the spray liquid agitated, but the valve to the boom is closed throughout the turn (Fig. 7.19). However, some farmers now have an untreated headland, which is managed separately to conserve wildlife in the hedgerows.

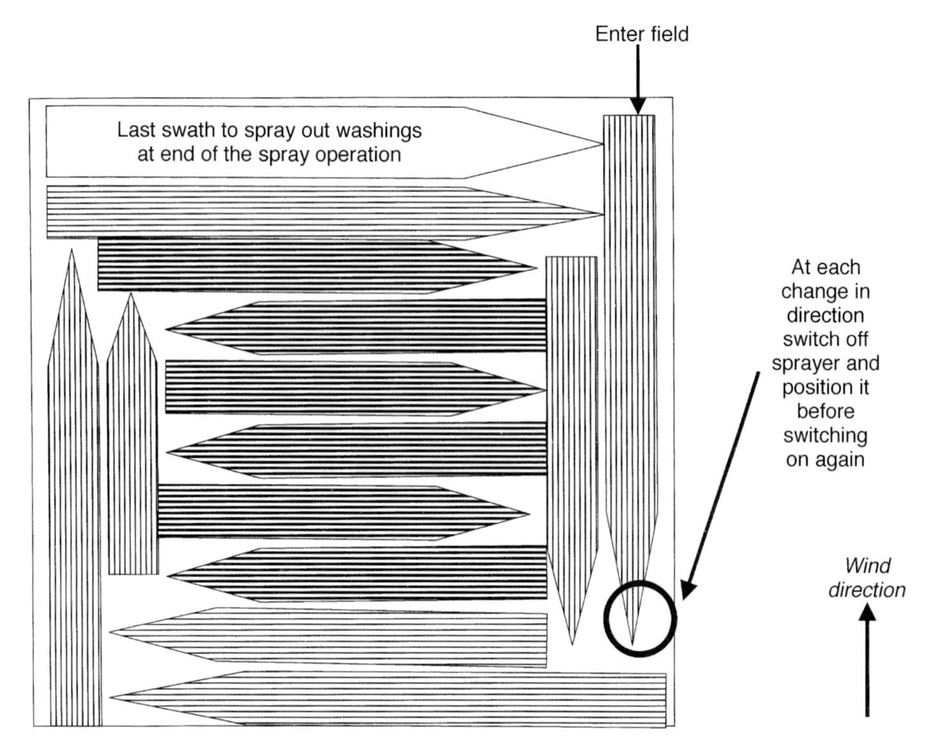

Fig. 7.19 Sequence of spraying a field. Never spray while doing a turn.

Refilling the sprayer

If the spray tank is not fitted with a sight gauge, a flickering of the pressure-gauge needle indicates that the sprayer is empty. The p.t.o should be disengaged immediately to avoid the pump running while 'dry'. Most operators prefer to empty the tank before refilling to avoid errors in calculating and measuring the amount of chemical for a different volume of liquid. Nevertheless, if the sprayer empties while in the middle or far side of a field, time is wasted in returning to the refilling point, and a smaller quantity may be required for the last load to avoid having any spray left over. If possible, the farmer should have detailed measurements of his field, so that with accurate calibration the appropriate amounts of chemical can be calculated beforehand for each load, thus reducing the time for ferrying to refill the sprayer.

To increase safety to the operator, the trend is to use the sprayer pump to fill the tank partially with water and then transfer and mix a measured quantity of pesticide using a closed-transfer system to reduce direct contact with the chemical (Brazelton and Akesson, 1987). Such systems include the use of a suction probe to use the sprayer pump to draw chemical from its container (Fig. 7.2). In Europe, an industry standard requires a closed coupling without any spillage, using equipment such as the MicroMatic. Most sprayers now have an induction bowl (Fig. 7.20) which is used especially with particulate formulations to provide mixing before the chemical is drawn into the sprayer tank (Frost and Miller, 1988). An International Standard for induction bowls is in preparation. This equipment is fitted with a system of using clean water to rinse containers to reduce residues and thus eliminate the hazards associated with disposal of contaminated containers. In one test 69 per cent of the participants were able to clean a 5 litre container so that it had less than 0.5 ml of pesticide residue after 20 s washing, and thus below the upper limit defined by the standard BS 6356 (Cooper and Taylor, 1998).

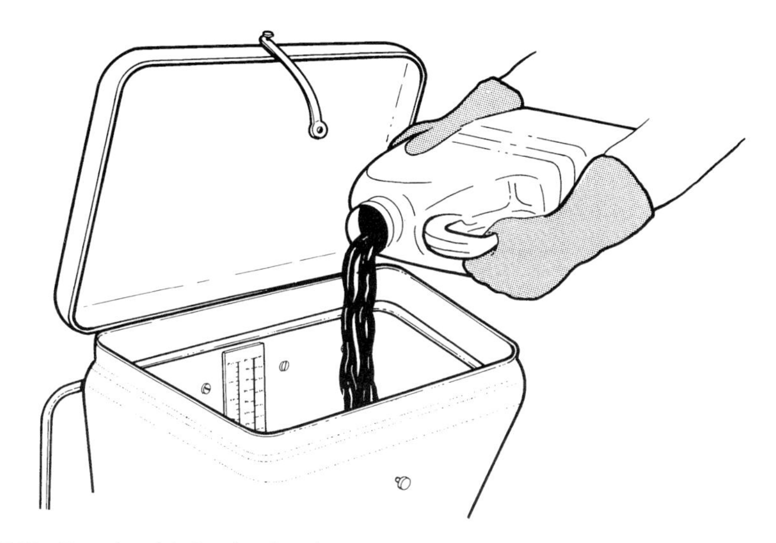

Fig. 7.20 Low-level induction bowl.

The chamber must be fitted with a system to rinse the container and in some the empty container can then be crushed to prevent re-use.

Alternatively, the farmer may have a separate water tank or mixing facility with its own pump to refill the sprayer. Great care must be taken to avoid contamination of the water source.

Metered spraying

Uniform application with the equipment described so far depends on a constant tractor speed and constant pressure. Forward speed may vary, so systems are needed to regulate the flow of liquid to the nozzles. A variation in speed from 0 to 80 km/h must be considered when herbicides are applied to railway tracks (Amsden, 1970). Some systems incorporate a metering pump which is linked to the p.t.o. or sprayer wheel, and a proportion of spray may or may not be returned to the tank. Pump output must be proportional to the forward speed, so a diaphragm or piston positive displacement pump is needed; gear or roller-vane pumps are unsuitable. When the pump – usually a piston pump with an adjustable stroke – is driven by the sprayer wheel, a second p.t.o. pump is needed for agitation and refilling the tank (Fig. 7.21). The main disadvantage is that the power required to drive the metering pump is high, 10 hp being needed to supply 500 litres/ha through a 12 m boom. This can be overcome by using the ground-wheel pump at low pressure and a separate pto pump to boost pressure to the nozzles. These systems are relatively simple to operate, but droplet size is also affected when flow rate is adjusted by pressure. The operator should try to keep within ± 25% of the selected speed so that the pressure is not greatly affected. Other systems include a centrifugal regulator linked to the sprayer wheel and metering pumps or valves operated electronically by the forward speed of the sprayer (Fig. 7.22).

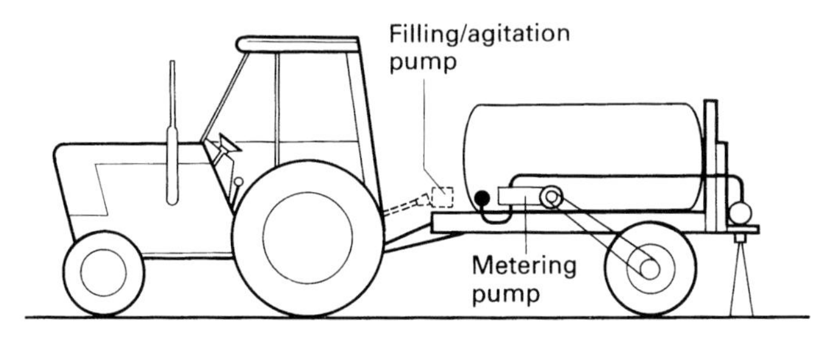

Fig. 7.21 Metering pump system.

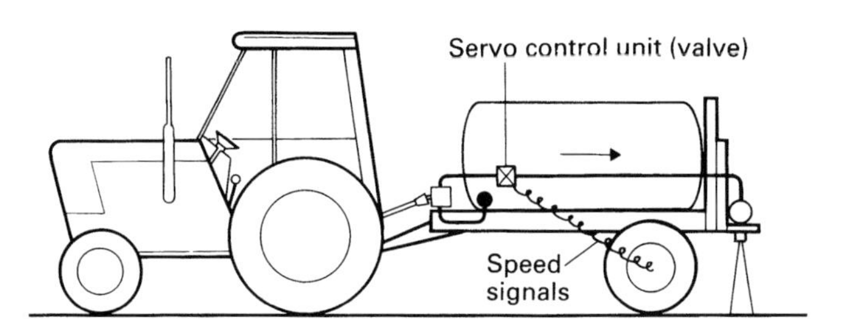

Fig. 7.22 Output controlled by electronic sensing of forward speed.

The more complex electronic systems are expensive, and their use is limited unless specialized maintenance facilities are available. All systems linked to the rotation of the p.t.o. or wheel may be affected by wheel slip causing underdosing or overdosing, so the metering device must be operated by a trailed wheel rather than a driving wheel (Amsden, 1970). The spray is already mixed in the sprayer tank with these automatic regulating systems. Ultimately, the chemical and diluent may be kept in separate tanks (Fig. 7.23) and using an in-line mixing system with the concentration of spray affected by forward speed (Hughes and Frost, 1985). Unused chemical can then be readily returned to the store. Frost (1990) has described a novel metering system in

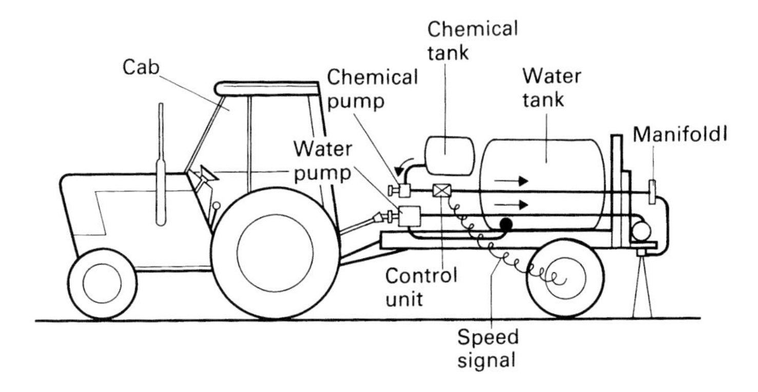

Fig. 7.23 Servo-operated system with separate chemical and diluent tanks and pumps.

which the flow of water is used to control the flow rate of the chemical, making the system independent of the characteristics of the chemical.

In another closed system, a piston pump with a ceramic piston to withstand the effects of the pesticide concentrate is used to meter the chemical into a mixing chamber. An electric stepper motor, controlled from the tractor cab, is used to adjust the length of pump stroke and thus the input of chemical into the water that is pumped separately into a mixing chamber and thence to the nozzles (Landers, 1988). Humphries and West (1984) describe a similar system that uses compressed air to force the pesticide to the mixing chamber. Zhu *et al.* (1998) describe how the lag time and uniformity of mixing can be assessed when an in-line injection system is operated.

Regardless of which system is used, the sprayer must be properly calibrated, and worn parts, especially nozzles, replaced regularly.

Precision (patch) spraying

Instead of treating the whole field, systems are being developed in precision agriculture to treat specific areas within fields according to the pests that are present. At present, patch spraying is mostly with reference to weeds, the type and position of which can be determined by walking the field and the locations recorded in a computer linked with geographical positioning systems (GPS) data (Rew *et al.*, 1997). The tractor can then be programmed to spray the patches. Paice *et al.* (1995) used a sprayer with an injection system, but an alternative system with a twin-boom and individual nozzles controlled by solenoid valves has been used commercially (Miller *et al.*, 1997). Womac and Bui (1999) have patented a device that will facilitate application at a variable flow rate and avoid the use of electrically complex equipment. Giles *et al.* (1996) have controlled the flow through nozzles using a pulsed solenoid independently of pressure, while controlling droplet size by adjusting the pressure of the spray liquid (see p. 123).

Portable-line sprayers

A flexible boom or hose can be used when a horizontal boom on a tractor cannot be used in orchards or forests, or because the land is undulating. Sufficient labour is needed to carry the hose. Various types of portable line have been used. In cotton fields, operators spaced at 4 m intervals carry an interconnecting hose on a short mast, supported in a waist strap. The interconnecting hose is liable to stretch unless strengthened by Terylene or a similar fibre in the hose wall. The portable line is connected to a spray tank and pump, mounted on a trailer driven or pushed along pathways across the field. The main difficulties with the system are the need for a pressure regulator at each operator to compensate for the pressure drop along the line, and the need for each operator to walk at the same speed as the tractor (sometimes difficult if an operator meets an obstacle or a snake!). With a line of spray operators, care must be taken to avoid contaminating each other with spray droplets drifting downwind.

Alternatively, a hose on a reel is paid out from a stationary pump as the operators move down the field and is wound in on their return. This method has been used in small orchards, as well as for cotton. These systems are generally no more expensive than using teams with knapsack sprayers, but require sufficient supervision to co-ordinate the operators and ensure that they do not get contaminated by the spray.

Incorporating herbicides

Some volatile herbicides, such as trifluralin and dinitramine, must be incorporated into the soil to prevent loss by volatilisation or photo-decomposition by sunlight. Incorporation also reduces the rainfall requirement and places the chemical in close contact with the weed seeds or roots for best control. The implements used for incorporation will vary with different herbicides, depending on the distribution of chemical required. Power-driven devices are required to incorporate herbicides evenly to a precise depth. Walker *et al.* (1976) reported that a single pass with a rotovator gave an even distribution of trifluralin in the top 5 cm of soil, i.e. its working depth. Incorporation to half its working depth was obtained by cross-cultivations with a rotary power harrow, reciprocating harrow, spring-tined harrow and disc cultivator. Single passes of these implements, and even cross-cultivation with a drag harrow, left much of the herbicide close to the soil surface.

Special booms for tree spraying

High-volume application (100 litres/tree) has been used to protect citrus trees. These sprays are applied with a vertical boom, which is mounted on the rear of a tractor travelling at 2–2.4 km/h. Narrow-cone nozzles (16°) with 2.4–3.5 mm orifices are placed at intervals of 30–40 cm, from 45 cm above the ground to no lower than the top of the average tree height. A high-capacity pump is needed to deliver up to 500 litres/min at pressures exceeding 30 bar. The nozzles are oscillated in a continuous cyclic rotation pattern once every 0.5–0.6 m of forward travel. The high pressures and volume with this system were considered essential to penetrate the peripheral 'shell' of foliage and achieve up to 88 per cent coverage, including the upper central parts of the trees (Carmen, 1975). However, Cunningham and Harden (1998, 1999) have pointed out that retention of spray on leaf surfaces, as a percentage of the spray applied, increases as the total volume applied decreased.

To treat small trees, it is possible to mount the boom within a tractor-mounted shield to reduce the effect of wind on spray dispersal (Cooke *et al.*, 1977). With the need to reduce spray drift in orchards there is likely to be an increase in the use of 'tunnel' sprayers (Fig. 10.16, p. 235).

High-volume fungicide sprays (800 litres/ha) to control coffee berry disease in East Africa have been applied with twin horizontal overhead booms covering eight rows (2.7 m apart) on a single pass (Pereira and Mapother, 1972). Deposits were predominantly on upper surfaces and decreased from the top to lower parts of the coffee, so that effective control of the disease depended on redistribution of the fungicide. Overhead spraying is less effective against coffee leaf rust because these spores germinate on the undersurface of leaves. Pereira (1972) has also described an inverted 'hockey-stick' boom at the rear of a tractor-mounted sprayer operated at 7 bar for both overhead and lateral spray application of coffee trees. Variable-cone nozzles, 250 mm apart, on the vertical section of boom were angled to spray up through the branches to improve coverage on the berries. Small fruit trees can be sprayed with an overhead boom.

Some experiments have been made with a crossflow fan driven by a hydraulic motor to provide an air flow to improve penetration of a tree canopy. The main advantage of the system is that a fan unit can be mounted on a hydraulically manoeuvred boom so that the nozzles are placed relatively close to the foliage (see p. 234).

Animal-drawn sprayers

Animal-drawn sprayers have been used where farmers have draught animals such as oxen. The tank, boom and pump are usually mounted on a suitable wheeled frame. A high-clearance frame is needed for some crops (Fig. 7.24). These sprayers can be operated even when conditions are too wet to allow the passage of a tractor, and the animals do not damage the crop. The pump can be driven by a small engine or by means of a chain drive from one of the wheels on the frame. When the latter is used, the pump has to be operated for a few metres to build up sufficient pressure at the nozzles before spraying starts. If wheel slip occurs, spray pressure will decrease.

Fig. 7.24 Animal-drawn sprayer with engine-driven pump.

8

Controlled droplet application

In contrast to the relatively wide range of droplet sizes produced by hydraulic nozzles, controlled droplet application (CDA) emphasises the application of a narrow spectrum, choosing the appropriate droplet size to optimise deposition on the intended spray target (Bals, 1975b; Matthews, 1977). Fraser (1958) and others had previously stressed the need for a nozzle that produces a *narrow* spectrum of droplet sizes to avoid losses caused by off-target spray drift or run-off, or both. Bals (1969) began a crusade to develop rotary nozzles to achieve a relatively narrow droplet spectrum and later the term CDA was developed and differentiated the question of droplet size from the use of ultralow volume formulations. The VMD/NMD ratio has been used as a criterion to assess whether a nozzle is producing a CDA spray, but with greater use of laser light instrumentation to measure droplet spectra and their measurement of spray volumes rather than numbers of droplets, the relative span has become more relevant. Bateman (1993) suggested that a high proportion of the spray should be within two size classes, the upper class being double the diameter of the lower class. This represents an eight-fold increase in spray volume. Bateman and Alves (2000) plot the droplet size expressed as the VMD together with the $D_{(9.0)}$ and $D_{(1.0)}$ to show the range of sizes obtained with different disc speeds. They also show where the droplet spectra fall within the recommended size range for applying a mycoinsecticide (see also Chapter 17).

CDA developed from the need for greater efficiency when applying ULV sprays. With usually only 0.5–3 litres of spray per hectare, it was important to avoid large droplets that wasted a high proportion of the pesticide. More recently there has been a need for greater flexibility in the choice of insecticides in IPM programmes, so with limited availability of different UL formulations and their greater cost, formulations suitable for dilution in water have been used increasingly for CDA at very-low volumes (VLV). While sprays with droplets having a VMD of 50–100 µm were suitable for ULV application, larger droplets (100–150 µm) are needed for applying insecticides and fungicides at VLV. This is to compensate for evaporation of the diluent during flight. Larger droplets (>200 μm) to minimise the risk of downwind drift have been used with CDA equipment to apply herbicides. Specialised CDA formulations have been developed for certain markets. As the reduced volumes provide a pattern of discrete droplets over the target surface, an adjuvant may also be needed to enhance redistribution, cuticle penetration or rainfastness. Molasses has been used to reduce the effect of evaporation of water from spray droplets and act as a feeding stimulant to enhance mortality of insect pests.

Controlled droplet application been widely adopted in semi-arid areas where water supplies are poor and prevent widespread adoption of more traditional spraying techniques. In particular, large areas of cotton in Africa have been treated with spinning disc sprayers (Matthews, 1989a, 1990). Cauquil and Vaissayre (1995) have reported on the extensive use of these sprayers to treat nearly two million hectares of cotton in West Africa. This equipment has also been recommended for locust and armyworm control, in forestry, and by local authorities for pest control in urban situations. CDA has also been adopted in amenity areas where drift of herbicides from pathways on to adjacent flower beds must be avoided. The reduced weight of the equipment and eliminating the need to carry up to 15 litres of spray in a knapsack have also helped to promote the use of CDA.

An important method of controlling size of droplets within fairly narrow limits is by using centrifugal-energy nozzles (e.g. spinning discs or cages), with which droplet size can be adjusted by varying their rotational speed.

Centrifugal-energy nozzle (e.g. spinning discs)

Centrifugal-energy nozzles have proved valuable in the laboratory as a means of obtaining a narrow spectrum of droplets, but early attempts to use them in the field were not successful. This was due to attempts to apply the same volumes of liquid as used with hydraulic nozzles causing flooding of the nozzle. Liquid is fed near the centre of a rotating surface so that centrifugal force spreads the liquid to the edge at or near which the droplets are formed. Fraser *et al.* (1963) defined three methods of droplet formation as the liquid flow rate is increased. These are:

(1) single droplets leave directly from the nozzle at low flow rates
(2) liquid leaves the nozzles in the form of long curved threads or ligaments which break down into droplets
(3) liquid leaves the nozzle in the form of an attenuating sheet which disintegrates – mostly caused by aerodynamic waves of increasing amplitude so that fragments of the sheet break up into ligaments and subsequently droplets (Fig. 8.1).

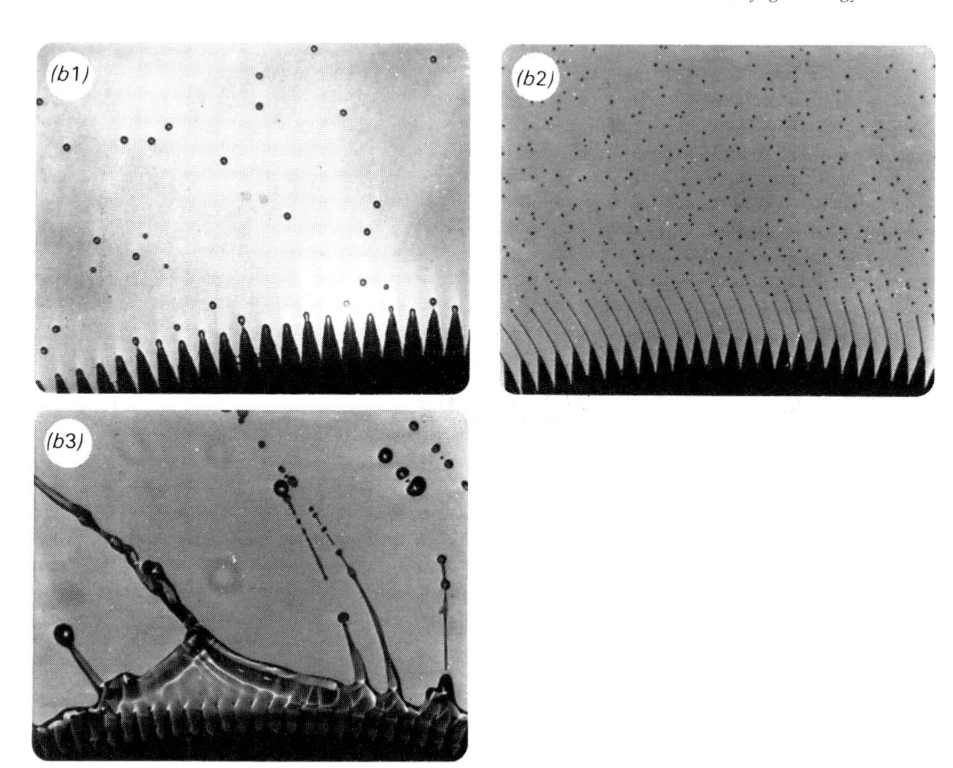

Fig. 8.1 Variation of droplet size – single droplet, ligament and sheet formation from a spinning disc: (b1) Herbi disc 2000 r.p.m., 60 ml/min; (b2) 2500 r.p.m., 100 ml/min; (b3) 1000 r.p.m., 800 ml/min (photos: Micron Sprayers).

Sheet formation occurs when the rotating surface is flooded; droplet formation is then similar to that with hydraulic nozzles and a wide range of droplet sizes is produced. The transition between droplet and ligament and between ligament and sheet formation occurs over a range of flow rates when droplets are formed by both mechanisms (Frost, 1974). Droplet size distributions from a rotary nozzle often have two principal droplet sizes, corresponding to the main and satellite droplets (Hinze and Milborn, 1950; Dombrowski and Lloyd, 1974). Satellite droplets are formed from a thread which connects the main droplet to the rest of the ligament or liquid on the nozzle. In the transition from single to ligament droplet formation, the size and number of satellites increases, causing a decrease in the mean diameter (Dombrowski and Lloyd, 1974).

$$d = K\frac{1}{\omega}\sqrt{\frac{\gamma}{D\rho}}$$

where d is the droplet diameter (μm), ω is the angular velocity (rad/s), D is the diameter of disc or cup (mm), γ is the surface tension of liquid (mN/m), ρ is the density of the liquid (g/cm^3), and K is a constant which has been found experimentally to average 3.76 (Fraser, 1958).

This can be written as

$$d = \frac{\text{Constant}}{\text{r.p.m.}}$$

The constant will be affected by disc design but is usually about 500 000.

The main types of centrifugal energy nozzles are discs, cups and cylindrical sleeves or wire mesh cages (Figs. 8.2, 13.14). Spinning brushes have also been used.

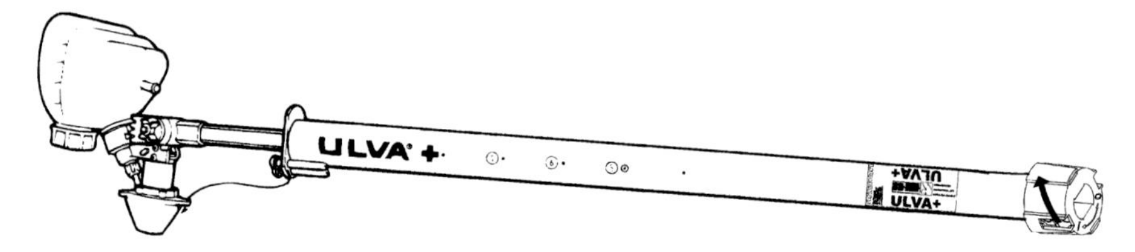

Fig. 8.2 Hand-carried battery-operated spinning disc sprayer (ULVA+).

Studies of disc design have used smooth-edged discs, but Fraser (1958) reduced droplet size by 13 per cent with a 45° chamber around the edge. Bals (1970) made discs with serrated edges to provide 360 half-pyramidical 'zero issuing points' to reduce the force required to overcome surface tension and breakaway droplets of a given size or, for a given force, produce smaller droplets. Bals (1976) introduced discs with a grooved inner surface to provide a reservoir of liquid to feed 'ligaments' of spray liquid to individual issuing points around the periphery. A very narrow range of droplet sizes is produced with discs having both grooves and teeth (Fig. 8.3), hence their suitability for controlled droplet application. There is an optimum flow rate for a given rotational speed; this decreases with increased speed (Matthews, 1996b). Application of a higher flow rate was possible when the larger cup-shaped disc was used with grooves to each of the peripheral teeth (Heijne, 1978). Similarly, a series of stacked discs on the rotary cages allows higher flow rates to be applied with sprayers on vehicles or aircraft.

Spinning discs, cups or cages are less liable to clog, but they are subject to different types of wear and breakdown compared to hydraulic nozzles.

Centrifugal-energy nozzles can be mounted in the airstream emitted from mistblowers (see Chapter 10). However, droplet spectra produced can be affected by the interaction of centrifugal and air shear forces. Large droplets produced from a rotary nozzle will be sheared at high air velocities; thus with rotary sleeve nozzles, a higher air velocity produced a wider droplet spectrum when the nozzle rotated at only 50 rev/s (Hewitt, 1991). However, with a

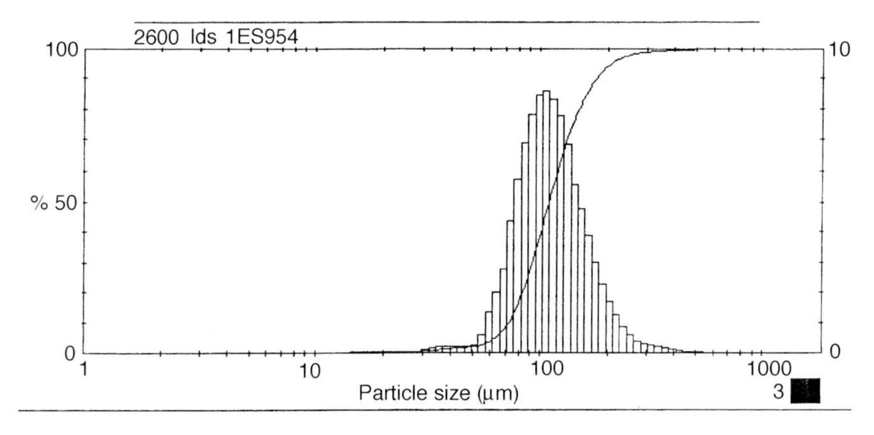

Fig. 8.3 Droplet spectrum from ULVA+ spinning disc sprayer.

knapsack mistblower, Bateman and Alves (2000) reported a narrower spectrum from a single disc (Micron X) mounted in the airstream. The distance over which droplets are thrown from the periphery of a disc is important when droplets have to be entrained in such an airstream. According to Byass and Charlton (1968), the upper limit of droplet size from a nozzle mounted in an airstream into which the droplets have to be turned can be determined by an equation given by Prandtl (1952)

$$d = \frac{K\gamma}{\frac{1}{2}\rho V^2}$$

where V is the velocity of the airstream (m/s), d is the diameter of the largest surviving droplet (μm), K is a constant depending on the droplet size range, γ is the surface tension (mN/m), and ρ is the density of air (g/cm^3).

The actual distance a droplet is thrown depends largely on the effect of air resistance, which is reduced if more droplets are produced. Courshee and Ireson (1961) showed that within certain limits the distance (S) single droplets were projected in ambient air is approximately proportional to the square root of the product of droplet size (d) and disc diameter (D).

$$S = 1.3\sqrt{dD}$$

where 1.3 is a constant. Thus a 250 μm droplet produced on a 9 cm disc should travel 61.7 cm.

These nozzles operate efficiently only when the volume of spray applied is restricted to prevent flooding resulting in sheet formation. Ideally, a suitable formulation and flow rate are selected, so that at a given rotational speed droplet formation is from ligaments with a minimal number of satellite droplets. Very uniform droplets can be produced if the flow rate is low enough to avoid ligaments being produced. A narrow droplet spectrum is also achieved with an electrodynamic nozzle described in Chapter 9.

The spray volume required depends not only on the selected droplet size, but also on the number of droplets required on a target surface. When a spray is evenly distributed over a flat surface, the same number of droplets per unit area (100/cm^2) is achieved with as little as 500 ml/ha, when 46 μm droplets are applied, in contrast to 1.8 litres/ha with 70 μm droplets or 200 litres/ha with 340 μm droplets (see p. 78). When fewer droplets are needed to control a pest, less liquid is needed per unit area. In practice as little as 18 litres/ha of certain herbicides has given good weed control, when 300 μm droplets provided an average of 14 droplets/cm^2. In some cases in the UK, as little as 2.5 litres/ha has been used in upland pastures and 8 litres/ha for bracken control using smaller droplets.

Hand-carried, battery-operated spinning-disc sprayers

These lightweight sprayers have a plastic spray head with small d.c. motor which drives a rotating disc, a liquid reservoir (a screw-on bottle), a handle and a power supply. Various designs are available to provide particular droplet spectra and to accommodate different types of battery (Fig. 8.2) (Clayton, 1992).

Disc design

Following the early designs with two discs joined together (Boize and Dombrowski, 1976), only a single disc, which is saucer- or cup-shaped, is used on the small equipment as its use requires less power. While some discs have a smooth edge, the droplet spectrum is better with more uniform distribution of liquid via grooves to teeth around the edge of the disc. This gives better ligament formation even at higher flow rates to produce a narrow droplet spectrum.

One of the problems when liquid is fed near the centre of a disc is to prevent liquid entering the motor along its shaft. A separate baffle plate or spinner may be fixed to the shaft between disc and motor. Alternatively, the centre of the disc can incorporate a cylindrical baffle which interleaves with corresponding channels moulded in the motor housing. The motor should always be run for a few seconds after stopping the flow of liquid, so that all the liquid is spun off the disc. To facilitate calibration, one CDA sprayer has small holes on the disc so that spray which drains through can be collected and measured.

Disc speed and power supply

Most of these sprayers operate with zinc–carbon (Leclanche-type batteries), but the number of batteries required has been significantly reduced by having a smaller single disc, and improved motor. The performance of batteries will vary depending on the manufacturer and their storage life. High power transistor-type batteries provide a longer service life than single-power cells (Matthews and Mowlam, 1974; Beeden, 1975). Rechargeable batteries are also available, and in some countries the sprayer has been adapted to use larger motor-cycle batteries. Solar energy could be used in conjunction with rechargeable batteries to avoid disc speed fluctuation due to variation in the amount of sunlight.

Droplet size will increase as voltage and thus the motor speed decline, so it is important to check the batteries before use. Resting the batteries allows re-polarization to occur and the voltage partially recovers. Normally, spraying should be confined to relatively short periods of continuous use of the motor. Thus a period of 15–20 min spraying can be followed by a rest of 5–7 min to change the bottle, and spraying of insecticides should be completed normally within 2 hours per day. Spraying for longer than 2 hours per day is possible with the herbicide sprayers. Where long periods of use are needed, different sets of batteries can be used, provided they are numbered for use in the correct sequence. Care must be taken when changing batteries that they are all inserted correctly and that wires and connections are not damaged.

A constant disc speed can also be achieved by using a motor with a mechanical governor when slow disc speeds are required; thus large droplets for herbicide application (250–350 μm) are produced by direct droplet formation with a disc speed of 2000 r.p.m. over a range of voltages (Bals, 1975a, 1976).

Disc speed can be checked with a tachometer. This is particularly important if phytotoxicity is liable to occur because the droplets are too large. A relatively inexpensive tachometer suitable for use in the field, the 'Vibratak', consists of a thin wire inside a metal cylinder. One end of the cylinder is held against the back plate surrounding the motor, and the wire is pushed out of the cylinder. When it vibrates at maximum amplitude the r.p.m. reading is taken directly. A direct reading of disc speed is preferable to measuring the voltage of the power supply, as motor efficiency and the amount of spray liquid fed on to the disc also affect disc speed.

Control of flow rate

Interchangeable restrictors control the flow of liquid from the reservoir by gravity to the disc. A vacuum inside the reservoir is avoided by incorporating an air bleed into the design so that constant pressure is obtained. The air bleed is along a small channel at the base of the thread in the socket. Apart from the size of the restrictor, flow rate is also affected by viscosity of the formulation, which may change with the temperature (Cowell and Lavers, 1988).

Flow rate should be checked, before spraying and during spraying if there is a marked change in temperature, by timing the period to spray a known quantity of liquid, preferably with the discs rotating. Comparison of the flow rate between different formulations can be made by using the restrictor separately from the sprayer. In general, the lowest effective flow rate is chosen to reduce the load on the motor and thus avoid increased power consumption and droplet size. Some restrictors are colour coded (Table 8.1). On some sprayers a filter is inserted between the reservoir and the restrictor to prevent blockages.

Table 8.1 Orifice diameter for one range of colour-coded restrictors

Colour	Diameter of orifice (mm)
White	0.50
Brown	0.65
Blue	0.78
Yellow	1.00
Orange	1.30
Red	1.56
Black	1.60
Grey	2.00
Green	2.90

A plastic bottle, usually of 1 litre capacity, is used as a reservoir for the spray liquid. If necessary the operator can carry a reserve supply of spray liquid in a plastic bottle mounted on the shoulder or on a knapsack frame (Fig. 8.4). On one sprayer designed for herbicide application, a 2.5 litre bottle is fitted to the battery case at the end opposite to the spinning disc and acts as a counterbalance to the rest of the sprayer. The bottle must be screwed in firmly to avoid leakage (Fig. 8.5). Contamination of the outside of a bottle must be avoided, particularly when sprays contain oil, as some bottles are difficult to hold when wearing rubber gloves. Some agrochemical manufacturers provide pre-packed containers to fit directly on the sprayer. The use of very-low or ultralow volumes makes the distribution and sale of pre-formulated, pre-packed chemicals economically feasible. This has advantages in eliminating mixing and filling operations which is where there is the greatest potential risk of operator contamination. Indeed, when this type of sprayer was introduced, this was seen as one of the potential advantages over conventional knapsack sprayers.

Swath width and track spacing

Movement of droplets after release from a nozzle depends on their size, wind velocity and direction and height of release above the crop (or ground). As

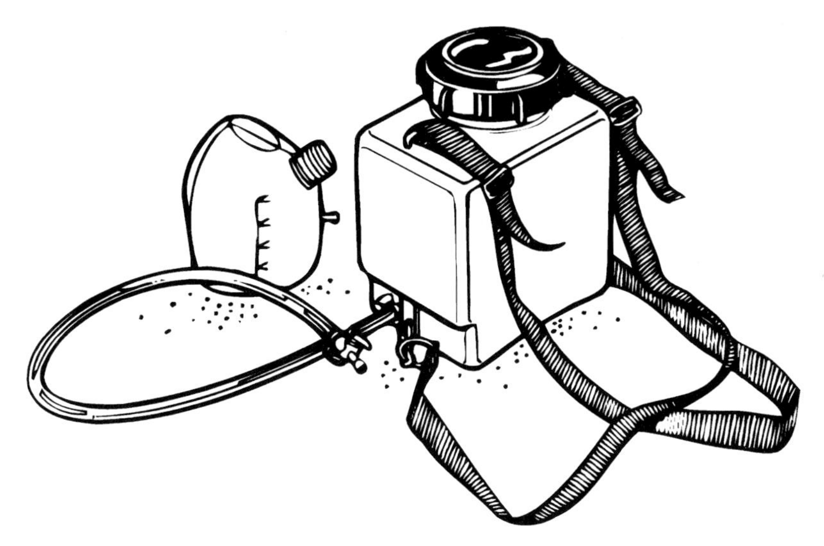

Fig. 8.4 Optional knapsack tank to refill ULVA+ sprayer.

discussed earlier in Chapter 4, large droplets deposit quickly by gravity with minimal displacement by the prevailing wind. An unshielded 80 mm disc producing 250 µm droplets has a swath of approximately 1.2 m. In contrast, 70 µm droplets may be blown more than 10 m downwind if released 1 m above the crop, even when wind velocity is less than 7 km/h. Convective air turbulence could carry such droplets much further. Swaths up to 20 m downwind of the operator can be treated effectively with droplets less than 100 µm diameter under certain circumstances, but there is a risk of thermal air movement taking such small droplets upwards away from a crop.

The overall distance downwind over which sufficient droplets are deposited is referred to as the swath, whereas the distance between successive passes across a field is more appropriately referred to as the track spacing. Often the term 'swath' is used synonymously with track spacing.

Choice of track spacing used in incremental spraying will depend on the behaviour of the pest and the type of foliage of a host crop, as well as wind velocity affecting the amount of downwind displacement across each swath. Spraying across wide swaths can control insects exposed on the tops of plants, for example the leafworm *Alabama argillacea*. Similarly, a 15 m track spacing has been used to spray aphids on wheat in the UK. However, when an insect is feeding on the lower part of plants and penetration of the foliage is needed, a narrow track spacing is essential so that more droplets are carried by turbulence between the rows. A 3 m track spacing has been used to spray tall bracken. Too wide a track spacing should be avoided as variations in wind velocity may result in uneven distribution of spray. Incremental spraying by overlapping swaths generally improves coverage (Fig. 8.6). Narrow track spacing (0.9 m) gave better control of *Helicoverpa* on cotton than when a wide track spacing (4.5 m) was used, even though a lower concentration of spray

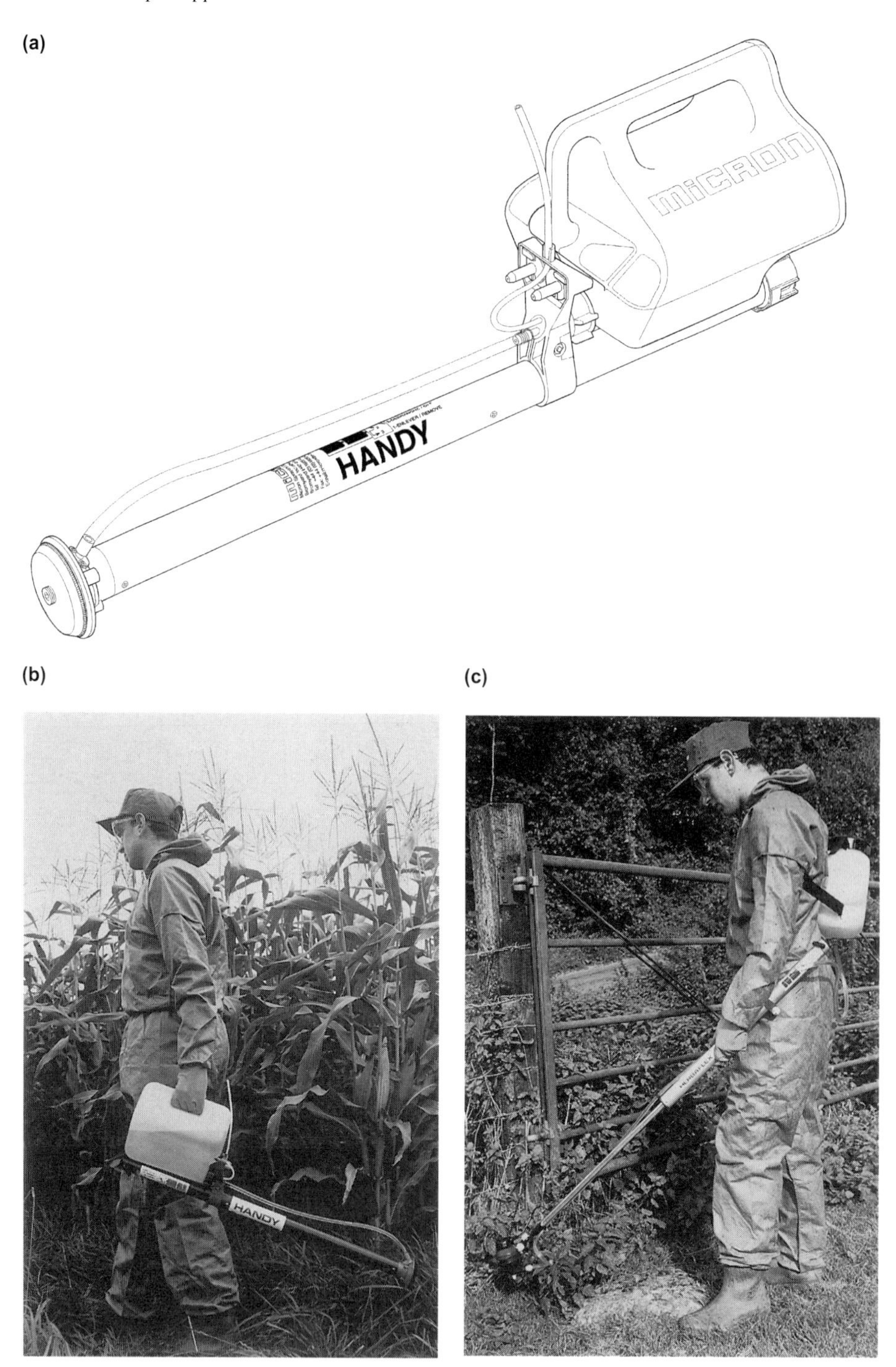

Fig. 8.5 (a) Handy herbicide applicator. (b) Using Handy sprayer in maize. (c) Herbi sprayer.

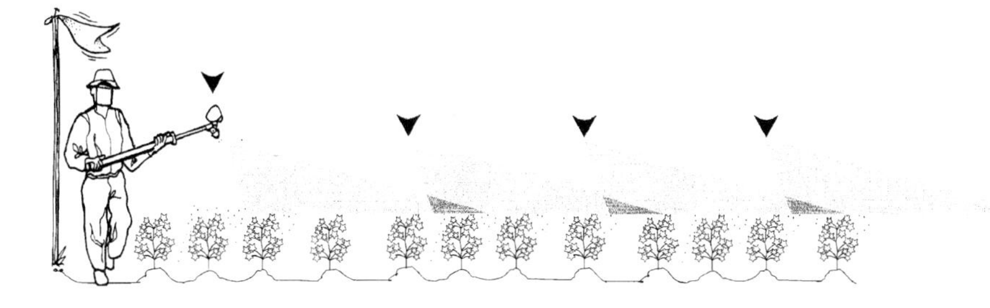

Fig. 8.6 Overlapping swaths when using downward movement of spray from spinning disc.

was used on the narrow track spacing (Matthews, 1973). Raheja (1976) obtained no difference in the yield of cowpeas when spraying at 2.5 litres/ha over 1.8 or 3.6 m wide track spacings, mainly against *Maruca*.

Sometimes adjacent swaths can be displaced not only in space but also in time (sequential spraying) (Joyce, 1975); thus twice-weekly sprays over a double-width swath may be preferred with a less persistent chemical or if rain reduces the effectiveness of a weekly spray. An increase in the frequency of application without increasing the total volume of spray per unit time may improve deposition, as the chance of sprays being applied under different wind conditions is increased, thus a change in wind direction and an amount of turbulence exposes other leaf surfaces. A spray repeated when the wind is from the opposite direction is ideal for improved spray coverage. When more frequent sprays are logistically possible, it is feasible for the farmer to use a lower dosage and repeat a spray if necessary to compensate for the effect of rain or vigorous plant growth.

For any given swath width and droplet size an increase in the volume of spray may not improve control, as the greater number of droplets produced are carried in the same volume of air to more or less the same positions within the crop. Matthews (1973) using ULV sprays obtained no differences in yield of cotton when 0.5 and 1.0 ml/s flow rates were examined with one-, two- and five-row track spacings.

On some crops the track spacings should be changed in relation to the size of the plants; thus, as the area of foliage increases, reduction in track spacings increases the volume per unit area. On cotton in Central Africa, UL formulations sprayed at 0.5 ml/s, the track spacing was reduced from six to four and finally to two rows as plant height increased from 25 cm to 25–50 cm and more than 50 cm, respectively (Matthews, 1971). With VLV sprays a two-row track spacing is used until plants are knee height (approximately 50 cm), and then a single row when wettable powder formulations are applied (Mowlam *et al.*, 1975; Nyirenda, 1991).

Penetration of a crop canopy is poor when plants have large, more or less horizontal leaves, as droplet dispersal is dependent on air movement. More droplets can penetrate the canopy if a suitable variety is selected; for example, okra leaf and frego bract have characteristics which breeders are endeavouring to incorporate into commercial varieties of cotton (Parrott *et al.*, 1973).

The time required to spray 1 hectare with different track spacings when walking at 1 m/s is shown in Table 8.2. With a narrow 1 m track spacing, less than half a person day is required per hectare to apply a herbicide as a placement spray at 10 litres/ha, in contrast to over 30 person days needed to hand-hoe weeds. Even less time is needed using incremental spraying with wider track spacings.

Table 8.2 Time required to treat a (1 ha) square field at a walking speed of 1 m/s. [a] Extra time is required to mix the spray, replace the bottles and carry the materials to the field

Track spacing (m)	Time (min)
1	168
2	85
5	35
10	18
15	13

[a]Time to treat area (min) $= \dfrac{\left(\dfrac{\text{Area to be treated (m}^2\text{)}}{\text{Area covered per second (m}^2\text{)}}\right) + \text{Field width across rows (m)}}{60}$

Spraying procedures

Incremental drift spraying

Wind direction is noted so that the spray operator can walk progressively upwind across the field through untreated crops. A piece of thread can be attached to a wire fixed to the spray head to check wind direction. Spraying commences 1 or 2 m inside the downwind edge of the field. The disc speed is checked before the spray liquid is prepared. The bottle is filled and then screwed on to the sprayer. When the operator is ready to spray, the motor is switched on and the disc allowed to reach full speed. The sprayer is held either with the handle across the front of the operator's body, or over the operator's shoulder with the disc above the crop, pointing downwind, so that droplets are carried away from the operator while walking through the crop. The bottle is then inverted as liquid is gravity fed to the disc, but if the operator stops for any reason or reaches the end of the row, the sprayer should be turned over again to stop the flow of liquid and avoid overdosing. The inversion of the bottle at the end of each row ensures that the spray remains well mixed. The motor is not switched off while the operator walks along the edge of the field to the start of the next swath. If there is more than one operator in a field, great care must be taken to avoid walking in each other's spray cloud. An extra swath outside the upwind edge of the field may be necessary. At the end of spraying, the motor is left running for a short period to remove any liquid from the discs. Ten litres/ha is applied when the operator walks at 1 m/s (÷) spraying a track spacing of 1 m wide (÷) with a flow rate of 1 ml/s (×). If any of these variables is changed the volume (10 litres/ha) is divided or multiplied as indicated by the sign in the brackets. Thus, with a 5 m swath: 10 ÷ 5 = 2 litres/ha.

The spinning disc is normally held about 1 m above the crop. It may be necessary to hold it lower while spraying the first swath along the leeward side of a field to reduce the amount of chemical which may drift outside the treated area. Similarly, the nozzle may be held lower during the final swath on the windward side of a field, to cover the edge of the field. Nozzle height can be lowered if necessary when the wind velocity increases but, if the area being treated is sufficiently large, a wider track spacing can be used to take advantage of the wind. Simple anemometers are available to check wind velocity which should be 2–15 km/h. One small simple anemometer has a pith ball which moves up a vertical tube according to the strength of the wind (Fig. 8.7). Extreme conditions, such as a dead calm or a strong, gusty wind, should be avoided.

Placement spraying

When spraying herbicides, the disc is held only a few centimetres above the weeds so that downwind displacement of the spray is negligible. The disc is

Fig. 8.7 Measuring wind velocity with a simple anemometer.

held behind the operator at 60° from the ground (Fig. 8.8a) to avoid the hollow-cone pattern from a horizontal disc. The operator does not walk over treated surfaces with this method but, if greater control of the position of the swath is needed, the less poisonous chemicals can be applied with the disc tilted slightly away and in front of or to the side of the operator (Fig. 8.8b). Also, a wider swath can be achieved by mounting two or more discs on a hand-held boom, a practice used in plantation agriculture. Sprayers with a shrouded disc have been developed to allow adjustment of the swath width as well as apply a narrow droplet spectrum.

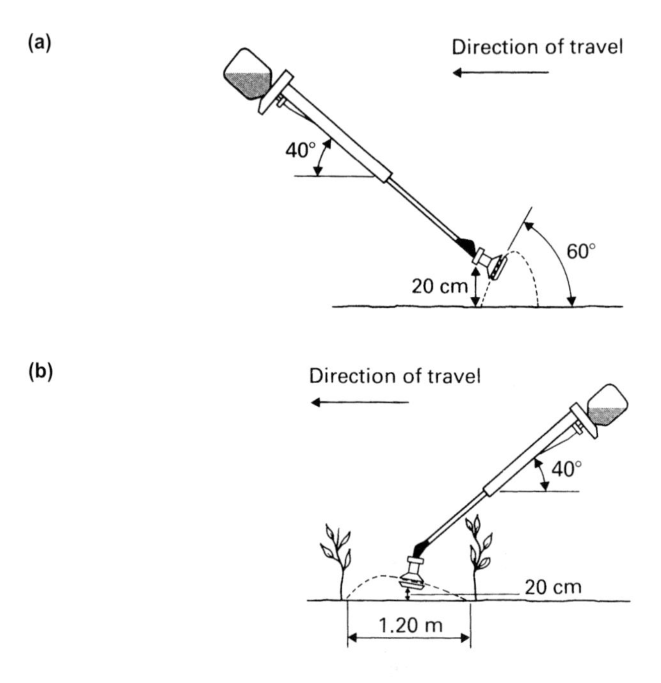

Fig. 8.8 (a) Position of 'Herbi' sprayer behind operator. (b) Position of 'Herbi' sprayer when walking towards spray.

Motorised fan-assisted spinning-disc sprayers

Discs can be mounted in front of a fan that provides a directional airstream so that insecticide and fungicide sprays can be applied in warehouses, glasshouses and other enclosed areas where natural air movement is insufficient to disperse the spray droplets. The power required to move air is much greater than that required to produce the droplets. Small battery-operated sprayers with a fan have been used in glasshouses (Fig. 8.9a), but the period of operation is limited, even when rechargeable batteries are used. An a.c. electric motor may be used if a mains electricity or a portable generator power supply is available, but a trailing cable is a disadvantage. For greater mobility the sprayer with a two-stroke engine is used (Fig. 8.9b).

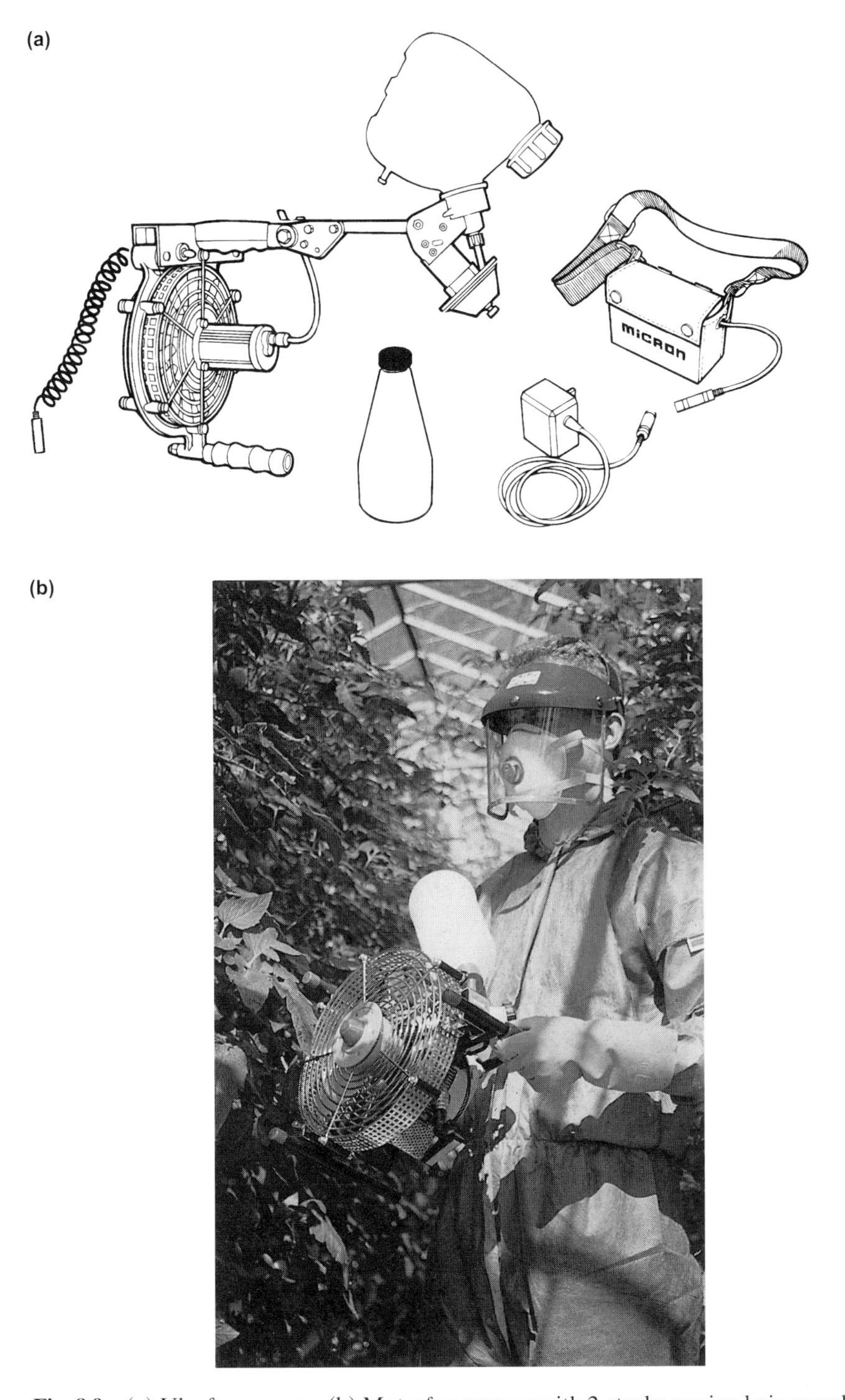

Fig. 8.9 (a) Ulvafan sprayer. (b) Motorfan sprayer with 2-stroke engine being used in glasshouse.

Discs mounted directly on the drive shaft of the engine operate at a relatively slow speed of about 6000 r.p.m., and some larger droplets may collect on operators, particularly on their hands, if sprayers are too close to their body. Hand-carried sprayers of this type are not very comfortable to use because of the vibrations from the engine; their weight is also a disadvantage. Machines with gravity feed of liquid are rotated to stop and start spraying; thus a modified carburettor is needed to allow the engine to run in different positions. Slight pressure in the spray tank to force liquid to the nozzle is achieved, if necessary, by positioning an air line from in front of the fan or engine to the tank. Changes in the direction of the airstream cause leaves to flutter and collect droplets more efficiently.

A CDA knapsack mistblower has been developed with a two stroke engine to drive a propeller fan, in front of which a single spinning disc is mounted (Fig. 8.10) (Povey *et al*., 1996).

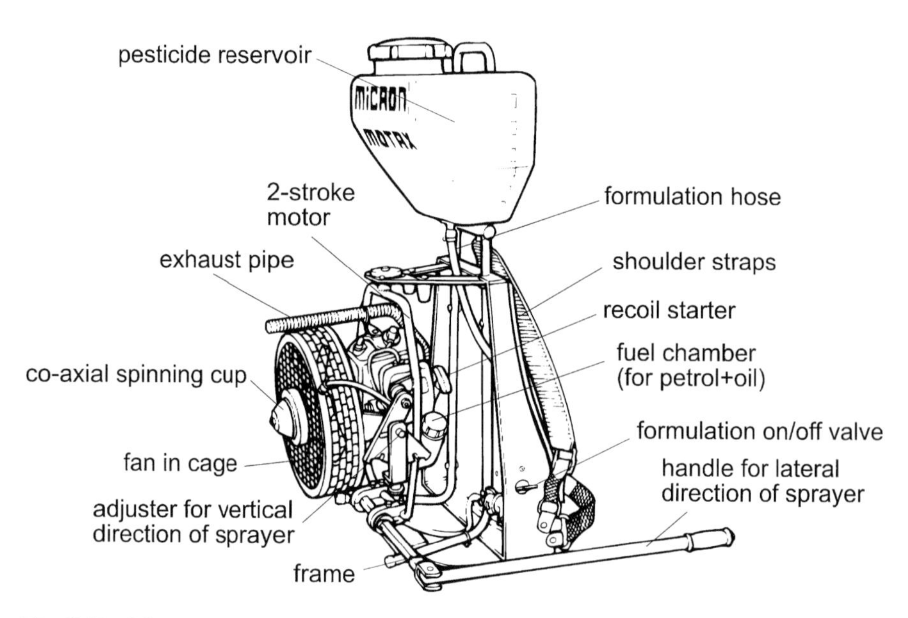

Fig. 8.10 Motax sprayer.

Vehicle-mounted sprayers with centrifugal-energy nozzles

Vehicle-mounted 'drift'sprayer

A series of spinning discs are mounted on a shaft through which spray liquid is pumped from a reservoir mounted on the vehicle. The discs and pump are driven by electric motors (Fig. 8.11). The Ulvamast sprayer has been used most widely for applying insecticides to control locusts and has become a replacement for the exhaust nozzle sprayer. Hewitt (1992) reported on the

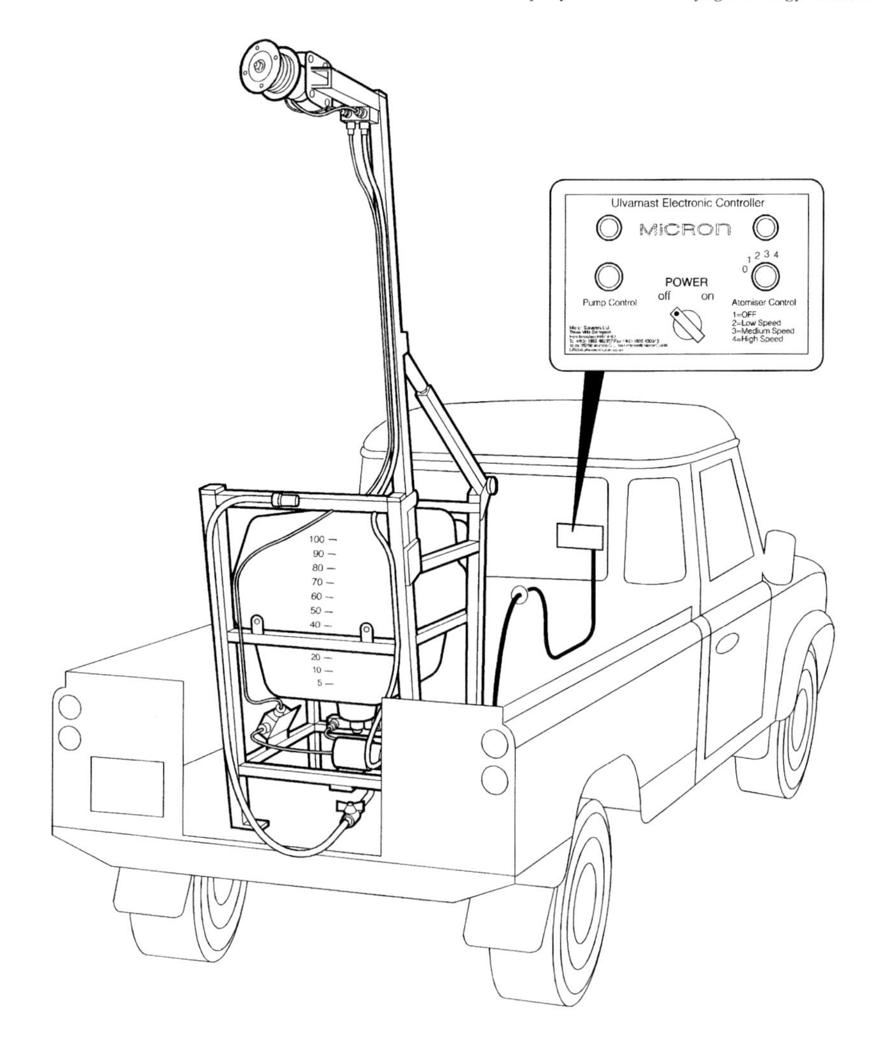

Fig. 8.11 Ulvamast sprayer (see also Fig. 2.4c).

droplet spectra obtained with the Mark II version. A new version provides a direct drive atomiser, increased tank capacity and electronic in-cab controller to regulate flow rate and atomiser disc speed.

Similarly, droplets can be drifted through the nesting sites of the red-billed quelea, *Quelea quelea* (Dorow, 1976). Low dosages of fungicides have also been applied to cereals with similar equipment.

Boom sprayers

Initial research with CDA sprays with boom sprayers was with multiple shrouded discs to obtain a spray pattern similar to that of the fan nozzle and avoid flooding a single disc. Instead of multiple shrouded discs on each unit, Heijne (1978) reported data for a single large spinning cup (Micron 'Micromax') with individual grooves to 180 teeth (Fig. 8.12). This rotary atomizer allowed herbicide, insecticide and fungicide application by alteration of the disc speed (see also Bode *et al.*, 1983). Several versions of this nozzle have

(a)

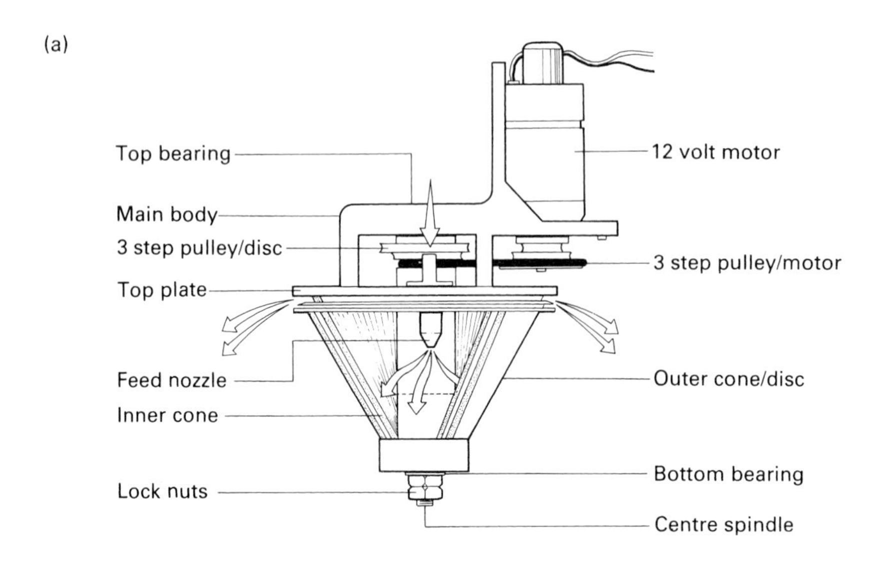

(b)

Fig. 8.12 (a) Spinning nozzle for use on tractor. (b) Tractor boom fitted with Micromax nozzles.

been supplied commercially with electrical or hydraulic drive systems. However, there has been limited acceptance of CDA sprayers despite the advantages of low volume and drift reduction when large droplets are applied with low disc speeds. This is mainly due to the lack of penetration into a crop canopy when spray is released in a horizontal plane. Certain studies suggested that lower dosages could be applied against aphids on wheat as aphid survivors remained as essential food for predators (Holland *et al.*, 1997). In Australia, spinning discs mounted on booms are usually in conjunction with a propeller fan to improve spray penetration.

A shrouded vertical large (14 cm) spinning disc was developed to improve penetration (Morel, 1985) and good results were achieved with 25 litres/ha. Other shrouded disc equipment has been developed for use on a vehicle in urban areas to treat pavements and gutters.

Air-assisted sprayers

Some of the sprayers with air assistance use a rotating cage atomiser. More information on these is given in Chapter 10.

9

Electrostatically charged sprays

The need to reduce the amount of spray drift from treated fields and improve dose transfer to target surfaces has led to research on the possibility of improving spray deposition by electrostatically charging droplets (Bailey, 1986; Law, 1987; Marchant, 1987; Matthews, 1989b; Carlton *et al.*, 1995). However, despite considerable research, the use of charged agricultural sprays has been very limited. Three methods of charging agricultural sprays have been used:

(1) induction charging of conductive liquids
(2) ionised field charging of either conductive or non-conductive liquids
(3) direct charging of semiconductive liquids.

In each case, the normal balance of positively charged protons and negatively charged electrons is disturbed by movement of electrons, so that additional electrons provide a negative spray droplet, while a deficit of electrons makes the spray positive.

Induction charging

When a high-voltage electrode, positioned close to where spray liquid is emitted from a nozzle, is positively charged, a conductive liquid, such as a water-based pesticide spray at earth potential, has a negative charge induced on its surface by the attraction of electrons. If the electrode is negative, the reverse occurs and electrons repelled from the liquid to earth provide a positively charged liquid. As the droplets are formed, the charge is retained on them. A conductive liquid is needed so that the charge transfers from earth to the liquid jet in the very short time while it passes the electrode. The level of charge induced per unit area of surface will be proportional to the voltage applied to the electrode.

The charge on the spray droplets is the opposite of that on the electrode, so some spray is liable to be attracted on to the electrode, which if wetted, is liable to short circuit the power supply. An air stream is used on some nozzles to blow droplets away from the electrode and keep it dry (Law, 1978).

Ionised field charging

A high voltage applied to a needle-point can create an intense electric field around it that is sufficient to ionise molecules of the surrounding air. A positively charged conductor will repel the positive ions created, while the electrons that are released in the ionisation process will be attracted to the conductor and neutralise some of its charge. With a negatively charged conductor, the reverse is true and positive ions are attracted back to the conductor. Great care is needed to protect the fragile needle and avoid reverse ionisation. The level of charge is dependent upon the dielectric constant of the spray, its surface area, the electrical characteristics of the corona discharge and the time within the ionised field.

When a stream of liquid passes near to the ionising tip of the needle, the charged ions produced are attracted to the liquid and carried away by it. The needle is usually negatively charged, as higher voltages are required to create an equivalent positive corona. Liquids with a wide range of conductivities can be charged with this method (Arnold and Pye, 1980).

Direct charging

When a semiconductive spray liquid, with a electrical resistivity in the range 10^4–10^6 ohm m, is exposed to a high voltage (15–40 kV) as the liquid emerges through a narrow slit, mutual repulsion between different portions of the liquid overcomes surface tension and ligaments are formed. These ligaments break up into droplets due to axisymmetric instabilities. The level of charge on the droplets represents the maximum that can be attained and is called the Rayleigh limit (Rayleigh, 1882). The initial droplet size distribution is initially bimodal, but the very small satellite droplets are attracted to an earthed electrode or 'field adjusting electrode' positioned close to the nozzle, so that essentially a monodisperse spray is produced. The size of droplets is reduced for a particular flow rate by increasing the applied voltage. Increasing the flow rate without changing the voltage will increase droplet size. There is therefore a complex interaction of electrical, viscous and surface tension forces, affecting droplet size with the resistivity and viscosity of the spray liquid being particularly important factors.

Electrostatically charged nozzles

Induction charging nozzles

Hydraulic nozzles

In one system, spray to hydraulic spray nozzles was charged at a potential of up to 10 kV and the electrode was earthed (Marchant and Green, 1982). The distance between the spray tank and individual nozzles was lengthened by using a long coil of narrow bore tube to increase the electrical resistance and reduce the hazard of a large tank at high voltage. The electrode was a perforated hollow tube through which any liquid wetting its surface was sucked by a vacuum and recycled to the spray delivery line. The nozzle was considered impractical under field conditions and was not developed for commercial use. In another system with direct charging of the liquid, the main sprayer tank is separate from the boom and liquid is pumped to a smaller tank from which the charged liquid is delivered to the nozzles.

Later charged electrodes were positioned on either side of the spray sheet emitted from a fan nozzle (Fig. 9.1). Supports for these electrodes were designed to prevent liquid accumulating on their surface, but in practice small charged droplets did collect and drip from the outer shroud. Marchant *et al.* (1985a,b) showed that the charge-to-mass ratio increased with voltage and spray angle, and reduced with nozzle size, electrode spacing and pressure. Few farmers used these nozzles as deposition on the crop was not significantly increased (Cooke and Hislop, 1987), although drift downwind was less, presumably due to a reduction in the volumes of small droplets in the spray cloud. Later studies showed no significant drift reduction and at higher wind speeds airborne drift 5 m downwind of the nozzle increased (Sharp, 1984; Miller, 1988).

Further research has led to the patenting of cone nozzles (Carlton, 1999) with cylindrical electrodes arranged so that one set has the opposite charge to the second set. This apparatus was designed for use on aircraft so that corona discharge of the airframe is substantially near zero. Charging is controlled so that each nozzle has an equal charge-to-mass ratio of at least 0.8 mC/kg. The cone nozzles were also used to overcome flow rate restrictions with spinning disc nozzles previously assessed.

Spinning disc atomisers

Marchant (1985) described mounting a high-voltage electrode around the periphery of an atomiser disc so that liquid was charged as it left the disc (Fig. 9.2). Charged droplets attracted back to the electrode were re-atomised as the electrode rotated at the same speed as the atomiser. Carlton and Bouse (1980) also designed an induction charged spinner to study the use of charged sprays from aircraft. Using these nozzles, higher spray deposits on cotton were obtained with a bipolar charging system (Carlton *et al.*, 1995).

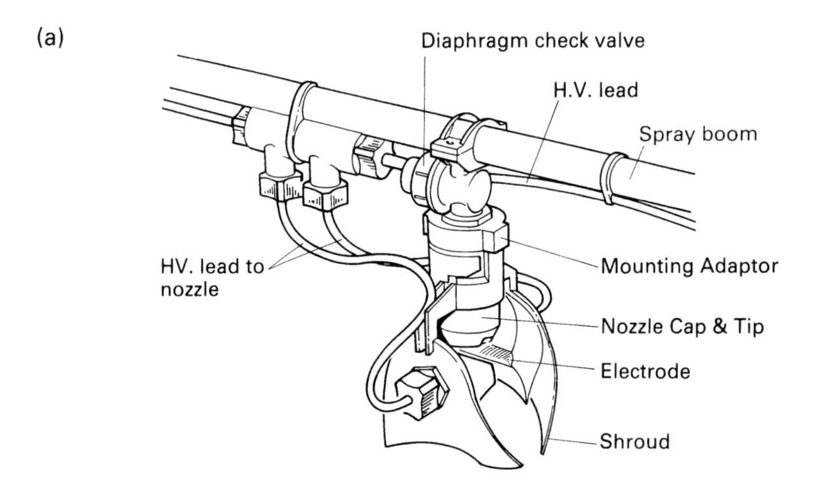

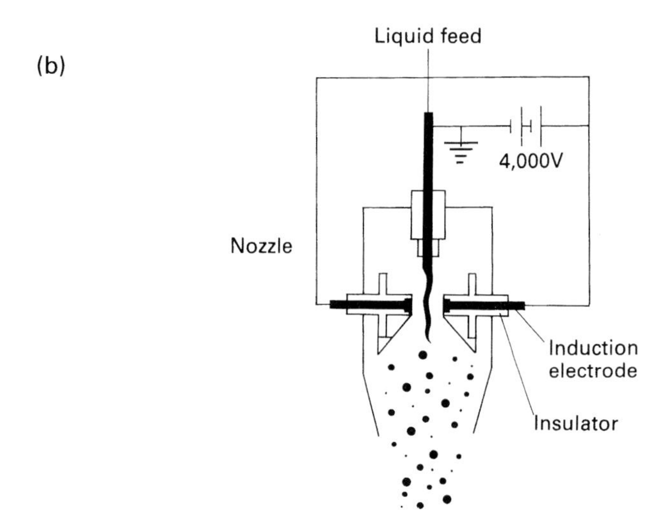

Fig. 9.1 (a) Hydraulic nozzle with induction charging. (b) Diagram to show position of electrodes relative to the spray.

Air-shear nozzles

The advantage of an air-shear nozzle (Fig. 9.3) for induction charging is that the airstream carries the charged spray away from the electrodes. Law (1978) embedded an electrode at the outlet of an air-shear nozzle and has done extensive studies of charged sprays with his equipment. The nozzle was optimized to produce 30–50 μm VMD droplets at flow rates of 75–100 ml/min with a −10 mC/kg charge-to-mass ratio by charging with voltages less than 1 kV and a power consumption less than 25–50 mW (Law, 1987). However, overall power consumption is at least 100 times greater, as a large compressor

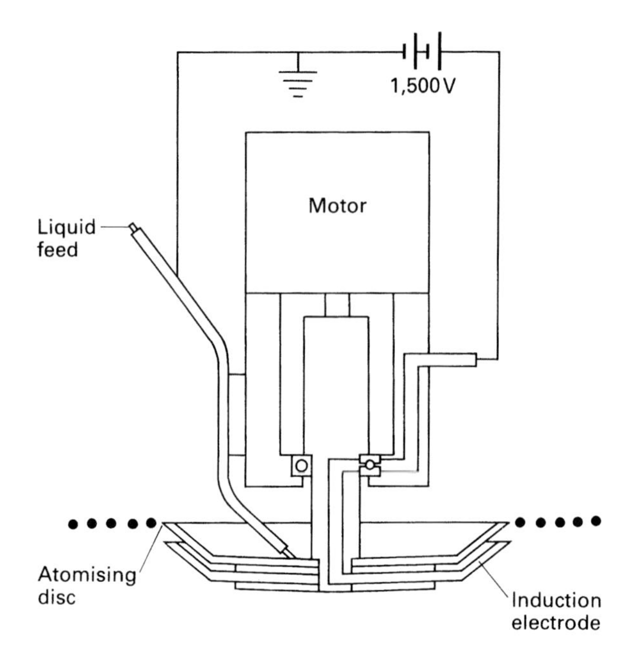

Fig. 9.2 Spinning disc nozzle with induction charging.

is needed to provide the air needed to atomise the spray. This nozzle is more suited for use on a tractor-mounted sprayer, but commercial use has been limited. Flow rate can be increased to 500 ml/min with suitable modification of the nozzle (Frost and Law, 1981). Another air-shear nozzle for use on an orchard sprayer was designed with a high voltage petal electrode mounted opposite the liquid outlet (Inculet *et al.*, 1981).

Piezoelectric nozzles

A piezoelectric nozzle designed to produce monosized droplets for experimental work was improved by incorporating an induction charging system to ensure that the droplets did not coalesce (Stent *et al.*, 1981). The apertures are extremely small, so only well-filtered solutions can be applied with this nozzle. Reichard *et al.* (1987) also used an electrostatic charge on a Bergland and Liu (1973) droplet generator to provide uniform sized droplets for laboratory studies.

Ionised field charging nozzles

Most studies have been with a rotary atomiser with a needle mounted so that it does not quite touch the liquid moving over the surface of a disc, which is coated with a thin metallic layer (Fig. 9.4). When a water-based spray (a good

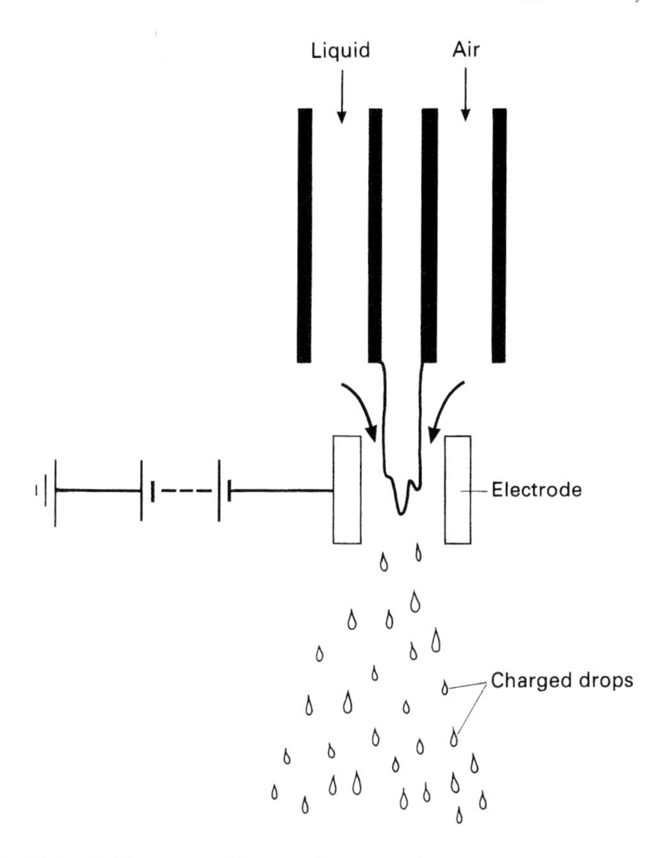

Fig. 9.3 Twin-fluid nozzle with induction charging.

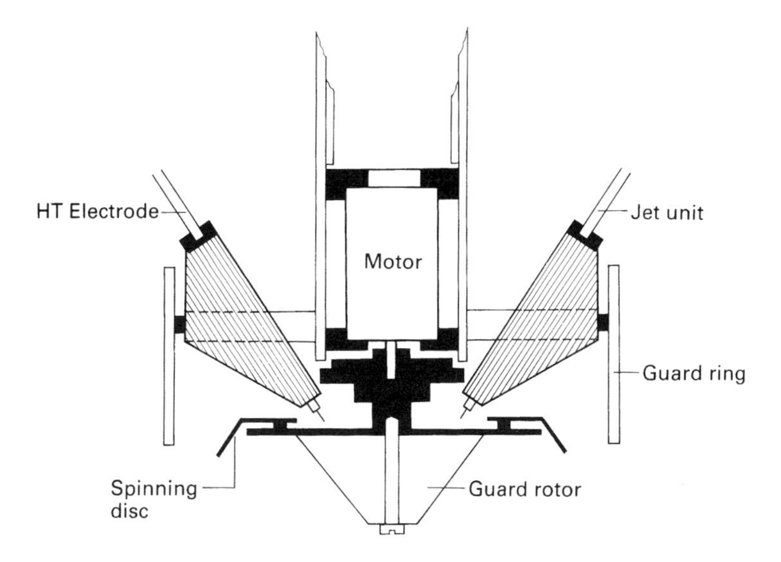

Fig. 9.4 Corona charging of spinning discs.

conductor) is used, the liquid is charged to a potential close to that of the needle, so the spray liquid must be isolated if a high voltage is used. A.J. Arnold (1983, 1984) used a peristaltic pump to maintain a constant flow from separate small spray containers to each individual nozzle, an inverted spinning grooved cup with toothed edge, the 'Micromax', which was referred to as the 'Jumbo' atomiser. In tests with a smaller disc, droplet size was significantly reduced as the applied voltage increased. As an example, with an oil spray fed at 20 ml/min and disc speed of 3500 r.p.m., droplet size decreased to about 50 μm at 30 kV from 200 μm with an uncharged disc (Arnold and Pye, 1980). Cayley *et al.* (1984) reviewed a series of trials with these nozzles.

Electrodynamic nozzles

Coffee (1979) described an electrical-energy nozzle in which a high voltage was applied to a semiconducting liquid emitted through a narrow annulus or slit. The charged liquid forms ligaments which break up into electrostatically charged droplets with a very narrow size range. This nozzle was incorporated into a hand-carried battery-operated sprayer, in which the 'electrodynamic' nozzle was manufactured as an integral part of the pesticide container, and known as a 'Bozzle' (Fig. 9.5). The nozzle, made of a special plastic material, enabled an electric charge to be conducted to the pesticide formulated in an oil, as the spray liquid was fed by gravity through the very narrow annulus. Four batteries in the handle provided 6 V that were converted by the generator to 24 kV fed to the nozzle. A restrictor and air-bleed in the 'Bozzle' determined the flow of liquid to the nozzle. A colour-coded protective cap over the 'Bozzle' indicated the flow rate. At the far end of the handle away from the 'Bozzle' a switch had to be kept depressed during spraying. As soon as the switch was released, the voltage to the 'Bozzle' was discontinued, although there was a small residual charge on the nozzle until it leaked away or the nozzle was deliberately earthed.

Deposition studies

Uncharged spray droplets may sediment onto mainly horizontal surfaces due to gravity or be impacted on vertical surfaces by their velocity by movement in air currents (see also Chapter 4). The latter applies to the smallest droplets with a low terminal velocity, but these droplets can be readily carried by the air movement around targets such as stems and leaves, or carried upwards on thermals due to convection. Improvement of the collection efficiency of these small droplets (< 100 μm) can be achieved by the addition of an electrostatic charge on the spray. Deposition of charged droplets is influenced in several ways. There is the electrical field between the nozzle and the nearest earthed object, the space cloud effect, an induced field effect (Fig. 9.6) and there is often a naturally occurring electropotential gradient near the surface of the

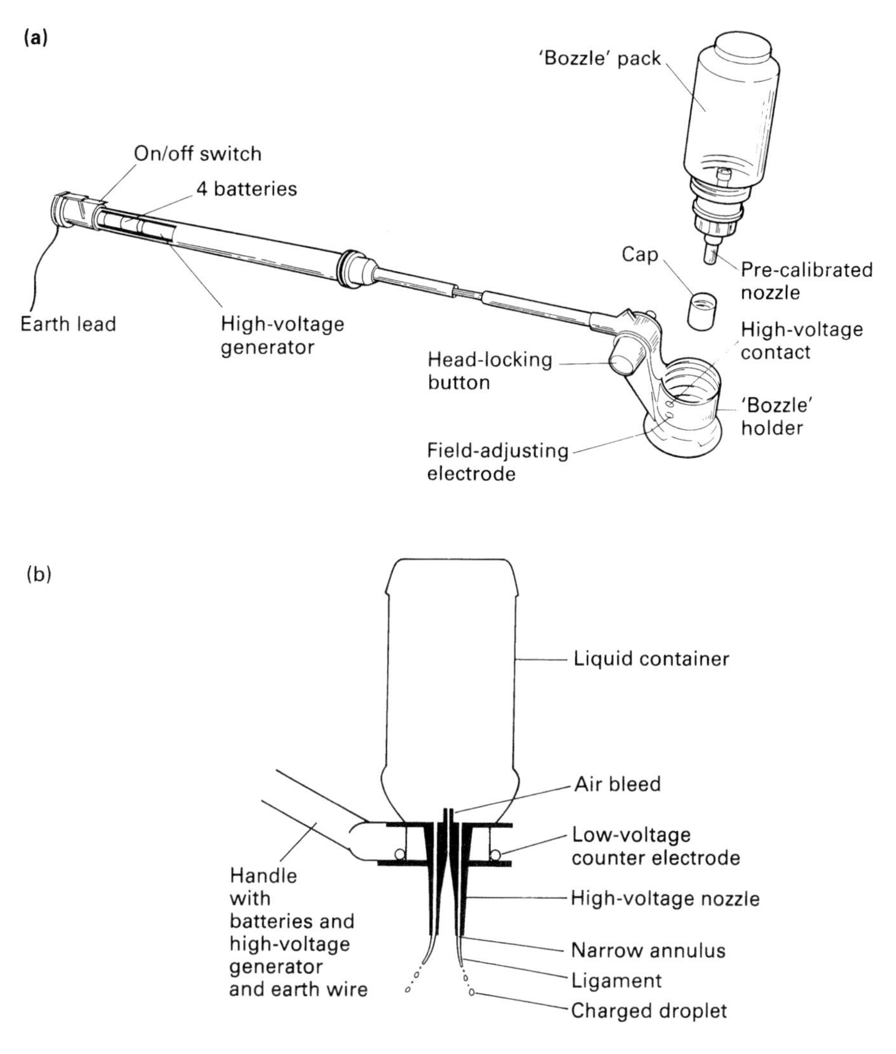

Fig. 9.5 (a) Electrodynamic sprayer (ICI Agrochemicals, now Zeneca). (b) Diagram to show 'Bozzle'.

ground because positive polarity in the upper atmosphere induces a negative charge on the earth's surface. Lake and Marchant (1984) have reported modelling of deposition of a charged aqueous spray in barley.

Shemanchuk *et al.* (1990) reported deposition of electrostatically charged sprays on cattle to protect them from mosquitoes.

Nozzle effect

When a charged nozzle is relatively close to a crop, the electrical forces exerted on the droplets are much greater than the gravitational force;

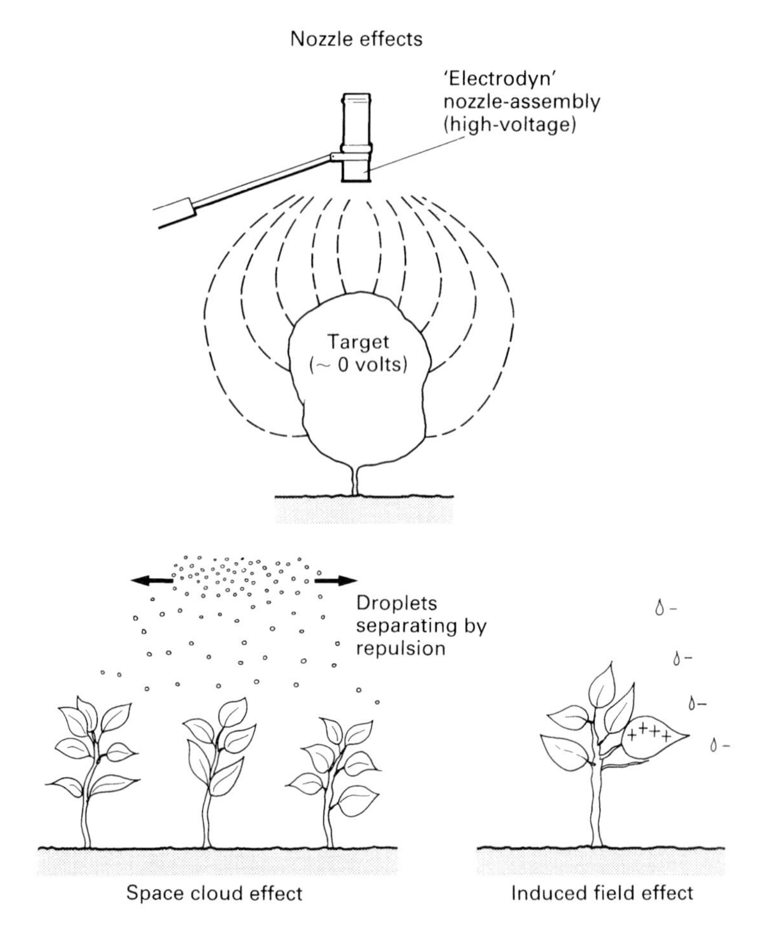

Fig. 9.6 Diagram to show nozzle, space cloud and induced charge effects on deposition.

thus Coffee (1979) calculated that for a 100 μm droplet, with a charge at about 75% of the Rayleigh limit, the initial electrical force acting on it would be about 50 times the gravitational force. Computer simulation suggests that the terminal velocity of a 100 μm droplet would be increased about 16 times to approximately 5 m/s, and that with an air velocity of 4 m/s the electrical force would be about 20% greater than the air drag force (Marchant, 1980). Thus droplets of the same charge as the nozzle would be repelled from the nozzle towards the nearest earthed object, and their trajectories would be less affected by the air movement above or within the crop canopy. In some cases droplets would travel upwards against gravity when the nearest earthed surface is the undersurface of a leaf.

Space cloud effect

The spray cloud containing a large number of droplets of the same polarity expands rapidly as each droplet repels its nearest neighbour. The spray cloud thus creates its own electrical field, which influences the trajectory of the individual droplets. While the effect of the electrical field from the nozzle is relatively short-lived and occurs only while the nozzle is above the crop, the space cloud effect continues after the passage of the sprayer as long as there is still a cloud of droplets. Some of the droplets, repelled outwards away from the centre of the spray cloud, move upwards and could be carried by convection away from the sprayed area, thus Miller (1989) and Western and Hislop (1991) have reported no reduction in spray drift with charged hydraulic spray nozzles, in contrast to data reported by Sharp (1984). However, much depends on droplet size and volatility of the formulation and, provided the droplets do not become too small, there is generally a downward movement of the spray cloud, and downwind drift is less than with uncharged sprays (Johnstone *et al.*, 1982). Penetration into a crop canopy will depend on the openness of the canopy structure so that the space cloud can force droplets into air spaces between branches and leaves.

Induced field

One effect of a charged space cloud is to induce an opposite charge on an earthed target surface; thus when the droplets are positively charged, electrons are attracted to the crop surface from earth. Due to the inverse square relationship of force to distance, the opposite induced or image charge will attract droplets to a surface only when they are very close to the surface (i.e. less than 1 cm). Thus deposition of the smallest droplets is enhanced as, if uncharged, these would be the most likely to be carried by air currents around some target surfaces. If, however, there is an excessively resistive pathway to earth, a charge due to deposition of droplets could raise the electrical potential on a target and consequently diminish further deposition (Law, 1989).

Factors affecting deposition

Pointed leaves

Laboratory experiments readily confirm increased deposition on artificial spherical targets (Fig. 9.7); thus Law and Lane (1981) reported a sevenfold increase compared with an uncharged spray. Similarly, Arnold and Pye (1980) obtained up to eightfold increases in deposition of oil sprays at 30 kV. However, when a needle-point was present on a target, deposition was significantly reduced (Law and Lane, 1982) due to a gaseous discharge which flows from the point to the charged spray cloud (Coffee, 1971). The electric field

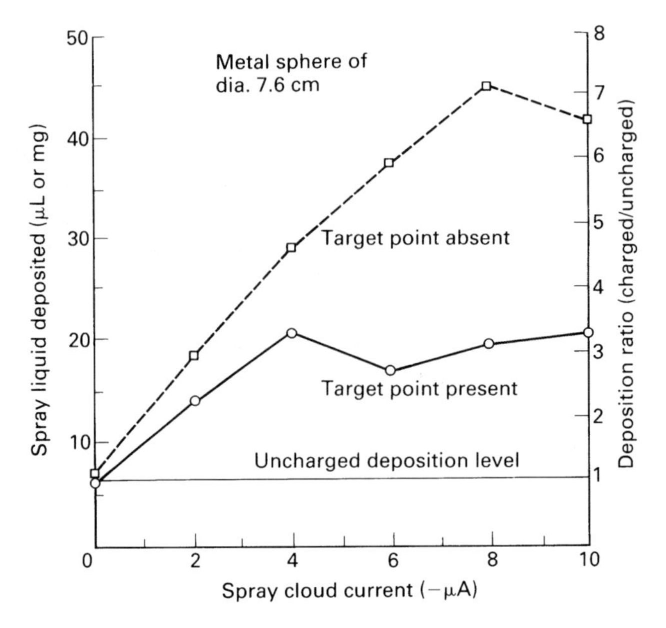

Fig. 9.7 Comparison of charged and uncharged spray deposition on a sphere with and without a needle point (from Law, 1980).

intensification and ionisation at the point caused by the spray cloud's space charge cause this. A single point can account for up to 80% of the total charge exchange between a target and the incoming spray cloud (Cooper and Law, 1985). The counterflow of positive ions drawn from an earthed target point by a negatively charged spray reduces the charge/mass of the adjacent spray cloud so that the effect of space charge (approx 25 μC/m^3) is so decreased that droplets are no longer forced into air spaces between foliage. Laser Doppler studies showed that the induced corona from a point affected the momentum and charge of approaching charged droplets, and in some locations repelled droplets from targets (Law and Bailey, 1984). Nevertheless, droplet deposition was 20% better if the spray cloud was negatively charged compared with a positively charged spray (Cooper and Law, 1987a). Attempts to overcome this problem by using a bipolar instead of unipolar charged spray, failed to improve deposition (Cooper and Law, 1987b).

The effect of the ionisation from pointed leaves will also vary between different types of foliage, and the way in which the plants are spaced apart in the field. Initial studies were with artificial targets. Giles and Law (1985), using cylinders of different diameters and spacings, achieved better deposition (a) closer to the top of the cylinders, (b) the wider the spacing between cylinders and (c) the larger the diameter of the cylinders. Later Law *et al.* (1985) used fluorometric analysis to examine deposition on different segments of cereal leaves, broadleaved weeds under the cereals and the soil. Charging droplets in the 1.5–4.5 mC/kg range increased deposition on all plant surfaces and reduced residues on the soil, but deposition was not uniform and was not improved by increasing air velocity from 2 to 4 m/s. When an external voltage was applied to a cylinder mounted just behind the spray cloud in an attempt to repel charged droplets further into the crop canopy, the electric field of 37 kV/m did not increase deposition significantly, but the gaseous charge exchange between the spray cloud and leaf surfaces was exacerbated via undesirable leaf-tip coronas (Fig. 9.8).

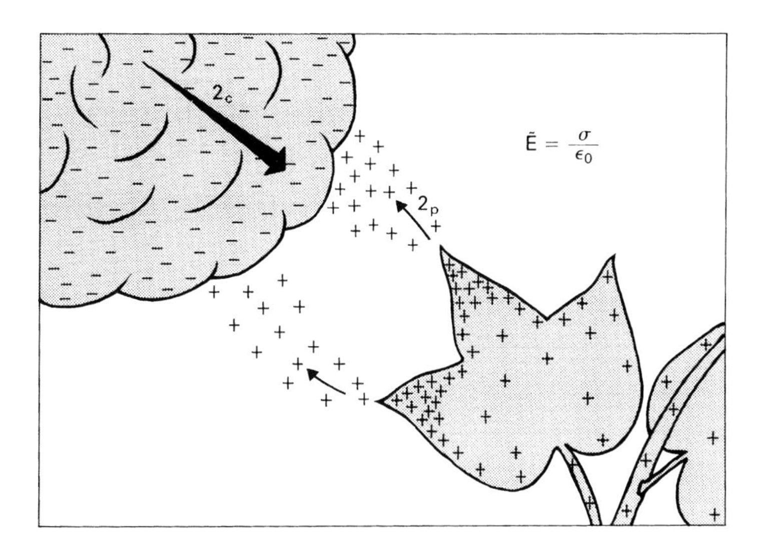

Fig. 9.8 Partial neutralisation of charged spray droplets by leaf-tip ionisation (from Law, 1980).

Lane and Law (1982) examined the level of deposit on cotton plants affected by drought stress, but plant moisture levels did not significantly affect the transient charge-transfer ability of the plants. Improved underleaf coverage was obtained on cotton with an electrostatically charged spray from a spinning disc when compared to an uncharged spray (Cooper *et al.*, 1998). In laboratory experiments Giles *et al.* (1991) improved deposition on earthed spheres by pre-charging a dielectric film underneath them.

Evaporation

As Law (1978) was using aqueous sprays in small droplets, he was concerned that evaporation would affect the charging of the sprays. By study of a 3 mm diameter droplet held within a closed cabinet in which humidity was controlled, Law (1986) concluded that surface charge did not affect vapour movement, and the evaporation liquid did not dissipate the charge. As non-conductive vegetable oils are sometimes added to sprays to reduce the effect of evaporation on droplet size and enhance rainfastness of deposits, Law and Cooper (1987) investigated the use of oil-based sprays through an induction charged nozzle. A combination of formulated vegetable oil with surfactant was suitable for induction charging and a charge-to-mass ratio of 4.1 mC/kg was achieved.

A hand-carried unit of the charged rotary atomiser was used in field trials to investigate spray coverage of cotton and soyabeans (Arnold and Pye, 1980). Spray deposition on these crops was increased with increasing voltage (0–30 kV), but the higher voltage did not improve canopy penetration. These trials indicated that with aqueous sprays the more rapid trajectory of charged droplets overcame, to some extent, the effect of evaporation at high ambient temperatures. However, Lake *et al.* (1980) had calculated that droplets of a given size produced by a hydraulic nozzle are airborne for less time than one in free fall from a horizontal spinning disc, even when the spray is charged.

Air-assisted spraying

Projection of charged sprays into crop canopies using an airstream has been investigated in glasshouses, tree crops and cotton. Abdelbagi and Adams (1987) obtained the most efficient distribution of droplets for whitefly control using 18 μm charged droplets with a small fan providing 2 m^3/min air flow, as abaxial leaf coverage was very good. Improved control of *Aphis gossypii* was achieved with electrostatic charged sprays of *Verticillium lecanii* with more spores deposited on the abaxial leaf surface (Sopp *et al.*, 1989; Sopp and Palmer, 1990). Dai *et al.*, (1992) have also reported that air-assisted charged sprays increased deposition on the undersurface of cotton leaves and within chrysanthemum canopies. Improved deposition on the outer part of apple trees was obtained with an electrostatically charged spinning disc mounted on a knapsack mistblower (Afreh-Nuamah and Matthews, 1987) and tractor mounted mistblowers (Allen *et al.*, 1991). Bjugstad (1994) reported that spray deposits in orchards could be improved by up to 46% using nozzles with induction charging. In China, improved control of cotton bollworms was reported by Shang and Li (1990), using an electrostatically charged mistblower.

Studies with a linear electrohydrodynamic nozzle showed that the addition of air assistance increased canopy penetration of highly charged sprays, and that the potential for increased drift with smaller droplets was decreased by over 90% with air assistance (Western *et al.*, 1994).

Commercial development of electrostatic spraying

Attempts to market electrostatically charged hydraulic nozzles on tractor-mounted boom sprayers or air-assisted orchard sprayers have met with little success due to the higher cost of equipment and the relatively small improvement in spray deposition obtained under field conditions. In the USA, a motorised unit with lance incorporating an air-atomising nozzle based on work by Law (1978) has been promoted, mainly for use in glasshouses (Lehtinen *et al.*, 1989). In the charged mode, bifenthrin gave better control of *Trialeurodes vaporariorum* than an uncharged spray or a high volume (1200 litres/ha) treatment (Adams and Lindquist, 1991; Adams *et al.*, 1991). Improved deposition achieved on strawberries indicated that the dosage of captan could be reduced by 50 per cent (to 1.12 hg a.i./ha) in a charged spray (80 litres/ha) and achieve similar persistence as the full rate of uncharged high volume spray (1870 litres/ha) (Giles and Blewett, 1991). One tractor mounted sprayer in the USA introduced a system of direct charging of spray in a separate small tank between the main spray tank and the nozzles.

For several years, the hand-carried 'Electrodyn' sprayer was used commercially on cotton in Africa and South America (Smith, 1988; Matthews, 1990). However, limitations on the number of pesticides that could be formulated successfully in an oil applied at ULV and the monopoly of supply of 'Bozzles' from one multinational company led to its withdrawal. Furthermore, while it was very effective on small plants, the penetration of the crop canopy of well-grown cotton was inadequate.

'Electrodyn' nozzles have been used to treat young trees placed on a conveyor with insecticide before planting to protect them from *Hylobius* damage. Similarly, potato tubers have been carried under a charged rotary nozzle to ensure distribution over their whole surface.

Some interest in electrostatic sprays has continued, but so far very little of the equipment that has been developed has been commercially acceptable. Perhaps with future legislation demanding reduction in dosages of pesticides, an electrostatic option to certain air-assisted sprays may be developed.

10

Air-assisted sprayers

Interest in the use of air-assisted sprayers has increased with a number of different systems now available. Potts (1958) was one of the earliest to recognise the ability to reduce spray volumes by using a flow of air to project droplets into a crop canopy. Air-assisted sprayers are often used to apply the same quantity of pesticide in one-tenth of the volume normally applied with hydraulic sprayers. They have also been adapted for ULV application by restricting the flow of liquid to the nozzle. They have traditionally been used primarily for treating tree crops, but air-assistance has now been adopted in many other types of crop. Downwardly directed air assists droplet penetration of arable field crop canopies and in reducing downwind drift (Fig. 10.1). However, in contrast, registration authorities are concerned that when spraying orchards, it is the small droplets that can drift above the tree canopy and outside orchards, so buffer zones are wider for this type of application.

Various terms have been used including 'concentrate', 'mistblower' and 'air-carrier' spraying. Hislop (1991) provides a review of air-assisted spraying. Mistblowers should strictly be confined to those sprayers which produce

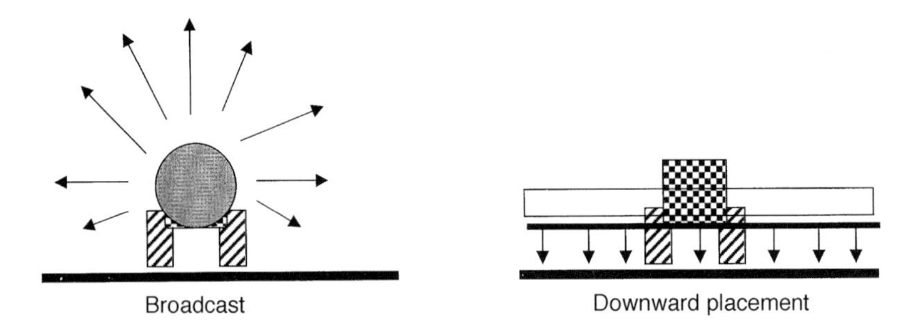

Fig. 10.1 Air-assistance pictograms.

droplets in the 50–100 μm size range as these droplets are most effectively conveyed within an airstream. Thus Potts (1958) found that in a particular airstream, droplets of 60–80 μm diameter were carried 46 m, while the larger 200–400 μm droplets travelled only 6–12 m. Larger droplets will be influenced more by gravity, while smaller droplets are less likely to impact on foliage and other targets because they remain within the airstream. This is particularly important when projecting spray upwards into a tree canopy; fall-out due to gravity can result in considerable wastage of pesticide on the ground as well as increasing the risk of operator exposure to the pesticide. However, with greater concern about spray drift out of orchards, coarser sprays are increasingly selected. The distance that large droplets are transported depends very much on the strength of the air assistance and the initial direction of trajectory of the droplets.

The airstream may be used to break up liquid into droplets, by using an air shear nozzle (Fig. 10.2). Alternatively, some sprayers have internal mix twin-fluid nozzles (see p. 125; Fig. 5.18), or droplets are produced by hydraulic, centrifugal-energy or other types of nozzle mounted in the airstream. When droplet production is independent of air-shear, emphasis can be given to the air volume rather than the air velocity at the nozzle. As pointed out later, the volume of air and turbulence within a crop canopy may be more important than having a high air velocity. Droplet size is affected by the position of the orifice of hydraulic nozzles in relation to the direction of the airstream. Both cone- and fan-type nozzles have been used on air-assisted sprayers. Wide-angle cone nozzles permit very efficient break-up of the spray, but when larger droplets are needed, a narrow cone is used. More recently, air-induction nozzles have been used where the larger droplets penetrate further into the crop canopy. Although the droplets are larger, they contain air inclusions, so the increase in volume to maintain similar droplet numbers is less than with a coarse spray without aerated droplets (Heijne, 2000). The position of the nozzles in relation to the air outlet is important for achieving proper mixing and projection of droplets in the airstream.

There is a similar requirement to other types of sprayers for routine inspection of air-assisted sprayers to ensure that they are operating efficiently (Koch, 1996).

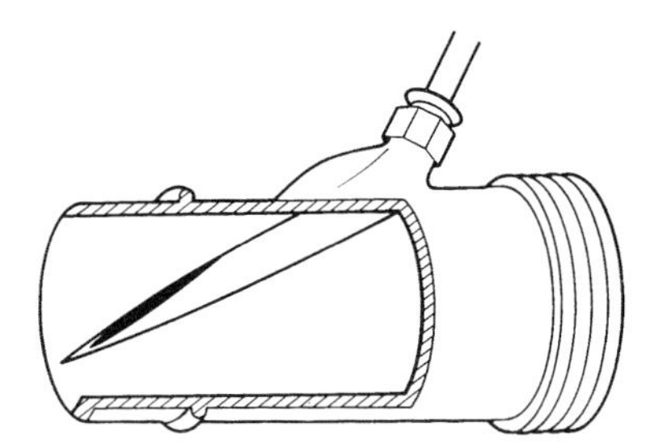

Fig. 10.2 Motorised mistblower showing simplest type of nozzle.

Fans

The central feature of an air-assisted sprayer is the fan unit, although a few sprayers rely on a compressor or rotary blower to provide air to twin-fluid nozzles. Four main types of fan described below are used, namely propellor fans, centrifugal fans, crossflow fans and axial fans. (Fig. 10.3). In choosing a suitable fan, consideration needs to be given to air volume, air velocity and the amount of turbulence created within the crop canopy. The propellor fan is simplest and is most frequently used in conjunction with a centrifugal energy nozzle (Figs 8.9 and 8.10).

An axial fan has blades of 'aerofoil' shape similar to an aeroplane wing with a blunt leading edge and a thin trailing edge. In an axial fan, air is accelerated in the same direction, whereas in a centrifugal fan air is drawn in at the centre and discharged at 90° to its entry. Axial fans are used to move large volumes of air at low pressure and the air velocity is usually insufficient to use with air shear nozzles. The performance of the fan depends on the shape and angle or 'pitch' of the blades in relation to the direction of rotation. Air pressure can be increased, within limits, by increasing the blade pitch or hub diameter, but this reduces the airflow. The clearance between the tip of the blade and the casing is also critical for optimum efficiency.

The centrifugal fan is similar to a centrifugal pump and consists of a wheel with blades rotating in a 'volute' or scroll casing. There are three types of these fans:

(1) those with the tip of the blade curved forwards (i.e. in the direction of rotation to provide a 'scoop' effect)
(2) straight radial blade fans
(3) those with the tip of the blade curved backwards to provide a smoother flow of air.

The forward curved fan is run at a slower speed (r.p.m.) and the backward curved fan faster than a radial blade when moving the same volume of air at the same velocity. The forward curved fan, although it may be less efficient, provides a higher velocity for a given rotational speed and is the most commonly used type. Centrifugal fans are used on knapsack mistblowers as well as some types of tractor-mounted equipment.

The crossflow fan has been used on sprayers designed for spraying blackcurrants and deciduous fruit. An impeller has long blades in the axial direction, similar to those on a forward curved centrifugal fan. Air entering on one side is accelerated out of the opposite side. Crossflow fans are less efficient than axial or centrifugal fans and operate at lower pressures. The length of the fan is limited, because an unsupported drive shaft will be prone to whirling and other out-of-balance effects at high speeds (Miller and Hobson, 1991). This linear fan can be driven by a small hydraulic motor, and thus a series of them can be positioned around a crop canopy to project spray into foliage from nozzles mounted in the airstream.

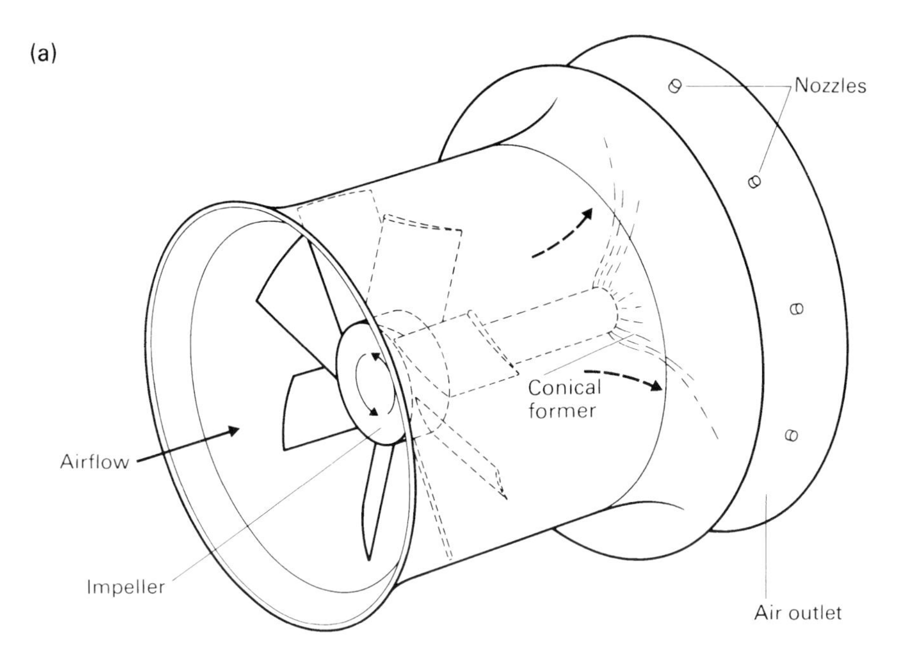

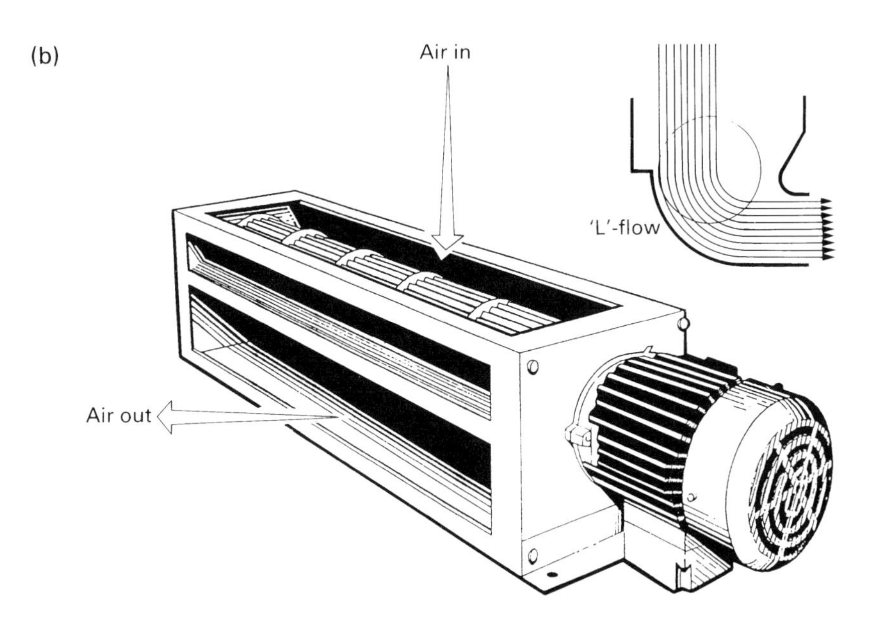

Fig. 10.3 (a) Axial fan; (b) crossflow fan.

The rate of flow (m^3/s) varies directly with rotational speed with each particular size and type of fan. Similarly, the air pressure developed varies with the square of the speed of rotation, and the power absorbed in relation to the cube of the speed. When fans of different size, but geometrically similar, operate at a particular rotational speed, then the rate of flow varies with the cube of the size, the pressure with the square of the size and the power absorbed with the fifth power of the size. Thus, generally an increase in fan diameter rather than fan speed is a more efficient way of increasing rate of flow. The fan is rotated either by a belt drive from the p.t.o. shaft or it is mounted directly onto the shaft of a separate motor.

Ideally, the airstream from a fan should continue in the same direction for at least two diameters of the impeller before any bend. Unfortunately, on most sprayers, vanes or a 90° elbow are positioned much closer, thus causing pressure losses before the air is discharged. For spraying tall trees, the outlet of the fan on a knapsack mistblower should be vertical rather than horizontal (MacFarlane and Matthews, 1978). When air is discharged into the atmosphere it loses velocity owing to friction with the atmosphere, and also entrains some air with the jet. Air velocity decreases from the fan outlet, depending on its initial velocity and the area and shape of the outlet. When a slot outlet is used, the equivalent round outlet diameter is determined from the equation

$$D = 1.3 \times W + \frac{\sqrt{L}}{4}$$

where D is the diameter of the round outlet, W is the width of the slot and L is the length of the slot.

The decrease in axial velocity of a circular low-velocity air jet with distance is illustrated in Fig. 10.4, which shows a decrease to 40 per cent of the initial velocity at 20 diameters and to 10 per cent at 90 diameters; thus, if the initial velocity from a 5 cm diameter nozzle is 50 m/s, then at 200 cm the velocity has decreased to 10 m/s. In practice, lower velocities are usually recorded under field conditions (Potts and German, 1950). The discharge tube should have the largest circular opening to achieve maximum throw of droplets, but there is an optimum diameter for a given air capacity and air pressure. The velocity field with contours of equal velocity from an air jet is illustrated in Fig. 10.5. At the mouth of the discharge tube, a turbulent mixing region surrounds the air core at the initial velocity, but at about five diameters this air core disappears.

When spraying tall trees, it is better to establish a column of air moving up into the canopy, then spray briefly and continue the flow of air to carry the droplets up to the target. Without an airstream, the droplets may drift and fail to reach the target, and the larger droplets are liable to fall out to ground level. On some sprayers, it is possible to extend the air delivery tube to a greater height before releasing the spray. A pump is needed to get the spray liquid to the elevated nozzle. Air velocity is also affected by ambient temperature, humidity, wind speed and its direction in relation to the blower, and thus the speed of travel of the sprayer.

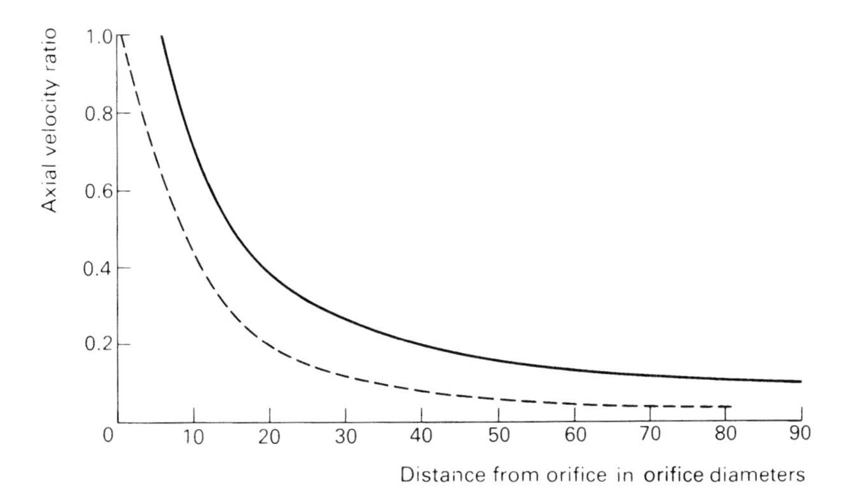

Fig. 10.4 Axial velocity of an airjet with distance. Theoretical velocity (solid line) data from Potts and German (1950) – obtained with sprayers of different diameters and air velocities (broken line) (after Fraser, 1958).

Air velocity can be measured with an anemometer or pitot tube (Fig. 10.6). Air velocity can be important when projecting spray up into trees, but displacement of the air within a crop canopy by air containing droplets is usually more important, so sprayers that deliver large volumes of air, or at least match the volume of air with the volume of the tree, are generally more suitable than those with a low volume of air at high velocity.

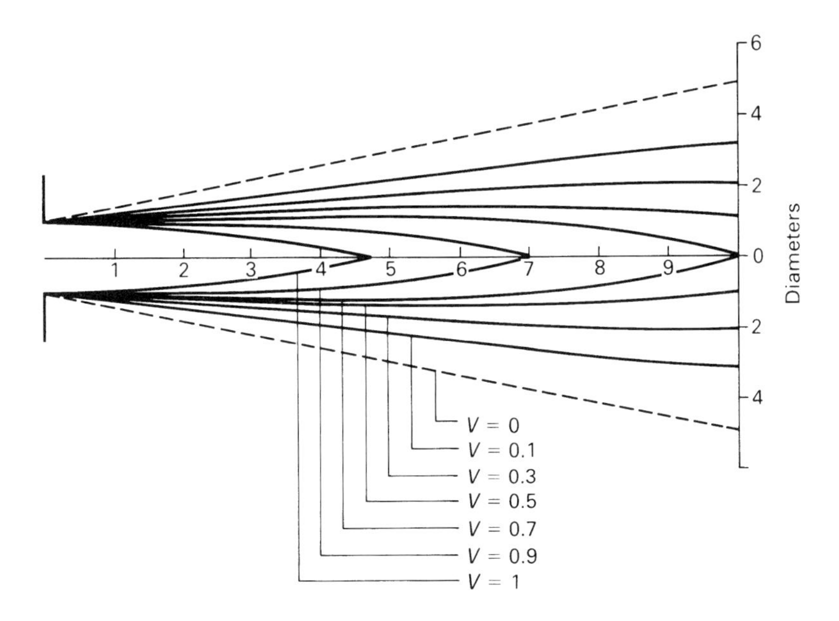

Fig. 10.5 Velocity field of a symmetrical airjet (after Fraser, 1958).

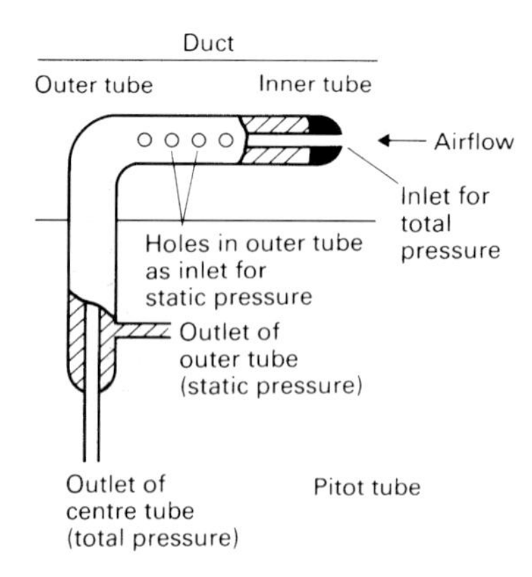

Fig. 10.6 Measuring air velocity from a mistblower.

The volume of air can be calculated from the equation

$$Q_a = VA$$

where Q_a is the volume of air, and V is the velocity of air at the end of the discharge tube which has an area A. In practice, the different velocities recorded across the area have to be integrated. In the laboratory, the volume of air moving through a duct can be calculated more accurately by measuring the differential pressure across two orifices, partially separated by a sharp-edged plate, and mounted in a smooth-bore pipe so that the upstream pipe is $20 \times$ pipe diameter and downstream $5 \times$ pipe diameter (Fig. 10.7). Another method is to deliver the volume of air into an enclosed space and, while maintaining no pressure change, measure the volume of air expelled using a previously calibrated standard fan.

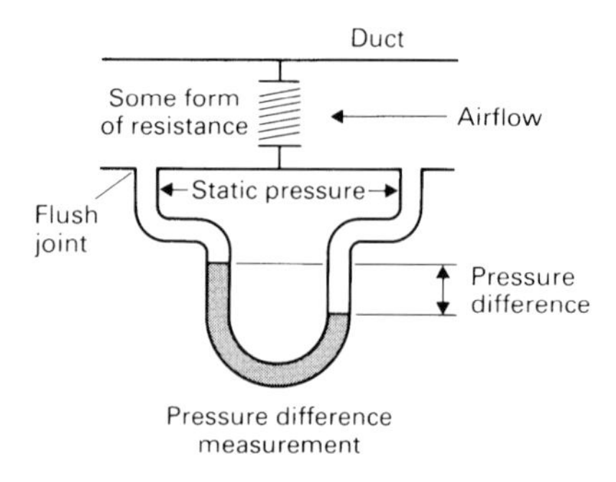

Fig. 10.7 Measuring air volume from a mistblower.

Pumps

Low liquid pressures are usually sufficient to feed spray to the nozzles, so any of the pumps described in Chapter 7 can also be used as well as peristaltic pumps. As a high speed drive is available, simple centrifugal pumps are suitable, but diaphragm and piston pumps are frequently used. On some sprayers, spray is fed into the airstream by gravity; on others air from the fan is used to pressurise the spray tank, in which case the lid of the tank must be airtight.

Motorised knapsack mistblowers

Portable air-assisted sprayers, invariably referred to as knapsack mistblowers, were developed initially to treat cocoa crops, but are used on a wide variety of crops and also in vector control. They have a lightweight two-stroke engine with a direct drive to a centrifugal fan, which is usually mounted vertically and attached by antivibration mountings to a strong L-shaped frame. The lighter versions have a 35 cm^3 engine, while the larger versions with a 60–70 cm^3 engine have a more powerful fan. The latter are more suitable when spray has to be projected up into tall trees. The frame is designed to allow the sprayer to stand upright on a horizontal surface. The spray tank is mounted above the engine/fan unit and normally has a capacity of 10 litres. A large opening facilitates filling, and this should have a large capacity filter with a fine mesh to prevent nozzle blockages. An on/off tap is fitted in the spray line, but unfortunately none of the machines has a trigger valve to facilitate intermittent spraying of individual targets. In recent designs, the controls for engine speed are now mounted in front of the operator rather than on the L-shaped frame, and include an on/off switch to shut off the engine when necessary (Fig. 10.8). The basic weight of a knapsack mistblower is often as much as 14 kg when empty so they are much heavier to carry than other types of knapsack sprayer. The straps are usually provided with a non-absorbent pad over the shoulder and a padded backrest to improve operator comfort and reduce the effect of engine vibration.

An alternative design has a propellor fan in the front of which is mounted a spinning disc nozzle. This sprayer, which was developed initially for treating coffee (Fig. 8.10), directs the spray behind the operator and is thus safer than when the operator walks into treated foliage. Low volumes of 30–70 litres/ha can be applied (Povey *et al.*, 1996).

Two-stroke engine

A brief description of the engine (Fig. 10.9) is provided, as its maintainence is essential when using this type of equipment. When the piston is moving up the cylinder to compress the fuel/air mixture, the inlet and outlet ports are covered

Fig. 10.8 Knapsack motorised mistblower.

initially but then, as the piston continues to travel upwards, it creates a partial vacuum and uncovers the inlet port. This vacuum causes a depression in the carburettor inlet and air passing over the fuel jet collects a metered quantity of the oil + petrol fuel mixture. This fuel/air mixture is mixed and drawn through the inlet port into the crankcase. Meanwhile, the previous charge of fuel/air mixture is compressed in the combustion chamber and ignition occurs before the piston has ascended to the top of the cylinder. The momentum of the piston carries it over top dead centre and the expansion of the burning gases

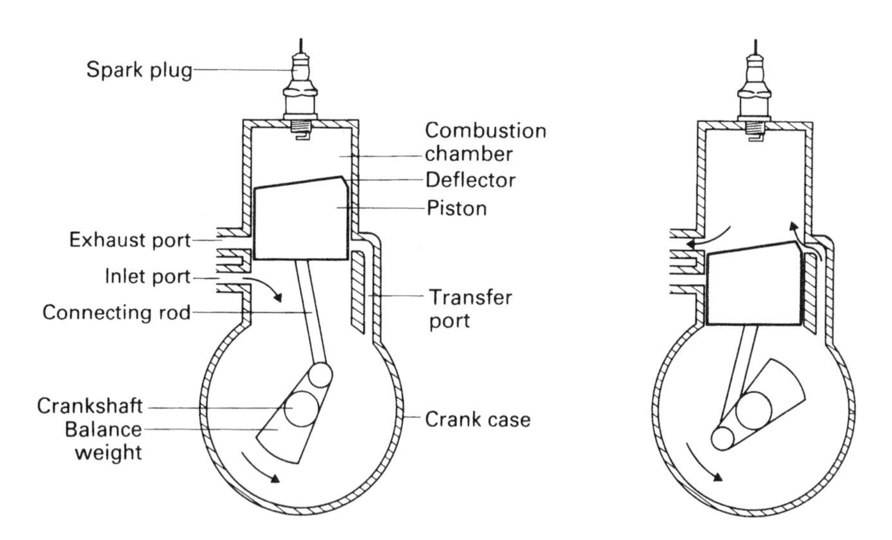

Fig. 10.9 Diagram showing operation of a two-stroke engine.

provides the power stroke, the downward movement of the piston. After a short distance, the exhaust port is uncovered and burnt gases escape. As the piston moves down, the fuel/air mix in the crankcase is compressed and when the transfer port is opened, it is forced into the combustion chamber, ready to be compressed by the next upward stroke of the piston.

The fuel for the two-stroke engine is a mixture of oil and petrol, usually in the ratio of 1:24. The correct mixture should be indicated clearly on the fuel tank or its cap. The most suitable oil is 30 SAE. Multigrade oil should never be used, because the additives it contains may cause engine failure. Similarly only lead-free petrol should be used.

The fuel is usually fed by gravity to the float chamber of the carburettor through a needle valve (Fig. 10.10). The float is designed to maintain the required level of fuel in the float chamber, but when starting the engine a tickler knob can be used to 'flood' the carburettor to provide a richer fuel/air mixture. Air is drawn through the filter which should be cleaned regularly to prevent dust and grit entering the engine. The flow of air is speeded up by a narrowing of the tube, known as a venturi. The increase in speed causes a decrease in air pressure which draws in fuel through a jet. A throttle valve controls the volume of fuel/air mixture entering the combustion chamber, and hence the speed and power of the engine. The throttle is operated by means of a flexible cable (Bowden cable) connected to a lever that is easily accesible to the operator. Often the throttle lever is placed behind the operator which makes it difficult to locate. A choke or 'strangler' restricts the flow of air through the venturi and is used to enrich the fuel mixture when starting the engine.

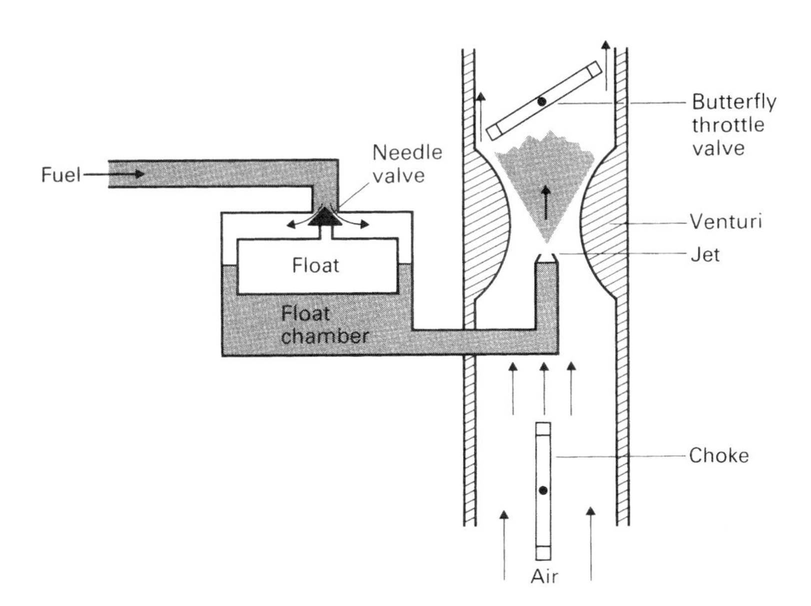

Fig. 10.10 Principles of a simple carburettor.

Ideally, fuel should be drained from the tank and carburettor when the sprayer is being stored, especially in hot climates, otherwise petrol may evaporate, affecting the petrol/oil ratio. Oil deposits in the carburettor may make it difficult to start the engine. If it is necessary to stop the engine in the field, even for short periods, this should be done by closing the fuel valve rather than by shorting the electrical circuit. The engine is usually easier to start if the carburettor has been left dry. Some of the new machines have an electric switch which is useful if there is an emergency requiring an instant stopping of the engine.

The engine is usually provided with a recoil starter but, when a pulley wheel is provided as part of the starter, the engine can also be started by using a rope or strap. The starter mechanism should be fully covered by a cap while the engine is running to prevent the operator touching a moving part. Electronic ignition is now provided on some engines.

Nozzle on mistblowers

Air from the fan is directed through a 90° bend through a flexible hose to a rigid duct on which the nozzle is mounted. On the majority of these machines, the high velocity airstream is used to shear the spray liquid into droplets (Fig. 10.2). The flow of liquid is controlled by a variable or fixed restrictor and then fed through a small tube into the airstream. A fixed restrictor is preferred so the user is unable to alter the flow rate in the field (Jollands, 1991). If too high a flow rate is used, the droplet size tends to be larger so more of the pesticide is wasted as these larger droplets do not remain entrained in the air projected from the nozzle. More uniform droplet size is obtained if the liquid is spread more thinly over a flat surface mounted in the airstream. On some mistblowers there is a fixed disc, while others have a spinning disc (Fig. 10.11) (Hewitt, 1991), one design of which has been shown to provide a narrower droplet spectrum (Bateman and Alves, 2000).

When the nozzle is held above the level of liquid in the spray tank, spraying will cease unless the lid on the tank has been fitted properly and the tank is slightly pressurised (0.2 bar) with air from the fan. On some machines there is a small pump fitted to the engine shaft. This is particularly important if the nozzle is positioned on an extended delivery tube to gain extra height for the spray. Unfortunately these pumps are not very durable in the field.

Assessment of knapsack mistblowers

The performance of different mistblowers can vary significantly despite the use of the same basic design (Table 10.1). As they were designed primarily for projecting spray upwards, the vertical throw of droplets should be examined. This can be done by fixing sample cards, such as water sensitive paper, horizontally usually at 30 cm intervals to a rope that can be raised over a pulley attached to a tower. The highest target card should be at least 12 m and the lowest 4 m above the ground. Each target should have an upper and lower

(a)

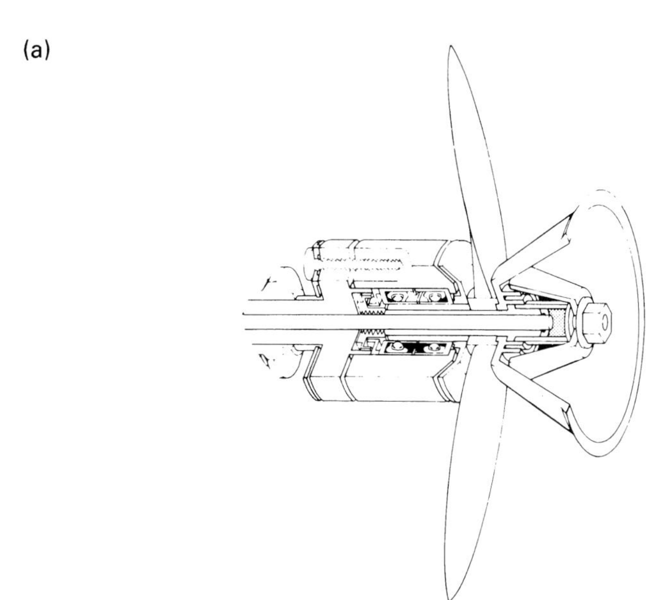

(b)

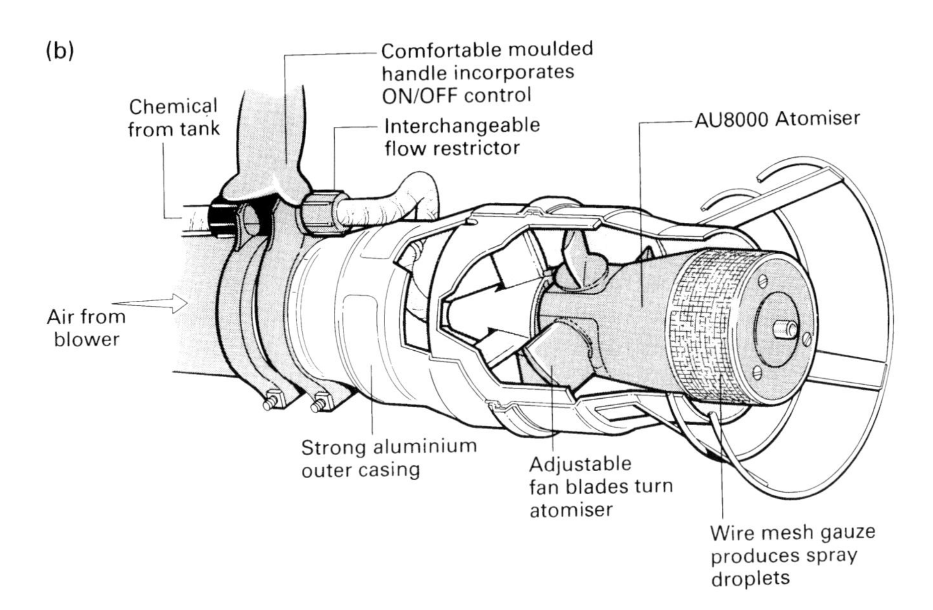

Fig. 10.11 (a) Spinning disc (Micron X-1); (b) Micronair AU8000 mounted on a knapsack mistblower.

surface to sample droplet density. The sprayer is operated using water so that the nozzle is held at an angle 1.5 m above the ground and 3 m from the rope. Spray is directed upwards at the targets for a brief known period with the minimum interference from natural air movements.

Table 10.1 Comparison of the performance of two knapsack mistblowers (adapted from Clayphon, 1971)

	Mistblower A	Mistblower B
Engine capacity (cm^3)	35	70
Fuel tank capacity (litres)	1.25	1.5
Fuel consumption (litres/ha)	0.9	1.6
Air velocity at nozzle (m/s)	66	74.6
Air volume		
at fan outlet (m^3/min)	7.9	14.7
at nozzle (m^3/min)	3.2	8.2
Flow rate (litres/min)	0.7–1.8	0.04–2.8
Horizontal throw[a] (m)	13.7	16.8
Vertical throw[a] (m)	6.1	9.75

[a] Measured at maximum flow rate.

In practice, many mistblowers are used to project spray horizontally over field crops. Horizontal throw can be determined in a similar manner, using water sensitive cards attached to the front and back surfaces of an array of stakes. A typical layout has ten rows, each with seven stakes placed at 1.5 m intervals between rows and 0.75 m intervals within rows. The first row is 3 m from the nozzle, which is directed down the centre line of the target layout when spraying for 5 s. The width of the airstream is indicated by the spread of the spray across the array of targets.

Using a knapsack mistblower

As with any equipment, it is important that the mistblower is calibrated before use. The correct petrol/oil mixture is poured through a fine-mesh filter into the fuel tank. Some water is put into the spray tank through the filter and the tank lid replaced tightly. Any on/off switch is turned on, the petrol tap opened and the carburettor allowed to fill with fuel. The choke lever is moved to the closed position and, with the throttle closed, the engine is started by pulling the recoil starter evenly. The starter rope should be allowed to rewind slowly and not released to snap back. When the engine starts, the choke can be moved to the open position and the throttle opened up to allow the engine to run at full throttle. Engine speed can be checked using a tachometer (Fig. 10.12). The engine should never be allowed to idle at slow speeds. To calibrate the equipment allow the small volume of water in the spray tank to be sprayed and stop the machine as soon as spray liquid has been used. While it is spraying, check visually for any leaks or other problems. A known volume of water is then put into the spray tank (sufficient for at least 1 minute of spraying) and on restarting the engine, the time taken to spray the known volume is measured using a stopwatch. The volume application can be calculated if the swath (track separation) is known and the walking speed of the operator has been

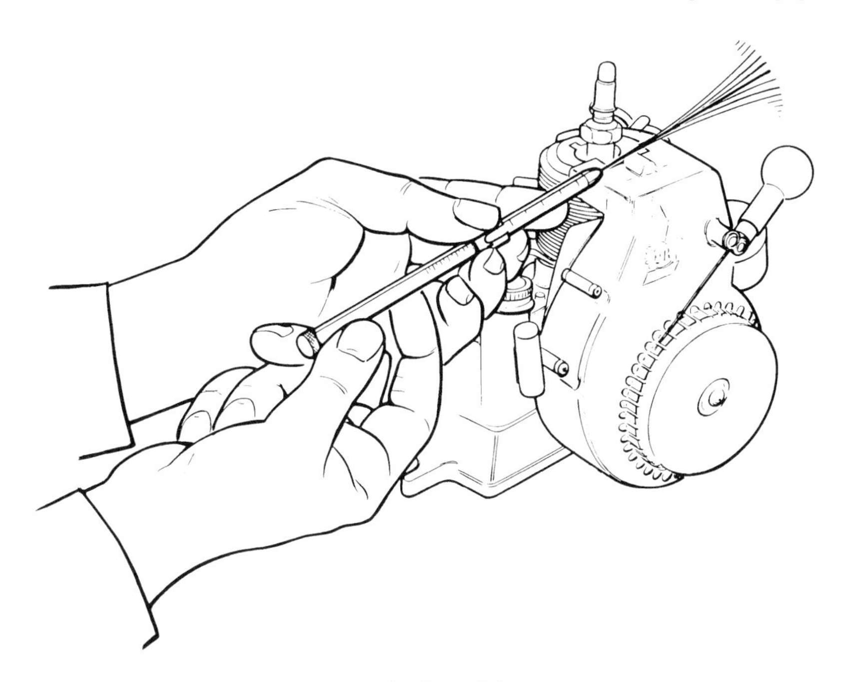

Fig. 10.12 Vibrating wire tachometer ('Vibratak').

measured. This calibration should be repeated to check that the volume application rate is consistent.

Once the calibration has been completed, the spray tank can then be filled with the pesticide liquid, and with maximum engine speed the nozzle is directed downwind so that any natural air movements assist dispersal of the droplets away from the operator. If the nozzle is pointed upwind, droplets are liable to be blown back onto the operator. The discharge tube should be held at least 2 m from the target to allow dispersal of the droplets, as the air velocity close to the nozzle may exceed 80 m/s. Operators should walk at an even pace through the crop and close the spray liquid tap whenever they stop to avoid overdosing part of the crop.

Knapsack mistblowers can be adapted to apply dry formulations as indicated in Chapter 12.

Tractor-operated equipment

Equipment is designed for use on arable or orchard type crops. Some air-assisted sprayers are used in glasshouses.

Arable crop sprayers – 'downwardly directed air assistance on boom sprayers'

The air movement caused by the forward speed of the standard tractor-mounted boom sprayer can significantly affect the subsequent dispersal of spray from hydraulic nozzles. This is especially evident as farmers increase the speed over larger, relatively flat fields. Concern about the smallest droplets being caught up in vortices and drifting downwind were soon recognised, and early attempts to reduce the proportion of downwind spray drift led to the covering of the boom (Edwards and Ripper, 1953). Later the boom design was modified with an aerofoil so that the forward movement of the sprayer (8–12 km/h) directed air downwards to reduce trailing vortices behind nozzles (Jegatheeswaran, 1978; Goehlich, 1979; Lake *et al.*, 1982; Rogers and Ford, 1985). Unfortunately, at the higher forward speeds at which the aerofoil performance is improved, there are problems of controlling the spray output while accelerating and slowing down at the field edges (see also Chapter 7).

Equipment with a downwardly directed air curtain (Fig. 10.1) has been used to increase penetration of droplets into crop canopies and reduce spray drift (Taylor *et al.*, 1989; Cooke *et al.*, 1990; Hadar, 1991; Taylor and Andersen, 1991). An axial fan delivers a very large volume of air through an inflatable sleeve mounted above the boom and nozzles (Fig. 10.13). These sprayers have proved to be popular on many arable farms, although the penetration of a crop canopy is better with cereals than in broadleaved crops, such as cotton. Wind tunnel studies with trays of plants have confirmed that when finer sprays were angled forwards with air assistance, total deposition of sprays on cereals increased and soil contamination was reduced (Hislop *et al.*, 1995). Nordbo (1992) reported less variability and enhanced deposition with air assistance, providing some scope for reducing spray volumes. Taylor *et al.* (1989) suggested that the reduction in spray drift permits the use of nozzles with a finer spray or allows a faster forward speed. However, the air curtain can increase drift in the absence of crop foliage on which the droplets can impact, due to the deflection of air by the ground.

Lack of penetration to provide deposition on the undersides of leaves in the lower canopy of cotton has led to several different designs based on air assisted droplegs. In one design of vertical booms somewhat similar to the tractor-mounted tailbooms (see p. 170), air is mixed with the spray liquid and the mixture fed to large orifice fan nozzles through which more air is projected. This system developed in Uzbekistan allows low volume applications to be made and creates a fine spray. In Israel, Manor (1999) used vanes within a thin vertical duct to create turbulent air within the canopy into which the spray is released from cone nozzles. Gan-Mor *et al.* (2000) and Grinstein *et al.* (2000) had problems with the passage of larger ducts; they therefore shortened them and allowed some of the air to be directed downwards to the soil so that on rebound it carried spray droplets to the undersurface of the lower leaves. Under some dry conditions, soil is also thrown up on the leaves. The Benest nozzle was designed particularly for treating potatoes with fungicides, but provides a thin dropleg with a twin fluid nozzle suitable for other row crops (Fig. 10.14).

Some farmers have also used twin-fluid nozzles, especially as these nozzles can be adjusted to provide a coarser spray, and thus reduce the drift potential and increase the number of days on which a spray may be applied for optimal timing of a pesticide (May, 1991; Nettleton, 1991).

(a)

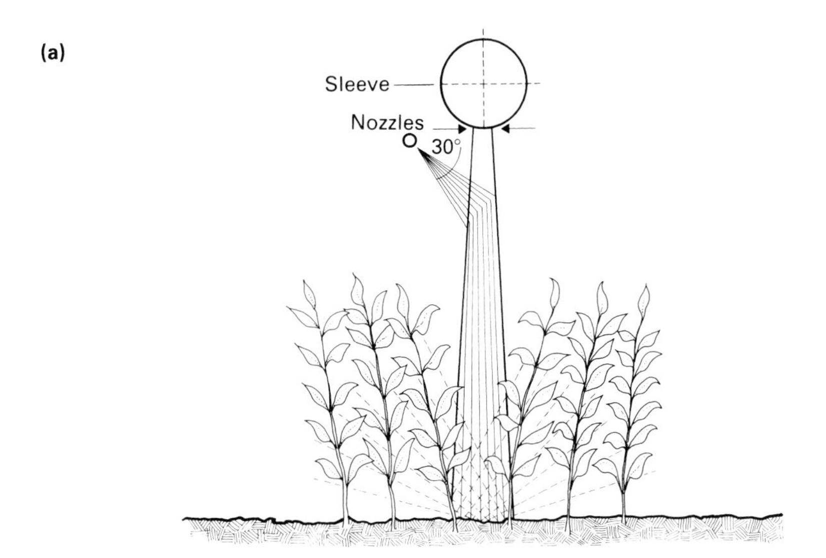

(b)

Fig. 10.13 (a) Diagram to show relative position of airflow from airsleeve and spray from nozzle. (b) Air-sleeve sprayer.

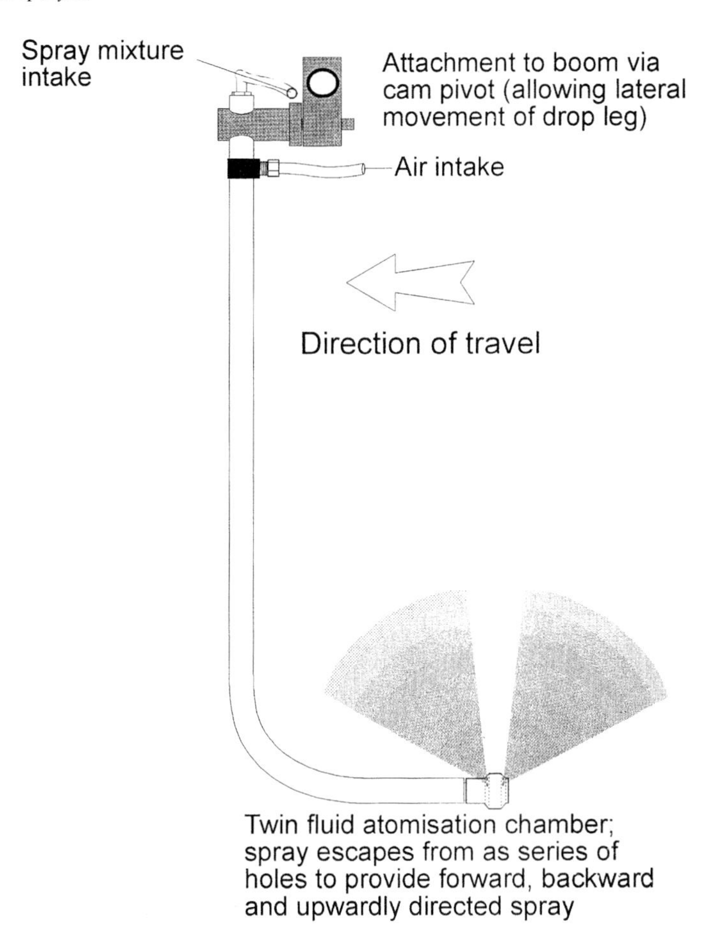

Fig. 10.14 Benest twin-fluid nozzle.

Orchard sprayers

A wide range of equipment is used to treat tree and bush crops (Fig. 10.15). On the majority of these the spray is produced by using hydraulic nozzles. On some sprayers with centrifugal fans, air shear nozzles are used, while rotary nozzles are more frequently mounted in front of propellor fans. The four basic types of orchard sprayer are

(1) the airstream from an axial fan is deflected through 90°, and a series of nozzles are mounted close to the outlet
(2) the airstream is provided by one or more centrifugal fans
(3) the airstream is provided from a crossflow fan, and is particularly suited for low, trellis or spindle pruned trees
(4) a small propellor fan has been used on equipment designed as an alternative to the knapsack mistblower (see above).

Modifications of these basic designs have been made to adjust for tree canopy size and shape, and especially to minimise downwind drift. According to Göhlich *et al.* (1996) air-assisted orchard and vineyard sprayers most commonly available were still those with an axial fan, but the projection of spray upwards to reach the top of a tall tree inevitably results in spray above the tree canopy, and this spray can be transported over long distances by the wind (Planas and Pons, 1991). Thus many of these have been modified by different ducting (Fig. 10.15b). Comparatively few sprayers use a centrifugal fan or a crossflow fan. Where a centrifugal fan type air-assisted mistblower has been used to treat wide swaths, spray deposition is generally greater close to the air outlet, thus Parkin *et al.* (1992) reported 90 per cent of the spray with certain sprayers was within 10 m of the vehicle track.

According to Doruchowski *et al.* (1996a) who compared sprayers with an axial fan, crossflow fan and a centrifugal fan with ducts, an increase in air velocity reduced losses to the ground, but more airborne spray was lost and spray deposition was not necessarily improved. Converging air jets used with crossflow fans gave some improvement in uniformity of the spray distribution of spray (Svensson, 1994). Using computational fluid dynamics, Walklate and Weiner (1994) modelled the relationship between penetration of the crop, crop structure, and sprayer variables, but as most orchard sprayers have a fixed air outlet and air velocity, users are limited in the adjustments that can be made. Apart from changing nozzles, forward speed is the main variable that can be changed. Care is needed to avoid blowing leaves together as this will restrict penetration, so a design that improves turbulent air flow within a crop canopy may be more advantagous.

The traditional tractor-powered mistblower has an axial fan, and spray is blown in an arc around the sides and top of the fan outlet. Much of the energy from an axial fan is lost when the air is deflected by vanes through 90° to aim spray at the trees. When studying the distribution of spray on large apple trees, Randall (1971) concluded that the optimum performance required a volume of 13.4 m^3/s at an outlet velocity of 31 m/s. Uniformity of the deposits was improved if the forward speed of the tractor was as slow as economically possible (i.e. 2.75 km/h was better than 6.5 km/h), but the actual speed will depend on wind conditions and the type of plantation. On each side of the sprayer there may be up to ten hydraulic nozzles, but often only five may be fitted. These may have a hollow cone spray pattern, but recent studies have included air inclusion nozzles. A valve may be fitted to separate spray lines on each side of the fan, but on some machines a valve on each nozzle enables specific nozzles to be shut off if necessary. Some users have fitted rotary atomisers or air-shear nozzles to reduce the volume applied, but machines with air-shear nozzles require a high velocity air jet (Hislop, 1991).

Since the 1970s, changes in orchard management have led to shorter trees, often grown along trellises, so the standard axial fan sprayer is no longer very suitable (Cross, 1991). Assessments of spray deposition and drift have been reported by Walklate *et al.* (1998, 2000a). In some countries, the ducting above the axial fan has been modified to release the air at a greater height but in a

Fig. 10.15 Orchard sprayers: (a) with axial fan; (b) with ducting over axial fan; (c) with crossflow fan; (d) with centrifugal fan; (e) with Turbocoll system.

lateral direction to minimise spray going above the tops of the trees. Where trees are in a single row rather than multiple row beds, the entire canopy can be enclosed by a mobile tunnel (Fig. 10.16) (Matthews *et al.*, 1992). This idea was initiated by Morgan (1981) and has been developed to reduce emissions into the environment (Huijmans *et al.*, 1993; Heijne *et al.*, 1993; Van de Werken, 1991; Siegfried and Holliger, 1996).

Fig. 10.16 Tunnel sprayer.

A number of different designs have been tried to improve the distribution of spray on the crop, yet avoid spray escaping from the rear of the tunnel. In one design incorporating a 'closed loop system' air with droplets is drawn from the rear of the tunnel and is blown out near the front. Spray that goes through the canopy and is deposited on the other side of the tunnel can run down into a gutter where it is collected and recycled. Less spray volume is recycled when trees develop a full canopy, but Holownicki *et al.* (1997) measured an average of 30% of the spray being recycled over the whole season. Apart from the number of leaves and their size, the amount of spray retained on individual leaf surfaces declined during the season, as leaf hair density decreased (Hall *et. al.*, 1997).

Although other nozzle designs have been tried, many tunnel sprayers have a vertical boom fitted with hydraulic nozzles. When the air flow is directed 40° upwards, deposition was improved (Holownicki *et al.*, 1996a) and this protected apples from scab, even with reduced dosages, although untreated trees had 90% of their leaves infected (Holownicki *et al.*, 1996b). With a tunnel sprayer Cross and Berrie (1995) obtained more efficient mildew and scab control by increasing the spray volume from 50–200 litres/ha with an approximately constant droplet size of 140 μm. As pointed out earlier, the crucial requirement is sufficient air to displace still air within the tree by air carrying spray droplets and with sufficient momentum to get the leaves to move and assist deposition.

Tunnel sprayers require relatively flat land and are more expensive than other types of sprayers, so their uptake has been relatively slow. Also, they are not suitable where hail nets are used to protect trees. Some manufacturers have attempted to make a cheaper version of a tunnel sprayer; the 'reflection' sprayer has a shield to reflect air and spray droplets back into the crop and collect spray that impacts on the shield to recycle it. This has not been very satisfactory as the row may need to be treated twice to ensure both sides of the canopy get sprayed (Göhlich *et al.*, 1996).

Some tractor mounted mistblowers have a centrifugal fan which delivers air at high velocity through a series of ducts. Air shear or hydraulic nozzles are mounted at the exit of the ducts which can be positioned at different heights and angles to direct spray at specific sites of the crop canopy (Fig. 10.15d). Where crossflow fans have been tried, they are generally mounted with hydraulic nozzles close to the crop canopy (Raisigl *et al.*, 1991). An alternative to this has been the 'Turbocoll' system (Fig. 10.15e) developed in France which uses a venturi system to entrain more air projected at the crop (Fig. 10.17) (Morel, 1997).

In Australia, the use of several propellor fan units with a rotary atomiser has

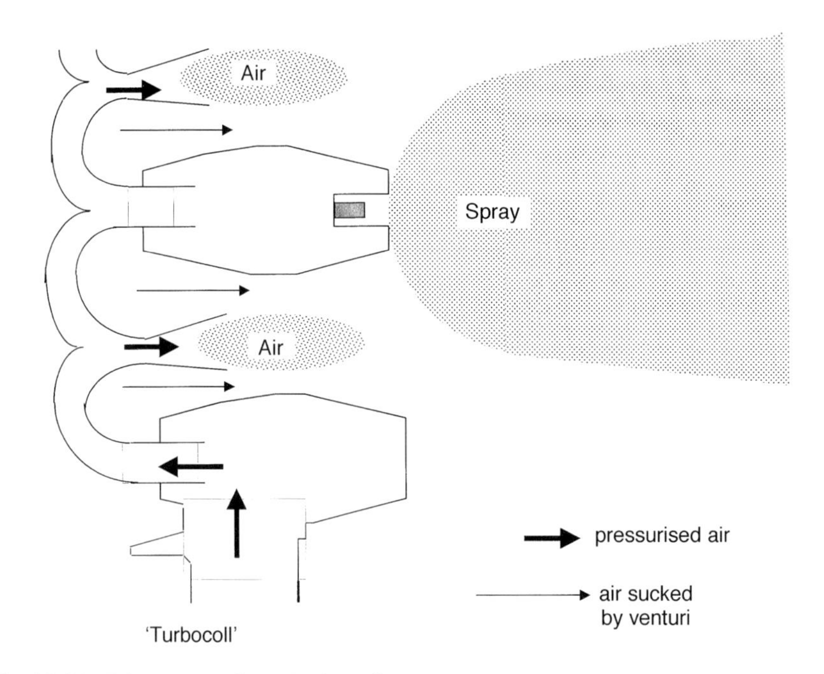

Fig. 10.17 Diagram to show Turbocoll.

been tried. Furness (1996) made changes to the fan design to achieve a large air volume of very turbulent air at relatively low velocity, which has greatly improved spray coverage. Hollow cone nozzles are mounted behind four axial fans, so that droplets are sheared by the air flow across the fan blades into a very fine spray (Furness *et al.*, 1997).

In some countries, an assessment of the vertical spray distribution has been made with a special patternator (Kaul *et al.*, 1996), but Koch (1996) points out the results of such tests are not appropriate for specific adjustment of a sprayer for a particular crop. In practice, farmers normally select the output of the nozzles with reference to the changes in canopy with tree height. As pointed out earlier, uniformity of deposit was considered better if the forward speed of the tractor was as slow as economically possible, because this exploited the air velocity to push spray deeper into the tree canopy. However, a higher forward speed which is now possible with smaller, narrower trees, reduces the proportion of the spray plume above the crop and reduces drift. Vercruysse *et al.* (1999a) have reported assessments of spray drift from a conventional axial fan sprayer up to 40 m downwind from an orchard with semidwarf trees.

Most growers with orchards have decreased spray volumes from >2000 to <600 litres/ha. Several systems of adjusting the volume have been advocated. One version is the 'tree-row-volume' (TRV) concept (Fig. 10.18) (Sutton and Unrath, 1984; Ras, 1986). The following is an example of one method of using the TRV system:

'Crown' height × width of tree at 1/2 crown height × length of row = Air volume to treat

For example, 2 m × 1 m × 10 000 m = 20 000 m^3. If the speed of travel is 6 km/h, the sprayer will take 10 000/6000 = 1.67 h to pass the length of row, thus the fan must deliver a minimum of 20 000/1.67 = 12 000 m^3 of air per hour. This can be expressed in terms of volume of trees to be treated. If the trees are in rows 4 m apart, then 2 m tall trees × 1 m wide foliage × 10 000 m^2/4 m = 5000 m^3/ha. Recommendations on the amount of liquid needed to achieve adequate coverage without run-off have varied between 10 and 100 litres per 1000 m^3 of foliage, so if 20 litres is selected for the above example, the required coverage will be 5000 × 20/1000 = 100 litres/ha.

In practice, farmers may not follow this system, but will close off individual nozzles depending on their perception of the need to adjust for different tree canopies.

Unit canopy row (UCR), proposed by Furness *et al.* (1998) is based on a volume per 100 m^3 of foliage (1 m high × 1 m wide × 100 m along row), omitting the inter-row spacing, but making adjustments for canopy density, type of foliage and sprayer being used.

Cross *et al.* (1998) considered that the TRV system is too simplistic, because it does not recognise the need to adjust air output, forward speed and other parameters, especially spray quality. A fine spray with a VMD of 100–150 μm has been used to achieve good coverage but, as mentioned above, some growers now use air induction nozzles on orchard sprayers to reduce spray

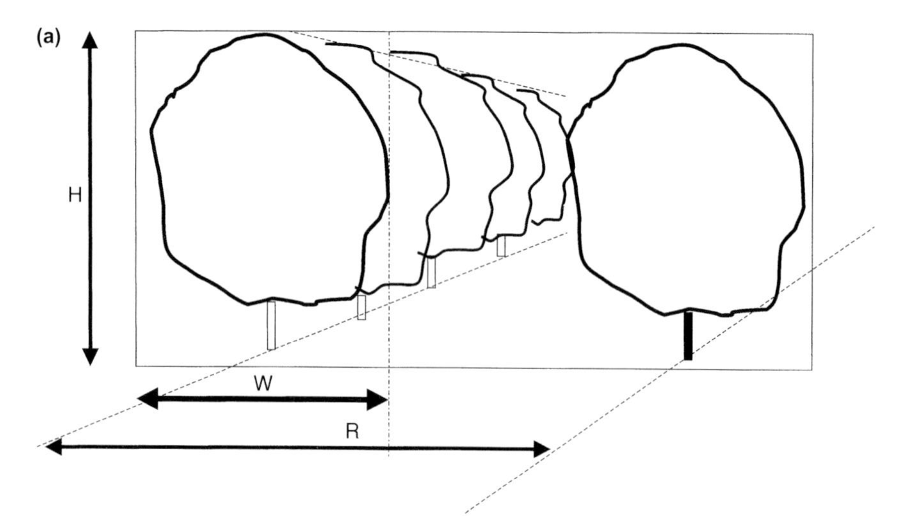

(b)

Fig. 10.18 (a) measurements of crop to calibrate sprayer application rate. (b) LIDAR equipment to assess tree canopy.

drift. The success of the coarser spray will depend very much on the extent to which the chemical is redistributed from the larger droplets, and the use of an appropriate adjuvant to increase spreading and rainfastness may be necessary. The volume of spray applied per 100 metre row is used in Norway without reference to row spacing. Bjugstad (1994) reported that 10 litres per 100 metres gave the best deposit on trees before blossom, after which the volume has to be increased. Doruchowski *et al.* (1996b) were able to reduce the amount of pesticide applied by decreasing the spray volume, because this increased deposition of the active ingredient and reduced run-off from small trees. Cross (1988) also reduced dosage and spray volume on fruit trees, but application of calcium chloride at 1/10th dosage did not increase the calcium content of fruit (Cross and Berrie, 1990). Where use of reduced volumes has been successful in orchards, it has required a higher level of management.

Further research is examining the crop structure using a LIDAR system (Fig. 10.18b) so that the effects of operational adjustments of an axial fan sprayer can be investigated on different sized trees (Richardson *et al.*, 2000).

At present in the UK the unsprayed buffer zone (UBZ) for orchards when using broadcast applications is 18 m. For tunnel sprayers this is reduced to 5 m. A LERAP system is to be introduced so that in some circumstances the UBZ can be reduced.

A potential new development for air-assisted spraying is the positioning of a fan-shaped air jet to impinge on the liquid sheet emerging from a hydraulic nozzle. This produces a finer spray entrained in the air flow directed at the crop (Matthews and Thomas, 2000).

11

Space treatment by fogging

Space treatments require droplets to remain airborne for as long as possible, so it is usual to apply a fog. Strictly, a fog is produced when aerosol droplets having a diameter of less than 15 μm fill a volume of air to such an extent that visibility is reduced. The obscuring power of a fog is greatest when droplets are 1 μm in diameter. However, in agricultural practice, a fog refers to a treatment with a VMD of less than 50 μm, but with more than 10% by volume smaller than 30 μm. This definition of a fog includes both thermal fogs produced in a very hot air flow and cold fogs produced by a vortex of air. Both thermal and cold fogs have droplets small enough to present an inhalation hazard, so particular care is needed to protect the person applying them and to provide sufficient ventilation before anyone enters an area treated by fogging. In some circumstances, in glasshouse treatments, a mist treatment (see Chapter 10) with droplets having a VMD between 50 and 100 μm and less than 10% by volume smaller than 30 μm would be preferred, because the inhalation hazard is significantly reduced. A mist will sediment more rapidly on foliage. Fogging at higher volume rates and with a greater proportion of larger droplets, sometimes referred to as a 'wet' fog will leave a heavier deposit on foliage. This may provide a longer residual effect, but at high flow rates foliage close to the nozzle is liable to be damaged by an overdose of large droplets. Although less effective as a space treatment, droplets are deposited on foliage more rapidly by a wet fog, so quicker re-entry into a glasshouse will be possible.

Fogging is particularly useful for the control of flying insects, especially mosquitoes (Fig. 11.1a), not only through contact with droplets, but also by the fumigant effect of a volatile pesticide. It is used mainly to treat unoccupied enclosed spaces, such as warehouses, glasshouses (Fig. 11.1b) (Matthews, 1997), ships' holds and farm sheds, where the fog will penetrate inaccessible cracks and crevices. Fogging has also been used to treat sewers. Insecticides and some fungicides are applied as fogs. Some air movement within a building is needed to disperse the fog evenly. Fog will then slowly settle onto the horizontal surfaces. Unlike most applications where the surface area of ground

Fig. 11.1 Thermal fogging: (a) for vector control; (b) in glasshouse.

or foliage needs to be known to determine dosage rates, the volume of the space to be treated should be calculated and the machine's output calibrated carefully so that the correct dosage is applied. Fog can rapidly escape through small openings in the structure of a building, especially when treating a glasshouse, so some allowance for this is needed in calculating the dosage. Optimum results with low dosages of insecticide are obtained against mosquitoes when they are actively flying in the evening and inversion conditions occur.

In thermal fogging machines, pesticide, usually dissolved in an oil of a suitable flashpoint, is injected into a hot gas, usually in excess of 500°C and vaporised (Fig. 11.2). A dense fog is formed by condensation of the vapour when discharged into the cooler atmosphere. Most fogging machines also produce droplets larger than 15 μm diameter, especially if the flow rate is too high to achieve complete vaporisation. When the droplet size is too large, a high proportion of the pesticide will be deposited on the ground within 10 m of the fogger (Wygoda and Rietz, 1996). In enclosed buildings, all naked flames must be extinguished and electrical appliances disconnected, preferably at the mains. In the case of pilot lights, sufficient time must be allowed for gas in the pipes to be used up. Thus, in glasshouses, automatic ventilation, irrigation and CO_2 systems should be switched off and the glasshouse kept closed as long as possible after fogging. Care must also be taken in buildings in which there may be a high concentration of fine dust particles in the air, such as flour mills. A single spark can set off an explosion when more than 1 litre of a formulation containing kerosene is fogged per 15 m^3. Overdosing may be confined to localised pockets of fog which exceed the explosive limit. Fogging rates are usually less than 1 litre/400 m^3, but this lower rate can be ignited by a naked flame.

When wettable powder formulations are applied using fogging machines

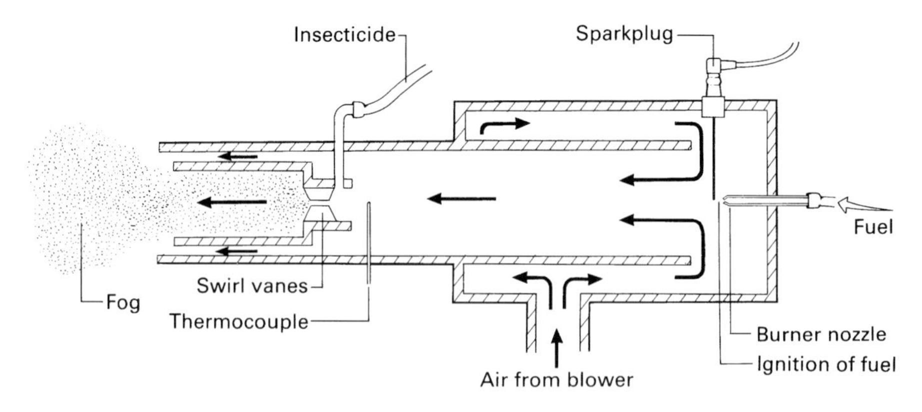

Fig. 11.2 Thermal fogging nozzle.

(see also Chapter 3), it is better to have agitation in the tank to keep the formulations in suspension.

Foliage should be dry, with temperatures between 18° and 29°C, and fogging should be avoided in high humidity conditions, and in direct sunlight to minimise risk of phytotoxic damage. Application is often better if made in the evening. Plants needing water should not be fogged. Fog is normally directed upwards over the crop at about 30°, while the equipment is moved from side to side to minimise any risk of phytotoxicity due to any localised over-treatment. The small droplets (<15 μm) eventually sediment on horizontal surfaces. Experiments with fogging using the microbial insecticide *Bacillus thuringiensis* confirmed that 95 per cent of the spray was deposited on the upper surfaces of leaves (Burges and Jarrett, 1979). Such deposits have little or no residual effect unless a persistent chemical has been fogged, so reinfestation can take place readily from neighbouring areas. Also, not all stages in the life cycle of a pest may be affected by a pesticidal fog; for example, whitefly adults are readily killed (Mboob, 1975), but egg and pupal stages on the undersurface of leaves are less affected. When fogging indoors, the lowest flow rate possible should be used to reduce the proportion of large droplets in the fog.

Fogging equipment is moved gradually through the space to be treated by an operator wearing appropriate protective clothing. Alternatively, the equipment can be mounted on a trolley and pulled through the building by a rope so that the operator can stay outside. While someone must be present when a thermal fogger is used, cold foggers can be operated using a remote control with a timer. This allows treatments to take place during an evening when the building is normally unoccupied. However, when operators attempt to fog large spaces from one position there will not be an even distribution of pesticide (Fig. 11.3) (Nielsen and Kirknel, 1992), unless fans provide sufficient air circulation to spread the fog away from the nozzle. Some buildings have a series of shuttered openings along the exterior walls so that treatments can be carried out at intervals from outside.

Fogs can be used outdoors, when advantage can be taken of temperature

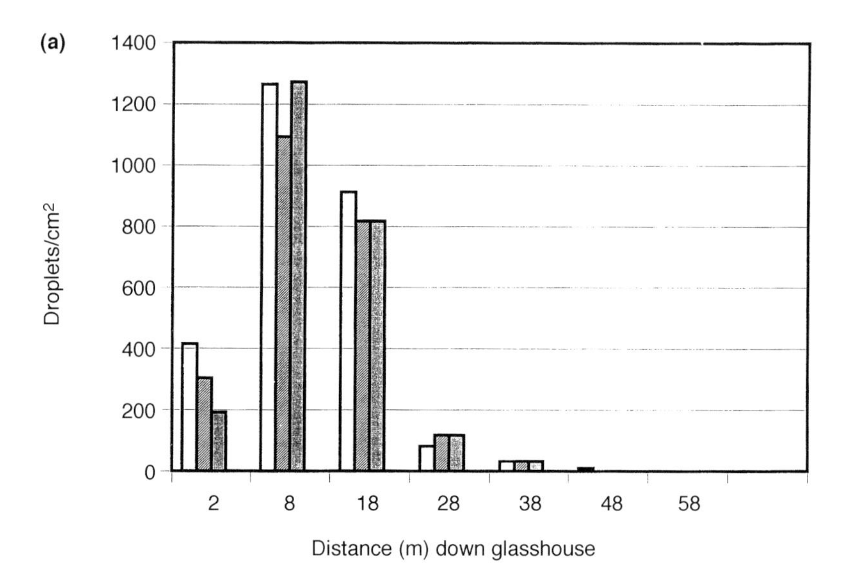

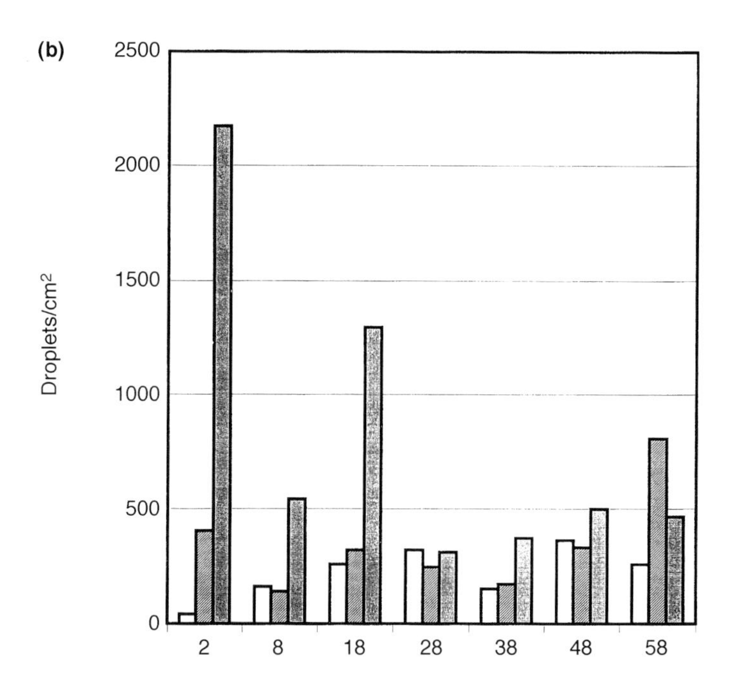

Fig. 11.3 Distribution of a pesticide applied as a fog with: (a) stationary fogger; (b) fogger moved through glasshouse pointing to each side from centre path (from Nielsen and Kirknel, 1992). Each set of histograms shows droplets/ cm^2 at 0.25, 4.6 and 9.9 m from the path.

inversion conditions, usually either early morning or early evening, so that the fog remains close to the ground. The fog is released as close to the ground as possible, or directed towards ground level and drifted across the area to be treated. Wind velocity should not exceed 6 km/h or the fog will disperse too quickly. Thermal fogs have been used extensively for the control of adult mosquitoes and other vector or nuisance insects, but in urban areas preference has been given in some countries to cold fogs using ULV aerosols, to avoid traffic hazards associated with the reduced visibility of thermal fogs and also to avoid the use of large volumes of petroleum products as diluents. On vehicle mounted equipment, the flow of pesticide can be controlled in relation to the vehicle speed and automatically switched off if the vehicle has to stop at traffic lights or is travelling too fast to achieve an effective flow rate. Some equipment fitted with GPS will also record where spraying has been carried out.

In forests, the retention of a thermal fog has been utilised to control cocoa mirids and to treat tall trees, such as rubber, as the chimney effect caused by the spaces between individual trees lifts the fog into the upper canopy (Khoo *et al.*, 1982). Unless there are inversion conditions, the fog is likely to be sucked rapidly out of the canopy by air movement above it. Water is injected into the hot gases closer to the combustion chamber (Fig 11.4) on some machines deliberately to cool the fogging temperature and help keep the fog closer to the ground.

Great care must be taken to avoid inhalation of fog, as the smallest droplets are not trapped in the nasal area and may be carried into the lungs. Research has shown that the particle sizes most likely to reach the lungs are 10 μm and smaller (Swift and Proctor, 1982; Clay and Clarke, 1987). Proper protective clothing must be worn; this includes a full-face respirator when many pesticides and special fogging carriers are applied. After application, all doors to enclosed spaces should be locked until after the required ventilation period, usually not less than 5–6 h. Treated areas should be marked with suitable warning notices. The concentration of airborne pesticide decreased by 60% in the first hour and by 95% during 12 hours after application (Giles *et al.*, 1995b).

Thermal fogging machines

Several types of thermal fogging equipment are available. Ideally, all types should be started outdoors and never in an area partially fogged.

Pulsejet

These machines are usually hand- or shoulder-carried (Fig. 11.5); a larger more powerful model is trolley mounted. These machines consist essentially of a fuel tank and pesticide tank, a hand-operated piston or bellows pump, spark plug, carburettor and long exhaust pipe. A few machines have an electrically

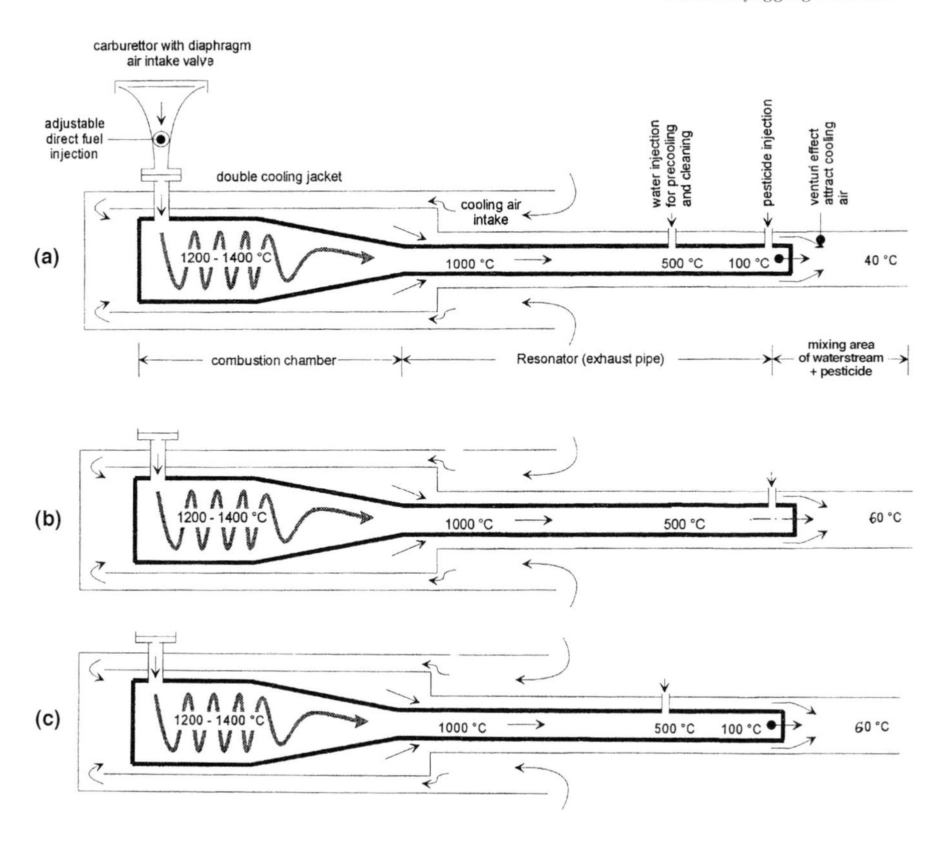

Fig. 11.4 Three basic models of pulse jet thermal fogger: (a) biosystem to cool fog; (b) standard type for oil-based formulations; (c) unit to produce finer dry fog with combustible liquids with a flash point greater than 75°C.

operated pump. Some machines have a translucent tank so that the quantity of pesticide remaining can be easily seen. Where the tank is detachable it is easy to change to a different chemical when necessary, if a spare tank is used. The provision of a large tank opening not only facilitates refilling but also cleaning.

To start the machine, the pump is operated to pressurise the fuel tank and force fuel through non-return valves to a metering valve. The initial mixture of fuel and air drawn through a filter is supplied into a combustion chamber, where it is ignited by a high-tension spark obtained from a battery-powered vibrator or mechanically operated magneto connected to the plug for a few seconds. The fuel is regular-grade petrol and about 1 litre/h is used on the smaller machines. Once the machine has started, the high-tension spark is no longer required and can be stopped. The exhaust gases from the combustion chamber escape as a pressure wave at high velocity through a long pipe of smaller diameter than the combustion chamber, and draw in a fresh charge of fuel and air into the combustion chamber. If operating with the correct mixture, there are about eighty pulsations per second, slightly irregular with maximum noise.

Fig. 11.5 Pulse-jet hand-carried thermal fogger (photo: Motan gmbh).

By means of a non-return valve the pesticide tank is also pressurised, and when the machine has warmed up, after about 2 min running, a valve is opened to permit the flow of pesticide solution through an interchangeable or variable restrictor into the end of the exhaust pipe. Temperatures on one small pulse-jet fogger are shown in Table 11.1 On some machines suitable for oil-based formulations only, the inlet is nearer to the combustion chamber to give more complete vaporisation of the liquid (Fig. 11.4). Some machines have a variable restrictor, but these are generally difficult to set repeatedly at the appropriate position. Droplet size is larger if the flow rate is increased, and the larger droplets will sediment usually within 2–3 m from the nozzle. On some machines the liquid is injected into the hot gases through two openings on opposite sides of the exhaust pipe. This gives a better distribution and break-

Table 11.1 Temperatures measured at different distances from a small pulse-jet fogging machine (from Munthali, 1976)

	Temperature (°C) at			
Flow rate (ml/min)	Injection of insecticide	End of pipe	0.8 m from nozzle	2.5 m from nozzle[a]
0	524	379	78	28.5
109	420	318	72	28.5
193	350	290	65	28
270	318	185	52	32
370	196	178	40	32

[a] Essentially ambient temperature.

up of the liquid. The temperature to which the liquid is exposed is lowered by an increase in flow rate, but even at low flow rates there is a very rapid decrease in temperature as soon as the fog is formed.

At the end of fogging, the valve should be closed with the engine running for at least 1 min to clear the exhaust and feed pipes of all liquid. Some machines have a safety valve, so that if the engine stops running a reduction in pressure to the spray tank causes a valve to close and stop liquid reaching the hot exhaust pipe. The machine is stopped by closing the fuel valve.

The volume which can be treated will depend on the capacity of the machine, but it is possible to treat a space of 200 m^3/min and cover an area of 3 ha in 1 h with fog of varying densities. The exhaust pipe can be tilted upwards, but particular care must be taken that insecticide solution does not leak from the restrictor and run down the hot exhaust. Ignition of fog has occurred on some machines, possibly because of an excess of unburnt petrol vapour in the exhaust gases. The problem is reduced by using a smaller fuel jet and renewing any faulty valves in the system.

As there is a fire hazard, this type of equipment should be operated only be well-trained personnel who should be supplied with a fire extinguisher in case of emergencies. Only formulations suitable for fogging should be used, and the fuel and chemical containers should be refilled very carefully in the open without spillage. In particular, refilling should be avoided when the unit is hot. The manufacturer's recommendations should be carefully studied by the operator before use, so that specific instructions for the particular model of the fogging machine are carried out. Ear muffs must be worn when using the larger machines and these are supplied by some machinery manufacturers for operator protection.

Engine exhaust fog generator

In this type of fog generator an engine, sometimes a two-stroke, drives two plates so that friction between them as they rotate preheats a pesticide solution fed from a separate knapsack container or from below the plates within the same container. The heated solution is metered into the hot gases of the engine exhaust. Although the temperature of the insecticide solution is lower compared with other fogging machines, breakdown of *Bacillus thuringiensis* formulations occurs because the duration of exposure to a high temperature is longer. Since the insecticide solution is separated from the spark plug, the fire hazard is considerably reduced.

Large thermal fog generators

A larger type of fog generator, known as the Todd insecticide fog applicator (TIFA), was developed originally for military use; a vehicle mounted machine is shown in Fig. 11.6. A petrol engine is used to operate an air blower and two

Fig. 11.6 TIFA fogger mounted on vehicle.

pumps. The blower supplies a large volume of air at low pressure to a combustion chamber, in which petrol pumped by a gear pump from a second fuel tank is ignited by a spark plug to heat the air. Some models do not have a pump and the fuel tank is pressurised from the blower. The hot gases at 500–600°C pass through a flame trap to a distributor head to which the insecticide solution is delivered by a centrifugal pump. A small proportion of the hot gases partly vaporises the liquid in a stainless steel cup in the distributor, while most of the hot gases pass to the outside of the cup, complete the formation of the fog and then carry it away from the machine. The temperature within the fogging head operating with odourless kerosene at 95 litres/h was 265°C (Rickett and Chadwick, 1972). Decomposition of some insecticides, including natural pyrethrins, occurs in temperatures over 230°C, but thermal degradation of certain pyrethroids was negligible in fogging machines, no doubt because they were so briefly exposed to high temperatures. Also, the hot gases contain little oxygen and so are less destructive chemically. The direction of the distributor head can be set in various positions. Normally, fog is drifted across a swath of 150 m, but it can be effective for 400 m (Brown and Watson, 1953). When mounted on a truck, some models have a self-starter and remote controls for operating the fogger, located in the cab. The vehicle is usually equipped with a low-speed speedometer, an hour meter to record the period of fogging and fire extinguishers.

A fogging machine can be used for ULV aerosol application by restricting the flow of insecticide and removing the heating section, and utilising the blower and distributor units: for example, Brooke *et al*. (1974) achieved over 85 per cent reduction of adult *Aedes taeniorhynchus* by applying only 1.5 g of biomesmethrin in 50 ml dieseline/ha with a modified thermal fogger. These fogging machines have also been used to treat sewers for cockroach control (Chadwick and Shaw, 1974). Thermal foggers can be used to apply fungicides to rubber trees, and to apply sprout suppressants in potato stores.

The same machines can be used as a blower, and also for conventional spraying by fixing a hose and lance to the insecticide pump and not using the heating or air blower sections.

Cold foggers

A cold fogger has a petrol- or propane-powered engine or an electric motor to drive a blower which forces air though a vortical nozzle (Figs 11.7, 11.8). A range of machines with different capacities is now available (Table 11.2). While some are suitable for trolley mounting in glasshouses and warehouses, equipment suitable for mounting on a flat-bed truck is used for mosquito control in urban areas. A typical truck-mounted unit used for mosquito control has a 12 kW four-stroke engine with direct drive to a Roots type blower. At the nozzle, spray liquid is fed into a vortex of air so that droplets generally smaller than 30 μm are formed. A major advantage of the vortical nozzle is the ability to apply ultralow volumes, compared with thermal foggers. This is why cold fogging is preferred in the USA, because pollution due to the diluents used in

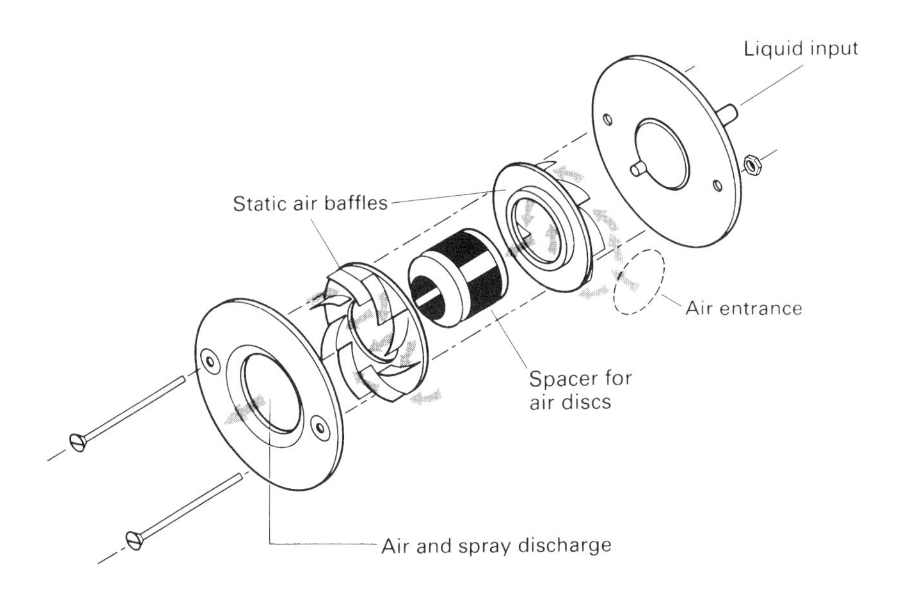

Fig. 11.7 Vortical nozzle – liquid fed into airstream, droplets fed into air vortex.

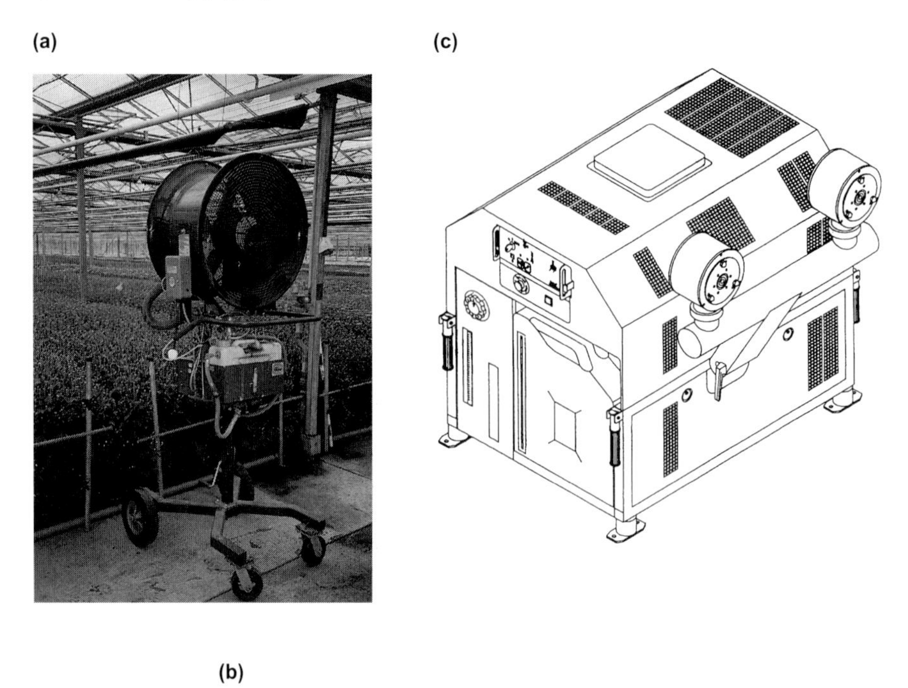

Fig. 11.8 (a) Cold fogger in glasshouse. (b) Close up of nozzle. (c) Vehicle mounted cold fogger for mosquito control.

thermal fog is avoided. Mount (1998) has reviewed the use of ULV aerosols for mosquito control.

Trolley mounted glasshouse equipment (Fig. 11.8a,b) is fitted with a fan to distribute the fog away from its fixed position. The fan should be operated for a period after treatment to maintain air circulation and obtain a more uniform distribution of the pesticide, but in some buildings separate fans should be used to circulate the fog throughout the space requiring treatment. Equipment adapted to use propane is suitable for applying pesticides in warehouses to

Table 11.2 Performance of two sprayers with vortical nozzles

	Sprayer A	Sprayer B
Power (kW)	1.86	12
Insecticide tank capacity (litres)	1.0	56
Weight (kg)	6.1	146
Air velocity (m/s)	109	196
Air volume (m^3/s)	0.0095	0.1
Droplet size (µm VMD)	17	17 (at 140 ml/min)
Max. flow rate of light oil (ml/min)	30	590

reduce the risk of carbon monoxide in the atmosphere. On the small machines, droplet size is affected by flow rate more than on the larger machines, presumably because the air volume emitted through the nozzle is insufficient to shear liquid effectively at high flow rates.

On some machines, air from the blower is used to provide a low pressure in the pesticide tank from which liquid is forced via a variable or fixed restrictor to the nozzle. The problem with the variable restrictor is that it is very difficult to reset at a particular position. A thermometer is required to note temperature changes which affect the viscosity of UL formulations, so on newer equipment preference is given to a positive displacement pump. On vehicle mounted equipment, this pump can be operated in conjunction with a speed sensor and the pump output displayed digitally in the vehicle cab.

Dosage recommendations are based on flow rate and vehicle speed and an intended swath width. The actual passage of the vehicle will be dictated by the layout of roads and wind direction. Spray droplets emitted while the truck is driven through an urban area will drift downwind between dwellings. New versions of this equipment with electronic controls will automatically stop treatment if the vehicle speed is too low, for example when stopping for traffic lights. Similar cut-off will occur if the vehicle speed is too high and flow rate to the nozzle is inadequate. Wind speed should not exceed 1.7 m/s for maximum efficiency. Penetration into houses is poor (Perich *et al.*, 1990) even if doors and windows are open, so when there is a major disease epidemic, treatment inside houses, especially in the latrines, is required with portable equipment to reduce populations of vectors such as *Aedes aegypti*. Goose (1991) used a knapsack mistblower adapted for ULV application.

Other fogging machines

A small, hand-carried, electric-powered machine has a fan which blows air over a heater so that hot air vaporises insecticide impregnated on a special cartridge. Other machines can fog a water based formulation which is pumped into the hot gas.

Aerosols can be produced with very small droplets equivalent to a fog by mechanical devices in which a series of baffles prevent large droplets escaping from the nozzle. A cloud of droplets less than 10 μm VMD is released. Warehouses can also be treated by space sprays using an insecticide formulated in a compressed gas (CO^2) supplied in cylinders that are fitted to a spray gun. Slatter *et al.* (1981) reported the use of non-residual synthetic pyrethroids applied through a cone nozzle having a 0.5 mm orifice at a nominal output of 6 g/s at 5000 kPa. Immediately after discharge, the droplets of insecticide plus solvent were approximately 9 μm VMD. Efficient insect control was obtained.

A high speed rotary atomiser has also been used to produce an aerosol for space treatments against mosquitoes. The rotary nozzle on a vehicle mounted unit used in the USA has a speed of about 24 000 r.p.m. in order to achieve an appropriate droplet spectrum. Interest in rotary atomisers for 'fog' sized droplets is due to the need to have a narrower droplet spectrum than from conventional fog nozzles and also to reduce the noise level.

12

Seed treatment, dust and granule application

While most pesticides are applied as liquid sprays, the application of dry formulations has one main advantage in that the product requires no dilution or mixing by the user. This is important in areas where the usual diluent, water, is expensive to transport. Nevertheless, the cost of transporting heavy and bulky diluent in the formulated product has to be paid for, and the relative cost of the active ingredient is increased.

Use of dusts has declined, largely because of the drift and inhalation hazards of fine particles less than 30 μm in diameter. Dusts are useful when treating small seedlings during transplanting, and in small buildings where farm produce is stored. Certain dusts, especially sulphur fungicide, are used on a few crops, notably grapevines when humid conditions improve retention of dust on foliage. The main use of dusts is now for seed treatment, although even in this application dusts have been replaced, mainly by particulate suspensions. Small and irregular shaped seeds are often pelleted, so that with a more uniform size they are easy to sow with greater precision (Clayton, 1988). The amount of pesticide applied is related to the pellet size (Dewar *et al.*, 1997).

Granular insecticides are used principally to control soil pests, aphids, stem borers on graminaceous crops and the larval stages of various flies, preferably where there is adequate rainfall or irrigation. They are sometimes added to compost used in peat blocks to raise seedlings such as brassicas (Suett, 1987). An increasing number of herbicides are also formulated as granules, certain of which are used widely in rice in the Far East. In the USA, aerial application of rice herbicides is common. Granules are very often applied by hand, especially in tropical countries, but the amount of active ingredient used is higher than with other application techniques when the granules are broadcast into irrigation water (Table 12.1) (Kiritani, 1974), whereas accurate placement of granules at their appropriate target with precision equipment means that less active ingredient is needed than with other application methods (Walker, 1976).

Table 12.1 Amounts of active ingredient in relation to different formulations used in rice paddy (after Kiritani, 1974)

	Active ingredient in formulation (%)	Active ingredient (kg/ha)
Spray	50	0.5
Dust	2	0.6
Microgranule	3	0.9
Granule	5	1.5

Seed treatment

The 'Rotostat' (Fig. 12.1) has been developed especially for treatment of batches of seed. It can be used to apply dusts, but is more generally used with liquid formulations. Smaller versions have been developed for use in developing countries. Fluidised-bed film coating systems are also used to deposit the pesticide in a thin durable polymer coating (Maude and Suett, 1986; Halmer, 1988). Seed treatment provides a valuable method of protecting young plants with minimal quantities of toxicant (Elsworth and Harris, 1973; Middleton, 1973). Seed treatment is discussed by Jeffs and Tuppen (1986) and Graham-Bryce (1988).

Equipment consists essentially of a hopper, preferably with an agitator, and a metering device to feed particles at a constant rate to the discharge outlet. A blower unit to produce an airstream to convey particles towards the target is essential on a duster, and may be used also on granule applicators, unless granules are allowed to fall by gravity directly from the metering mechanism. Walker (1976) lists the requirements of a good applicator (Table 12.2), and Bruge (1975, 1976) details characteristics of a range of granule applicators (Table 12.3), the main features of which are discussed below.

Features of dust and granule applicators

Hopper design

Ideally, the hopper should have smooth sides sloping down to the outlet, thus conical-shaped hoppers are better than those which are square-box shaped, unless the floor slopes (Fig. 12.2). Conversion of spray tanks to hoppers is unsatisfactory when the floor is level. An agitator is useful to prevent packing of the contents and to ensure an even delivery of the contents directly to the metering device or through a constant-level device. The latter is particularly useful where an agitator damages friable materials, such as attapulgite. Mechanical agitators are linked to the drive shaft of the blower unit. On some machines, air is ducted through the hopper from the blower unit. Certain agitators are less effective when dust particles bind together, as they merely

(a)

(b)

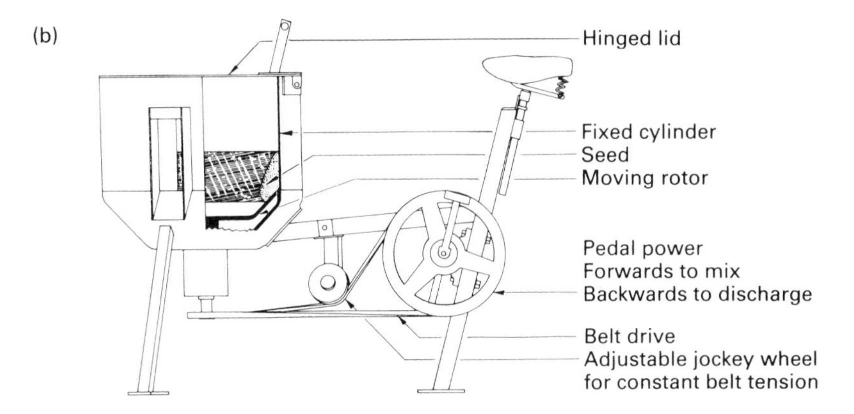

Fig. 12.1 (a) 'Rotostat' seed treatment machine (ICI Agrochemicals – now Zeneca). (b) Low-cost, pedal-powered seed treater.

cut a channel in the dust. Some machines have an auger in the hopper to move the contents to the metering device.

The hopper should have a large opening to facilitate filling; great care is needed to avoid fine particles 'puffing' up when the hopper is filled. Some granule products are now supplied in containers that allow direct transfer, so eliminating operator exposure at this stage. Where granules are not in a closed transfer system, a sieve over the hopper opening is essential to eliminate

Table 12.2 Requirements of a good granule applicator (after Walker, 1976)

(1) Deliver accurately amount calibrated, either continuous or intermittently
(2) Spread particles evenly
(3) Avoid damage by grinding or impaction
(4) Adequate mixing and feeding of material to metering device
(5) Easy to use, calibrate, repair and replace worn parts
(6) Light hand-carried and knapsack versions need to be comfortable to carry on the back
(7) Robust
(8) Corrosion, moisture and abrasion proof
(9) Inexpensive
(10) Output directly related to distance travelled

foreign matter and large aggregates. A lid must provide a seal to protect the contents from moisture. Ideally, hoppers and components should be made from corrosion-resistant materials, so various types of plastic and light alloys are preferred to ferrous metals, although the latter are cheaper. Granules should never be left in the hopper, otherwise corrosion will occur, so the hopper should be designed to be easily emptied. One knapsack granule

Table 12.3 Three examples of tractor-mounted granule applicators (some data selected from Bruge, 1975)

Manufacturer	Horstine Farmery	Merriau	SMC
Model	'Microband'	Granyl	Bimigrasol
Rows/hopper	2	2	2
Hopper capacity	16 or 35 litres	25 litres	18 litres
Hopper	Polythene	Polythene	Metal sheet
Indicator of level of granules	Yes, translucent hopper	Through translucent hopper	No
Metering device	Grooved rotors in special plastic or aluminium	Rotors with oblique lateral holes	Rubber belt
Method of regulation	By speed, using different sets of pulleys and width of rotor	By variation of speed	Variation of hopper outlet aperture by moving slide with micrometer screw
Drive	By spider wheel or by belt and pulleys	Direct from seeder by chain and sprockets	Direct from seeder or by wheel
Outlet tubing	Transparent plastic	Rubber	Transparent plastic
Internal diameter of outlet (mm)	25	20	20
Adaptability	Fits all seeders	Fits all seeders	Fits all seeders

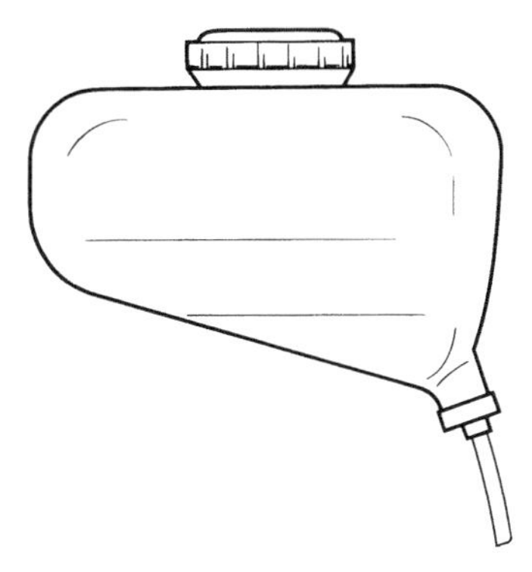

Fig. 12.2 Hopper design.

applicator was designed to incorporate a collapsible hopper to facilitate storage and transport.

Metering system

Various systems of metering dust and granules are used. The amount of product emitted by some machines is adjusted by altering the cross-sectional area of a chute by means of a lever or screw. For most applications the chute must be at least half open. Alternatively, particles drop through one or more holes, the size or number of which can be regulated. Both these systems are liable to block, especially if the particles are hygroscopic. Even collection of a small quantity of particles on the sides of the orifices is liable to reduce their flow and ultimately block the metering system. These systems will not give an accurate delivery unless the forward speed is constant.

Metering is improved by using various types of positive-displacement rotor (Fig. 12.3) which deliver a more or less constant volume of product for each revolution. Output is varied by changing the speed of rotation or capacity of the rotors or both, as on the Horstine Farmery 'Microband' equipment. Great

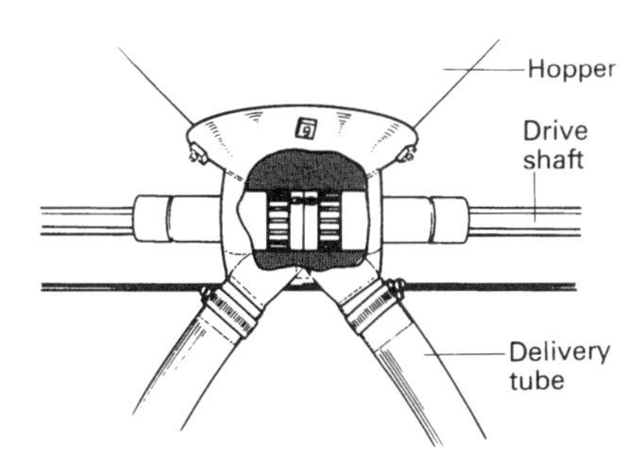

Fig. 12.3 Displacement rotor.

care must be taken in the design and construction of the metering system to avoid it acting as a very efficient grinder or compressor of granules (Amsden, 1970). Variations in size, specific gravity and fluidity characteristics of particles affect the efficiency of the metering system, so each machine requires calibration for a particular product. Farmery (1976) pointed out that the characteristics of a particular product may differ from one season to the next owing to a change in the granule base. Amsden (1970) suggested that the bulk density and flow rate of granules should be given on the label of the granule packaging as a guide to calibration. As some granules contain the most hazardous pesticides, the main development has been in closed transfer systems, such as 'Sure Fill' (Fig. 12.9). In the closed transfer system known as 'Smartbox', metering is controlled with a flow sensing device that allows a precise dosage to be applied, even at very low rates. As with other granule applicators the rate of application is determined by the forward speed of the tractor (Fig. 12.7). Monitoring application continuously and a positive shut-off at the end of rows minimise misapplication and wastage. On many granule applicators a trailing wheel is used to control a positive displacement meter in relation to the forward speed, but on more modern versions ground speed is monitored by radar. The granules are applied in-furrow or in a T-band to protect the roots of the young seedlings.

Calibration under field conditions at the appropriate forward speed is recommended because the flow of granules can be influenced by the amount of vibration caused by passage over uneven ground. Calibration can be done by collecting granules separately from all delivery tubes in suitable receptacles while travelling over 100 m and checking their weight (Bruge, 1975). The amount within each section of the target area has to be considered during calibration, as well as the total amount of product applied per hectare (Table 12.4). Thus, the metering system must provide as even a flow of particles as possible and avoid irregular clumping of particles. This is achieved when rotors have many small cavities to hold the particles, rather than a few larger ones. The speed of rotation can be reduced to minimize attrition of the product. Bruge (1976) emphasised that the moving rotor must be set close to its housing to avoid particles being crushed. On some machines a scraper is positioned at the bottom of the hopper to ensure that the contents in the rotor remain intact. The metering system should not be adjusted to a lower flow rate while particles are present, as packing of the product is liable to jam the unit.

Table 12.4 Amount of formulation required in small areas

Rate (kg/ha)	Area covered by 100 g (m^2)	Rate per m^2(g)
10	100	1.0
15	66.7	1.5
25	40	2.5

Blower unit

Small hand dusters usually have a simple piston or bellows pump. Bellows have been used in knapsack dusters as they are useful for spot treatments, but rotary blowers provide a more even delivery. The fan may be driven by hand through a reduction gear (about 25:1) or by a small engine. Compressed-air cylinders have also been used to discharge small quantities of dust.

Delivery system

Particles drop from the metering unit into a discharge tube connected to a blower unit, if present. When a blower unit is not used, the discharge tube should be mounted as vertically as possible to avoid impeding the fall of particles. If it must be curved, a large radius of curvature is essential. The internal diameter of the tube should be sufficiently large, ideally not less than 2 cm, and uniform throughout its length. Some tubes are divergent at the outlet end, or subdivided to permit treatment of two rows. At the outlet a fish-tail or deflector plate may be fitted to spread the particles. The position of the discharge tube should be fixed, especially when granules are applied in the soil, and the outlet has to be 10–30 mm from the soil at the back of a coulter. Clear plastic tubes are often used as they are less liable to condensation and blockages can be easily seen, but they are sometimes affected by static electricity. Instead of a blower and discharge tube, some machines have a spinner to throw particles over a wide swath.

Examples of equipment

Package applicators

Some dusts and granules are packaged in a container with a series of holes which are exposed on removal of a tape cover. The contents are shaken through the holes, so the quantity emitted will vary depending on the operator and amount remaining in the container. The main advantage is that the contents do not require transferring to other equipment, but the container has to be carefully disposed of after use. Similar 'pepperpot'-type applicators can easily be made by punching holes in the lid of a small tin.

Hand-carried dusters

Various types of bellows dusters are available with capacities from 20 to 500 g. Plastic materials are used now in preference to leather or rubber, which deteriorate more rapidly under hot and humid tropical conditions. They were used to apply sulphur dusts in vineyards, and Mercer (1974) used bellows dusters to treat small plots of groundnuts with fungicide. The design was similar to that described by Swaine (1954) who improvised a bellows duster by cutting a small tin in half and joining the two halves with a piece of car tyre inner tube. A metal handle was soldered to both ends and an outlet tube was fixed to the bottom half of the tin. These dusters were used to apply a puff of dust in the funnel of maize plants for stalk-borer control.

Simple plunger air-pump dusters have a bicycle-type pump, which blows air into a small container. Some have double-action pumps to provide a continuous airstream. The air agitates the contents and expels a small quantity through an orifice. This type of duster was used extensively to treat humans with DDT to prevent an outbreak of typhus in the 1940s. The World Health Organization has a specification for this type of duster (WHO/EQP/4.R2). They are also useful to spot-treat small areas in gardens and around houses for controlling ants and other pests. Similar dusters with a foothold, a cut-off valve to close the dust chamber, and a flexible discharge pipe, have been used to blow a dust into burrows for rodent control. When sufficient dust has been emitted the valve is closed and more air blows in to drive the dust deep into the burrow (Bindra and Singh, 1971). Small dusters with a rotary blower are also made for garden use.

Pest control operators sometimes use a dust applicator, which is very similar in appearance to a compression sprayer. The duster can be pressurised from an air supply through a schrader valve. Dusters with an electrically powered fan are also available. A duster can be improvised by using a loosely woven linen or fabric bag, sock or stocking as a container which is shaken or struck with a stick. The amount applied is extremely variable and most of the dust is wasted.

Hand-carried granule applicators

These have a tube container (approximately 100 cm long, 1–1.5 litres capacity) with a metering outlet operated by a trigger or wrist action rotation of the container (Fig. 12.4). On one machine a small meter is positioned on each side of the outlet. The output of granules depends on the position of the cones, which can be altered by adjusting a connecting-rod. Robinson and Rutherford (1988) found that many applicators which rely on gravity flow are slow, and trigger-operated systems were tiring to operate and more expensive to manufacture. They developed a ‘rotary valve’ using a wrist-action for granule application in transplanted tobacco. These applicators are particularly useful for spot treatment at the base of individual plants, and have been used in cabbage root fly control and in selective weed control, but are not suitable for burrowing nematode control on bananas for which larger doses are required. Granules are normally left on the soil surface but, by modifying the outlet with a spike, subsurface application is also possible.

Fig. 12.4 Hand-operated granule dispenser.

Shoulder-slung applicator

An applicator, known as a 'horn seeder', consists of a tapered metal discharge tube containing a variable opening which is inserted into the lowest point of a rubberised or neoprene-treated cloth bag. This bag has a zipped opening and is carried by a strap over the operator's shoulder. A swath of up to 7 m can be obtained when the discharge tube is swung from side to side in a figure-of-eight pattern.

Knapsack and chest-mounted dusters and granule applicators

A blower is usually mounted to the side and base of a hopper of 8–10 litres capacity. On hand-operated knapsack versions, a crank handle is situated in front of the body and is connected to a gear-box by a driving chain which is protected by a metal case. This drive is eliminated on some machines by hanging the hopper on the operator's chest, but the front position is less comfortable, restricts the quantity of material which can be carried in the hopper, and may be hazardous if fine particles escape and contaminate the operator's face.

The volume of air emitted by hand-operated machines will depend on the operator, but is at least 0.8 m^3/min at a speed of 14 m/s with the crank handle turned at 96 r.p.m. The discharge tube is normally on the opposite side of the hopper to the gear-case, which must be protected as much as possible from particles liable to cause wear of the gears. Compared with knapsack sprayers, dusters are relatively expensive, owing to the cost of the blower unit.

Most motorised mistblowers described in Chapter 10 can be converted for dust and granule application by removing the spray hose from the tank and inserting a wider tube to feed particles directly down from the hopper through a metering orifice into the airstream (Fig. 12.5). The outlet tube is rotated to stop the flow of material. Machines with a tank having a sloping floor are more easily adapted for application of dry materials. In Japan, a 30 m long plastic tube, carried at each end, has been fitted to these machines. Dusts and microgranules are dispersed through a series of holes along its length. Tabs next to the holes improve distribution (Fig. 12.6) (Takenaga, 1971). Hankawa and Kohguchi (1989) reported that good control of brown planthopper (*Nilapavata lugens*) was obtained using buprofezin or BPMC insecticides applied with this type of applicator fitted to a large capacity fan to increase airflow at each hole.

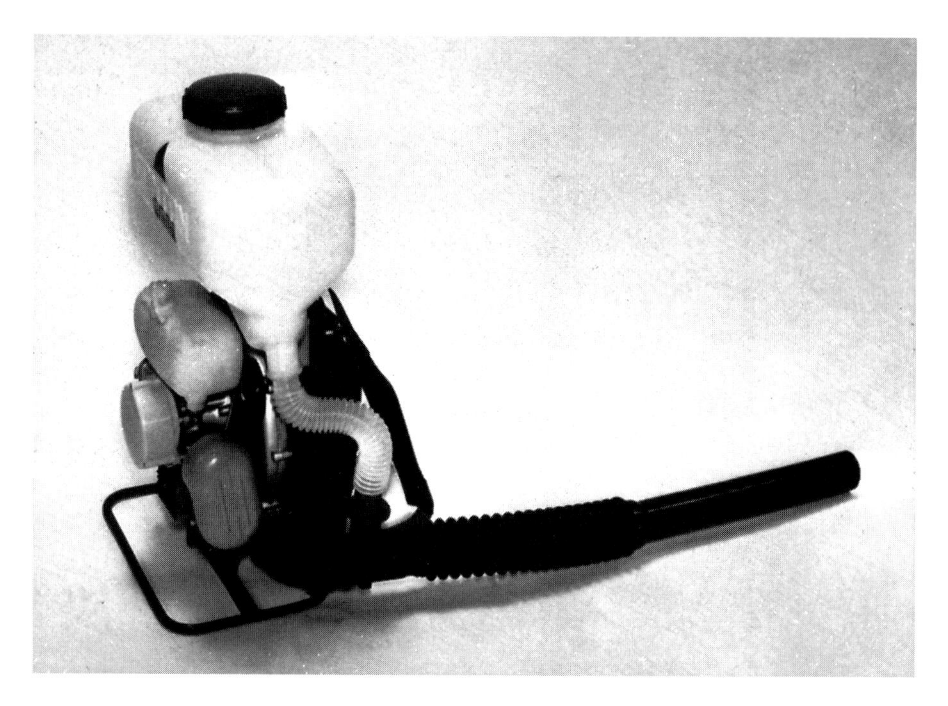

Fig. 12.5 Motorised mistblower converted to direct-air granule application.

Some knapsack equipment is made specifically for granule application with or without an airflow. Machines designed to apply granules by gravity only can sometimes be modified to spot-treat with a measured dose. A knapsack in which a cup is moved by a lever mechanism from the hopper outlet to the

(a)

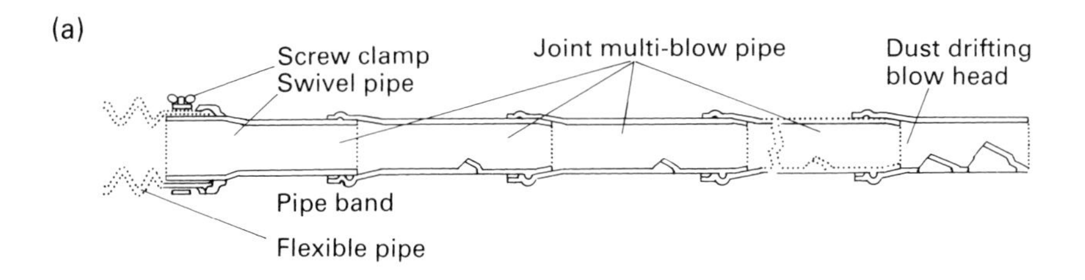

(b)

Fig. 12.6 Examples of extension tubes to apply microgranules: (a) detail of rigid tube; (b) flexible lay-flat tube inflated by the airstream applies microgranules over a 30 m swath.

discharge tube each time a dose is applied can be used for spot application on bananas.

Power dusters

Relatively few larger power-operated dusters are manufactured. Some are coupled to the p.t.o. of a tractor to operate a fan with an output of approximately $50\,m^3/min$. Units with an independent 3–5 hp engine are heavy (> 50 kg), but are sometimes mounted on a two-person stretcher to facilitate transport to areas inaccessible to vehicles. Such units have been used to apply fungicides, especially sulphur dust in vineyards, using machines with twin outlets. In rubber plantations, dust has been projected vertically to a height of up to 24 m under calm conditions.

Tractor-mounted granule applicators

A few machines have been designed with a boom fed from a single blower unit (Palmer, 1970). Some are adapted from fertiliser spreaders and have a spinner, but most have a series of separate units fixed on a horizontal frame (Fig. 12.7) which can be mounted either at the rear or front of a tractor. The containers of granules can now be fitted directly to the hoppers to avoid exposure of the user to the pesticide (Fig. 12.9). Normally, up to four units each with two outlets are used together, but wider booms are possible. Displacement rotors are ideally driven at 0–8 km/h from a rimless spoked land wheel designed to follow the ground contours accurately. A p.t.o. drive replaces the land wheel on high-clearance frames.

Fig. 12.7 'Microband' applicator (Horstine Farmery).

An airflow, provided by a compressor fan driven from the p.t.o. or by a hydraulic motor, increases precision when broadcasting as little as 3 kg/ha (Fig. 12.8). A deflector plate at the discharge tube outlet, if properly angled for a particular granule size and density (Goehlich, 1970), will spread granules over a swath up to 1–2 m wide. Thus an applicator with four hoppers each

Fig. 12.8 Airflow granule applicator (Horstine Farmery).

Fig. 12.9 'Sure Fill' closed transfer of granules.

carrying up to 30 kg granules can treat up to 40 ha without refilling, and at 8 km/h cover the ground in under 6 h. Farmery (1970) discusses the problems of achieving an even air volume and speed with airflow applicators.

Applicators can be mounted on the tractor alongside seed drills, special coulters, fertiliser applicators and other farm implements, depending on where the granules are needed (Bailey, 1988; Pettifor, 1988). In the 'bow-wave' technique, granules are metered in a 100 mm band on the soil surface in front of a coulter which then mixes them in the soil (Whitehead, 1988). Dunning and Winder (1963) found that the bow-wave technique decreased phytotoxicity.

Some pesticides can be applied as granules placed as an in-seed furrow treatment at the time of sowing. The discharge tube can be positioned close to the coulter for accurate placement of the granules in relation to the seed. Wheatley (1972) studied placement and distribution on the performance of granular formulations of insecticides for carrot-fly control. He concluded that the performance of these granules could not be easily improved by minor modifications of the granule-placement equipment, provided an even distribution along the row is achieved. Granule-application criteria for certain pest-control problems are summarised in Table 12.5 (Wheatley, 1976).

Table 12.5 Dose required for control of cabbage root fly (after Wheatley, 1976)

Application method	Amount of granules	Particle distribution		Dose/plant	
		cm^2 soil/granule	cm^3 soil/granule	No. of granules	mg a.i.
Spot surface (15 cm dia.)	0.17–1 g/plant	0.08–0.6	–	340–2100	17–45
Band (5 × 10 cm deep)	0.7–1.2 g/m row	0.09–0.71	0.8–3.6	210–880	10
Broadcast					
surface	45 kg/ha	1.1	–	160	8
mixed to 5 cm depth	30 kg/ha	0.71	3.5	250	2.7

In contrast to 17–45 mg a.i./plant required with spot treatment against cabbage root fly, the dose can be reduced to 10 mg a.i./plant for banding treatments, and to as little as 2.7 mg a.i./plant when granules are broadcast and mixed to a depth of 5 cm. Against aphids, the amount per plant depends on the size of the plant, larger plants requiring a heavier dose to maintain a similar lethal concentration in the sap (Table 12.6). Side dressing of some insecticides is effective later in the season, especially if moisture is sufficient to redistribute the chemical.

A major problem with the development of ground equipment for granular

Table 12.6 Dose required for aphid control with granules on different crops (after Wheatley, 1976)

Crop	Crop density (plants/ha)	Treatment (amount of granules)	Dose/plant	
			Number of granules	mg a.i.
Carrot	500000	Band 7.5 cm × 1 cm (11–25 kg/ha)	32–160	1.1–2.5
Sugar beet	40000	Soil or foliage (5.6–11 kg/ha)	38–880	7–14
Brussels sprouts	10000	Soil or foliage (14–39 kg/ha)	3100–12000	33–190

application has been the lack of fundamental studies on particle distribution and the wide range of granule sizes available, since each manufacturer has sought the cheapest and most easily obtained supply of suitable base carriers. Sand is readily available, but is very abrasive. Heavy carriers are suitable for aerial application to reduce the risk of drift, and for soil application, but lighter and softer carriers such as the clay bentonite, are more suitable for rapid release of herbicides used in rice-weed control. A greater control of particle size within defined limits, namely controlled particle application, is possible with new coating systems.

13

Aerial application

Aerial application of pesticides is important, especially over forests and where large areas of crops need to be treated rapidly and access is difficult for ground equipment. Irrigated fields and areas invaded by locusts are often treated using aircraft. Aircraft were also used in large-scale vector control programmes, especially for tsetse control (Allsopp, 1984) and in the Onchocerciasis Control Programme in West Africa (Gratz, 1985). Their use has declined in some countries due to public concern about spray drift as pesticides are released at a greater height above the crop canopy. In Europe the extensive use of tramlines to allow access for tractor mounted equipment has also reduced the need for aerial treatment. Earlier references to aerial application were given by Akesson and Yates (1974) and Quantick (1985a,b). Bouse (1987) discusses aerial herbicide application.

Types of aircraft

Fixed and rotary-wing aircraft (Figs 13.1 and 13.2) are used for applying pesticides. Information on the performance for international standard atmosphere conditions at mean sea level of certain aircraft is listed in Tables 13.1 and 13.2. Such data need to be converted to estimate correctly the performance under local operating conditions. Microlight aircraft have been used for ULV application; however, all but larger (2 seat) models are generally too small. The pilot is too exposed to spray contamination unless the microlight has a closed cockpit, and it is difficult to maintain an accurate track for adjacent swaths. Autogyros (Hill, 1963; Johnstone *et al.*, 1971; Johnstone *et al.*, 1972) and remote-controlled model aircraft (Embree *et al.*, 1976; Johnstone, 1981) have also been used on a very limited scale, although remote controlled helicopters are now used commercially in Japan (Fig.13.3) (Hasegawa, 1995). Larger multi-engined aircraft such as the DC-3 (Dakota) and DC-6 have been used in vector-control programmes (Lofgren, 1970; Lofgren *et al.*, 1970a; Lee

Fig. 13.1 Fixed-wing TurboThrush aircraft.

Fig. 13.2 Helicopter spraying.

et al., 1975b), and for spraying forests (Randall, 1975; Quantick, 1985a). Many different types of aircraft have been converted for pesticide application, but there are several which have been specifically designed for crop spraying. Drift potential was greater with biplanes (Teske and Thistle, 1998). The main features of these single-engined low-wing monoplanes are

Table 13.1 Data on certain fixed wing aircraft used in agriculture (performance details are given for international standard atmosphere (ISA) conditions at mean sea level)

Aircraft	Pawnee Brave	Air Tractor AT-502B	Turbo Thrush
Engine power (kW)	230	550	550
Fuel capacity (litres)	329	817	863
Spray tank capacity (litres)	1000	1892	2498
Weight empty (kg)	930	1952	2381
Gross weight (kg)	1770	4403	5600
Ag load weight (kg)	839	2451	3219
Wing span (m)	11.9	15.86	15.25
Wing area (m^2)	20.9	29	34.8
Stall speed (km/h)	114	109	92
Spraying speed (km/h)	163	233	161–282
Take off distance (m)	267	348	457
Landing ground run (m)	213	–	183
Rate of climb (m/min)	241	329	–

(1) a high-performance engine to lift a heavy payload from earth or gravel strips to a height of 15 m in less than 400 m at sea level
(2) an airframe stressed to withstand frequent landings and take-offs and to provide protection for the pilot in the event of an accident
(3) an operational speed of 130–230 km/ha
(4) a low stalling speed of 65–100 km/h
(5) a high payload to low gross weight ratio
(6) light and responsive controls to reduce pilot fatigue

Table 13.2 Data on certain rotary-wing aircraft used in agriculture (performance details are given for international standard atmosphere (ISA) conditions at mean sea level

Aircraft	Hughes 300	Bell Ag-5	Hiller UH-12E
Engine power (kW)	130	193	230
Main rotor diameter (m)	7.7	11.3	10.8
Overall height (m)	2.5	2.83	3.1
Length (m)	8.8	13.3	12.4
Capacity fuel (litres)	114	227	174
Capacity hopper (litres)	304	454	635
Weight empty (kg)	433	770	770
Weight max. AUW[a] (kg)	755	1293	1220
Weight Ag load (kg)	204	544	239
Speed cruising (km/h)	97	135	140
Rate of climb (m/min)	350	262	463
Range (km)	355	547	298

[a] AUW, all-up weight.

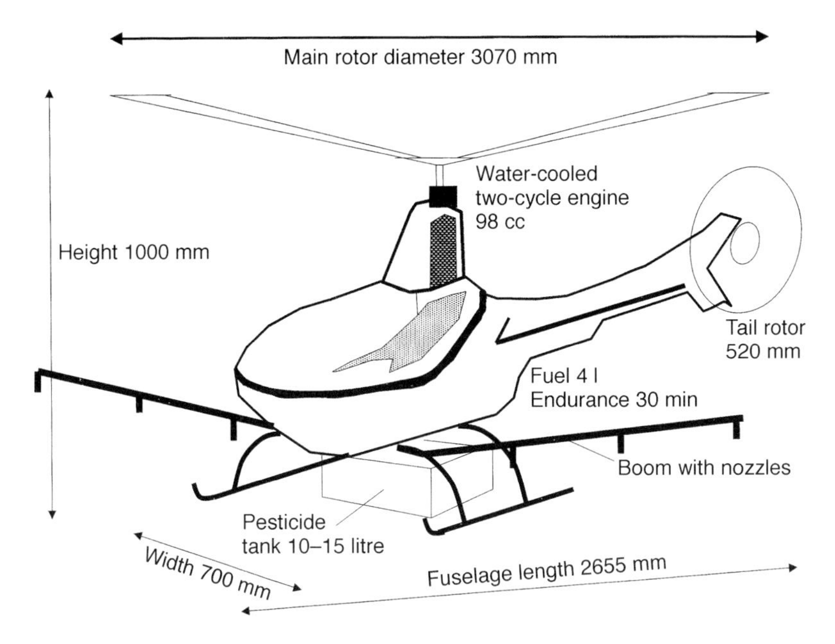

Fig. 13.3 Remote controlled helicopter used in Japan.

(7) distinct separation of flight controls from application equipment
(8) cockpit with good all-round visibility
(9) landing gear and the canopy with sharp leading edge to minimise the hazard of hitting power lines or wires
(10) a deflector cable fitted between the top of the canopy and the tail
(11) a pressurised and air-conditioned cockpit to reduce the risk of contaminating the pilot with pesticide
(12) a recoil-type harness and safety helmet to protect the pilot
(13) a pesticide tank or hopper located in front of the cockpit and aft of the engine and over the centre of lift, so that aircraft trim is minimised by changes in weight during spraying
(14) the maximum permissible weight is indicated clearly near the filler opening
(15) a tank designed for rapid loading, easy cleaning and maintenance with provision for rapid dumping of a load in an emergency
(16) provision for loading by pumping the spray into the bottom of the tank through a filler opening to the rear of the wing
(17) provision of top-loading of dry particulates through large dust-tight doors
(18) fuel tanks placed as far away from the pilot as possible, preferably as wing-tanks.

The basic design should facilitate inspection, cleaning and maintenance of all parts of the aircraft and application equipment. Corrosion-resistant materials and coatings should be used with readily removable panels to permit easy access to the fuselage.

Multi-engined aircraft (Fig. 13.4) are needed over populated areas and forests and swamps, where opportunities for a safe emergency landing in the event of an engine failure are limited. Such aircraft generally require well-constructed runways and can operate over long distances, even at night when it is necessary to take advantage of inversion conditions.

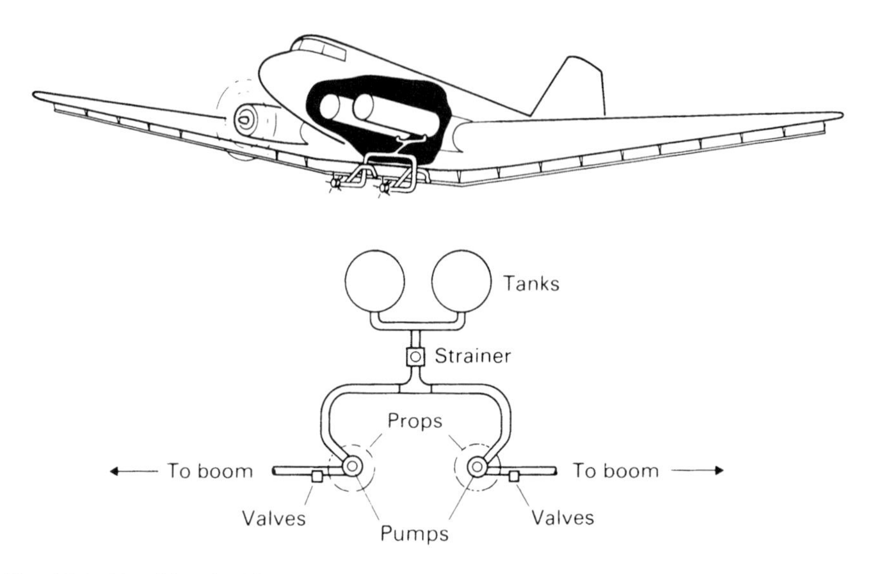

Fig. 13.4 Double air-driven pump option mounted in twin-engined aircraft (after FAO, 1974).

Helicopters provide an alternative to fixed-wing aircraft where reduced flight speed and greater manoeuvrability within fields is desirable to increase penetration, or is necessary due to the presence of trees or other obstacles, and where landing strips are not available. They are used to treat vinyards on steep slopes in Germany (Schmidt, 1996). Helicopters may be landed in any suitable clear area or on special platforms such as the top of a vehicle. Helicopters are particularly useful where on-the-spot survey and treatment need to be combined, for example in mosquito- and blackfly control programmes. The contribution made by helicopters to agricultural aviation has been reviewed by Voss (1976).

Improved penetration of a crop canopy with spray droplets in the strong downwash of air created by the rotor is not achieved unless the helicopter is flown at less than 25 km/h (Parkin, 1979). Unfortunately, the initial cost and maintenance costs are much greater than with fixed-wing aircraft and extra flying skills are needed by the pilot, so spray application at a low speed is not economic. Discriminative residual placement spraying of a 20 m swath, along the edge of fringing woodland and riverine forest for tsetse control at 25–

40 km/h was tried in West Africa (Spielberger and Abdurrahim, 1971; Baldry *et al.*, 1978; Lee *et al.*, 1978). The rotor downwash pattern changes from a closed toroid to a horseshoe vortex as the helicopter increases forward speed. At operational speeds above 40–50 km/h, distribution of spray in the wake of a helicopter is similar to that of a fixed-wing aircraft (Fig. 13.5). Productivity with helicopter spraying can be improved with booms up to 15 m wide, although care must be taken to ensure that spray droplets do not enter the rotor vortex if the booms are too wide in relation to the rotor diameter.

Spray gear

The arrangement of pump, tank and other components of spray application equipment on a fixed-wing aircraft is shown diagrammatically in Fig. 13.6. Equipment for dispersal of solids is discussed on pp. 288.

Spray tank or hopper

The tank may be constructed with stainless steel; one aircraft has a titanium tank integral with the fuselage. Subject to government regulations concerning the structure of aircraft components, fibreglass tanks are acceptable for application of most pesticides. They have a translucent zone at the rear of the tank mounted in the fuselage to permit the pilot to check the volume of liquid remaining in the tank, otherwise a contents gauge is provided. The shape is designed so that it will drain completely, either in flight or on the ground. A dump valve is fitted so that a full load can be jettisoned within 5 s in an emergency. Small planes converted for spray work may have a belly tank fitted to the bottom of the aircraft (Fig. 13.7). In this case, the whole tank may be jettisoned if the need arises, although almost all belly tanks incorporate a conventional dump and most pilots prefer not to drop the tank. Cockpit contamination and pilot exposure to pesticides is minimised with a belly tank as it is outside the fuselage. Internally mounted tanks and pumps are installed in large aircraft when required. In helicopters, saddle tanks can be mounted on either side of the engine with a large-diameter interconnecting pipe to maintain a level load. On turbine helicopters, a belly tank or sometimes an internal tank is normally used. An electrically driven agitator or liquid recirculating system is normally fitted to tanks used with particulate suspensions such as *Bacillus thuringiensis* for mosquito larviciding.

The tank opening is provided with a basket-type filter, but loading is quicker and safer when the load is pumped through a bottom loading point from a ground mixing unit. A filter incorporated in this feed line usually has a sufficiently fine mesh to protect the nozzle orifices, although each nozzle should be provided with its own filter. The mesh size is therefore 25–100 mesh, depending on the type of nozzle used; 50 mesh is suitable for most spray work and is the first which should be used for wettable powder formulations. An in-line strainer is fitted to the outlet of the tank to protect the pump. A larger mesh size (6–8 mesh) is usually desirable to reduce pressure drop at this point. All filters should be readily accessible to allow a change of mesh size if necessary and to facilitate regular cleaning; a valve is necessary upstream of the filter to allow cleaning, even when the spray tank is full. An overflow pipe and air vents are ducted from the spray tank to the rear and bottom of the fuselage to prevent contamination of the cockpit. An air vent prevents a vacuum being created in the tank, as this would affect the flow of liquid to the pump.

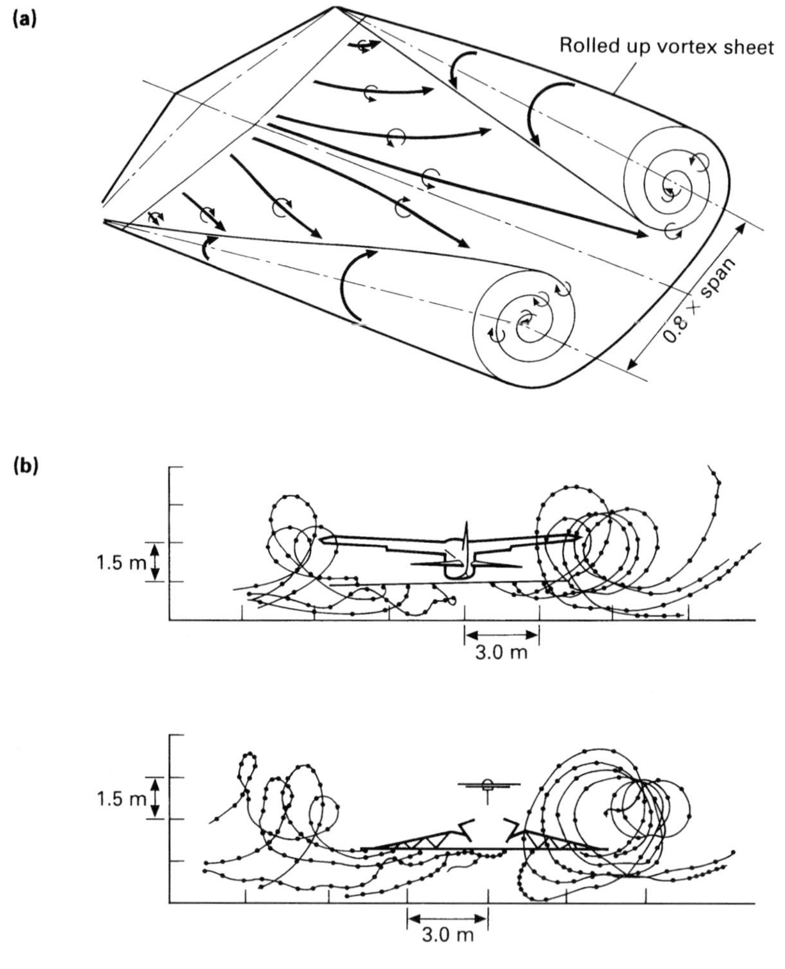

Fig. 13.5 (a) Trailing vortex system behind an aircraft (after Spillman, 1977). (b) Aerodynamic trailing wake of a high-wing monoplane traced by gravitationally balanced balloons and of a helicopter (after FAO, 1974). (c) Droplet trajectories from mid span of wing in relation to gravity, vortices, ground effects and size of droplets (after Spillman, 1977). (d) Photograph to show wing tip vortices.

(c)

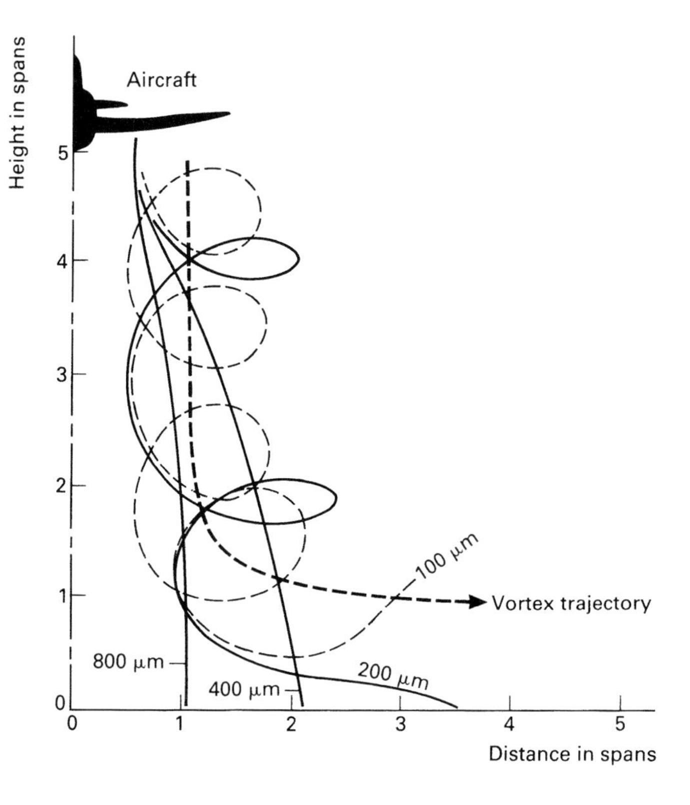
Aircraft
Height in spans
5
4
3
2
1
0
100 µm
Vortex trajectory
800 µm
200 µm
400 µm
0
1
2
3
4
5
Distance in spans

(d)

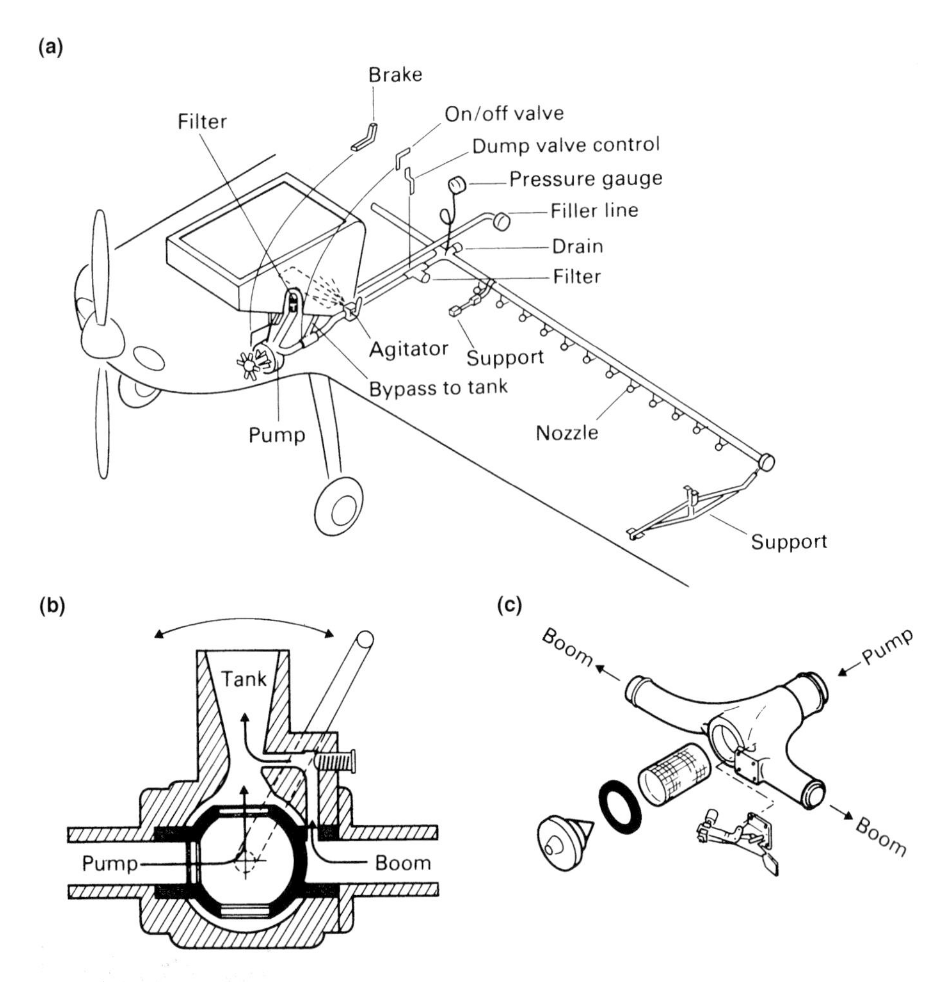

Fig. 13.6 (a) Cutaway diagram of spraying system for a small fixed-wing aircraft (after FAO, 1974). (b) Main control valve for aircraft sprayer showing boom vacuum positions for positive shut-off; check valves are needed at each nozzle. (c) Liquid screen filter between pump and boom.

Pump

A centrifugal pump is usually used. It may be driven directly by a fan mounted in the slipstream of the propeller of the aircraft engine, usually between the landing wheels. However, to reduce the drag on the aircraft, the pump can be driven by a hydraulic drive system that operates at up to 200 bar and can provide 10–18 kW at the pump. Electrically operated pumps are sometimes used, particularly for LV application when only 1–2 kW or less is required. This improves power utilisation, so the stalling speed is reduced and climb performance and cruising speed increase (Boving et al., 1972). The overall efficiency of pumps is only about 10%, due partly to poor transfer of energy

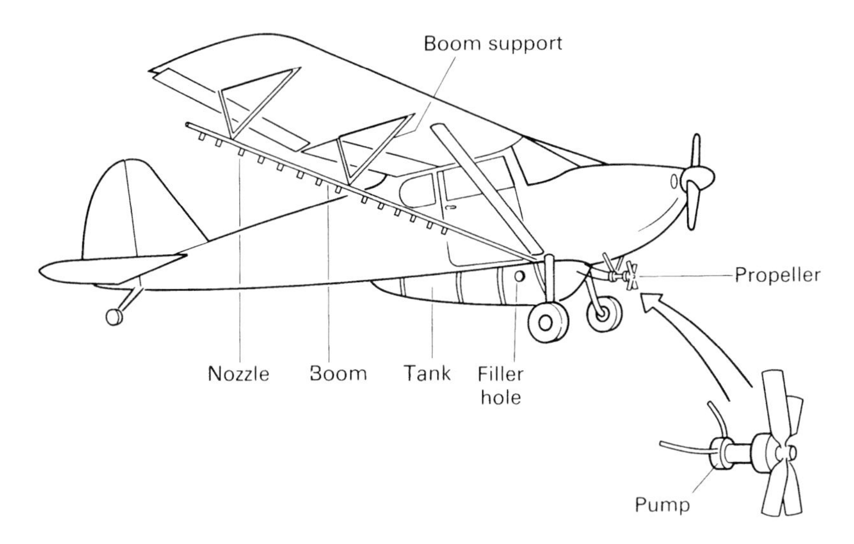

Fig. 13.7 Spray system with quick detachable belly tank on a small passenger aircraft (after FAO, 1974).

from the engine to the propeller drive. Tests with a 220 kW Stearman aircraft cruising at 137 km/h indicated that the pump developed approximately 3.3 kW (Akesson and Yates, 1974). The pump is fitted below tank level to ensure that it remains primed. Piston, gear, or roller-vane pumps may be used if higher pressures are required, although these types are vulnerable to damage by solids in suspension. A valve should be fitted close to the pump inlet so that if any maintenance is needed or the pump has to be replaced, it can be removed without draining the system.

On helicopters, a drive from an ancillary pad on the gearbox can be connected through a clutch directly or via a belt drive to a pump or hydraulic pump driving the chemical pump by a hydraulic motor. Many helicopter systems now use one or two electrically driven centrifugal pumps.

The pump must have sufficient capacity to recirculate a proportion of its output to the tank to provide hydraulic agitation. Some agitation, even during actual spraying, is desirable. A bleed line from the top of the pump may be required to remove airlocks.

Spray boom

On fixed-wing aircraft the boom extends for most of the wingspan, but usually avoids the wing-tip area where a vortex could carry droplets upwards. Teske *et al.* (1998), using computer simulation, concluded that the suggestion that boom length should be less than 75 per cent of the wing span or rotor diameter is based on anticipated position of rolled-up vortices rather than solid experimental evidence. Parkin and Spillman (1980) showed that the amount of spray carried off-target by wing-tip vortices could be reduced by extending the wing horizontally by fitting 'sails'. These wing-tip sails, originally designed to reduce drag, have been used only in experiments. The boom is often mounted at the trailing edge of the wing, but actual positions depend on the wing structure. Young *et al.* (1965) studied the spray-distribution patterns with different boom lengths and positions. Generally, a better distribution has been found when the spray bar is mounted below the wing. A round pipe may be used, but some booms with an internal diameter of up to 50 mm to cope with high flow rates (>500 litres/min) are streamlined to reduce drag. At lower volumes the boom can be decreased to 13–20 mm ID. Larger-diameter booms (64 mm ID) have been used for very viscous materials such as invert emulsions.

On helicopters, the central section of the boom may be fixed aft of the engine, but nozzles are then in an up-draught of air near the centre of the rotor. Wooley (1963) suggested nozzle positions outside the trailing vortices. Ideally the nozzle should be below and in front of the cockpit, where there is a down-draught of air. An alternative system with helicopters is to avoid fitting the spray gear directly to the aircraft by using a separate underslung unit (Fig. 13.8). The unit is attached to the cargo hook and is detachable in only a few seconds, thus permitting maximum utilisation of the helicopter. Chemical contamination of the helicopter is less, thus reducing the cost of maintenance. The unit is self-contained and has an engine-driven pump controlled from the cockpit by means of an electrical connection. Ideally, with two units – one spraying while the other is being refilled – the need for the helicopter to land

Fig. 13.8 Underslung unit on helicopter (photo: Simplex).

and wait for another load is eliminated. A standard unit with a 6 kW, four-cycle engine has a 560-litre capacity tank weighing 134 kg when empty. Outputs of up to 350 litres/min can be applied.

Spray nozzles

Droplet spectra of the various spray nozzles used in aircraft are affected by air shear caused by the high-speed slipstream (Yeo, 1961). The angle at which liquid is discharged from the nozzle relative to the slipstream changes the droplet spectra. Hydraulic nozzles angled forwards and downwards into the slipstream produce smaller droplets and a wider range of sizes than nozzles directed downwards or backwards (Figs 13.9 and 13.10) (Kruse *et al.*, 1949; Spillman, 1982). Thus, backwardly directed fan nozzles (8005) were considered suitable for herbicide application. Higher air speeds will increase the proportion of small droplets, but the higher aircraft velocity also changes the effect of the wing tip vortices, thus offsetting the potential drift due to small droplets (Womac *et al.*, 1992).

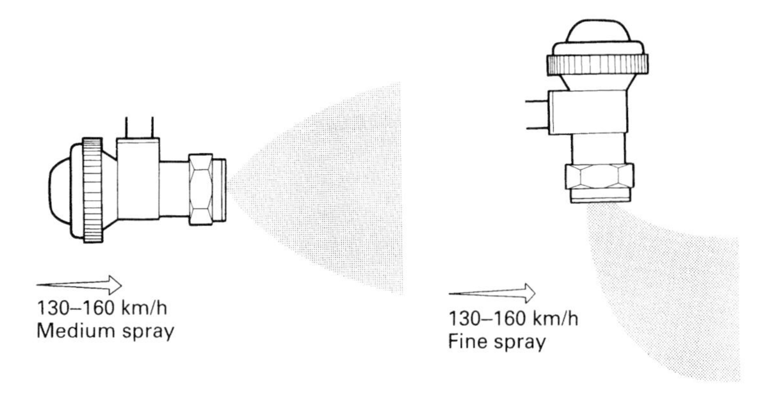

Fig. 13.9 Position of nozzles relative to aircraft slipstream.

Although fan- and cone-type hydraulic nozzles are widely used, a type of deflector nozzle known as the CP nozzle (Fig. 13.11) has been extensively used in the USA. Like other hydraulic nozzles, it is fitted into special nozzle bodies incorporating a diaphragm check valve to provide positive shut-off of the spray even when boom suck-back is not provided. When pressure along the spray boom exceeds 0.2–0.5 bar, a spring-loaded chemically resistant diaphragm is lifted to allow liquid to pass through a filter to the nozzle tip. A PTFE disc should be used to protect the diaphragm when some aggressive solvents are used in the spray. This is particularly important with some ULV formulations.

The CP nozzle is available with three different angles of deflection and easily changed flow rates. A second version of this nozzle has the option of a

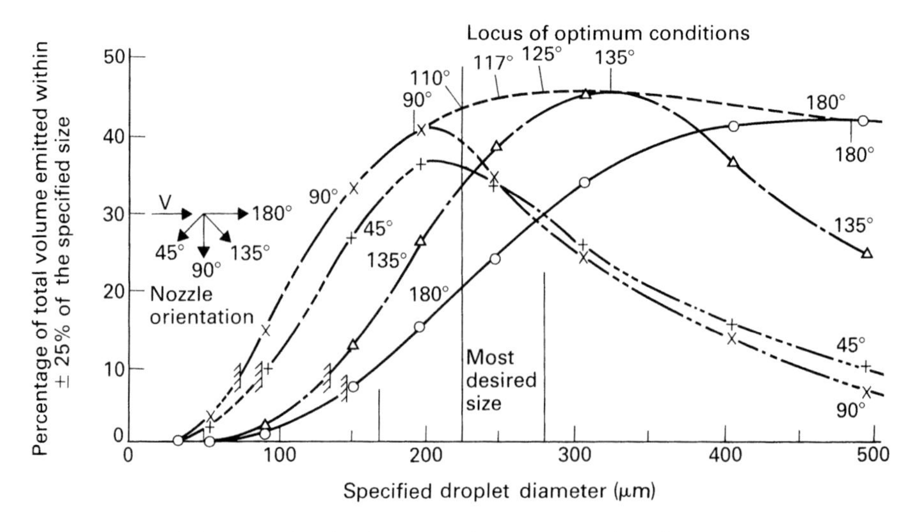

Fig. 13.10 Efficiency of droplet size control with nozzles (8005) set at different angles when aircraft is flying at approximately 46 m/s (from Spillman, 1982).

rear-directed spray without any deflection for larger droplets that are less affected by air shear. A computer model has been developed to advise users of the setting for different spray qualities (Fig. 13.11b). Other special deflector nozzles include the Reglo-Jet, which has a curved plate on which the liquid is fed before atomisation (Fig. 13.12). This has been used primarily for herbicide application. A very coarse spray can be obtained using a cone nozzle with an additional orifice, the 'Rainjet' nozzle. Air deflectors or 'winglets' have also

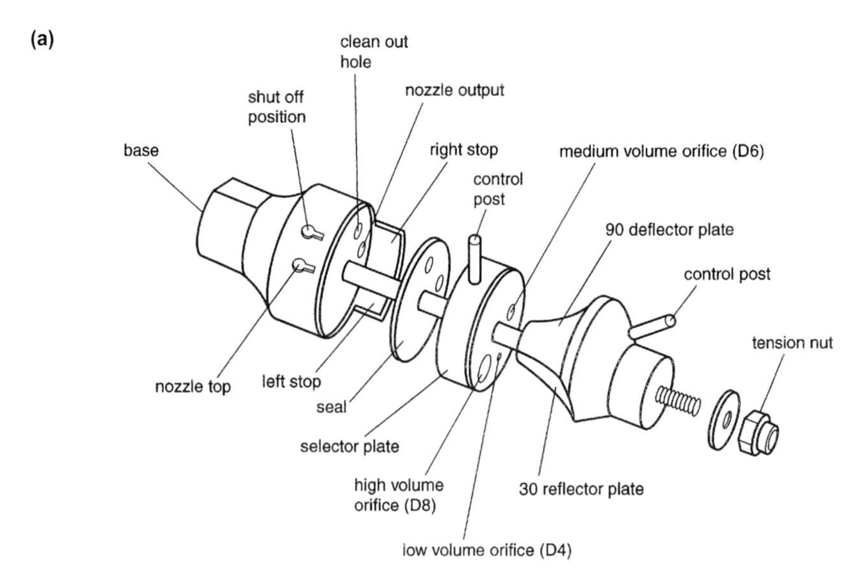

Fig. 13.11 (a) CP selectable nozzle.

(b)

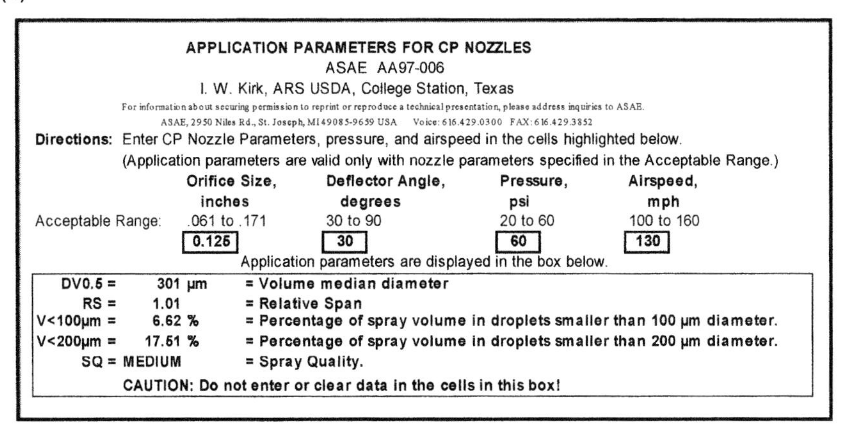

APPLICATION PARAMETERS FOR CP NOZZLES
ASAE AA97-006
I. W. Kirk, ARS USDA, College Station, Texas

For information about securing permission to reprint or reproduce a technical presentation, please address inquiries to ASAE.
ASAE, 2950 Niles Rd., St. Joseph, MI 49085-9659 USA Voice: 616.429.0300 FAX: 616.429.3852

Directions: Enter CP Nozzle Parameters, pressure, and airspeed in the cells highlighted below.
(Application parameters are valid only with nozzle parameters specified in the Acceptable Range.)

	Orifice Size, inches	Deflector Angle, degrees	Pressure, psi	Airspeed, mph
Acceptable Range:	.061 to .171	30 to 90	20 to 60	100 to 160
	0.125	30	60	130

Application parameters are displayed in the box below.

DV0.5 =	301 µm	= Volume median diameter
RS =	1.01	= Relative Span
V<100µm =	6.62 %	= Percentage of spray volume in droplets smaller than 100 µm diameter.
V<200µm =	17.51 %	= Percentage of spray volume in droplets smaller than 200 µm diameter.
SQ =	MEDIUM	= Spray Quality.

CAUTION: Do not enter or clear data in the cells in this box!

Fig. 3.11 (b) Application parameters for CP nozzles, based on computer model.

been used to increase the downward projection of droplets to minimise drift (Womac *et al.*, 1994).

Nozzles are sometimes irregularly spaced along the boom to try to counteract the effect of the propeller or rotor vortex which shifts spray from one side to the other, especially when the flying altitude is less than about 3 m. At greater heights, the maximum horizontal velocity of the down-wash due to the wing-tip vortex is less and turbulence causes sufficient mixing of the spray droplets (Trayford and Welch, 1977). Johnstone and Matthews (1965)

Fig. 13.12 Reglo-Jet deflector nozzle on aircraft to apply coarse spray.

describe experiments with a helicopter to determine the optimum nozzle arrangement. On fixed-wing aircraft extra nozzles are fitted about 1 m to one side of the fuselage and fewer nozzles on the other side, depending on the direction of rotation of the propeller.

An extensive range of hydraulic nozzles is available; nozzle tips can be easily interchanged for different flow rates and mean droplet sizes (Payne, 1998). The disadvantage is that spray booms often have 30–60 individual nozzles, so cleaning and changing tips is a lengthy task. Moreover, droplet-size range is so great that inevitably some spray drifts, even when spray tips are selected for a coarse spray. Viscosity additives have been used with sprays to try to reduce the number of small droplets produced. When production of large droplets (> 500 μm) is essential, the 'Microfoil' nozzle (Fig. 13.13) can be used, but air speed must be less than 95 km/h to avoid droplets being shattered (Table 13.3), so choice of aircraft is limited. Production of droplets of 250 μm is also possible with a transducer-driven, low-turbulence nozzle, but this has such small orifices (125 μm) that 400-mesh filters are needed and wettable powders cannot be used (Wilce *et al.*, 1974; Yates and Akesson, 1976). 'Raindrop' nozzles have been used on helicopters to apply particulate suspensions for mosquito larviciding.

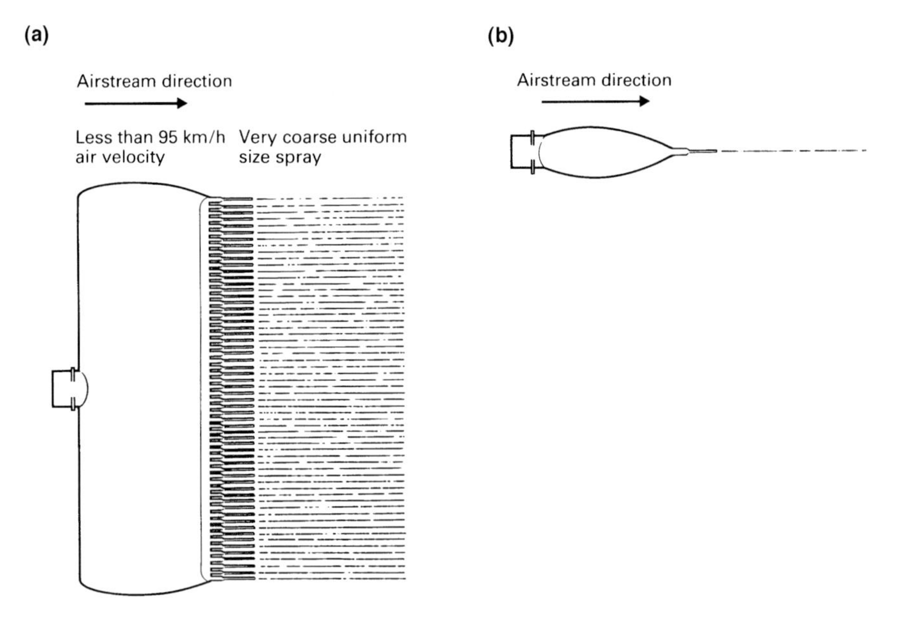

Fig. 13.13 (a) Microfoil nozzle. (b) The nozzle from the side.

A versatile aircraft nozzle is the centrifugal-energy Micronair equipment. An advantage of this type of nozzle is that greater control of droplet size can be achieved (Parkin and Siddiqui, 1990; Hooper and Spurgin, 1995). Any adjustments can be made very rapidly, as there are only a few units on each

Table 13.3 Critical air velocity for various droplet sizes (water)

Droplet diameter (mm)	Shatter velocity (km/h)
100	322
170	241
385	161
535	137
900	105

aircraft. This nozzle consists of a cylindrical, corrosion-resistant, monel metal wire gauze rotating around a fixed hollow spindle mounted on the aircraft wing (Fig. 13.14). Speed of rotation is controlled by adjustment of the pitch of a series of balanced blades, which form a fan. The blades are clamped in a hub, which carries the bearings. To adjust the angle, bolts are slackened on the clamping ring; the blades are twisted to the correct angle setting on the clamp ring and the bolts retightened evenly to nip the blades. Spray liquid is pumped through a boom via a variable restrictor unit to the hollow spindle in which there is a shut-off valve. Opening this valve allows liquid to hit a deflector to spread it in a diffuser tube. An initial break-up here provides even distribution of liquid on the gauze.

The number of units fitted to aircraft will depend on the wing span, intended swath and volume of spray being applied. A similar number of units can be

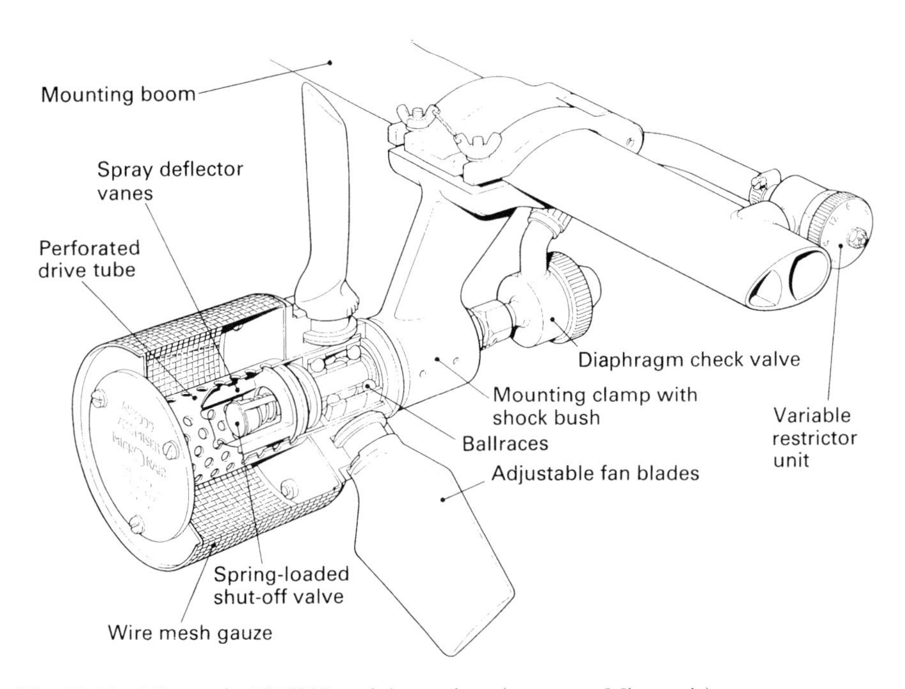

Fig. 13.14 Micronair AU5000 aerial atomizer (courtesy: Micronair).

fitted to helicopters, but larger propeller blades have been used. The layout of Micronair units is shown in Fig. 13.15. The earlier large AU3000 unit was fitted with a hydraulically operated brake for use in an emergency or during ferrying. The newer AU5000 atomiser is now preferred for normal agricultural spraying; 6–10 of these smaller units are normally installed instead of 4–6 AU3000 units. The AU4000 is recommended for high rotational speeds on fast aircraft, while the AU7000 is intended for small helicopters and slower fixed wing aircraft. Blockages are rare with these atomisers as small orifices are not required to break up the liquid, and wettable powders and suspensions are more easily applied than with hydraulic nozzles.

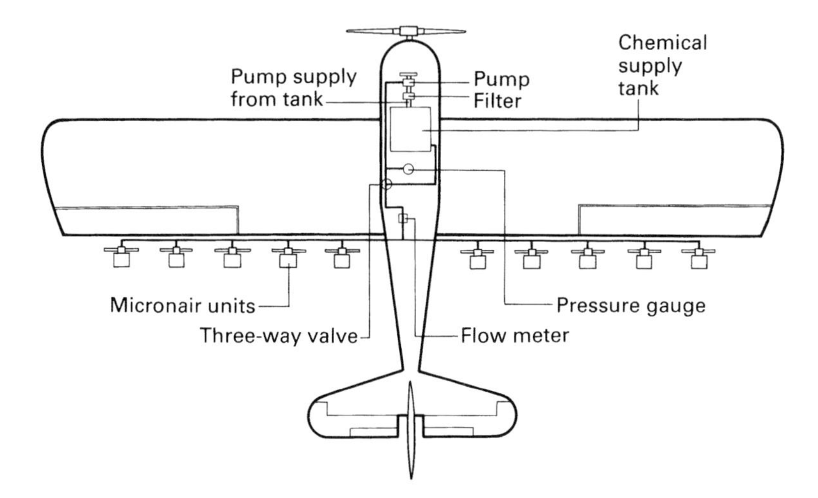

Fig. 13.15 Typical layout of Micronair AU5000 installation.

The variable restrictor unit (VRU) has a single orifice plate with a series of orifices. Numbers 1–7 (0.77–2.4 mm) are intended for ULV application and 8–14 (2.65–6.35 mm) for conventional LV spraying. The standard plate has all the odd number restrictor sizes 1–13 (see Table 13.4). Alternative plates are available. Care must be taken to install the unit so that liquid flows through it in the correct direction.

The angle of the fan blades is determined by first selecting the speed of rotation that is expected to produce the required droplet size (Fig. 13.16). Then, knowing the air speed of the aircraft and flow rate through the atomiser, charts as shown in Fig. 13.17 are examined to determine the angle of the blades. The blades are usually set at 40–70° on the AU5000 (Fig. 13.18). A check should be carried out with the particular chemical formulation being applied to determine the droplet sizes obtained, as the manufacturers' charts are intended only as a guide. An electronic application monitor can be fitted to provide the pilot with an accurate record of flow rate, quantity of liquid emitted and atomiser rotational speed.

Micronair equipment is particularly suitable for producing droplets of less

Table 13.4 Range of flow rates with Micronair units

Micronair restrictor	Flow rate (litres/min) at different pressures (bar)		
	2	2.8	3
1	0.15	0.27	0.42
2	0.20	0.30	0.56
3	0.35	0.63	0.99
4	0.51	0.90	1.42
5	0.71	1.27	1.98
6	1.02	1.81	2.8
7	1.43	2.54	3.97
8	1.84	3.27	5.1
9	2.66	4.72	7.38
10	3.47	6.18	9.66
11	4.90	8.7	13.6
12	7.56	13.5	21.0
13	8.79	15.6	24.4
14	14.3	25.5	39.8

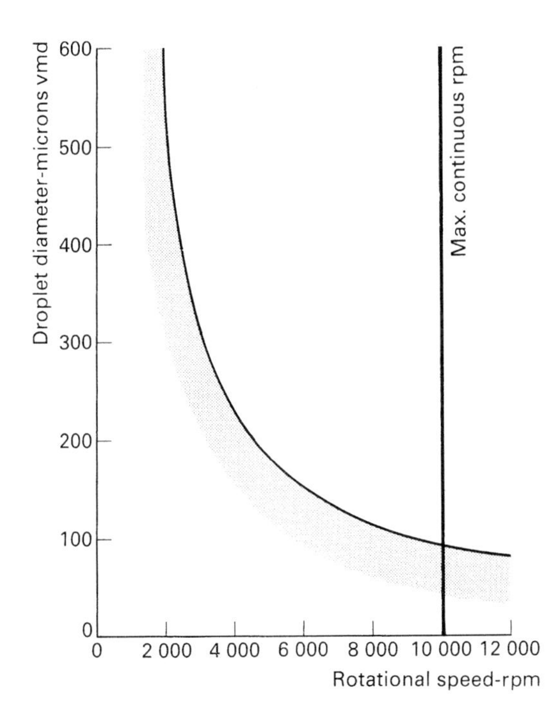

Fig. 13.16 Droplet size in relation to the rotational speed of the Micronair AU5000 unit (courtesy: Micronair).

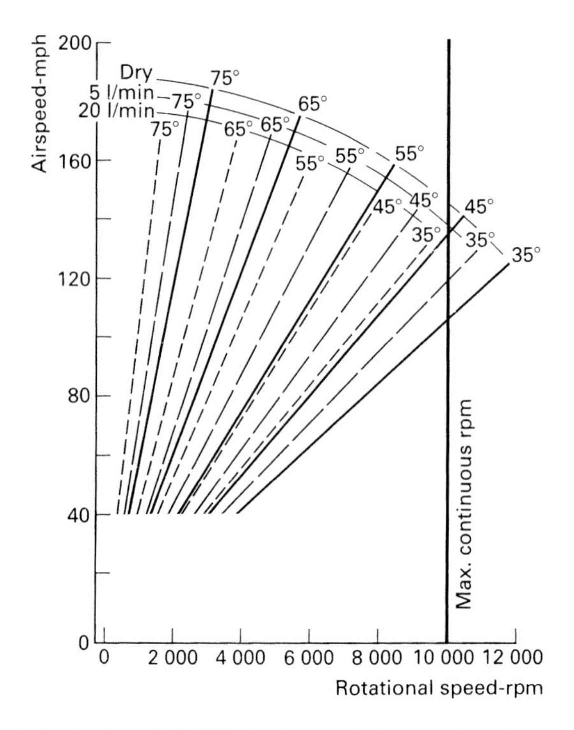

Fig. 13.17 Speed of rotation of AU5000 unit in relation to aircraft speed, blade angle and flow rate (courtesy: Micronair).

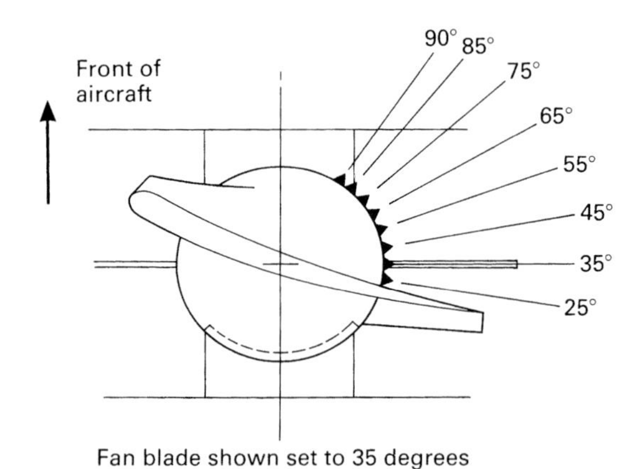

Fig. 13.18 Fan blade angles for AU5000 unit.

than 100 μm, owing to the elimination of the need for higher pressures (> 10 bar) required if hydraulic nozzles were used. The use of Micronair equipment increases aerodynamic drag, and the spray distribution is uneven if large droplets are applied and the aircraft flies too low because each aircraft has relatively few nozzles. Various attempts have been made to produce a small spinning spray nozzle, which fits the check valve of a conventional hydraulic nozzle (Spillman and Sanderson, 1983). Electrically driven nozzles have been designed to overcome the need to use a propeller (Skoog *et al.*, 1976). The Micronair AU6539 electric atomiser uses a cylindrical gauze cage mounted directly on an electric motor. This unit is intended mainly for use on helicopters, and the power of the motor has been selected to be compatible with the limited capacity of many helicopter electrical systems. An advantage of the system is that the pilot can control rotational speed and easily adjust droplet size when required, irrespective of the forward speed of the aircraft. Such nozzles are particularly useful on the slower aircraft, including helicopters. For spraying against tsetse flies in Nigeria, a helicopter was fitted with up to six spinning discs, the speed of which was controlled by means of a computer mounted to the rear of the engine (Spielberger and Abdurrahim, 1971). Multiple spinning discs have also been used to apply baculoviruses to control pine beauty moth (Entwistle, 1986, Entwistle *et al.*, 1990) and pine sawfly in forests (Doyle, 1988). To generate sufficient numbers of droplets, a spray unit based on the X-15 stacked disc atomiser was developed. Six units with 30 discs driven by a direct drive electric motor were used at a flying speed of 45 knots with a lane separation of 40–50 m.

Parker *et al.* (1971) and Parkin and Newman (1977) have examined prototype venturi nozzles for ULV application. Thermal nozzles are used to produce aerosols for vector control (Park *et al.*, 1972), but there is less control of droplet size than with centrifugal-energy nozzles. Specialised equipment for use on aircraft has been developed to release known volumes of liquid into rivers. The 'vide-vite' system was used to apply a larvicide, temephos, into West African rivers for the control of *Simulium damnosum*, vector of onchocerciasis (Lee *et al.*, 1975a; Baldry *et al.*, 1985).

Aerial application of dry materials

Hazardous pesticides formulated as dusts should not be applied from aircraft owing to the high drift loss potential, although fertilisers are applied. Microgranules and granules can be applied either by ram-air spreaders or spinners (Brazelton *et al.*, 1971; Spillman, 1980; Bouse *et al.*, 1981). An air speed as high as 240 km/h directly from the propeller slipstream is used in the ram-air type on fixed-wing aircraft to distribute the material in the wake of the aircraft. Air enters the front of a tunnel sloped like a venturi tube, and with internal guide vanes or channels. A control gate fitted under the hopper can be opened to a preselected position determined by flight calibration, and is also the shut-off valve (Fig. 13.19). Metering can be improved by using a vaned rotor system

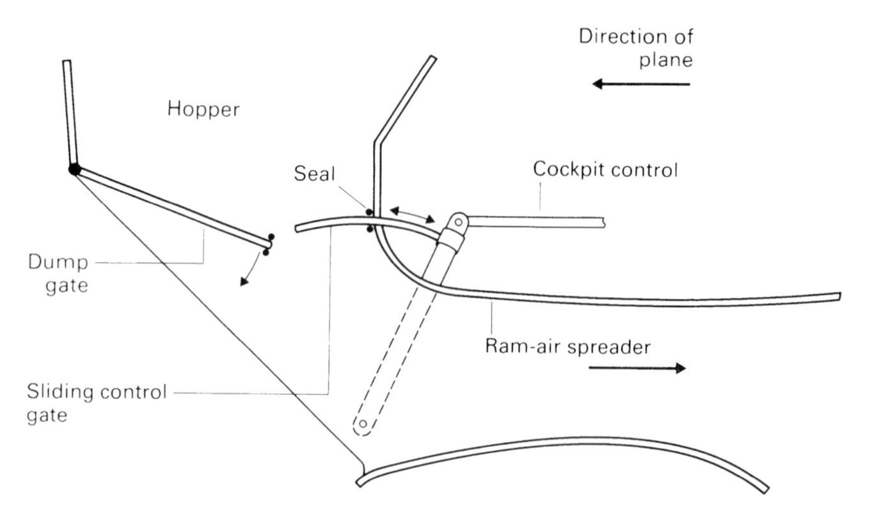

Fig. 13.19 Granule application from aircraft: simple diagram to show sliding metering gate (after FAO, 1974).

(Bouse *et al.*, 1981). A revolving agitator may be fitted above the throat of the metering gate. A windmill placed in the propeller slipstream may drive this agitator via a reduction gear. Ram-air devices suffer from high drag and low spreading power. Stephenson (1976) has studied the effect of the spreader on aircraft performance. The weight penalty is offset by gains from a wider swath, except for light aircraft. Drag is less with a tetrahedron spreader, which gives a wider swath (Trayford and Taylor, 1976). Ducting a limited flow of fine particles to the wing tips has been investigated by Lee (1976), who showed that wide swaths were not possible unless granules were released remote from the aircraft centre line. Experimentally, a conveyor duct was built into the trailing edge of the wing along which were a series of discharge outlets (Harazny, 1976).

On helicopters a separate blower unit driven by the engine forces air along two ducts positioned at the base of the side tanks and out on short booms.

An alternative system to the ram-air spreader is to have two spinners, each driven by an electric motor or hydraulically activated (Fig. 13.20). These revolve in opposite directions, throwing granules outward from the front of the spinner, so that swaths up to 15 m wide can be obtained. Deflector plates protect the landing gear and propeller. Distribution of granular material by a spinner is reported by Courshee and Ireson (1962) and by Hill and Johnstone (1962). The metering gate is the same type as that used in ram-air spreaders. Lee and Stephenson (1969) developed a rotary-cylinder spreader. Two were mounted on an aircraft with a central spreader so that the swath was widened from 10 to 18 m, but the rotary-cylinder spreader is not suitable for low-density or small particles.

Apart from spreading pesticides, aircraft have been used to distribute biological agents as discussed in Chapter 17.

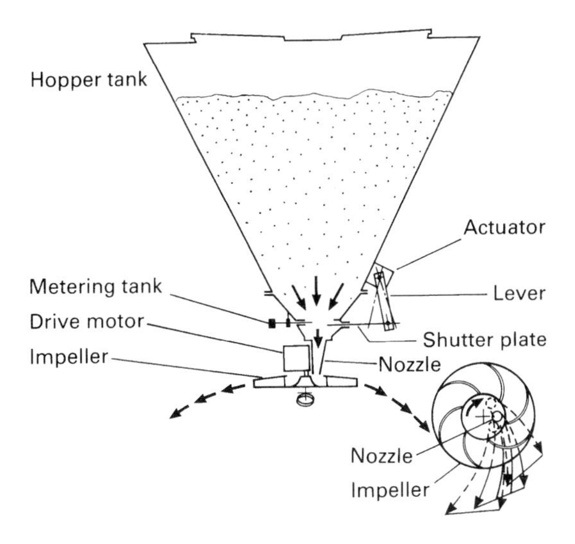

Fig. 13.20 Kawasaki granule distributor using a spinner on a helicopter (from Quantick, 1985a).

Flight planning

Aircraft flying height

Amsden (1972) has listed the factors which determine the height from which pesticides should be applied. These are

(1) the velocity of the crosswind component relative to the flight path
(2) the aircraft design characteristics
(3) the composition of the spray spectrum being produced
(4) the specific gravity of the spray liquid (or particles)
(5) the rate of evaporation from the spray droplets.

All these factors may vary from one operation to the next, and even within a single flight. As discussed earlier in Chapter 4, the relationship between spraying height and the crosswind component can be expressed as

$$HU = C$$

where H is the height of the wing or rotor above the crop (m) and U is the crosswind velocity in km/h.

The constant cannot be calculated, but is estimated by observing the biological effectiveness of sprays applied under a number of known conditions with a swath width determined at a particular height. Overdosing and underdosing is liable to occur if the aircraft is too low. Excessive drift is liable to occur if the aircraft is flown too high. Amsden (1972), illustrating the effect of a crosswind on aircraft height (Table 13.5), points out that the maximum wind speed is dictated by the safety of the pilot. At speeds in excess of 25 km/h,

Table 13.5 Variation in flying height with wind speed where $HU = 80$ (after Amsden, 1972)

Height (H) (m)	Wind speed (U) (km/h)	
2.0	40	
2.67	30	
3.2	25	↑
4.0	20	operating
5.0	16	limits
5.7	14	↓
8	10	

conditions are normally so turbulent that the pilot will find it too uncomfortable to fly at the very low altitude required. If the pilot continues to spray and flies higher, excessive drift will occur. At the other extreme, the distance between the aircraft and the crop should not exceed half the wing span if full use is to be made of the downwash of turbulent air, so there is a minimum wind speed for a given *HU* factor. In the example in Table 13.5, these limits are 14–25 km/h for an aircraft with an 11 m wing span. According to Amsden (1972) *HU* values are usually between 40 and 90. The effective crosswind speed must be calculated if wind direction is at an angle to the flight path (Table 13.6).

Table 13.6 Calculation of crosswind velocity U for winds at different degrees off true crosswind (after Amsden, 1972)

Degrees	Correction factor	Effective crosswind (km/h) where $U = 20$
0	× 1.0	20
20	× 0.94	18.8
40	× 0.77	15.4
60	× 0.5	10
80	× 0.17	3.4
90	× 0.0	–

In contrast, Bache (1975) suggests that distribution of small droplets (< 60 μm) is relatively insensitive to changes in windspeed; therefore consistent deposition downwind over particular crops can be obtained by adjusting flying heights with time of day. Thus, in one example he suggests a flying height of 7 m above the crop during the morning, reducing to 5 m at dawn and dusk to achieve maximum deposition 50 m downwind. The latter technique is probably only suitable when large areas of a single crop are being treated. Johnstone and Johnstone (1977) recommend that spraying at 5 litres/ha or less with involatile droplets smaller than 120 μm VMD should cease if wind velocity exceeds 12.6 km/h. The theory of downwind dispersion is discussed in greater detail by Bache and Johnstone (1992).

Sensitivity analyses using the FSCBG model developed for the US Forest

Service (Teske *et al.*, 1993) confirmed that small changes of release height can significantly affect spray deposition and drift (Teske and Barry, 1993). Hooper and French (1998) used the model to examine ULV spray deposits for locust control. In contrast to the more complex models such as FSCBG and AgDrift, Craig *et al.* (1998) used a simple Gaussian diffusion model to predict aerial spray drift deposition. Koo *et al.* (1994) developed a laser system to measure aircraft height accurately. The main problem was that the crop canopy can interfere with the laser pulse to offset the above- ground level (AGL) altitude.

In general, aerial sprays are more effective when there is a crosswind and the aircraft is flying at the appropriate height.

Track separation (swath width)

The swath treated by an aircraft will depend on the type of aircraft, its flying height, droplet or particle size and wind conditions prevailing at the time of application (Parkin, 1979; Kuhlman, 1981; Parkin and Wyatt, 1982; Woods 1986) (Fig. 13.21). The minimum swath may be determined when the aircraft flies into wind, although use is made of a crosswind during normal commercial applications so that adjacent swaths overlap, even if there is little wind. The swath obtained with two aircraft of different dimensions can be compared by flying each one into wind with the wing at a height exactly one-half span above a line of targets. This height can also be assumed to be about the maximum height likely to be used for crop spraying, although greater heights are used, for example under carefully monitored conditions when spraying forests. Atkinson *et al.* (1968) studied the distribution of pasture seed and recommended that swath width should be expressed as a multiple of the standard

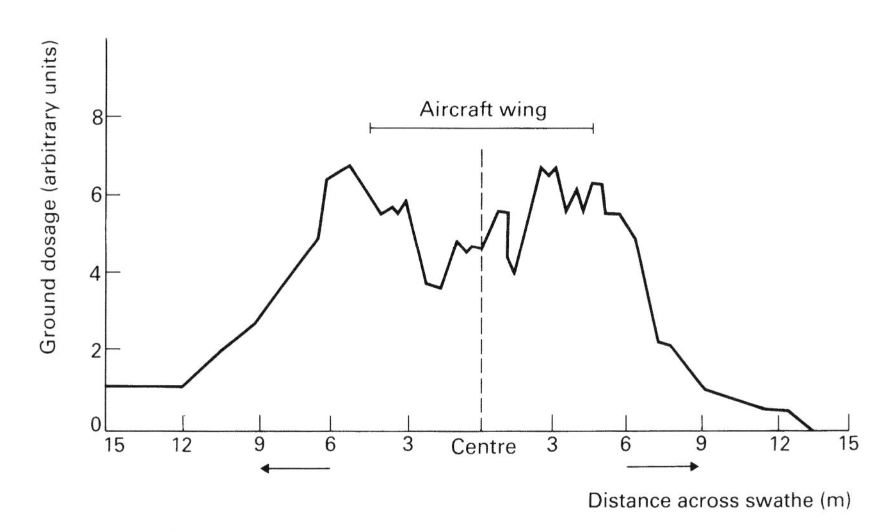

Fig. 13.21 Deposit distribution achieved with aircraft flying into wind.

deviation of the mean deposit achieved across a single swath. They recommended using a track spacing which is three and a half times the standard deviation, but a greater overlap is achieved if the track spacing is only twice the standard deviation, as 95% of the spray deposit would normally fall within these limits. Indeed, half the overall single swath width is normally used when marking out a field to ensure adequate incremental dosing, and thus sufficiently even application. A narrower track spacing (one-third or less of the overall swath) may be used when applying unselective chemicals. Wider swaths are obtained by applying smaller droplets or particles. Wider booms on larger aircraft operated at a higher altitude will increase the swath. Sometimes, instead of determining the swath, too wide a track spacing is selected so that an area can be covered more quickly, thereby reducing application costs. Inadequate coverage or, conversely, excessive overlap, may result in poor control or crop damage, and the pilot needs guidance so that successive flights over the area being treated are correctly spaced to ensure as uniform a coverage as possible. This is particularly true with herbicide applications when an overdose may damage the treated crop and, conversely, an underdose fail to control the weeds. Greater accuracy is needed with coarse sprays and application of granular materials, as downwind movement is minimal compared with aerosols and fine sprays.

Particular attention is being given to assessments of spray drift beyond the treated area (e.g. Riley *et al.*, 1989) to determine the width of a buffer zone required around sensitive areas during pesticide application (Payne *et al.*, 1988). A number of models are being developed to predict aerial spray drift (Atias and Weihs, 1985; Mickle, 1987; Parkin, 1987; Barry *et al.*, 1990; Teske *et al.*, 1990). Validation of these models is needed.

Track guidance

Several navigational systems have been used for track guidance. Generally these can be divided into two categories: those using an external reference and those using only on-board equipment. Early systems such as the Decca 'Agri-Fix' (Walker, 1973) relied on two mobile ground stations transmitting signals which intersect to produce a family of hyperbolic position lines. A receiver and left/right indicator in the aircraft enable the pilot to follow the hyperbolic tracks. This system has the disadvantage that the ground stations must be correctly aligned relative to the required track. Another system used two or more ground-based transponders placed in or around the area to be sprayed. Equipment on the aircraft interrogates these transponders and a computer calculates its position relative to them. The computed position is compared with a grid corresponding to the required tracks and a left/right indicator guides the pilot. These systems are typically accurate to within 1 metre, and have the advantage that the layout and spacing of the tracks can be defined independently of the position of the transponders. However, both transponders and airborne equipment are expensive, and a high utilisation is necessary for this approach to be cost-effective. Commercially available systems include the Flying Flagman from Del Norte and the Maxiran from Maxiran Inc.

Established chains of ground stations have also been used for track guidance. The VLF/Omega system is primarily intended for long range navigation, but has been used for track guidance in drift spraying with wide swath widths for tsetse control and similar work. The resolution and accuracy of this system is marginal even at very wide swath widths, and it is difficult to achieve reliable and repeatable performance. A more satisfactory system is Loran, using chains originally intended for marine and long-range aircraft navigation. These provide much better accuracy and resolution than VLF/Omega but the chains are available only in certain areas. This system has provided effective track guidance for mosquito spraying in the United States where there are extensive Loran chains.

Most current track guidance systems are based upon the US Department of Defence Global Positioning System (GPS). This uses a constellation of 24 satellites orbiting the earth and transmitting very accurately timed signals. The position of the aircraft is computed several times a second on the basis of signals received from the satellites by an on-board receiver. The aircraft system incorporates a computer that compares the actual position of the aircraft to its intended track (based on pre-programmed track spacing, swath pattern, etc.) and gives the pilot a visual indication to fly left or right so as to remain on track. This indication is usually by a 'light bar' consisting of rows of lights indicating cross-track distance and (in many systems) the angle of intercept of the aircraft's actual track to the required track. Several systems also incorporate a LCD screen providing a display of the area sprayed, required and actual tracks and other information (Fig. 13.22). Many of these systems also allow the job to be planned on the ground (using a digitised map or co-ordinates of the spray area) and loaded into the aircraft computer before flight. This facility is particularly useful for spray operations over unmarked or poorly defined areas such as in forest and locust spraying, etc.

Most agricultural GPS systems provide a data-logging facility that records the track of the aircraft and, often, the performance of the spray system throughout the flight. This data can be replayed on a standard office personal computer, overlaid on a digital map, printed out, archived or loaded into a GIS database. This provides conclusive proof of work done and also assists in the analysis of any claim regarding off-target application.

Unfortunately, the accuracy of a basic GPS system is limited to about ± 20 m (or worse when the US Department of Defence activates 'selective availability'). This is not adequate for most agricultural track guidance, and it is necessary to improve the overall system accuracy by using differentially corrected GPS (DGPS). This requires one or more ground reference stations in accurately determined locations in the same general area as the aircraft. Each ground station computes the instantaneous error of the signal from the GPS satellites and transmits it to a receiver on the aircraft. This error signal is then used to correct the 'raw' GPS signal being received from the aircraft. DGPS typically achieves accuracies of 1–2 m in an agricultural environment.

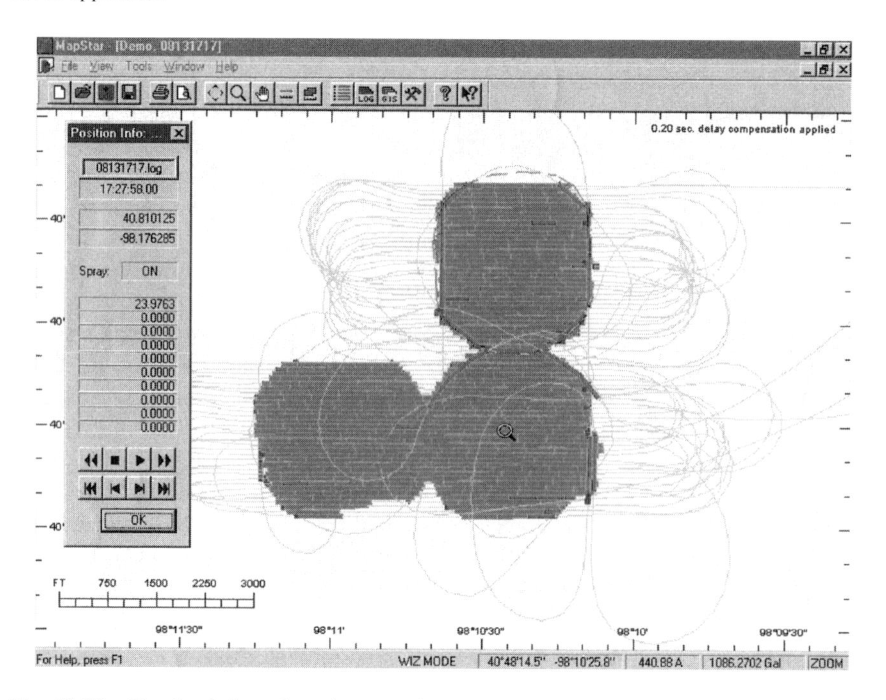

Fig. 13.22 Track of aircraft and sprayed area recorded by GPS system (SATLOC).

The reference station may be a portable unit sited near the spray area, but most users now prefer to subscribe to commercially available correction signals broadcast from communications satellites. Differential correction signals are also available from coastguard transmitters and FM radio-based networks in some areas. Differentially corrected agricultural GPS track guidance systems are available from several manufacturers, including AgNav, Del Norte, Satloc, UTS, WAG and others.

Flagging

In the absence of modern GPS equipment, the old method of marking field crops is by having two or more people carrying flags of a brightly coloured material. Haley (1973) described traditional methods of flagging. Flags are often yellow, orange or white to contrast with green vegetation and should be about 1 m square so that they are easily seen. The pilot prefers to fly along the rows of a field crop, and has little difficulty in keeping on a straight course between flag-people standing at each end of the field, indicating the position of the swath to be treated. However, this can be done only if the wind direction is across the rows. When the land is undulating or the field is excessively long (over 3 km), additional flag-people may be needed at intermediate positions through the field. Ideally, the position for the flag-people is measured earlier, and short pegs hammered into the ground so that they can move quickly to their next position as soon as the aircraft has flown by. The pilot may fly directly over the flag but, when highly toxic materials are used, the persons holding the flag should be positioned on the upwind side of the swath, or they should move rapidly upwind as the aircraft approaches to minimise chemical contamination (Whittam, 1962). If possible, they should be positioned back from the edge of the field, but a hedge or other obstruction may prevent this. The flag-people should be well protected with suitable clothing, depending on the pesticide being applied. Other people should be kept well clear of treated fields, not only during spraying operations but also afterwards for a period, the length of which will depend on the chemical applied. When a crop is treated several times with a relatively safe pesticide, the flag may be directly under the centre of the aircraft on one application and the wing or rotor tip at the next, as this may improve the uniformity of coverage.

A bright battery-operated signal light can be used instead of a flag when pesticides are applied at night. Xenon-gas-filled tubes flashing at 60–80 times per minute have also been used in daytime. Lamps have been used in conjunction with a compass for tsetse spraying (Miller and Chadwick, 1963).

When fields are treated many times, it is often more convenient to use a series of fixed markers. In annual crops the main problem with fixed markers is the need to adjust their height in relation to crop height. Markers may be hazardous if they are set too high at the start of the season when the aircraft is flying low, whereas later in the season they may be obscured by foliage. Successive fixed markers across a field need to be painted with contrasting colours (e.g. orange, white, yellow, black) and shapes (e.g. triangular, circular, square, rectangular) so that the pilot can readily identify which is the next swath to be treated. Bright fluorescent pigment paints of different colours are ideal. Ground markers such as an upside-down white or yellow plastic laundry basket, about 0.6 m high, have also been used to warn pilots of hazards such as power and telephone lines crossing fields (Keller, 1971). A serious drawback to fixed markers is that they cannot readily be moved. This may be necessary if the wind direction has changed from the time of one application to the next. This problem may be overcome by marking the field in two directions at right angles to each other. On each visit the pilot selects the more appropriate set of markers.

Flagging is more complicated in hilly or mountainous areas and also in forest or swampy areas, so modern systems using GPS should be used. Helium-filled balloons of different colours sometimes up to 1 m diameter have been used as fixed markers in forest areas where there is a sufficient opening in the canopy. However, they often get entangled in the foliage, burst or deflate, especially if set at appropriate places one or two days before treatment. A cluster of small balloons to allow for some deflating can be mounted at the top of a pole attached to a tractor. The more aerodynamic-shaped balloons such as the 'Kytoon', which flies like a kite, are less likely to be dragged downwards in a high wind, but is impractical as it must be lowered immediately the aircraft approaches to prevent the plane's vortex from hitting it. Otherwise the Kytoon may be damaged as it gyrates wildly above the trees (Whittam, 1962). Smoke flares have been used when there is radio contact between the pilot and ground crew, enabling flares to be used in the correct sequence and eliminating the fire hazard.

Logistics

The quantity of spray applied Q will depend on the throughput of each nozzle, the number of nozzles, swath width and flying speed thus:

$$Q = \frac{10Nq_n}{SV_s}$$

where Q is the output required (litres/ha), N is the number of nozzles, q_n is the throughput of a nozzle (ml/s), S is the swath width (m) and V_s is the flying speed while spraying (m/s).

The output and number of nozzles are selected to give the required output for a given aircraft for which the swath width and flying speed have been determined.

The time to fly 1 ha is 600/SV minutes, so the volume of spray applied per minute (QT) is

$$Q_t = \frac{QSV}{600}$$

where S is the swath width (m) and V is the flying speed (km/h). This should be checked by putting a known quantity of liquid in the aircraft and spraying at the correct pressure for a definite time. Normally this involves a flight check, although ground checks are possible if the pump can be operated at the correct speed. The tank is then drained to determine the quantity sprayed. Alternatively, the level of the tank before and after spraying can be measured if the tank has been properly calibrated. As with tractor sprayers, small adjustments of volume can be made by changing the pressure, otherwise different nozzles will be required. The number of hectares treated with one load is Q_f/Q where Q_f is the quantity of spray mix loaded per flight and the output is Q litres/ha. The number of hectares covered for different swath widths and field lengths is indicated in Table 13.7, so the load can be adjusted to avoid the aircraft running out of spray in the middle of a swath.

The approximate time needed to spray an area can be determined by reference to Table 13.8 in which the coverage (ha/min) is given for different combinations of swath width and flying speed.

Flight pattern

Pilots will normally fly a series of passes, gradually moving upwind across the area requiring treatment. At the end of each pass, pilots have to complete a

Table 13.7 Hectares covered for given field lengths and track spacings

Field length (m)	Track spacing (m)			
	7.5	10	15	20
250	0.19	0.25	0.38	0.5
500	0.38	0.5	0.75	1.0
750	0.56	0.75	1.13	1.5
1 000	0.75	1.0	1.5	2.0
2 000	1.5	2.0	3.0	4.0
3 000	2.3	3.0	4.5	6.0
4 000	3.0	4.0	6.0	8.0
5 000	3.75	5.0	7.5	10.0

Table 13.8 Hectares/min covered with different velocities and track spacings

Velocity (km/h)	Track spacing (m)			
	7.5	10	15	20
100	1.3	1.7	2.5	3.3
120	1.5	2.0	3.0	4.0
140	1.8	2.3	3.5	4.7
160	2.0	2.7	4.0	5.3
180	2.3	3.0	4.5	6.0

procedure turn. Initially, as they approach the end of a pass, they increase power, shut off the spray, pull up sharply to 15–30 m, turn away about 45° and then bring the aircraft round to approach the next swath. The power required will depend on the load and the height of obstacles, but adequate speed and power are essential to guard against stalls or incipient spins. Sometimes a pilot may prefer to fly a 'race-track' pattern which allows a wider turn, but would necessitate additional flag-people. The race-track or round-robin pattern is useful when a number of small fields are located close to each other and can give a more even spray coverage as the aircraft is flying successive passes in the same direction. When obstructions are close to the edge of the field, the pilot will normally fly one or two passes along the edge and along each headland to 'finish off' after completing the main part of the field. Useful advice to pilots is given in the *Handbook for Agricultural Pilots* (Quantick, 1985b).

Airstrips

Agricultural flying is often from unprepared airstrips or ordinary grass fields. Some governments have specific regulations on the size and condition of airstrips which can be used. In general, a strip should be about 30 m wide with a slight camber to permit drainage and at least twice as long as the distance taken by the aircraft to take-off. Longer strips are essential if there is an obstacle such as a low hedge at the end. The whole area around the strip should be as clear as possible of trees and bushes. Ideally the surface should be dry, smooth and with grass cut shorter than 100 mm, otherwise it clings to the wheels and delays take-off. When dry earth strips are used, the engine air filter must be cleaned frequently. The surface of the airstrip can be sprayed with used oil or water to reduce the dust problem. All strips should be checked regularly by driving over them in a vehicle at 40 km/h or more when excessive bumpiness is soon apparent.

The strip is widened at one point to allow the aircraft to turn around and load. The loading bay is usually at the end from which take-off normally commences, but operations are speeded up if there is sufficient space to have a long strip with a central loading area. The loading area must be accessible to ground vehicles without it being necessary for them to encroach on the strip. Aircraft should be refilled as rapidly as possible, and a mixing unit with a high-capacity engine-driven pump (up to 300 litres/min) is essential, and may be mounted on the support vehicle. When open tanks are used, foam can be a problem if air is trapped in the spray liquid during mixing. Adding a small quantity of a silicone antifoam agent can reduce such foam. Some countries have regulations concerning the mixing of pesticides (Brazelton and Akesson, 1976). Using closed-system mixing units with which a precise quantity of pesticide is transferred from its commercial container to the mixing tank and later pumped directly into the aircraft can reduce the hazards of handling concentrated materials. Dry materials are normally handled by special equipment and can be loaded through the opening on the top of the hopper.

The load will depend on the design of the aircraft, quality of the airstrip, altitude and air temperature. The pilot is responsible for determining what is a safe load for a given airstrip; the first few take-offs at a new strip will require a light load until the pilot is used to the local conditions. The pilot may need to reduce the load normally taken at a particular airstrip if the surface is softened or otherwise affected.

Great care must be taken to avoid overloading the aircraft. The hopper should be checked to ensure it is empty before reloading. Putting an excessive load into an aircraft accidentally by using the same volume of a higher-density material must also be avoided. This could occur if a technical material of density greater than water is used without mixing with water or hydrocarbon diluents.

Normally an aircraft will be used to apply a wide range of pesticides, so it is vital that as soon as the aircraft has completed treatment of an area or at the end of each day's (or night's) work, the whole aircraft should be cleaned, because chemical contamination can cause serious damage to the fabric. The hopper, pump and nozzles should be flushed clean with an approved detergent and clean water. Any spray gear which has been used for herbicides should not be used to apply fungicides or insecticides on susceptible crops such as cotton. If this is impractical, rigorous cleaning to a carefully devised schedule should be followed by complete replacement of all hoses and plastic components, which could have absorbed herbicide. Household ammonia may be added to the washing water, provided there are no brass components in the spray gear. All the washings must be carefully disposed of according to local regulations. Certain spray liquids may require special cleaning materials. Proper cleaning of the aircraft is essential to minimise problems due to corrosion.

Aircraft operations

The productivity of an aircraft depends on a number of parameters, including the size of the aircraft load, swath width, aircraft speed, size of the fields and distance to the refilling point. The Baltin formula (Baltin, 1959) expresses these parameters as follows:

$$t = 10^4\left(\frac{T_r q}{Q_f} + \frac{1}{V_s S} + \frac{T_w}{SL} + \frac{2aq}{V_f Q_f} + \frac{C}{V_f F}\right)$$

where t is the work time per hectare (s/ha), T_r is the time for loading and taxiing (s), q is the application rate (litres or kg/m^2), Q_f is the quantity of chemicals loaded per flight (litres or kg), V is the flying speed when spraying (m/s) and V_f the flying speed when ferrying (m/s), S is the swath width (m), T_w is the time for one turn at the end of spray run (s), L is the average length of fields (m), C is the average distance between fields (m), F is the average size of fields (m^2), a is the average distance from the airstrip to the fields (m).

A similar equation (the Baltin–Amsden formula) is available in imperial units (Amsden, 1959). Interflug (1975) has given a more detailed formula. More detailed discussion on the productivity of aircraft is given by Quantick (1985a). Akesson and Yates (1974) show the effect of variations in swath width, field length, ferry distance, loading time, application rate and payload on productivity in hectares per hour (Fig. 13.23). Each factor was varied separately, while median values were used as constant values for the other factors. These values were as follows: swath 12.2 m; field length 8.05 km; ferry distance 3.2 km; loading time 2 min; application rate 112 kg/ha; payload 907 kg; flying speed 144.8 km/h; and turn time 0.5 min.

Highest productivity is obviously favoured by long fields, wide swaths, low application rates, short ferry distance and a large load. Agricultural planners should consider field shape if aerial application is anticipated, and provide long runs for aircraft. Higher flying speeds favour fixed-wing aircraft in contrast to helicopters, but the ferry distance for the latter may be negligible. In most aerial application work a positioning time must be considered unless there is sufficient work to keep the aircraft occupied in one area for a period of several weeks. Lovro (1975) has developed a technique to calculate the optimum area to be treated from one or more airstrips. When an aircraft has to be moved to different widely spaced farms, sufficient time must be allowed for positioning, particularly as inclement weather can delay the arrival of aircraft.

The cost of operating an aircraft includes not only fuel and maintenance costs related to the number of flying hours, but also insurance and salaries,

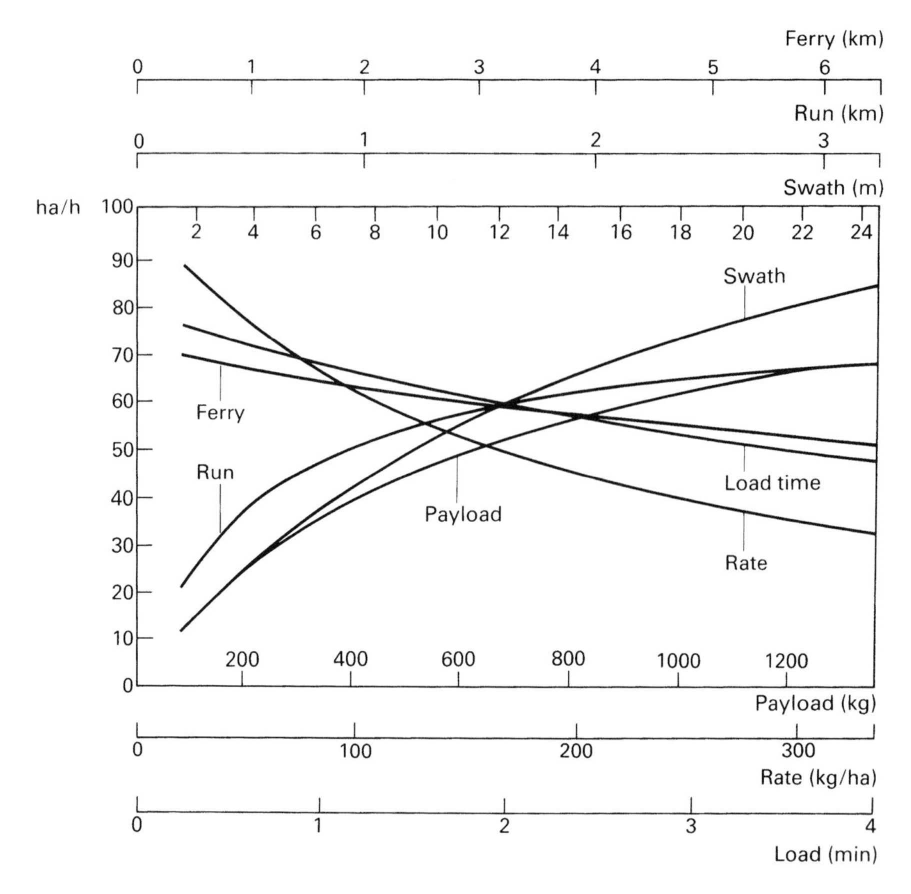

Fig. 13.23 Operational analysis for a fixed-wing aircraft at a normal application rate (after FAO, 1974).

which may be fixed irrespective of the proportion of the year the aircraft can operate (Schuster, 1974). Operations in remote areas also cost more because of the difficulties with maintenance and the need to transfer engines, equipment and spares over long distances. When an aerial operator has an hourly charge, the cost of application can be determined from the productivity data shown above. Obviously, large-area contract spraying on a regular basis is more attractive to the operator who can base pilots and engineers in one locality for a definite period. Routine cotton spraying in many areas of the world is a good example of this. Unfortunately, aircraft are not always available for the control of sudden or isolated outbreaks of a pest, although aerial application may be the most suitable method, because it is too expensive to keep aircraft waiting on the ground. Aircraft waiting for locust-control operations in Africa were redeployed when there were several years with low numbers of locusts. Sometimes aircraft are transferred to cope with a pest attack, but more of the damage may have already been done before the aircraft can reach the target area. Maximum use of aircraft is essential to keep costs as low as possible, so where outbreaks of pests are sporadic the aircraft is used to apply seeds and fertiliser, or is even used for transport work and firefighting (Pickler, 1976; Simard, 1976).

Aircraft regulations

The use of aircraft for the application of pesticides is controlled by legislation. Some countries merely require aircraft to be registered with, and inspected by, a civil aviation organisation, which has power to issue certificates of airworthiness where appropriate and to control the period of flying between routine maintenance checks. Pilots must undergo frequent checks on their physical fitness and competence to retain a licence issued by the same organisation, which also controls the number of hours a pilot is permitted to fly. Other countries have wider legislation to control which chemicals may be applied from aircraft. This legislation is essential in order to restrict the use of various herbicides, and the most toxic or most persistent pesticides. In the United Kingdom, agricultural aviation operators must comply with the 'Aerial Application Permission' issued by the Civil Aviation Authority. Previously, such operators had been exempted from the relevant parts of air legislation, which restricted both low flying (less than 152 m) and the dropping of articles from aircraft (Birchall, 1976). The Aerial Application Permission requires comprehensive standards for all safety aspects of aerial application operations, including avoidance of spray drift, marking of fields, reconnaissance pre-flight briefing and mapping of areas requiring treatment to indicate obstructions. Aerial spraying operators may apply only pesticides selected from a 'permitted list' compiled by the Ministry of Agriculture, Fisheries and Food under the Food and Environmental Protection legislation. Details of the regulations in the United States and other selected countries were published by the World Health Organization (WHO, 1970), but considerable changes in legislation are taking place, especially in the United States where the Environmental Protection Agency (EPA) has an overall responsibility for implementing the Federal Environmental Pesticide Control Act. An increase in the number of regulations and their scope can be expected, particularly in relation to the equipment which may be fitted to aircraft. Selection of nozzles and droplet size may be restricted in relation to meteorological conditions.

Where large-scale aerial spray operations are proposed, it is essential to carry out an environmental impact study. Ultralow volumes of insecticide have been applied successfully with minimal impact on non-target species by ensuring that with appropriate droplet sizes and spray concentration, most of the spray is collected on the foliage. Detailed studies have been carried out in Scotland (Holden and Bevan, 1979, 1981) and in Canada (e.g. Sundaram *et al.*, 1988; Sundaram, 1991). The trials in Scotland provided an interesting comparison between LV and ULV applications, the latter doubling the recovery on pine needles and the pine beauty moth larvae with a much higher work rate (Table 13.9) (Spillman, 1987).

Table 13.9 Comparison of two aerial spray treatments in a forest

Type of application	Ultralow volume application	Low volume application
Formulation	6 parts fenitrothion 50 EC 4 parts butyl dioxytol	3 parts fenitrothion 50 EC 97 parts water
Atomiser	Two Micronair AU3000 Standard 5″ cage, 13.5″ flat blades set at 25°	Six Micronair AU3000 Standard 5″ cage, 13.5″ flat bades set at 25°
Application rate	1 litre/ha	20 litres/ha
Active ingredient rate	0.3 litre/ha	0.3 litre/ha
Release height above canopy	6 m	3 m
Lane separation	50 m (two applications at 100 m on successive days)	25 m
Emission rate	15 l/min	150 l/min
Droplet sizes	VMD 97 μm nmd 24 μm	VMD 104.5 μm NMD 22 μm
Percentage volume between 10 μm and 40 μm	8.0%	8.4%
Spraying speed	170 km/h (50 m/s)	180 km/h (50 m/s)
Wind speed	7.8 knots day 1 13 knots day 2	1.6 knots
Area sprayed	50 ha (2 km by 250 m)	100 ha
Ferry distance	100 km	30 km
Overall work rate	309 ha/hour	88 ha/hour
Destination of active ingredient		
(a) Collected by needles or larvae	94.5%	41.7%
(b) Lost to ground within block	4.5%	38.3%
(c) Lost outside block	1%	20%
Average larval weight	28 mg	108 mg
Mean deposit on 20 needles	41.2 ng	23.6 ng
Mean deposit on single buds	18.8 ng	23.6 ng
Mean deposit on larvae per gram of larval weight	1 285 ng/g	407 ng/g
Mortality (%)	97.5	97.5

In several countries an untreated buffer zone has been proposed to protect ecologically sensitive areas, especially ponds and streams to prevent a significant impact on fish and their food populations. Payne *et al.* (1988) used a motorised mistblower with spinning disc to apply a synthetic pyrethroid as a fine spray to assess a worst-case scenario. Using a model they predicted that a buffer width of 20 m caused less than 0.02 per cent mortality of *Salmo gairdnei* rainbow trout in water depths greater than 0.1 m. For aerial application, Riley *et al.* (1989) have considered a 100 m buffer zone would ensure that there would be at least a tenfold decrease from the deposit observed at the edge of the target area, even when wind speeds exceed those currently recommended for agricultural sprays. As indicated earlier, much depends on the vegetation filtering out the spray droplets and, with higher wind speeds and turbulence, more of the spray will be impacted on foliage.

Following problems of translating laboratory data to field control of a forest pest, Payne *et al.*, (1997) followed a novel approach to determine the aerial spray parameters for applying an insect moulting hormone agonist, tebufenozide, against the spruce budworm. The aim was to optimise the dosage and deposit density where the larvae were feeding. Later studies investigated spraying only from the upwind wing to reduce downwind drift (Cadogan *et al.*, (1998).

14

Injection, fumigation and other techniques

Chemigation

Where crops are grown with irrigation, farmers have increasingly applied certain pesticides by injecting them into the irrigation water using a positive displacement pump, a technique known as chemigation. Irrigation water is applied on the surface as a furrow treatment and by drip irrigation to the soil, as well as with sprinkler systems. The sprinkler system allows both foliar and soil treatments; it is popular, especially where the continuously moving centre pivot or linear moving systems have been developed to improve uniformity of water distribution. These can achieve a coefficient of uniformity (CU) of 0.9 when properly calibrated, in contrast to many lateral move systems that achieve a CU of 0.70–0.75, and a travelling gun with a CU of less than 0.70 (Threadgill *et al.*, 1990). Scherer *et al.* (1998) describe the evaluation of a sprayer attached to a pivot system. Once the investment has been made, the farmer needs to utilise fully the irrigation equipment which is a viable alternative to conventional sprayers and requires less labout. The use of chemigation avoids tractor passage across the field.

Some herbicides washed into the soil by the irrigation water can be more effective, but concern about leaking of pesticides to ground water has been expressed. Care must also be taken to avoid back-flow which might contaminate the water source, so safety devices are essential (Fig. 14.1). Foliar applied chemicals may be less effective due to the extreme dilution (some centre-pivot systems use 25000 litres/ha, i.e. 25 times the maximum used in conventional systems of spraying), but herbicides may be less phytotoxic on crop foliage. Extreme dilution may be detrimental to pesticide activity if the active ingredient is readily hydrolysed or affected by the pH. Few pesticides are suitably formulated for chemigation, and particulate formulations may settle out during the application period (Chalfont and Young, 1982). The long time needed to complete a cycle is a disadvantage with insecticides, as damage may occur before a section of the crop is treated. Chemigation must be

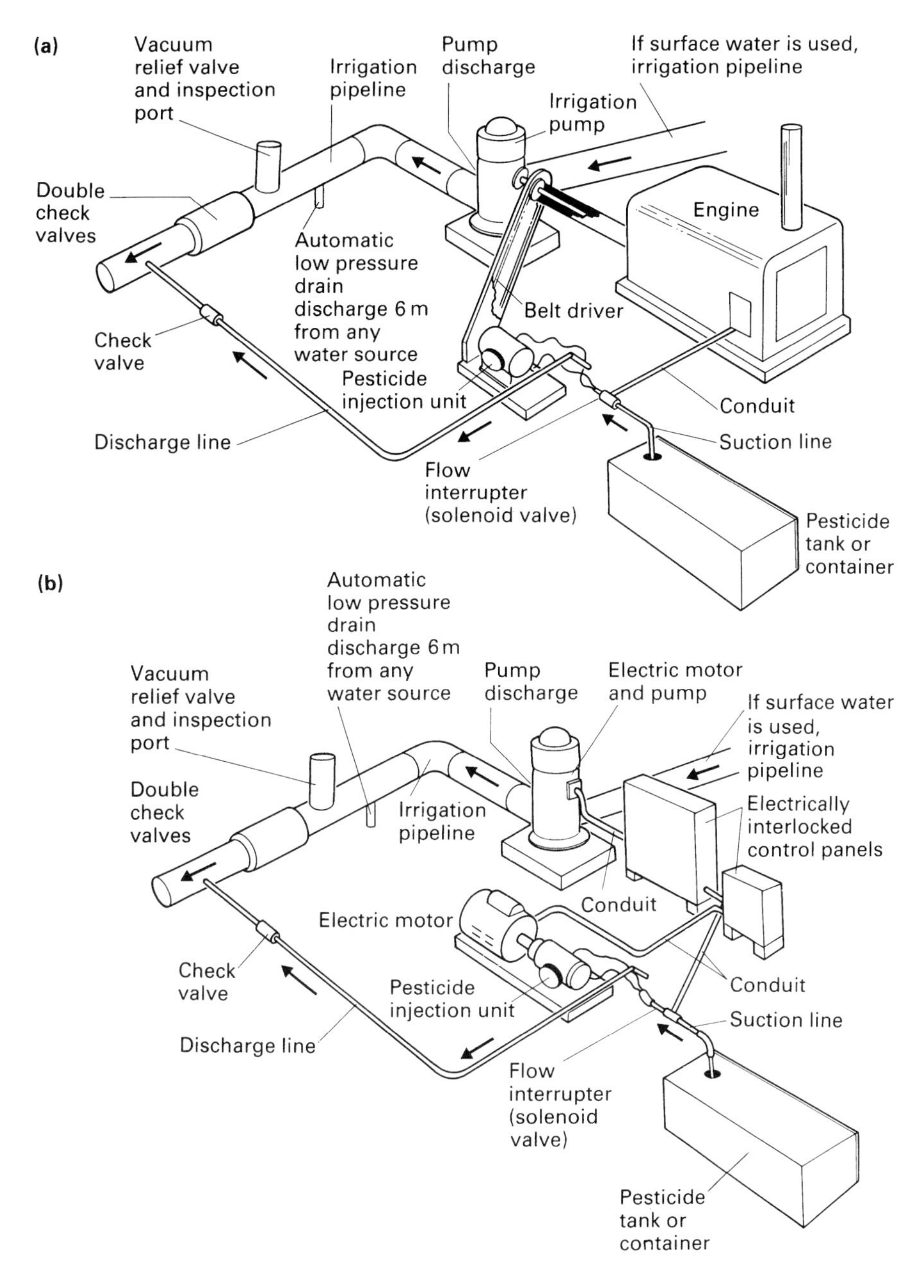

Fig. 14.1 Equipment for chemigation: (a) with engine driven pump; (b) with electrically driven pump.

avoided if wind conditions favour drift from the sprinklers. Drift is worse with travelling gun type irrigation equipment.

Ogg (1986) has summarised the response of crops and weeds to herbicides applied through sprinkler irrigation systems, but has pointed out the need for more research to elucidate the principles governing the behaviour of herbicides in irrigation water. Vieira and Sumner (1999) have reviewed the application of fungicides through overhead sprinkler irrigation. They concluded that chemigation with fungicides can be less, equally, or more effective, depending on the crop, pathogen, disease severity, fungicide and volume of water applied.

Chemigation has also been used successfully with drip-irrigation systems, for example to apply nematicides in pineapple culture (Apt and Caswell, 1988). To evaluate nematicides applied in a drip irrigation system, Radwald *et al.* (1986) have described a simple portable system for small plots. Due to problems of chemical contamination of groundwater, the application of entomopathogenic nematodes has also been investigated using drip-irrigation systems to control soil pests of horticultural crops (Reed *et al.*, 1986; Curran and Patel, 1988). Roger *et al.*, (1989) compared drip irrigation with foliar sprays and, at best, results of irrigation were equal to foliar sprays with a systemic insecticide.

If needed in the soil, the chemical is added at the start of an irrigation cycle so that the irrigation water washes it in, but for foliar treatment, application is at the end of an irrigation to minimise run-off. The technique has been used particularly where automatic irrigation systems are available as it reduces labour requirements. Distribution of water is not always sufficiently uniform to provide satisfactory coverage of foliage, so effective treatment depends on the selection of suitable chemicals which are readily redistributed. Users of chemigation should follow specific label instructions and any local regulations.

Other dispensers into water

Certain pests live in water; for example, the snails which are the intermediate hosts of *Schistosoma* spp. and the larvae of *Simulium*, vector of the disease onchocerciasis. There are also a number of important aquatic weed species.

Various types of equipment have been used to apply pesticides in sprays and granules to water, but where there is flowing water special dispensers can be used to avoid using labour involved in spraying. Complete mixing of a chemical with the water can be obtained if the dispenser is set at a narrow point or where the water is turbulent. The dose required and its distribution will depend on the volume of water and the rate of flow, as in sluggish water the chemical will be distributed only a short distance downstream. The quantity of water flowing per second past a given point can be determined from the cross-sectional area of flow and the average velocity. In many watercourses the velocity needs to be determined at various depths.

Some dispensers use a simple gravity feed, but unless the valve is adjusted or a constant-head device is used, the application rate will decrease as the reservoir empties because of the decrease in the head of liquid. The variation in flow can be minimised by mounting a squat tank as high as possible above the discharge point. A 200-litre drum is often used as a reservoir and can be mounted in a boat if necessary. More sophisticated systems adjust the dosage proportional to water flow (Klock, 1956). Alternatively, the chemical is bled into the suction line of a pump circulating water from a stream. In general, a pump delivering 100–200 litre/s and driven by a 1 kW engine is adequate. The flow of chemical needs to be checked with a suitable flow meter and adjusted with a regulating valve.

To minimise effects on non-target organisms, particularly fish, great care must be exercised in the selection of chemical and dosage required for the control of aquatic pests and weeds (Anon., 1995).

Weedwipers: rope wick applicator

Selective treatment of isolated clumps of weeds, especially if the weed is taller than an adjacent crop, is possible with a translocated herbicide such a glyphosate, using a rope wick applicator. This type of weed wiper has a container for the herbicide, which is fed through a restrictor to an absorbent surface. The aim is to wet the surface without any liquid dripping as this might damage non-target plants. The applicator can be a hand-held stick (Fig. 14.2) or a rope wick attached to a horizontal boom (Dale, 1979). The latter has been useful when treating tall weeds in 'set-aside' fields to avoid any possibility of drift to adjacent fields. The main problems with the equipment are difficulties in

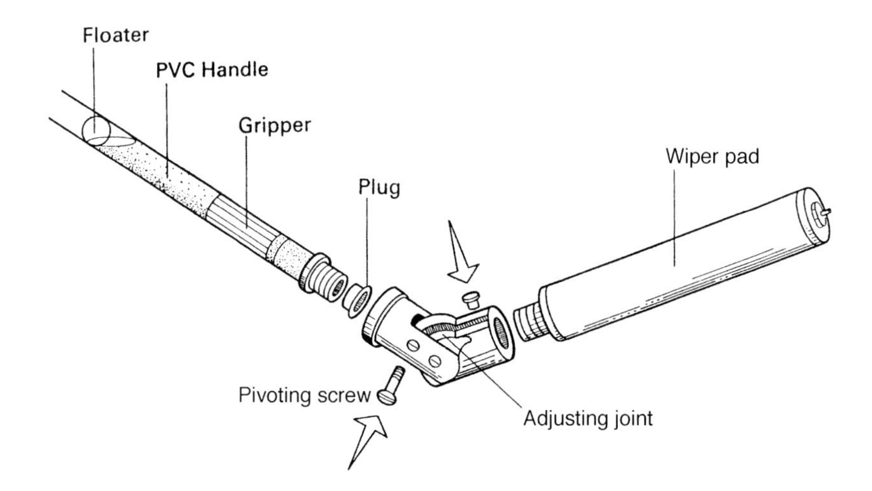

Fig. 14.2 Rope wick applicator.

avoiding dripping or, conversely, having too dry a wick, and accumulation of dirt on the surface. The hand carried rope wick applicator is mainly used for spot treatment. Some authorities have expressed concern about the wick after use. Care has to be taken to wash the herbicide from the wick and to cover it so that the surface cannot be touched. Individual plants can also be smeared with an oily rag treated with a suitable herbicide.

Soil injection

Generally, soil injection has been replaced by the application of granules, because most of the volatile pesticides, including nematicides and herbicides, are no longer registered. Granules can be applied to the surface and incorporated into the soil with a suitable harrow or plough. When a volatile chemical is applied, it requires specialised equipment and is justified only with high-value crops, especially as the treated surface may have to be covered with a polythene sheet to retain the fumigant for a sufficiently long period to be effective.

Tractor-mounted soil-injection equipment (Fig. 14.3) usually has a series of chisel tines fixed to a toolbar so that the depth to which the tines penetrate the soil can be adjusted. Pesticide is delivered through a tube mounted directly behind each tine. It may be supplied in a cylinder under pressure; otherwise a small low-pressure pump draws liquid from a spray tank through an in-line strainer, pressure regulator and distributor to a metering jet. This jet is mounted close to the tine but above the soil surface, so it is readily accessible for cleaning. It is impossible to see liquid flowing out behind the tine, so each outlet from the distributor has a flow meter, which can be checked visually by the operator. Good filtration of the liquid is essential to avoid blocking the jet. Injection to the correct depth is particularly important to distribute chemical into the most appropriate zone, i.e. where the roots of the subsequent crop will be liable to infestation. Often soil injection is too shallow, allowing a deep-rooted crop to be attacked by soil pests below the treated zone, but injecting below 150–200 mm is extremely difficult, except on some soil types. No chemical should be left in the equipment, otherwise corrosion may occur.

A hand-operated soil injector can be used when a small area needs treatment. This consists essentially of a pointed injector fitted to a piston pump, at the top end of which is a handle and adjustable metering cup. The handle may be fitted into the top of a container that surrounds the pump (Fig. 14.4). The operator holding the handle thrusts the injector into the soil to the required depth. Depth of injection is determined by a plate, which can be fixed at various positions to the injector head. A sharp tap on the cap or injector handle actuates the pump and forces a single measured dose, usually 1–15 ml, through an orifice near the end and to the side of the injector head. Forcing the injector into the soil is very arduous, especially on some soils, and less than 0.5 ha can be treated in one day with one injector. Injections need to be made

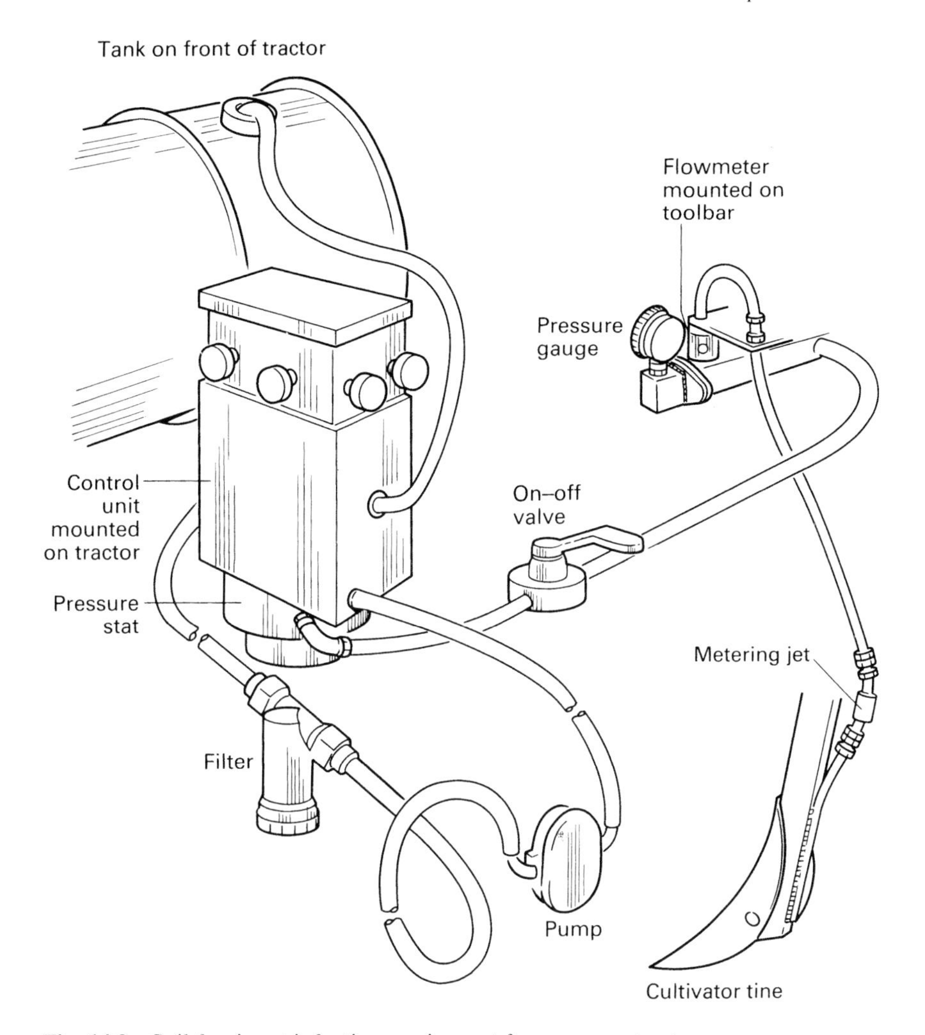

Fig. 14.3 Soil fumigant infection equipment for use on a tractor.

at regular intervals over the whole field or along rows in which the crop will be subsequently sown or transplanted. The distance between injection points will depend on the crop spacing, soil type and dose applied. A delay between treatment and planting may be needed to allow the chemical to disperse to avoid any phytotoxic effects on seeds or seedlings.

When filling the tank, a fine-mesh filter must be used in its opening to reduce wear of the pump and the risk of any blockage in the system. Great care is also needed to avoid trapping air in the injector, as this will affect the dose applied. After use, a soil injector should be thoroughly cleaned, and both the inside and outside treated with a mixture of equal parts of lubricating oil and kerosene before storage.

Fig. 14.4 Hand-operated soil injector.

Tree injection

Injection is most successful for treating coconuts and palm trees (Wood *et al.*, 1974; Khoo *et al.*, 1983). A simple brace and bit (Ng and Chong, 1982), or preferably a power-operated drill, is used to make a hole between the base of two frond butts, angled 45° downwards into the stem using a 10–15 cm long bit. Insecticide is metered into the hole as soon as possible using a small hand-operated injector (Fig. 14.5) and then the hole opening is covered with a fungicide paste to prevent loss of the insecticide. Lim (1997) describes a hand-operated pressure injector. Later callus tissue will cover the hole. A systemic

(a)

(b)

Fig. 14.5 Tree injection: (a) drilling hole. (b) injection of insecticide.

insecticide such as monocrotophos is used as it is readily taken by the xylem to the crown of the tree. Protection of the hands and face is needed during treatment as a concentrate is used to apply 5–7 g a.i./per tree. With a suitable power drill and injector about 3 ha/person-day can be treated.

Control of Dutch elm disease, caused by the fungus *Ceratocystilis ulmi* with benomyl or carbendazim injected into trees has also been attempted (Gibbs and Dickinson, 1975). The concentration of chemical used is usually a compromise to avoid too high a concentration which may cause phytotoxicity, and at the other extreme too low a concentration may result in an impracticable volume of liquid to be injected. The technique is slow and labour intensive, so application costs per tree are high and justified only where high-value amenity trees need to be saved.

Simpler techniques involving injection with a simple syringe to a bored hole or cut surface have been successful against some diseases. Specially adapted hatchets and secateurs have been made for repetitive treatments, for example *Trichoderma viride* has been applied to a cut shoot surface as a protection from invasion by pathogens such as silver leaf (Jones *et al*., 1974).

A trunk-implantation technique to apply systemic insecticides to trees was described by Ripper (1955) and used by Hanna and Nicol (1954) to control cocoa mealybug with minimal risk to pollinators, parasites and predators. Arboricides have been applied with special axes to kill trees.

Fumigation

Application of methyl bromide and other fumigants is a specialised method of controlling pests (Munro, 1961). Fumigants have been used to treat plants and the soil, for example in plant quarantine work, but are particularly useful when insect and other animal pests have to be controlled inside stored grain in silos, warehouses, ships and other enclosed areas, or in stacks of produce in the open. Concern about the release of methyl bromide into the atmosphere has led governments to introduce legislation that will phase out its use (UNEP, 1995). Apart from seeking suitable alternative fumigants, different techniques of controlling pests in the soil are being examined, e.g. soil solarisation and steam sterilisation, as is storage of produce in a controlled atmosphere. Where a fumigant is used, it is needed in the gaseous state in sufficient concentration for a given time; thus the dosage of fumigant is usually referred to as a concentration × time product ($c \times t$ product). A long time at a low concentration can be as effective as a short fumigation at a high concentration, but neither excessively long exposure periods nor high concentrations are practical. Some buildings are specially designed to permit fumigation of stored grain in bulk, or specially constructed fumigation chambers can be used, but if neither of these are available, a lightweight plastic sheet or gas-tight tarpaulin is used to retain the fumigant for the required exposure period. Persons with proper training and equipment must always carry out fumigation.

Soil fumigation

The technique is used mainly to control nematodes and weeds in seedbeds; for example tobacco seedbeds have always been fumigated before sowing. The area to be treated is covered with a plastic sheet or tarpaulin and the edges buried and sealed with soil. The centre of the sheet is held off the soil by a suitable support such as a bag of grass. The fumigant, usually methyl bromide, is discharged into a shallow pan (no more than 12 mm deep) under the sheet through a length of plastic tubing connected to a special device to puncture the container. Alternatively, the can is placed under the plastic sheet with a nail in a piece of wood. Without damaging the sheet, the wood is hit with a hammer to puncture the container. A can cannot be resealed once it has been opened. The dosage (500 g/10 m^2) is usually calculated in terms of the number of cans required. When several cans are required to treat an area the fumigant may be released at different places under the sheet to provide more even distribution. The gas is heavy so will diffuse down into the soil, but the amount of fumigant escaping to the atmosphere can be minimised by using less permeable plastic sheets, which also means less fumigant is needed (Gamliel *et al.*, 1998).

The tubing will be cooled when the fumigant is discharged, so in cold weather the can should be immersed in hot water up to 77°C so that the fumigant is discharged more evenly. The fumigant often contains a small proportion of a warning gas such as chloropicrin so that an escape of gas can be readily detected. Apart from avoiding breathing the fumigant gas, great care must be taken to avoid contact with liquid methyl bromide as it is liable to cause skin burns, although gloves should not be worn. A barrier cream can give protection to the hands. Any spillage will evaporate quickly, but hands should be washed immediately with soap and water.

The sheet can be removed after the prescribed exposure time (usually 48 h), and seed can be sown directly after removal. The soil should not be too wet, as the gas will not penetrate the soil, nor too dry, otherwise weed control will be poor. Some field crops such as strawberries have been fumigated in a similar manner with dosages of 24–48 g/m^3 of air space between the sheet and the ground, for 2–3 h during early morning or late evening to avoid high temperatures while the crop is covered.

Fumigation of stored produce

Unless a special fumigation chamber is available, the technique is basically the same for produce inside or outside buildings. Fumigation is useless if stored produce is kept in a building containing debris, or has cracks and crevices in the walls and roof, harbouring pests which can reinfest produce immediately after treatment. Thorough cleaning of stores is essential before treatment. A residual insecticide spray may be applied to the walls, although insects flying from elsewhere can infest produce without contacting the walls. A routine post-fumigation treatment with an aerosol spray may be needed to reduce the risk of reinfestation.

Before fumigation, an inspector should ensure that there is no risk of gas escaping to nearby offices, factories or living quarters, or that such areas are evacuated during treatment. Ready access to the produce is essential to permit the positioning of the cover sheets and seals. At the end of the treatment, the site must be well ventilated to clear any gas. If the floor is not gas-proof, the stack must be rebuilt on top of a gas-tight sheet. The dimensions of the stack of produce need to be checked so that the appropriate amount of fumigant needed can be calculated; for example 500 g methyl bromide per 10 tonnes of produce. An increased dose may be required for larger stacks; thus Hall (1970) indicated a dose of 5 kg/150 m^3 for larger stacks. Methyl bromide has been commonly used, as the gas will penetrate large stacks and diffuse away rapidly after the treatment is over. Large cylinders of fumigant need to be check-weighed to ensure that sufficient is available. The stack of produce in bags, boxes or cartons can be covered with sheets made of polyethylene (0.1 mm thick) which are light (100 g/m^2) and easily handled, but are liable to tear easily. Other types of sheet are made from nylon or cotton fabric and coated with neoprene, polyvinyl chloride (PVC) or butyl rubber. Pipes are laid from the gas cylinder to trays situated at the top of the stack where the fumigant is discharged. The top of the stack may need to be rearranged to accommodate the trays, and extra piping may be needed if the stack is inside a building so that the gas cylinders and scales can be kept outside. If available, fans can be placed under the sheets and operated for the first 15 min at the beginning of treatment to assist distribution of the fumigant. The fans should not be operated for a longer period as their use may tend to force fumigant out at the base of the stack, but can be used again during the subsequent ventilation period.

The sheets must be carried to the site and lifted to the top of the stack in a pre-rolled state to avoid puncturing them. Any holes must be repaired with adhesive or masking tape. The sheets are carefully opened to avoid dragging them over the surface. The edges of individual sheets need to be overlapped by 1 m, and rolled together to form an air-tight seal. Bags containing dry sand (to give flexibility) cover the joins between the sheets to prevent the joins unrolling. G clamps are also used to secure the joints, and sandbags are used to keep the edges on the ground. Plastic tubes filled with sand or water can be used as ‘snakes’ instead of sandbags.

Before the fumigant is released in the stack, a halide detector lamp is prepared and checked for satisfactory working. These lamps, which have a flame that turns to green and then intense blue as the concentration of methyl bromide in the air increases, are used to check any gas leakage during the fumigation. A final check is made to ensure that everyone has left the danger area, which should be cordoned off and patrolled by watchmen. Warning signs must also be placed at appropriate sites, such as entrances to the area. The operators then put on their gas masks with a correct filter-type canister and check that they are fitting correctly by placing a hand firmly over the air intake or pinching any hose connecting the canister with the face-piece. If the face-piece is fitting tightly the wearer will not be able to breathe, and the face-piece will be drawn into the face. Supervisors should also check everyone's gas mask as well as their own to ensure that it has the correct type of canister, that the canister is not out of date or exhausted, and that a first-aid kit and torches are readily available.

While the cylinder is opened to allow fumigant into the stack, frequent checks are made with the halide detector lamp around the cylinder connections, pipes and joins in the sheet. Extra sandbags may be needed to seal the sheets at ground level. The valve on the cylinder is firmly closed as soon as the required quantity of fumigant has been released. The cylinders can then be disconnected and removed, checking that no fumigant which may be in the pipes splashes onto the operators. When the stack is inside a building, all doors are closed and locked for the entire fumigation period, usually 24 or 48 h. Then the person supervising the fumigation and the assistants replace their gas masks and inspect the premises for gas with the halide detector lamp. Gas present in the building before the sheets are removed will indicate that leakage has occurred. The sheets are then removed methodically and as quickly as possible, so that staff are in the area with gas for as short a time as possible. Removal of sheets at the corner of the stack to allow partial aeration of the stack before returning to remove the remainder may be necessary when many sheets are involved on large stacks. Doors and windows are then left open for at least 24 h for ventilation to disperse the gas. The area is checked again with the halide detector lamp until declared safe for people without gas masks to enter. All the warning signs can then be removed.

Bond (1984) and Hall (1970) give more details of a range of fumigants and techniques, using special fumigation chambers. The fumigant, phosphine, applied as tablets of aluminium phosphide (see Chapter 3), can be incorporated into stacks at the rate of one tablet per two bags as each layer is built. The whole stack must be covered by a gas-proof sheet, as described above, within 2 h. Tablets can also be added to a conveyor-belt moving grain into a silo, or through special probes inserted into bulk grain. Alternatively, tablets sealed in a paper envelope can be placed in individual gas-proof bags and removed by the user so that no residue of aluminium hydroxide is left in the grain. Small quantities of grain can also be fumigated inside an empty oil drum or similar container, the top of which is sealed with a polythene sheet fixed with masking tape.

15

Maintenance of equipment

The need for well maintained equipment is now emphasised by the legal requirement at least in several European countries for an official inspection of sprayers at regular intervals. A European Standard for inspection of sprayers (CEN/TC 144) is being developed. Hitherto, farmers often considered maintenance when a part failed, but regular preventative maintenance is now recognised as essential to meet the increasingly demanding legislation to avoid leakages and contamination of the environment with pesticides. The cost of pesticides also necessitates that they are applied efficiently.

A problem still exists in many parts of the world where spare parts are less readily available and service manuals may not be in the local language or in sufficient detail to provide users with a clear step-by-step guide to what maintenance is required. Manufacturers of motorised equipment still often distribute a separate manual for the engine, instead of integrating the relevant information into a comprehensive manual, describing how to use and repair the sprayer or applicator. Manufacturers and their local agents also need to ensure appropriate spare parts are readily available.

Users of pesticides need to have some training in both the biological and chemical aspects of controlling pests, together with training in the correct and safe use, calibration, maintenance and storage of equipment. In some countries, such as the United Kingdom, all users of pesticides have to pass a practical test. In the United Kingdom this is organised under the auspices of the National Proficiency Test Council; in addition to a foundation course, there are courses provided at agricultural colleges to prepare participants who need to pass the relevant NPTC modules (Table 15.1) for the equipment they will be using, and this training is supported by manuals produced by the British Crop Protection Council.

In other parts of the world, training is not so readily available, and in some countries greater stress is placed on testing the machinery. Some of the agrochemical companies provide training as part of their product stewardship programmes. This is supported by the Global Crop Protection Federation

Table 15.1 Faults with two-stroke engines and their remedies (from Clayphon and Matthews, 1973; Thornhill, 1984)

Fault	Remedies
Engine does not start	
Fault in fuel system	
Fuel cock not opened or blocked	Ensure fuel is present in tank. Open cock. If no flow, remove clock, clean and replace
Air vent in fuel tank filter is blocked	Clean vent
Thimble filter in carburettor is blocked	Remove filter, clean and replace
Main jet in carburettor is blocked	Remove, clean and replace
Water in carburettor float bowl	Remove and clean. Check also that fuel in tank is not contaminated with water
Float needle sticking and stopping petrol supply	Remove needle, check for burrs or rough surface. Clean off rough surface, if not possible, replace with a new one
Too much fuel in engine	Close fuel cock, remove spark plug, open throttle, pull recoil starter rope to turn engine over a few times, clean, replace
Fault in ignition system	
High-tension lead to spark plug loose or disconnected or insulation broken or burned	Fasten lead securely to plug, if badly damaged, replace
Dirty spark plug, carbon or oil deposits on electrodes	Remove plug and clean; set gap as recommended by manufacturer. If porcelain insulation is damaged, replace with new plug
Contact breaker points dirty or pitted	Clean and adjust to correct clearance when points are open. If honing fails to remove pitting, replace with a new set
Exhaust blocked	Remove exhaust and clean or replace with a new part
Engine runs erratically or stops	
Dirt or floating debris in fuel system	Clean all fuel lines, filters and carburettor bowl and check there is no air in fuel line
Main jet blocked	Remove, clean and replace. Do not use nail, pin or wire to clear obstruction
High-tension ignition lead loose or 'shorting' on metal parts of the engine	Check that lead is firmly affixed to spark plug. Where lead has been chafing on bare metal, either cover bare wire with insulation tape or replace with a new lead
Fuel running low in tank. Engine vibration or irregular movement of operator leaves output pipe uncovered, resulting in fuel starvation	Refill tank with correctly mixed fuel

Table 15.1 (*Cont.*)

Fault	Remedies
Engine lacks power	
Choke is closed	Open choke
Fuel starvation	Partially blocked pipes or filter should be removed and cleared
Air cleaner blocked with debris	Remove, clean by washing in petrol and squirt a little light oil on the cleaner element. Conform with manufacturer's recommendations
Dirty carburettor	Remove from engine, dismantle carefully, clean and examine all parts. Any worn parts such as float needle valve, etc. must be replaced with new parts
Loose or leaking joint at carburettor flange to cylinder	Check gasket. Replace if worn or damaged and tighten nuts or studs
If whistling noise is heard from cylinder when engine is running, there is a possibility of the cylinder head gasket being worn or damaged	Check carefully by feel when engine is running. If gases are escaping, remove head, fit new gasket, tighten nuts evenly. On a new machine, it may be necessary to tighten the nuts evenly without fitting a new gasket. If heavy carbon deposits are seen on piston crown or when cylinder head is removed, these should be scraped away carefully. The ring of hard carbon should not be disturbed in the cylinder.
Dirty exhaust	Remove exhaust, clean carbon deposits from exhaust if possible, or replace with new part
Engine backfires	
Ignition may be badly retarded	Should be attempted only by trained or qualified personnel. Magneto should be checked and reset to manufacturer's specification
Carbon whisker bridging gap in spark plug	Remove plug, clean, adjust gap to correct clearance and replace
Overheating of engine	
Incorrect mixture of petrol and oil in fuel tank	Drain off tank. Refill with fuel in the correct ratio (see handbook or markings on tank)
Incorrect size of main jet	Remove and refit one that complies with manufacturer's specification
Ignition retarded too far	To be checked and reset by a competent person
Exhaust and silencer choked with carbon	Remove, dismantle, clean and reassemble

(GCPF), formerly the Groupement International des Associations Nationales de Fabricants de Produits Agrochimiques (GIFAP). In countries where farmers have insufficient mechanical knowledge to maintain application equipment, practical field training courses are essential for both individuals and those supervising spray teams. Such training must be supported by the availability of suitable instruction manuals which need to be well-illustrated and written in simple and clear terms to facilitate translation into the vernacular. International organisations such as WHO and FAO have published specifications of different types of equipment to ensure that minimum standards can be attained. They have also prepared some booklets that are aimed at promoting better use of equipment.

Problems with the spray system

Nozzle or restrictor blockage

Improvements in particulate formulations have been made so that any nozzle blockage is most likely to be due to extremely small particles in the water used as diluent, especially if it has been taken from a stream or borehole on a farm. There is also the possibility of particles flaking from the inside of the pipework of a sprayer if it has not been washed properly after use or some corrosion has taken place. Such blockages can be minimised by adequate filtration. When a closed system of loading is used, there should be a large filter between the mixing unit and the sprayer tank. The mesh size and area of the filter need to be selected to cope with the volume of liquid being used and in relation to the nozzles used. The filter mesh at the nozzle must be smaller than the orifice diameter; for most agricultural work a 50-mesh filter is adequate. When spraying has been completed, there may be several litres of spray remaining in the machine; the actual quantity will depend on the type and size of the sprayer (Taylor and Cooper, 1998). Ideally, the operator should only mix sufficient pesticide so that the spray liquid is used up just by treating a field; sufficient clean water is then used to wash the sprayer tank and pipework, and the washings are sprayed out within the last part of the treated field. An EC Directive 414 Annex III requires that an appropriate decontamination routine is defined to obtain approval for an agrochemical. This is important if traces of chemical from inside the tank were to contaminate the next pesticide used and cause phytotoxicity on the crop.

When washing out a sprayer, several washings of a small volume of water are better than filling the tank once with clean water. Cleaning the sprayer can be improved using a 0.2 per cent suspension of activated charcoal, but this is expensive. Some manufacturers now market products specifically for cleaning sprayers. Household ammonia diluted at 10 ml per 5 litres of water is also a useful cleaning agent, provided there are no brass components in the equipment. On motorised equipment, the volume of water must be sufficient to operate the agitation system. The final rinse must be with plain water.

Each nozzle should be dismantled and the individual components – filter, tip and cap – cleaned and replaced. All other filters on the sprayer should be removed, cleaned and replaced. In general, it is never possible to clean a sprayer completely, as some of the chemical can become impregnated in hoses. If possible, separate equipment should be used to apply herbicides such as 2,4-D, which could affect other crops when different pesticides are subsequently applied. Alternatively, equipment must be decontaminated with charcoal or other recommended procedure and the hoses replaced. The suitability of the sprayer can be checked by treating a few plants susceptible to the herbicide used in the equipment, for example tomato plants are susceptible to 2,4-D herbicide.

Care must be taken to avoid the washings contaminating any drinking or other water supply. Analysis of water in a section of a river in Europe indicated that pesticide was detected when sprayers were washed out rather than from spray drift (Fig. 15.1) (Beernaerts *et al.*, 1999). Some countries have issued guidelines or a code of practice concerning the cleaning of equipment, and this should be consulted. Protective clothing should not be removed until after the equipment has been cleaned and returned to the store.

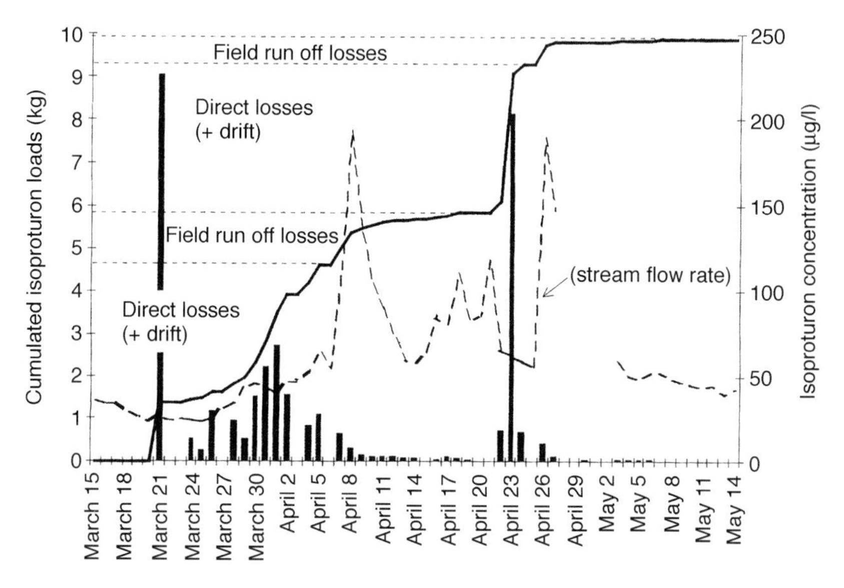

Fig. 15.1 Peak contaminants of a river related to direct losses when spraying – mostly due to washings of sprayer.

If special formulations have been used, a particular solvent may be needed to clean the equipment. Information on the suitability of solvents for cleaning should be obtained from the supplier of the pesticide or equipment, to check that there is no risk of detrimental effects on plastics and other materials used in the construction of the machinery.

If a nozzle blockage does occur while spraying, the nozzle tip and filter should be removed and replaced by clean parts. The blocked nozzle is more easily cleaned back in the workshop, so sufficient spares should be taken to the field. When spare nozzles are not available, sufficient water or solvent should be taken to the site of operations for cleaning a blockage. If washing does not remove the obstruction, giving the nozzle a sharp tap with the inner surface downwards may be sufficient to dislodge it. Alternatively, air pressure from a car or bicycle pump can be used to blow it from the nozzle orifice. Nozzles should **never** be placed in the user's mouth to blow through the orifice as their surface is inevitably contaminated with pesticide. A hard object such a pin, nail or stiff wire should **never** be used, as the orifice can be so easily damaged. When ceramic nozzle tips are used, extra care is needed as the slightest damage to the nozzle orifice can affect the distribution of the spray liquid. If several blockages occur, the whole system should be checked to determine the source of the material causing the blockage. Corrosion, especially inside the metallic parts of the sprayer, may result in small particles which can accumulate on the filters. With some of the particulate formulations, deterioration during storage can result in poor suspensibility, so particles settle out and can be the cause of a blockage.

The flow-control valve or restrictor may become blocked on sprayers which do not have hydraulic nozzles. As mentioned above, the occurrence of blockages can be reduced by proper filtration, but if a blockage does occur, it is usually quicker to replace the restrictor rather than attempt to clean it in the field.

Inefficient pumps

Piston pumps are fitted with 'O' ring seals or cup washers of synthetic material or sometimes leather, although this is less likely now. As ths seal can be damaged by particles suspended in the spray liquid, it should be checked regularly to keep the pump operating well. Where leather seals are used, they require regular treatment with a vegetable oil to prevent them drying out and shrinking. Some synthetic materials used in pump seals or as diaphragms may be affected by solvents and swell, making the pump harder to operate, but the majority of pesticidal liquids are diluted in water. Poor pump performance may also be due to faulty valves. Ball valves and their seating can be pitted or coated with sediment, debris from the water supply or pesticide. Apart from cleaning and replacing damaged parts, it may be necessary to change the formulation used or to improve the filtration of the water before use.

Leaks

'O'rings, washers and other types of seal are liable to wear or be damaged when hose connections, trigger valves and other components are unscrewed. Similarly, seals around the tank lid and in the pump assembly can be damaged whenever the connection is broken. The damaged part should always be replaced to avoid occurrence of leaks. Some connections such as nozzle caps may not have a washer and rely on direct contact of smooth surfaces to seal. Any dirt on the nozzle or cap, or damage to the threads, may prevent a proper seal.

Proper functioning of some spray equipment, such as compression and certain motorised knapsack mistblowers, depends on an airtight seal of the container or spray tank (Fig. 15.2). For example, it is impossible to spray liquid upwards with some mistblowers when the nozzle is above the level of the spray tank because there is insufficient pressure to force liquid to the nozzle. Small air leaks from the lid or other fittings to the tank, for example a pressure gauge on the container, can be detected by smearing a soap solution over the joint. Soap bubbles should be readily detected where air is escaping.

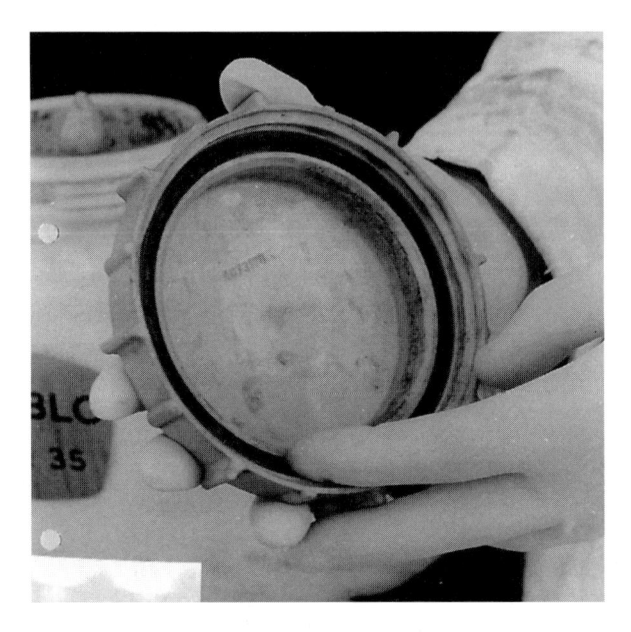

Fig. 15.2 Checking seal on lid of spray tank of motorized knapsack mistblower.

Problems with motorised equipment

Two-stroke engines

Users of motorised knapsack mistblowers frequently complain that the engine is difficult to start. Various causes for the failure to start and other problems are listed in Table 15.1, together with remedies. Many of the starting problems could be avoided if the carburettor and engine were drained of fuel after use to avoid gumming up the machine with oil when the petrol has evaporated. This can be done simply by turning off the fuel tap and allowing the engine to continue running until starved of fuel. Preferably, the fuel tank itself should also be drained to avoid the ratio of oil to petrol increasing, especially in hot climates. Starting problems are definitely reduced by ensuring the correct type of oil is used (see p. 225) and that the fuel is properly mixed.

The fuel line from the tank to the carburettor is often made of plastic, which becomes hardened by the action of the petrol and is sometimes loosened by the engine vibration. This plastic tube should be regularly inspected and replaced, if necessary, to avoid fuel leaking onto a hot engine and causing a fire. The sprayer's straps should be designed to allow the machine to be removed very easily in case a fire starts. On some machines, the fuel tank is now situated below the engine so that fuel cannot leak down on the engine.

The spark plug should be inspected regularly and cleaned if necessary, so it should be readily accessible (Fig. 15.3). The spark plug gap may need adjusting to obtain a good spark before replacing. The plug should be replaced with a new spark plug after 250 h as a routine. The air filter should also be examined at the end of each day's spraying and cleaned according to the routine recommended by the manufacturer.

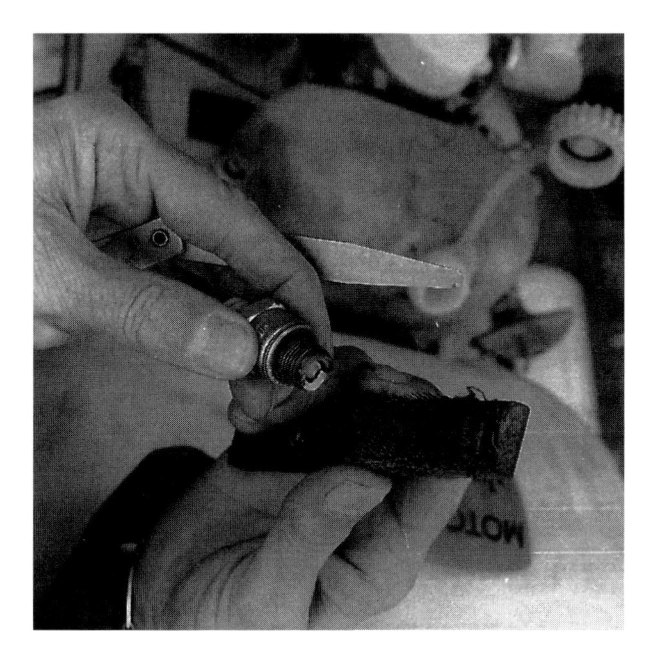

Fig. 15.3 Cleaning spark plug and checking gap.

Fault finding

Some of the faults commonly found when using hydraulic sprayers are indicated in Tables 15.2, 15.3 and 15.4, together with possible remedies. Similarly, faults with the small hand-carried spinning disc sprayers are given in Table 15.5.

Table 15.2 Faults with slide pumps (single- or double-acting continuous operation)

Fault	Remedies
No spray at nozzle	Check nozzle and clean if necessary Check that container is full Check pump, especially non-return valves Check for leaks on hose and connections
No suction	Check pump, especially pump seal Check valves and seatings Check strainer in container
Leaks from pump	Check gland and packing, replace if worn or damaged

Table 15.3 Faults with knapsack, lever-operated (piston or diaphragm) pumps

Fault	Remedies
No spray	If resistance is felt on downward movement of lever with cut-off valve open, check nozzle for blockage, and clean if necessary. Check and clean filter or strainer in handle of cut-off valve. If no resistance is felt, check tank contents and fill if necessary. Ensure that operating lever is tight, together with all the connections to the pump. Check that when the lever is operated, the shaft or connecting mechanism and the piston or diaphragm all move together. Pump valves and valve seat should be checked. If worn or damaged these should be replaced. Dirt and debris should be removed
No suction	Ensure that liquid is present in the container. Check that the suction and discharge valves are not sticking. Make certain that the liquid ports that permit flow from tank to pump are not blocked. If a piston-type pump is employed, check that the piston seal is not excessively worn or damaged, as this will permit the liquid to pass between the piston and cylinder wall.
No pressure	Check liquid contents of container. Fill if necessary. After several strokes of the operating lever, look in the tank to see if air bubbles are rising to the surface. If so, this could indicate a leak in the pressure chamber. Where pressure chamber is screwed into the pump body, check that the seal is not damaged. Replace if necessary. Check both suction and discharge valves. Remove any accumulated dirt or debris from discs or balls and valve seats. If discs are worn or damaged or the rubber is perished, replace. If ball valves and

Table 15.3 (*Cont.*)

Fault	Remedies
	seats are pitted or balls are no longer spherical, replace with new ones. If resistance is felt when pumping and no reading is seen on pressure gauge, replace gauge. If pump is of diaphragm type, check that it is seating correctly, that it is not damaged or split and that the rubber is not porous. Where a pressure-relief valve is embodied in the pressure chamber, check that it is adjusted correctly and that the spring-loaded valve is seating properly. Ensure that the openings between the pump inlet and outlet ports and the liquid container are not blocked. Check that the air vent in the filler cap is not blocked, as this could be the means of a vacuum forming in the container
Pressure drops quickly	Check pressure chamber for leaks. Air bubbles seen rising to the liquid surface are a good indication. Check valves for discharge. The discharge rate may be higher than pump capacity
Liquid leaks onto operator	Where pump is mounted in base of sprayer, a ruptured diaphragm, or one incorrectly assembled, will permit liquid under pressure to leak. For a piston type pump, a worn piston seal or deep scratches in the cylinder wall will also permit the liquid to escape and wet the operator. Check the container for cracks or leaking joints. Metal tanks can be soldered or brazed. Check that the lid of the container is fitting tightly

Table 15.4 Faults with compression sprayers

Fault	Remedies
No spray	Ensure container has liquid. If pressure gauge shows a reading and there is no spray when cut-off valve is opened, close valve and check nozzle. If nozzle is blocked, follow procedure for clearing blocked nozzles. Check strainer in cut-off valve. Clean and replace. Check hose connections and tighten. If no reading is shown on the pressure gauge, ensure that the gasket between the pump body and the liquid container is not leaking. Replace if leaks are present. Remove pump from container and check by giving a few smart strokes on the pump handle to test the valve. On each pressure stroke, the valve should 'grunt' or make a noise of escaping air. If the valve disc or ball is malfunctioning it should be replaced. Where a dip-tube is part of assembly, check that this is not blocked with debris

Table 15.4 (*Cont.*)

Fault	Remedies
Leaks from pump	After the container has been filled with spray liquid to the required level, if on the first or second downward strokes of the pump handle liquid is forced up past the shaft and out through the guide, this is a good indication that the valve requires attention. Furthermore, if strong resistance is felt on the downward stroke, again the valve is faulty and has permitted liquid to enter the pump barrel and, as liquid cannot be compressed, resistance is encountered
Pressure drops quickly	Check that the filter cap or lid gaskets are serviceable and that the cap is properly secured. Check also where a safety valve is fitted that it is not leaking and is in a working condition. Some compression sprayers have a constant-pressure valve fitted. Check that this is adjusted correctly and that there are no leaks from the point of entry to the tank. Ensure that all connections to the tank are tight and that all gaskets and washers are serviceable. Check tank for leaking seams by pressurising and immersing completely in water. Air bubbles rising to the surface will indicate the presence of a leak. Leaking tanks cannot be repaired in the field. All repaired compression sprayers must be pressure-tested to at least twice the working pressure before being used on spraying operations.
Other faults	If nozzle dribbles with cut-off valve closed, the 'O' ring seal or the valve seat is damaged. Dismantle and check. Replace with new parts if unserviceable. With some of the plastic-type pressure gauges, the indicator pointer sometimes becomes loose on its pivot. This can give a false pressure reading. By tapping the gauge against the hand it can be seen whether or not it is loose. If it is, remove the protective glass front, replace the needle on the pivot loosely and, with it pointing to zero, press it firmly on to its mounting. Replace the glass with a master gauge

Table 15.5 Faults with spinning disc sprayers

Fault	Remedies
No spray	Restrictor may be blocked. Clean with solvent or piece of very fine wire or grass stem. Check whether air vent is blocked
Leaks	Check that spray container is fitted correctly
Spinning disc not rotating or rotating intermittently or slowly	Check that enough batteries are fitted in containers. Check that all batteries are inserted the correct way. Check battery connections. Check switch (if any). Check connections to motor, clean connections with a dry cloth or sandpaper and fit new wires if necessary. Check that the '+' terminal of the batteries is connected to the '+' terminal marked on the motor. Replace batteries if necessary. Where large numbers of the sprayers are in use, it is advisable to provide a voltmeter and tachometer to check the revolutions per minute of the disc. Check whether disc is fitted correctly to motor shaft; it may be pushed on too far and touch the backing plate. If necessary, replace motor

Maintenance in the field

One or two tools should always be taken into the field while spraying, together with extra nozzles, washers and other spare parts. The non-mechanically minded user will find one pair of pliers, at least one screwdriver, or preferably two of different sizes, one small adjustable wrench, a knife and a length of string invaluable. Spare washers for the trigger valve, nozzle body or even the filler caps should be available, but if not, a length of oiled or greased string can be used as a substitute in some circumstances. Some washers may be cut from the inner tube of a car or cycle tyre and used temporarily until the proper spare washer can be fitted.

Quick repairs to leaking plastic containers which are not pressurised can be made by drawing the edges of a small hole with a black-hot nail, and smoothing it over with a wetted cloth. A 15 cm nail is suitable and can be heated in a fire, even out in a field, but it must not be made too hot, otherwise the plastic may melt and the hole enlarged beyond repair.

Those using engine-driven equipment, such as a knapsack mistblower, will also need to carry a spare spark plug and a plug spanner, while those using small battery-operated sprayers need a 'Philips' screwdriver, as well as a tachometer. Tools and spares can be conveniently carried in a small tool box. If the spray programme entails the use of several machines simultaneously, one or two complete machines could be taken to the field as spares so that work may continue when weather conditions are favourable, rather than delay spraying while repairs are attempted.

All stoppages and breakdowns that occur in the field should be reported to the workshop personnel, so that repairs and maintenance can be done without delay. Where several machines are used by a team of operators, it is a good policy to allocate a specific machine to each individual who then becomes responsible for its care and maintenance.

Storage of equipment

After each day's field work, and at the end of the season, complete checks should be made of the pump and, where necessary, the engine, before storing the sprayer in a dry place. All sprayers should be kept locked away from children, food and farm animals, and measures taken to prevent rats from chewing hoses and other parts. Many small hydraulic sprayers are preferably stored upside-down with the lids removed to allow complete drainage of the container. If engines are to be stored without use for a prolonged period, the spark plug should be removed and a small quantity of oil, preferably formulated with an anti-rust additive, poured into the crankcase. The engine should be turned over a couple of times to ensure the oil spreads. Similarly, at the end of each day it is advisable to add some oil to pumps on any type of sprayer. This is not necessary if the sprayer is used again the next day, but adverse weather conditions or some other factor may prolong the period of storage.

16

Safety precautions

An important part of any pesticide application is the assessment of the hazards involved in transportation, storage and use of particular pesticides, the toxicity of which varies according to their chemical structure, purity and formulation. A major trend has been away from emulsifiable concentrate formulations to particulate suspensions or dispersible granules, and an improvement in packaging to reduce exposure of the user to the pesticide. As indicated in earlier chapters, there has also been increased emphasis on engineering controls by using closed transfer systems to minimise the need for personal protective clothing. The hands are the key part of the body most exposed to pesticides, when opening containers or adjusting nozzles or other components of equipment. Concern about environmental pollution has led to greater emphasis on avoiding spillages, and specific recommendations regarding the washing down of spraying equipment and the disposal of washings and used containers.

Pesticides, like medicines and other chemicals, must be stored and used according to instructions, so the first requirement for all users of pesticides is to '**Read the label**'. A pesticide may be taken into the body by mouth (orally), through the skin (dermally) or through the lungs (inhalation). The uptake orally is minimal during pesticide application, unless operators unwisely eat, drink or smoke before washing their hands or face. Oral poisoning has occurred when pesticides have been improperly stored in food containers, especially soft drink or beer bottles, where recently sprayed fruit has been eaten or a person has committed suicide.

Contamination of the body is principally by absorption through the skin, which is particularly vulnerable where there has been a cut or graze. The backs of the hands and wrists absorb more than do the palms. Similarly, the neck, feet, armpits and groin are areas which need protection, and great care must be taken to avoid contamination of the eyes. The risk of skin absorption is increased in hot weather, when sweating occurs with the minimal amount of effort and conditions are not conducive to wearing protective clothing.

Unfortunately in many tropical countries, protective clothing is worn only by a few of the farm workers (Gomes *et al.*, 1999).

A pesticide can enter the lungs by inhalation of droplets or particles, principally those less than 10 μm diameter (Clay and Clarke, 1987), but the amount is usually less than 1 per cent of that absorbed through the skin. The greatest risk occurs when mixing concentrated formulations and applying dusts, fogs or smokes, especially in a poorly ventilated area. When treating inside buildings such as glasshouses, proper ventilation is needed before people can re-enter the treated area. The chances that these small droplets or particles will be inhaled is reduced under field conditions because the wind blows them away.

The relative hazards of these routes of exposure need to be evaluated with different operational procedures and protective clothing. Chester (1993) has reviewed methods of measuring exposure to and absorption of pesticides by workers involved in their use, and has given guidance in conducting field trials studying exposure (Chester, 1995). Exposure is generally much greater with manually operated equipment due to the close proximity of the spray to the user (Abbott *et al.*, 1987). Video imaging techniques have also been used to assess dermal exposure (Archibald *et al.*, 1994). Such videos are useful in training spray operators to reduce their exposure (Archibald *et al.*, 1995).

Parkin *et al.*, (1994) proposed a hierarchical scheme to classify pesticide application equipment according to its potential contamination hazard to the user and the environment. The scheme provides a framework that could be used by registration authorities. Figure 16.1 illustrates part of a decision tree

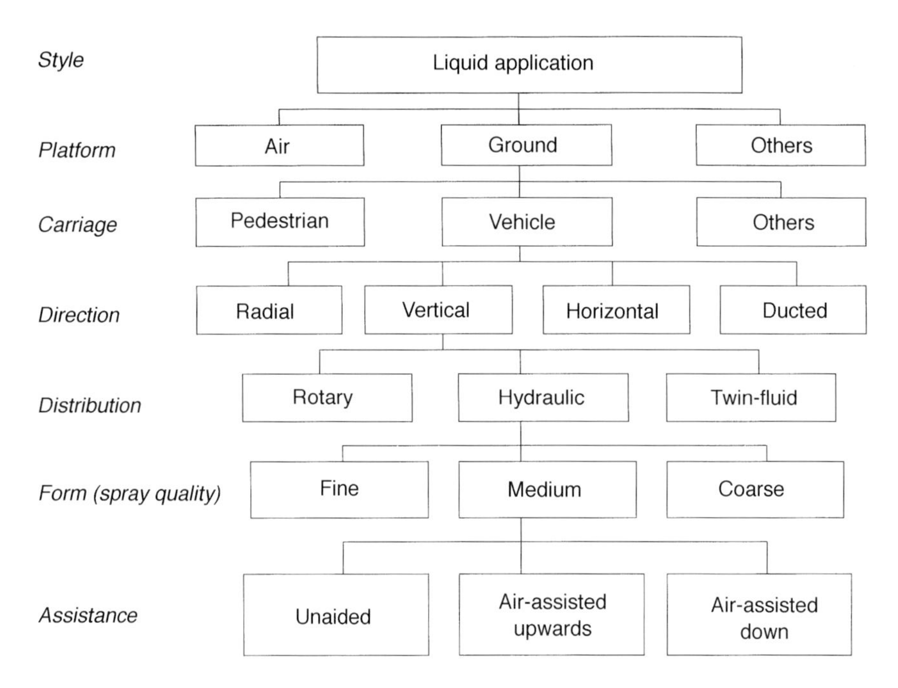

Fig. 16.1 Hierarchical classification of spray equipment.

used for liquid spray applications, the platform approach being included in the USA AgDRIFT model, although the EPA version of a classification scheme is likely to be based on droplet size with application height and wind speed (Hewitt and Valcore, 1998).

Thornhill *et al.* (1995) showed that by using low volume CDA manually carried sprayers, operator contamination was less than when herbicide sprays were applied with knapsack sprayers. Most of the contamination occurred on the lower legs and feet, irrespective of the sprayers used. Subsequent studies with smaller droplets for insecticide sprays confirmed lower contamination with CDA sprays. Most contamination of the operator using knapsack sprayers was due to walking through airborne spray and treated foliage (Thornhill *et al.*, 1996). High volume spraying resulted in more dermal exposure to operators than a low volume mistblower technique used in grenhouses (Moreira *et al.*, 2000). Wicke *et al.* (1999) had less operator exposure when using manually operated equipment in an orchard, if air induction nozzles were used. Fabrics with a water-repellent finish reduced penetration of a spray, and exposure was reduced by using a pressure control valve on a knapsack sprayer (Shaw *et al.*, 1999).

The hands are most exposed to pesticides during the preparation of sprays, as the concentrated chemical may splash or be touched by the user when opening the container or pouring pesticide into the sprayer. Hands are also exposed if the nozzles are touched to replace or adjust the nozzle tip. The hands of other workers are exposed to dislodgeable residues on treated crops if touched too soon after an application.

Apart from changes in formulation and packaging, exposure can be significantly reduced by engineering controls. The introduction of closed transfer systems is a major way of reducing exposure. Where such systems are not used, the addition of a low level induction hopper (Power and Miller, 1998) can facilitate pouring liquid from a drum and decrease the risk of splashing. Tractor cabs will significantly reduce operator exposure, particularly where air assistance is used to distribute sprays in orchards (Lunchick *et al.*, 1988). Personal protective clothing is a deterrent to dermal exposure, but needs to be comfortable to wear (Batel and Hinz, 1988; Cowan *et al.*, 1988; Fraser and Keeble, 1988). Other personal protective equipment (PPE) (see below) should be selected in relation to operational procedures.

Irrespective of how a pesticide enters the body, acute poisoning may occur after one dose or exposure, while chronic poisoning is caused by repeated small doses absorbed over a longer period of time. The latter is especially important when spray operators apply pesticides frequently, but others such as those scouting crops for pests, weeding or harvesting crops, may be at risk in treated areas. Many pesticides have a 'harvest interval' which specifies the minimum period which must elapse between the last pesticide application and harvesting.

The toxicity of a pesticide is usually measured in milligrams of active ingredient for each kilogram of body weight (i.e. parts per million) of the test organism. This is measured as the dose required to kill 50 per cent of a sample of test animals in a specified time (often 24 hours) and is referred to as the LD_{50} dose. This dose can be measured more accurately than the dose required to kill a higher or lower proportion of a sample of test animals. Concern about using animals to test pesticides has continued, but for registration the LD_{50} dose for rats is still used. Acute toxicity is much easier to assess, but subacute toxicity is measured initially over 90-day periods and chronic toxicity subsequently over 1 year or more. Inhalation toxicity is determined as the LC_{50} (lethal concentration) measured in milligrams per litre of air. Generally, pesticides which have been commercialised more recently have a much lower acute and dermal toxicity than many of the oldest pesticides. Further information on individual pesticides is available in the *Pesticide Manual* (Tomlin, 1997) while individual countries usually publish lists of pesticides which have been registered in that country, together with information on the crops and pests for which their use is permitted.

Much of the high cost of developing a new pesticide is due to the need for extensive toxicity testing of the chemical, its formulation and breakdown products, to determine its effect on representative non-target organisms (fish, birds, bees) and establish residue levels, before a new product can be marketed (Table 16.1). Kuiper (1996) discusses the current testing strategies to assess potential health risks. Extensive environmental impact studies are now required to obtain registration of a pesticide. The toxicity exposure ratio (TER) is one of the tools used in risk assessment to meet EC and EPA directives. Operator risk assessment is increasingly being based upon predictive operator exposure modelling such as 'EUROPOEM' used to harmonise the system throughout Europe (Gilbert, 1995; Van Hemmen and Brouwer, 1997). However, tests revealed higher values than those predicted by one exposure model (Vercruysse *et al.*, 1999b). Human exposures to pesticides have also been discussed by Krieger *et al.* (1992).

Table 16.1 Usual toxicological studies required before a pesticide can be marketed

(1)	Acute oral toxicity
(2)	Dermal toxicity
(3)	Eye irritation
(4)	Inhalation
(5)	Subacute studies – 90-day and 2-year feeding tests
(6)	Demyeliation
(7)	Carcinogenicity (tumour-susceptible strain)
(8)	Teratogenicity (pregnant rats)
(9)	Three-generation studies (mice)
(10)	Estimation of acceptible daily intake
(11)	Wildlife and fish studies
(12)	Studies on metabolism in plants and mammals
(13)	Residue studies
(14)	Potentiation

Classification of pesticides

The World Health Organisation has classified the commercially available pesticides according to the LD_{50} data for solid and liquid formulations (Table 16.2). Granular formulations are generally regarded as less hazardous to apply than sprays of the same chemical. Examples of the classification for selected pesticides are given in Table 16.3. When selecting a pesticide, preference should be given to the least hazardous pesticide which is effective and, if possible, to the least persistent. Preference is also given to the water dispersible formulation, if available, rather than liquid formulations such as the emulsifiable concentrate.

Table 16.2 WHO classification

Class	Hazard level	Oral toxicity[a]		Dermal toxicity[a]	
		Solids[b]	Liquids[b]	Solids[b]	Liquids[b]
Ia	Extremely hazardous	< 5	< 20	< 10	< 40
Ib	Highly hazardous	5–50	20–200	10–100	40–400
II	Moderately hazardous	50–500	200–2000	100–1000	400–4000
III	Slightly hazardous	> 500	> 2000	> 1000	> 4000

[a] Based on LD_{50} for the rat (mg/kg body weight).
[b] The terms 'solids' and 'liquids' refer to the physical state of the product or formulation being classified.

Users of pesticides should familiarise themselves with the appropriate legislation in their country. In the United Kingdom, the use of pesticides in agriculture is included in the *Food and Environmental Protection Act* (FEPA) and *Control of Substances Hazardous to Health (COSHH) Regulations.* A code of conduct for the safe use of pesticides on farms and holdings (the *Green Code*) (MAFF, 1998) sets out the responsibilities and requirements of those using pesticides to satisfy the legislation. Part of the requirement is for users of pesticides to obtain training and pass a practical test to obtain a Certificate of Proficiency appropriate for the equipment being used. In several countries, the equipment must be examined regularly to ensure it meets minimum standards of efficiency (Heestermans, 1996).

The protective clothing which must be worn will depend on the pesticide, its formulation and/or method of application (Table 16.4). In the United Kingdom, the personal protective equipment (PPE) required according to an EC Directive is shown in the annual *UK Pesticide Guide.* This publication provides a list of all the approved products for farmers and growers, and indicates which pesticides may be used on a particular crop. As chemical manufacturers may not seek approval for their product for all possible uses, for example on minor crops, other organisations or individuals can seek approval for off-label use. The *UK Pesticide Guide* lists these 'off-label' approvals, maximum residue levels (MRLs) permitted in food crops, and

Table 16.3 Examples of the acute oral toxicity of selected technical pesticides, based on data in the WHO document WHO/PCS/98.21/Rev. 1 with additional reference to the *Pesticide Manual*, 11th edn (Tomlin, 1997)

Classification	Common name	Trade name	Type[a]	Toxicity[b] (mg/kg)
Class Ia Extremely hazardous	aldicarb	Temik	I-C	0.93
	phorate	Thimet	I-OP	2
	disufoton	Disyston	I-OP	2.6
	methyl parathion		I-OP	14
Class Ib Highly hazardous	carbofuran	Furadan	I-C	8
	monocrotophos	Nuvacron	I-OP	14
	azinphos methyl	Guthion	I-OP	16
	methomyl	Lannate	I-C	17
	methamidophos	Tamaron	I-OP	30
Class II Moderately hazardous	endosulfan	Thiodan	I-OC	80
	fipronil	Regent	I-PPY	92
	lambdacyhalothrin	Karate	I-P	56
	DDT	–	I-OC	113
	rotenone	–	I-B	132[c]
	deltamethrin	Decis	I-P	135
	pirimicarb	Aphox	I-C	147
	paraquat	Gramoxone	H	150
	carbaryl	Sevin	I-C	300
	imidacloprid	Gaucho	I-CN	450
Class III Slightly hazardous	trichlorphon	Dipterex	I-OP	560
	tebufenpyrad	Oscar	A	595
	triadimefon	Bayleton	F	602
	malathion	–	I-OP	2100
	copper oxychloride	–	F	1440
	propanil	Stam F-34	H	1400
	isoproturon	Arelon	H	1800
Unclassified	atrazine	Gesaprim	H	2000
	buprofezin	Applaud	I-IGR	2200
	flufenoxuron	Cascade	I-IGR	> 3000
	spinosad	Tracer	I-S	3738
	teflubenzuron	Nomolt	I-IGR	> 5000
	tebufenozide	Mimic	I-HRA	> 5000
	temephos	Abate	I-OP	8600
	benomyl	Benlate	F	> 10 000
	carbendazim	Bavistin	F	> 10 000

[a] A, acaricide; B, botanical; C, carbamate; CN, chloronicotinyl; OC, organochlorine; F, fungicide; H, herbicide; HRA, hormone receptor agonist; I, insecticide; IGR, chitin synthesis inhibitor (insect growth regulation); OP, organophosphate; P, pyrethroid; PPY, phenylpyrazole; S, isolated from fermentation of the fungus *Saccharopolyspora spinosa*.

[b] Toxicity values refer to active ingredient, but classification is dependent on toxicity of formulation. If data on the toxicity of a formulation are not available, then an approximate value can be calculated from

$$\frac{LD_{50} \text{ of active ingedient} \times 100}{\% \text{ active ingredient in formulation}}.$$

Thus for 5% carbofuran granules the approximate toxicity is

$$\frac{(8 \times 100)/}{5} = 160\,\text{mg/kg}.$$

[c] Value depends on extract of plant used.

Table 16.4 Summary of the protective clothing which must be worn when applying scheduled substances

Jobs for which protective clothing must be worn	Protective clothing needed
Spraying pesticides	
(1) Opening container Diluting and mixing Transferring from one container to other Washing containers Washing out equipment, including aerial equipment	Overall and rubber apron[a] or mackintosh[a] Rubber boots[a], rubber gloves Face shield (or respirator[b])
(2) Spraying ground or glasshouse crops Acting as a ground marker with aerial spraying	As (1) above, except overalls should have a hood, and omit rubber apron and mackintosh
(3) Spraying bushes and climbing plants	As (1) above, but wear a rubber coat and sou'wester and omit rubber apron
Granule application	
(4) Opening container	As (1) above, but wear rubber gauntlet gloves with sleeves over their cuffs
(5) Application of granules by hand or hand-operated apparatus	As (1) above, but wear sleeves of overall over cuffs or rubber gauntlet gloves and omit apron
(6) Application of granules by tractor	Overall or mackintosh, but if a Part I substance, see (5) above
(7) Acting as a ground marker with aerial application	As (6) above, but add hood and face shield
Other applications	
(8) Sprays applied to soil Soil injection	Overall, rubber gloves rubber apron[c], rubber boots and respirator
(9) Bulb dipping	Overall, rubber gauntlet gloves, rubber boots and rubber apron
(10) Application of nicotine to roosts, perches and other surfaces in a livestock house	Overall, rubber gloves, face shield

[a] Not required with Part III substances.
[b] Respirator must be used (a) with all jobs involving Part I substances, except when diluted dimefox is applied to the soil, and (b) with any scheduled substances (Parts I, II and III) applied inside enclosed spaces, e.g. glasshouses, warehouses, livestock houses, as an aerosol or smoke.
[c] In enclosed spaces.

occupational exposure standards (OES), where appropriate. The *Poisons Act* and subsequent rules provide for the labelling, storage and sale of scheduled poisons. Aldicarb, chloropicrin, methyl bromide, demeton-S-methyl, endosulfan, paraquat and several other toxic pesticides are not scheduled chemicals, but are included in the *Poisons List*.

The container label must have an indication of a hazard 'VERY TOXIC', 'TOXIC', 'HARMFUL' or 'CAUTION' in relation to Class Ia, Ib, II and III category pesticides, respectively, shown in Table 16.3, and the statement 'Keep out of reach of children'. Such words need to be translated into the vernacular to be meaningful. Use of distinctive colours for labels has been used in some countries to denote the level of hazard, but this system is criticised because some users are colour blind. In addition, some manufactuers may use distinctive colours to advertise their products, and the significance of different colours does vary between different areas of the world. Some of the information on labels is now provided as pictograms, but these need to be sufficiently large and clear for the message to be appreciated by the user. A simplified guide for safe use is given in Table 16.5.

Table 16.5 Guidelines for safe use of a pesticide (Matthews and Clayphon, 1973)

Before applying pesticide – general instructions

(1) Know the pest, and how much damage is really being done
(2) Use pesticides only when really needed
(3) Seek advice on the proper method of control
(4) Use only the recommended pesticide for the problem. If several pesticides are recommended, choose the least toxic to mammals and if possible the least persistent
(5) **READ THE LABEL**, including the small print
(6) Make sure the appropriate protective clothing is available and is used, and that all concerned with the application also understand the recommendations, and are fully trained in how to apply pesticides
(7) Commercial operators using large quantities of organophosphate pesticides should visit their doctor and have a blood cholinesterase test, and have repeat checks during the season
(8) Check application equipment for leaks, calibrate with water and ensure it is in proper working order
(9) Check that plenty of water is available with soap and towel, and that a change of clean clothing is available
(10) Check that pesticides on the farm are in the dry, locked store. Avoid inhaling pesticide mists or dusts, especially in confined spaces such as the pesticide store.
(11) Warn neighbours of your spray programme, especially if they have apiaries.
(12) Take only sufficient pesticide for the day's application from the store to the site of application. Do **NOT** transfer pesticides into other containers, especially beer and soft drink bottles.

While mixing pesticides and during application

(1) Wear appropriate protective clothing. If it is contaminated, remove and replace with clean clothing
(2) Never work alone when handling the most toxic pesticides
(3) Never allow children or other unauthorised persons near the mixing
(4) Recheck the instructions on the label
(5) Avoid contamination of the skin, especially the eyes and mouth. Liquid formulations should be poured carefully to avoid splashing. Avoid powder formulations 'puffing up' into the face. If contaminated with the concentrate wash immediately. Use a closed system to transfer chemical to the sprayer where possible. Add washings of containers to spray tank
(6) Never eat, drink or smoke when mixing or applying pesticides
(7) Always have plenty of water available for washing

Table 16.5 (*Cont.*)

(8) Always stand upwind when mixing
(9) Make sure pesticides are mixed in the correct quantities
(10) Avoid inhalation of chemical, dust or fumes
(11) Start spraying near the downwind edge of the field and proceed upwind so that operators move into unsprayed areas
(12) **NEVER** blow out clogged nozzles or hoses with your mouth
(13) **AVOID** spraying when crops are in flower. Risk to bees is reduced if sprays are applied in evening when they are no longer foraging. Never spray if the wind is blowing towards grazing livestock or regularly used pastures. **OBSERVE** no-spray buffer zone.
(14) **NEVER** leave pesticides unattended in the field
(15) Provide proper supervision of those assisting with the pesticide application, and have adequate rest periods
(16) When blood tests are being conducted, do not work with pesticides if your cholinesterase level is below normal
(17) Wash sprayer in the field and apply invates to last section of field but do not exceed recommended dosage

After application
(1) **RETURN** unused pesticide to the store
(2) Safely dispose of all empty containers. As it may be difficult to dispose of empty containers after each day's spraying operations, they should be kept in the pesticide store until a convenient number are ready for disposal. **IT IS ABSOLUTELY IMPOSSIBLE** to clean out a container sufficiently well to make it safe for use for storage of food, water or as a cooking utensil. If any containers are incinerated, **NEVER** stand in the smoke
(3) **NEVER** leave pesticides in application equipment. Clean equipment and return to store
(4) Remove and clean protective clothing
(5) Wash well and put on clean clothing. Where there is a considerable amount of spraying, the operators should be provided with a shower room
(6) Keep a record of the use of pesticides
(7) Do not allow other people to enter the treated area for the required period if restrictions apply to the pesticide used

Protective clothing

Appropriate protective clothing must be worn whenever a Class I pesticide is applied or when application equipment contaminated with such pesticides is repaired (Fig. 16.2). The minimum protective clothing is a coverall defined as a single garment (or combination of garments which offers no less protection as a single garment) with fastenings at the neck and wrists which

(1) covers the whole body and all clothing other than that which is covered by other protective clothing such as face shield, goggles, respirator, footwear or gloves
(2) has its sleeves over the top of gauntlet gloves, unless elbow-length gloves are needed for dipping plants in pesticide

(3) is resistant to penetration by liquid or solid particles in the circumstances in which it is worn.

Test methods have been devised to assist with the selection of coveralls suitable for work while applying pesticides (Gilbert and Bell, 1990). The garments are rated according to the penetration of the material by solvents and by water + surfactant applied under pressure. Roff (1994) developed a

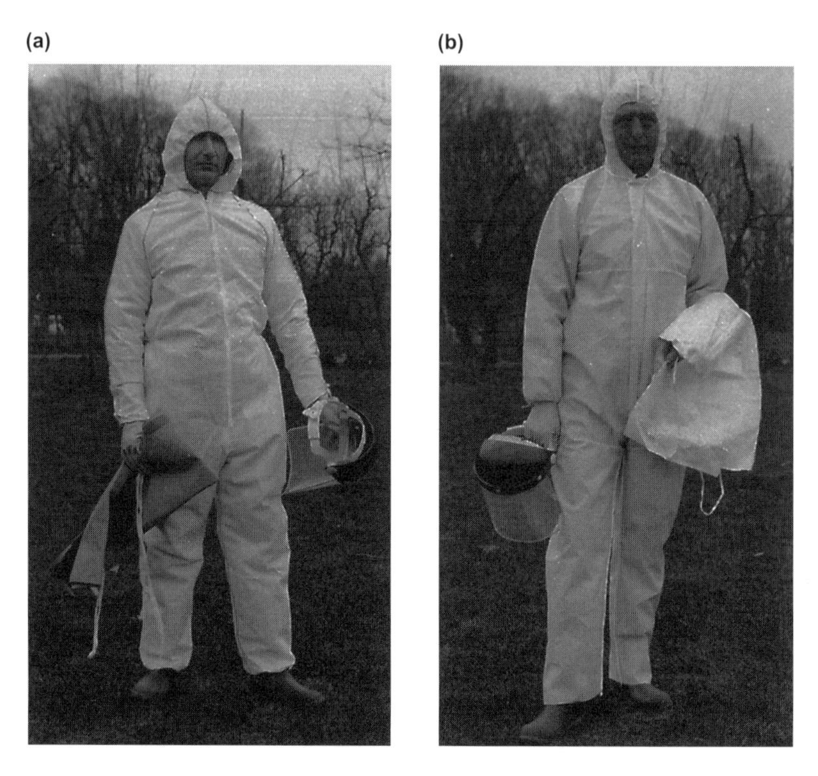

Fig. 16.2 Protective clothing (photos courtesy of: Application Hazards Unit, MAFF Central Science Laboratory). (a) Garment type 1. EP spun-bonded polypropylene boilersuit. (note: 'EP' stands for 'extra protection'.) This is a 'breathable' garment; hence no permeation resistance; hence the need for an impermeable apron (and faceshield) if handling concentrates or if expecting to get heavily contaminated by dilute liquid pesticide in the course of planned application. (b) Garment type 2. PE (polyethylene) coated spunbonded polypropylene boilersuit. Not breathable, but relatively low resistance to permeation. This implies the need for an impermeable apron (and faceshield) if handling concentrates. The suit is likely to withstand fairly heavy contamination by dilute liquid pesticide. (c) Garment type 3. Saranex laminated 'Tyvek' boilersuit. (Note: 'Tyvek' is the proprietary name for spunbonded polyethylene made by DuPont, who also make Saranex. All garments are made up by other intermediary companies, however.) Not breathable. Fairly high permeation resistance; hence no need for impermeable apron when handling most concentrates. The operator is thus

system of assessing dermal exposure using a fluorescent dye photographed with special UV lighting.

PPE required for Class I pesticides is not practical under hot tropical conditions and most small-scale farmers cannot afford it; hence there is the need to use the less hazardous pesticides. Unfortunately some farmers do not believe that they need to wear protective clothing (see, for example, Grieshop, 1988). The minimum requirement is a durable woven cotton fabric overall, or

(c)

(d)

pictured wearing only the faceshield, which would be required for concentrate handling. (It would be prudent to wear an apron if handling very hazardous products, to protect the coverall from contamination and extend its protective properties). (d) Garment type 4. PVC coated nylon boilersuit. Not breathable. High resistance to permeation (for most liquids); hence no need for impermeable apron when handling most concentrates. The operator is thus pictured wearing only the face-shield, which would be required for concentrate handling. (It would be prudent to wear an apron if handling very hazardous products, to protect the coverall from contamination and extend its protective properties.) Please note that all subjects are wearing protective gloves and rubber boots, with coverall arm cuffs and trouser cuffs worn outside. The wearing or holding of impermeable apron and/or faceshield is intended to denote whether wearing these particular items would be necessary to supplement the coverall alone when handling concentrates.

equivalent long sleeved shirt and trousers, without turn-ups where granules and dust particles can collect. Unwoven synthetic fabrics have been made up as disposable coveralls, which are effective in many situations, but they are unsuitable in the tropics as temperatures are too high to wear clothing made with impermeable materials. Cotton, polyester and cotton-polyester blend fabrics are more comfortable to wear, and sorption and penetration can be reduced by treatment with a fluoroalkyl methacrylate polymer (Shaw *et al.*, 1996). Particular attention to choice of protective clothing is needed when the period of exposure is likely to be prolonged or the concentration of chemical exceeds 10 per cent.

Operators are at the greatest risk when mixing concentrated formulations. As mentioned elsewhere, this risk is reduced by closed transfer systems, but where these are not available, the user should wear in addition to gloves, a plastic apron and faceshield to avoid splashes on the coveralls and face. Once the pesticide has been diluted this extra PPE can be removed. If gloves become contaminated with concentrate pesticide, they should be washed immediately, as deposits of some pesticide if left on the surface of the glove will penetrate quite rapidly through it. In any case, gloves exposed to dilute pesticides should also be washed with detergent and water before removal to avoid contaminating the hands, and also because dried deposits may adversely affect the glove material during storage. When gloves are worn they should be made of neoprene. These should be checked previously for pin-holes by filling the glove with water, gently squeezing it and then drying it before use. Gloves should be long enough to protect the wrist, and the cuffs of overalls should be outside the top of the gloves to reduce seepage of spray down inside the gloves.

During application, risk of exposure of the operator is reduced if the spray is directed downwind away from the body, even with power-operated equipment. A wide-brimmed hat, preferably waterproof, or a hood attached to the coverall to protect the back of the neck as well as the face is useful, not only to reduce spray deposition on the body, but also to minimise the effects of the sun.

After completion of spraying, equipment should be cleaned and returned to the store. Then the PPE should also be removed and cleaned. Regular washing of coveralls with soap or detergent should take place at the end of each day's use, and a second set used the following day. As not all pesticide deposits are removed by washing (Nelson, *et al.*, 1992), every effort must be made to minimise contamination of clothing. Throughout the spraying operation, there should be a good supply of water for washing the skin immediately if it is contaminated with pesticide. Treatment of overalls with starch is reported to enhance removal of pesticide during laundering (Ko and Obendorf, 1997).

Handling small objects such as nozzle tips is difficult when wearing rubber gloves, but operators should not be tempted to remove the glove; this can be extremely dangerous as some pesticides are easily absorbed through skin which is wet with sweat. People working in workshops where spraying equipment is being repaired should be particularly careful. When dismantling equipment they may touch chemical deposits which have not been removed by normal washing.

Special protective clothing includes eye and face shields, respirators, and impermeable overalls. Two types of respirators are available: the cartridge respirator which covers the nose and mouth and the gas-mask which also covers the eyes and may be incorporated in a complete head shield. Both types have one or two 'cartridges' which absorb toxic fumes and vapour, and are suitable for use when fogging (see Chapter 11). Gas-masks usually have more efficient fittings for more prolonged use. Both types must be worn tightly so that they are sealed around the face to prevent leakage around the edge, and are generally uncomfortable to wear in hot weather. All items need to be regularly cleaned, including the inside of gloves and masks. Any special filters on respirators must be changed according to the manufacturer's instructions. One of the dangers is that some operators wear a respirator while mixing sprays, but then remove it so that the inside is liable to get contaminated. Operators are liable to inhale the poison when they replace the respirator to mix another batch of chemical. Simple disposable masks are sometimes safer to use to reduce inhalation of droplets of the less hazardous chemicals and also minimize deposition of chemical around the mouth. An eye shield is needed when pesticide formulations containing certain solvents such as isophorone are sprayed.

All clothes, including protective clothing and the user's normal clothes, should be kept well away from the storage and mixing area in a separate changing room. If pesticides are used extensively, the changing room should ideally be fitted with a shower. In any case, soap and water should be available for operators to wash after work. Some tractor-mounted equipment has a separate water tank so any contamination of gloves can be washed off immediately (Fig. 16.3) The tractor sprayer should also have compartments for clean and used PPE.

Symptoms of poisoning

Symptoms of poisoning will vary according to the pesticide involved. Where pesticides are used regularly, advice from the local Health Authorities should be sought. In the United Kingdom, the Department of Health has published a book *Pesticide Poisoning – Notes for the Guidance of Medical Practitioners*, which gives relevant symptoms of each group. There is also a National Poisons Centre which provides a 24 hour information service.

Signs of organophosphate poisoning include headache, fatigue, weakness, dizziness, anxiety, perspiration, nausea and vomiting, diarrhoea and a loss of appetite. An increase in the severity of the symptoms leads to excessive saliva and perspiration, stomach cramps, trembling with poor muscle co-ordination and twitching. The patient may have blurred vision, a rapid pulse and some difficulty in breathing. Severe poisoning leads to convulsions, eyes with pinpoint pupils, inability to breathe and eventually unconsciousness.

Fig. 16.3 Using extra clean water tank on sprayer to wash hands.

Some of these symptoms can occur with other types of poisoning or other illnesses such as heat exhaustion, food poisoning or a 'hangover'. A person who becomes ill after using or being near pesticides may not necessarily have been poisoned, but the suspected poisoning is seldom verified by suitable tests. People using pesticides may develop dermatitis, especially on the hands. This may be due to the solvent rather than the pesticide itself, and may also be a reaction to wearing rubber gloves and sweating.

First aid

Immediate medical attention by a doctor or at a hospital is essential when a person using pesticides becomes ill. First-aid can be given before the patient reaches a doctor. The patient should be kept quiet and warm, away from the sprayed area and, if possible, in a sheltered place in the shade. All protective and contaminated clothing should be removed. All other clothes should be loosened and, taking care not to contaminate your own skin or clothes, the patient's contaminated skin should be washed thoroughly with soap and as much water as possible. If the person is affected by poisoning, keep the patient lying flat and at absolute rest; if conscious and able to swallow, the patient should drink as much water as possible.

When poisoning with the most toxic and rapidly acting substances has been by mouth, attempts should be made by specially trained medical personnel to induce vomiting within 4 h if the patient is conscious. Administration of salt solution is not now recommended. If the patient can be attended by a doctor or nurse, the use of ipecacuanha emetic is the preferred method of inducing vomiting. Note that its use should not be recommended to first aiders. Vomiting should not be induced if the person has swallowed an acid, alkaline or petroleum product. If the chemical has got into the eye, clean water is required quickly to flush the eye several times for at least 15 min. If breathing ceases or weakens, artificial respiration must be started, making sure that the breathing passages are clear. If the patient is in convulsion, a strong piece of wood or a folded handkerchief should be placed between their teeth to prevent them biting their tongue.

The doctor must be informed of the name of the active ingredient and given as much information as possible by showing them a leaflet or label about the chemical. Treatment by a doctor will depend very much on the type of poisoning. An injection of atropine is useful for organophosphate and carbamate (anticholinesterase) poisoning.

Suitable antidotes for organochlorine poisoning are not available. Large quantities of fuller's earth is used if a person is affected by paraquat. Morphine should not be given to patients affected by pesticide poisoning. A first-aid kit should be readily available and a supply of clean water for drinking and washing any contaminated areas of the body. On large-scale spraying programmes first-aid kits should be carried in vehicles and aircraft. People regularly involved in applying organophosphate pesticides should have a routine medical examination to check the cholinesterase levels in their blood plasma.

Combination of chemicals

When a crop is affected by more than one pest, a farmer may mix two or more pesticides together to control them with a single application or to increase the effectiveness of an individual pesticide. Apart from the problem of whether the chemicals are compatible, their toxicity to humans and other organisms may be increased. For example the LD_{50} of malathion is 1500 mg/kg and of fenitrothion 400 mg/kg, but the LD_{50} of the mixture is less than 200 mg/kg. Residues of mixtures may persist longer. Because of their potential toxicity, combinations of pesticides and various additives should not be used unless the specific combination has been tested and is recommended by the appropriate authorities (Godson *et al.*, 1999). There is also the danger of a cumulative effect of different pesticides used separately in a spray programme.

Pesticide packaging and labelling

Exposure to pesticides can be reduced by improved methods of packaging (Curle *et al.*, 1998). Most farmers in the tropics have a small acreage, and need small quantities of pesticide to avoid storage of partially opened packets; for example, 5 kg packets of granules are available for direct filling of hoppers, and several pesticides have been marketed in sachets containing sufficient wettable powder to mix 15 litres of spray, to fill one knapsack sprayer. Some of these sachets are covered by a water-soluble film, so the whole sachet is placed directly into the sprayer, but individual sachets must be kept dry until used. Similarly, some pesticides in dry tablet form are also easier to dispense without exposure to the user.

Apart from reducing the risk of pesticides contaminating the farmer's store, which is often in the house, small sachets eliminate the need to measure out the quantity required to obtain the correct spray concentration. Liquid containers may now incorporate a built-in measure to avoid pouring into a small measure (Fig. 16.4). Savings through reduction of loss by spillage and ensuring correct mixing, in addition to the improved safety aspects, more than repay the extra cost of packaging. Efforts have been made to standardise containers to facilitate the use of closed filling systems (Gilbert, 1989).

Fig. 16.4 Small pesticide container with built-in measure.

Appropriate labelling is essential and should be in the vernacular language. Apart from the brand name, the label should have details of the active ingredient and inert materials used in the formulation, the intended use of the product, full directions on the safe and correct procedure for mixing and application of the product (i.e. protective clothing required, which pests are to be controlled, dosage, time and method of application), and how to dispose of the container. Cautionary notices to protect the user, consumer (if the treated area is a food crop) and beneficial plants and animals should be clearly given on the label. The label should also indicate the minimum period between application and harvesting appropriate for the various crops on which the pesticide can be used. To assist some users, important information is also given in the form of pictograms (Fig. 16.5), but where these are used on labels, they need to be large enough to have sufficient visual impact on the user.

Fig. 16.5 Examples of pictograms.

Farmers should avoid storing chemicals for more than about 18 months. Containers left longer than this may corrode, or the active ingredient may be less effective.

Container and washings disposal

At the end of an application, a significant amount of pesticide spray mixture may be in the sprayer. Taylor *et al.* (1988) reported that over 9 litres may remain in the pump and associated pipework of a 600 litre sprayer, while Balsari and Airoldi (1998) found that air-assisted orchard sprayers had more spray mixture in them than boom sprayers. An approved decontamination method is required as a condition of approval of pesticides within the European Union, and de-activating agents may be required for a particular pesticide (Taylor and Cooper, 1998). Ideally, clean water is flushed through the equipment and sprayed out in the field, as legislation in some countries restricts the disposal of these washings which will contain an unknown quantity of pesticide. If washings are not sprayed in the area being treated with the pesticide, one option could be to install a water effluent treatment plant. Small-scale plants, albeit expensive, are now available which can remove organic substances. Treated effluent may be retained for reuse, such as cleaning of equipment, or may be discharged to a sewer with the consent of the appropriate authorities. The small quantity of sludge produced can be buried or disposed of by a waste disposal company. These effluent plants (Fig. 16.6) are especially important for spray contractors, including aerial operators, and large-scale farmers (Harris *et al.*, 1991).

Clean, rinsed empty agrochemical containers, plus outer packaging and related materials, should preferably be delivered to a licensed incinerator or to an approved landfill site. However in situations where no approved or licensed

Fig. 16.6 'Sentinel' system for decontaminating water used to wash sprayers and containers.

facilities exist, on-farm disposal is a possible solution. Options for this in order of priority are, (a) on-farm burning using a hot fire or (b) on-farm burial. It is essential that any method adopted does not conflict with any relevant local laws and regulations. This is important since burning containers on a farm is prohibited in some countries such as Germany and Canada. It is possible to return some empty purpose-designed containers to the supplier for refilling. The principle of a comprehensive container management strategy (Smith, 1998) has now been used by agrochemical companies, and the use of returnable containers has now increased in some areas. Dohnert (1998) compares costs of single and multi-trip containers.

As indicated earlier, developments in formulation have led to changes in packaging to reduce the problem of disposal of used containers. In developing countries, the empty pesticide container is still valuable for other uses. Large metal containers have been flattened to provide building materials, especially for roofs, and drums have been used to collect and carry water. As pesticide containers can never be adequately cleaned for other purposes, some fatalities have occurred. Strictly, where there is no pressure washing equipment in an induction hopper to clean containers, the user must triple rinse all liquid containers manually. A well-drained container is triple rinsed by adding sufficient clean water to fill the container to about 20–25% of its capacity, replacing the lid securely and shaking the contents vigorously so that all inside surfaces, including the lid, are cleaned; the contents are then poured into the spray tank and the container allowed to drain for at least 30 s before the process is repeated at least twice until the container is visually clean. In a survey with these washing methods, over 90% of the washed containers had less than 0.01% of the original contents (Table 16.6) (Smith, 1998). The rinsates are added to the spray. Power and Miller (1998) reported results of rinsing three sizes of containers from two manufacturers using a low-level induction hopper with a built-in rinsing system. After three rinses less than 0.1 ml of a simulant chemical residue remained, but the container had to be moved during the rinsing routine to ensure that flushing jets of water impinged on most of the inner surface of the container.

Table 16.6 Results of a survey of rinsing containers (from Smith, 1998)

	Rinsing method	
No. of tests	41	156
No. of results < 0.01%	38	142
Percentage of tests with < 0.01%	92.7	91.0

Cleaned containers can be punctured and buried in a deep pit (at least 1.5 m in depth), avoiding sandy sites and heavy clay or other areas liable to cracking, well away from any river or stream to avoid pollution of water supplies. The pit must be covered with at least 0.5 m of soil. Containers of special formulations which are not mixed with water, including those for ULV applications, should be cleaned with a suitable solvent such as kerosene before being buried. Boxes and other types of packaging may be burnt, provided air pollution does not become an additional significant concern and it is lawful. A suitable small-scale incinerator was developed for on farm use (Fig. 16.7) (Carter, 1998). Herbicide containers can be burnt, where on-farm burning is permitted. Aerosol cans should never be punctured or burnt. Regulations concerning disposal of containers and unused chemicals have been introduced in a number of countries to minimise the risk of human poisoning and environmental pollution. Local regulations should always be followed. Waste management is discussed by Johnson (1998).

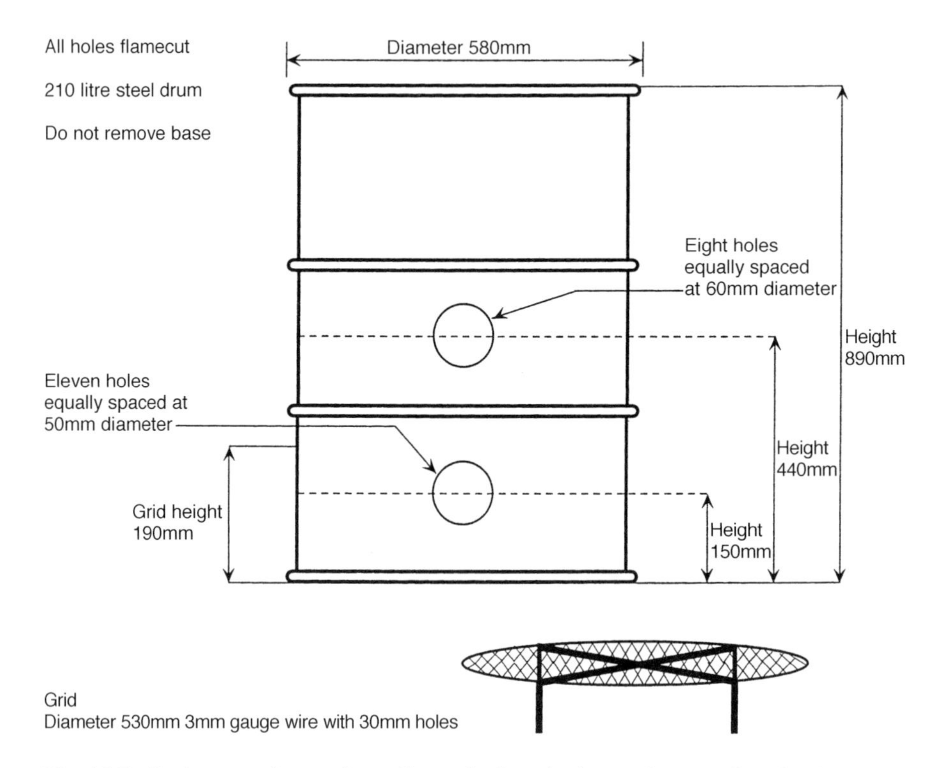

Fig. 16.7 Incinerator for on-farm disposal of washed containers and packaging.

Noise

Noise ratings greater than 85 decibels (dB) in any octave band in the speech range 250–4000 Hz can cause permanent hearing impairment. Human pain threshold is 120 dB. The noise level within a radius of 7 m from motorised sprayers often exceeds 85 dB, so the effect of noise should be considered in relation to the safe use of pesticides. Exposure to continuous noise should be restricted by interchanging spray operators or having definite rests between short periods of spraying. Ideally, hearing protection should be provided, especially when sprayers are operated inside buildings.

The safety of agricultural pilots is also affected by loud engine and other noises and vibrations on an aircraft.

Code of conduct

Concern about the hazards of using certain pesticides led to the *FAO Code of Conduct on the Distribution and Use of Pesticides*. Subsequently, this code has incorporated the principle of Prior Informed Consent (PIC) by which exporters of pesticides have to inform importers in developing countries about the toxicity and hazards associated with the use of products included on the PIC list (Table 16.7), and receive their authority before the products can be exported. The *FAO Code* is voluntary, but the requirements for PIC have been included in an EU regulation applicable by law in the member states. A PIC database is maintained at the International Register of Potentially Toxic Chemicals held at Geneva, where the International Programme on Chemical Safety (IPCS) is located (Younes and Sonich-Mullin, 1997).

Advice on the disposal of unwanted pesticide stocks is now available in a booklet published by the Global Crop Protection Federation (GPCF; formerly GIFAP), Brussels. The FAO has a programme to assist developing countries with the disposal of obsolete stocks of pesticides.

Table 16.7 Pesticides on the FAO Prior Informed Consent List (see http/ www.fao.org/pic/substances.htm for further information)

Pesticides subject to PIC regulations
2, 4, 5-T, aldrin, captafol, chlordane, chlordimeform, chlorobenzilate, crocidolite, DDT, dieldrin, dinoseb, EDB, fluoracetamide, HCH, heptachlor, hexachlorbenzene, lindane, mercury compounds, methamidophos, monocrotophos, parathion, pentachlorophenol, phosphamidon, PBBs, PCBs, PCTs, tris 2,3 dibromopropylphosphate

17

Application of biopesticides

There is considerable interest in developing the use of biological agents as an alternative to chemical pesticides (Menn and Hall, 1999). This chapter will consider the special requirements of biopesticides that can be applied with equipment designed for the application of chemical pesticides. These include baculoviruses, bacteria, fungi and entomopathogenic nematodes. In contrast to chemical pesticides, they are all particulate living organisms, although the effect of the most widely used biorational pesticide, *Bacillus thuringiensis*, is due to the endotoxin produced by the spores. They range in size from extremely small viral particles to the large nematodes. Many of the fungal spores may be relatively small but have irregular shapes. The use of botanical pesticides is not considered here as in most cases the derived product is used in the same way as synthetically manufactured pesticides.

While a number of biopesticides have shown promise in the laboratory, translation of effectiveness to practical field conditions has generally been elusive. Biopesticides are much more susceptible to the actual conditions – temperature, humidity, pH of the leaf surface and ultraviolet light – than chemical pesticides. Penetration of a biopesticide into plant or insect surfaces has to overcome not only the physical barrier of the surface which may be a hard cuticle or a very waxy layer, but also the natural resistance of the pest species to infection. Some of the solvents and formulants used with chemicals to increase suspensibility and surfactants to spread deposits over leaf surfaces may be incompatible, and in some cases lethal, to a biopesticide. Some of the aspects of plant surfaces in relation to viral deposits are discussed by Hunter-Fujita *et al.* (1998).

Determining the amount of biopesticide that needs to be applied is also conceptionally different from chemical pesticides, as activity will depend on the viability of the deposit in contact with the pest, whereas some chemicals can rely on redistribution by systemic or translaminar activity. Some biopesticides must be ingested by the pest. Thus coverage is essential in the areas of a crop or soil specifically inhabited by the pest, unless some attractant can be incorporated to bring the pest to the biopesticide, or other subtle means of transfer to the pest can be devised. In addition to the spatial distribution of the deposits, it will be essential to consider the concentration of the biopesticide in the carrier to ensure sufficient is delivered to affect the pest. Where sprays are applied it is vital that there is a high probability of droplets containing sufficient biopesticide; conversely, due to the cost of production of biopesticides, using droplets larger than the optimum for deposition on the target will lead to excessive wastage of material. One of the present constraints to greater use of mycopesticides is linked to production economics as dosages of the order of 10^{13}–10^{14} spores per hectare are often required to attain acceptable control (Wraight and Carruthers, 1999).

Many argue that biopesticides must be applied with hydraulic sprayers (e.g. Chapple *et al.*, 1996) as these are widely used by farmers, despite the poor record of success with different types of biopesticide in the field. The viability of both *Verticillium lecanii* and the entomopathogenic nematode *Steinernema feltiae* were decreased as the length of pumping time and pressure increased in hydraulic sprayers (Nillson and Gripwall, 1999). Hydraulic sprayers have been useful where it is possible to apply high volumes at low pressure, as in the application of certain nematodes to the soil. Chapple *et al.* (1997) and Hall *et al.* (1997) report using two hydraulic nozzles in combination (Fig. 17.1) and suggest that a coarse spray of water can entrain a fine spray of the biopesticide. However, such a 'double nozzle' requires a large volume of water and does not diminish the wide range of droplet sizes produced. There are other existing application systems, such as rotary atomisers, albeit mostly used in niche

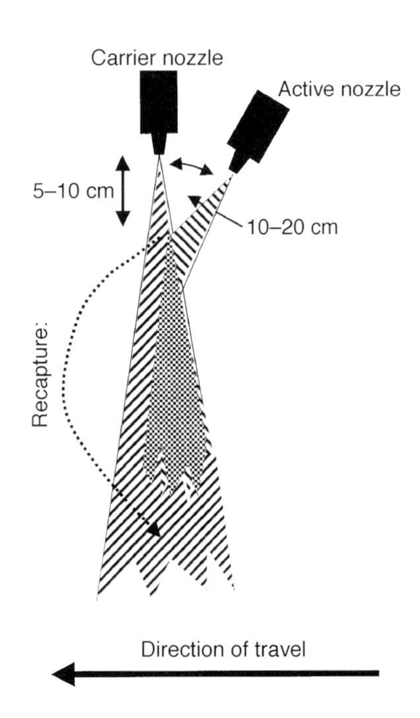

Fig. 17.1 Twin nozzle system (from Chapple *et al.*, 1997).

markets for chemical pesticides, which are in some cases more appropriate for optimising delivery of a biopesticide. The alternative of a specialised system of application is generally considered to be inappropriate due to the additional costs involved. The different types of ground-based equipment are discussed in relation to biopesticides by Bateman *et al.* (2000).

The wide range of droplet sizes produced with standard types of hydraulic nozzles is a key problem, as the smallest droplets produced in large numbers are unable to transport an effective dose of the biopesticide. Chapple and Bateman (1997) have reported that 63–72% of droplets from hydraulic nozzles applying between 159 and 560 litres/ha did not contain any spores of the biofungicide *Ampelomyces quisqualis*, while Lello *et al.* (1996) showed that although higher output hydraulic nozzles deposited the most infective juveniles of *Steinernema carpocapsae* on leaves, over 89% of droplets contained none of the infective juveniles. Secondly, due to the particulate nature of biopesticides, small orifice hydraulic nozzles can easily be blocked, so dictating a higher volume than may be required for a chemical pesticide. While this may be satisfactory for soil applied treatments, application of a higher volume implies a reduced concentration of the actual biopesticide; thus careful consideration has to be given to the formulations intended for foliar sprays to ensure adequate wetting of treated surfaces without detrimental effects on the biopesticide, caused by using more surfactant or other formulant.

These constraints suggest that it is important to assess whether the droplet size can be optimised, not only to improve efficiency of deposition at key target sites, but also to ensure delivery of sufficient biopesticide in the number of droplets likely to be deposited on a pest or on foliage inhabited by the pest. Droplet spectra illustrate the contrast between two air shear and two rotary nozzles (Fig. 17.2). Problems associated with too small a droplet liable to downwind drift and too large a droplet sedimenting on the soil rather than within a crop canopy, are equally, if not more important, with the application of biopesticides.

Metarhizium for locust control

In the development of the mycoinsecticide 'Green Muscle' based on the fungus *Metarhizium anisopliae* var *acridum* for locust control, it was possible at the outset to consider application with rotary atomisers as ULV technology had been adopted with chemical insecticides. Earlier studies had shown the advantage of applying droplets in the 50–100 μm range to enhance capture on locusts and the foliage on which they were feeding. Laboratory assays had shown a need for a particular dosage of conidial spores suspended in an oil to overcome the need for high ambient humidity (Bateman, 1994). Subsequently, through small scale (Bateman *et al.*, 1992) and then larger field trials (Bateman *et al.*, 1994; Bateman, 1997, Price *et al.*, 1997), effective control in the field was demonstrated by applying a dosage of 5×10^{12} conidia per hectare (Langewald *et al.*, 1999). Important differences in the field are the effect of secondary recycling of the fungal pathogen (Thomas *et al.*, 1995), which can offset the lack of persistence of the initial deposits, and survival of non-target organisms (Peveling *et al.*, 1999a). As part of the environmental impact studies, detailed assessment of *Metarhizium* on bees was conducted (Ball *et al.*, 1994). A crucial part of the delivery system was increasing the proportion of spray delivered with the optimum droplet size (Bateman, 1999) and using an oil carrier to protect spores in a harsh hot, dry environment. To assess the potential for the new insecticide in a recession period, 'pre-field' trials were carried out with laboratory reared insects under field conditions (Bateman *et al.*, 1998). Griffith and Bateman (1997) checked whether the oil-based formulation could be applied with the exhaust gas nozzle sprayer previously recommended for locust control.

Other studies with *Metarhizium* have examined the use of aqueous sprays for control of insect pests where ULV application is not used. Thus Alves (1998) examined the impact of a range of adjuvants to assess whether any could enhance tolerance to solar radiation. Generally, vegetable oil adjuvants did enhance conidial tolerance to UV light for up to 6 h exposure. Bateman and Alves (2000) report on the narrow droplet spectra obtained by adding a rotary nozzle to a motorised knapsack sprayer to apply oil/water emulsions containing a mycoinsecticide and enhance the proportion of spray droplets containing an optimum number of spores (Fig. 17.2). However, it must be emphasised that this applies only when the flow rate to the disc is kept low. Starch extrusion technology has also been examined as a means of preparing baits for entomopathogens for locust control (Caudwell and Gatehouse, 1996).

Biopesticide application in forests

Optimisation of droplet size has been adopted extensively for treatment of forests to minimise impact in streams within the forest ecosystem. In each case the aim has been to increase deposition within the canopy and avoid large droplets that can fall to ground level. Sundaram *et al.* (1997a) used rotary atomisers to apply Bt to control spruce budworm (*Choristoneura fumiferana*) and assessed deposits on canopy foliage. They reported later the effects of sunlight on Bt deposits affecting persistence and emphasised using an optimum droplet size (Sundaram *et al.*, 1997b). In studies on the feeding behaviour of gypsy moth larvae *Lymantria dispar* L., feeding was inhibited most by applying Bt at 9 droplets/cm^2 with 10 BIU/litre. The LD_{50} decreased from 14.1 to 3.1 BIU/litre between 48 and 144 h after application (Falchieri *et al.*, 1995). Bryant and Yendol (1988) considered that droplets in the 50–150 μm range would increase efficacy from a given volume of spray, but this would be difficult in practice if flying more than 15 m above the crop canopy due to off-target drift. Similar experiments by Maczuga and Mierzejewski (1995) indicated more than 90% mortality of second and third instar gypsy moth larvae with 5–10 droplets/cm^2, with good mortality of fourth instar with 200 and 300 μm droplets. If only 1 droplet/cm^2 was applied, Bt was ineffective against the later instars.

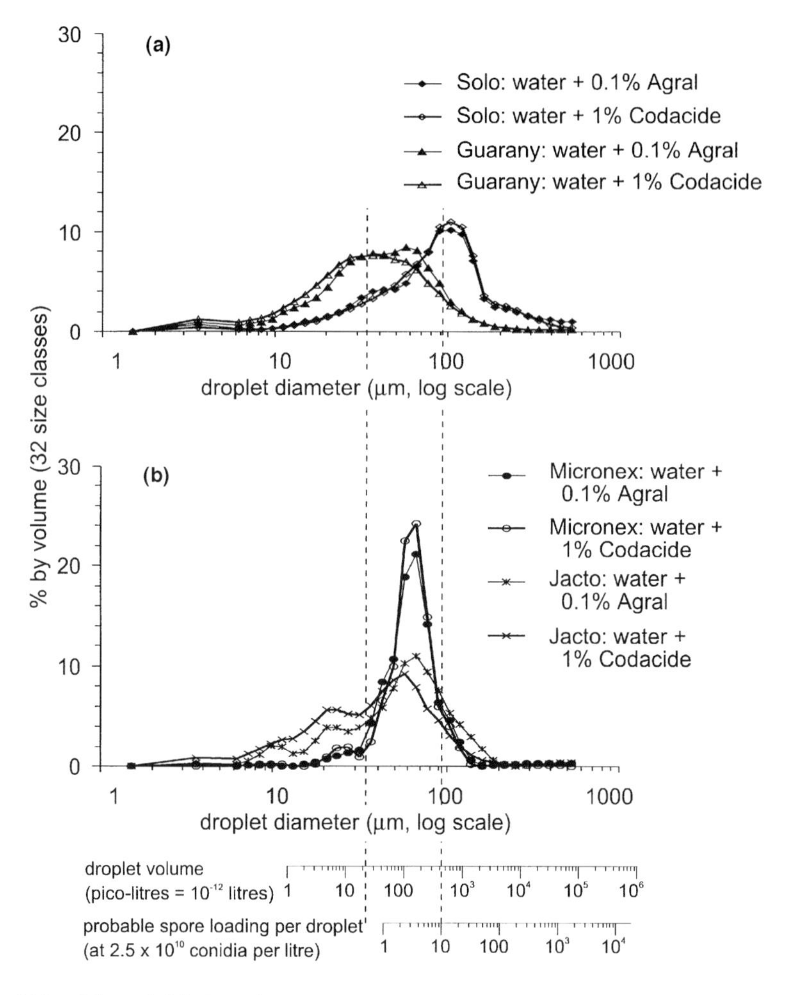

Fig. 17.2 (a) and (b) Droplet size spectra with four motorised mistblower nozzles showing the narrower spectra rotary atomisers. Data show effects of applying two liquids (water + 1% Codacide (an adjuvant containing rapeseed oil and water) + 0.1% Agral surfactant). All treatments applied at 200 ml/min. (c) Volume median diameter plus $D_{[9.0]}$–$D_{[1.0]}$ range for three mistblowers operating at different flow rates applying water +1% Codacide, showing narrowest droplet size range with the rotary atomiser at low flow rates.

A major example of an operational control programme was the treatment of the nun moth *Lymantria monacha*, a serious pest in parts of Europe, which was controlled with *Bacillus thuringiensis kurstaki* specially formulated for ULV application. Rotary atomisers (Micronair) were used on helicopters and fixed-wing aircraft (Butt *et al.*, 1999).

Another type of rotary atomiser, namely multiple spinning discs mounted on an aircraft, was also used to optimise delivery of the baculovirus of

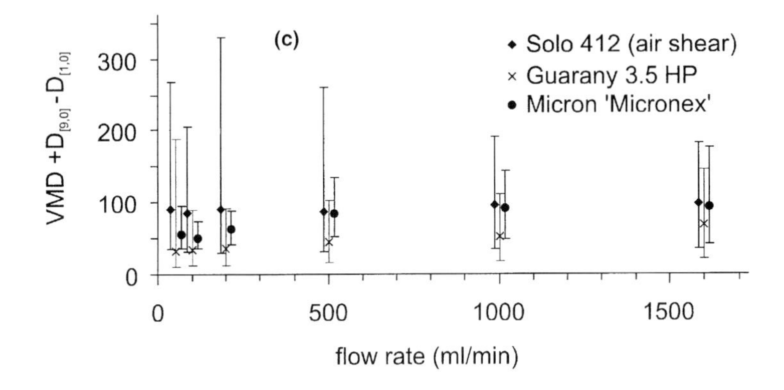

Panolis flammea on pine trees. Studies had shown a need to apply a minimum of five droplets per centimetre length of leaf in the top 30% of the tree canopy. With 40–50 μm droplets effectively filtered by pine needles and accurate timing, effective control was obtained with dosages as low as 2×10^{11} PIBs/ha (Cory and Entwistle, 1990; Evans, 1999). The need for small droplets and more turbulence within the tree canopy was pointed out by Payne *et al.* (1996) following applications of a nuclear polyhedrosis virus (NPV) and Bt. The dosage of NPV could be reduced by the addition of an optical brightener (Cunningham *et al.*, 1997). Parnell *et al.*, (1999) also report that application of *Helicoverpa armigera* NPV on cotton was better with a spinning disc sprayer compared to a mistblower. Silvie *et al.* (1993) had also used a manually carried spinning disc sprayer to apply a low dose of insecticide with a virus.

Calculations to determine the amount of bioinsecticide to apply, given by Evans (1999), are the following

The capture efficiency (CE) to ensure at least one droplet per host feeding area is given by

$$\mathrm{CE} = \frac{A \times \mathrm{LAI}}{sr^{-1}}$$

where LAI is the leaf area index, A is the area expressed in mm^2 ($1\,\mathrm{ha} = 1 \times 10^{10}\,\mathrm{mm}^2$), s = is the loss of spray to non-target area, and r is the area of feed (mm^2) consumed in the time taken to acquire a toxic dose. The volume (V) to be applied (litres/ha) is given by

$$V = \mathrm{CE}/N$$

where N is the number of droplets per litre; thus with 50 μm droplets $N = 1.53 \times 10^{10}$.

The dose per hectare (D_{ha}) is related to the incident dose (D_{i}) by the expression

$$D_{\mathrm{ha}} = \mathrm{CE} \times D_{\mathrm{i}}$$

where D_{ha} and D_i are expressed in IU/ha, while the concentration in the spray tank expressed in IU/litre is

$$D_l = N \times D_i$$

where $D_i = d/ar$, d being the LD_{95} (IU/target host stage) and a the proportionate loss of activity over time required for the host to acquire the target number of droplets. Therefore, if $D_i = 1000$, then $D_l = 1.53 \times 10^{10} \times 1000 = 1.53 \times 10^{13}$ IU/litre and $V = (4 \times 10^{10}/1.53 \times 10^{10}) = 2.6$ litres/ha.

These examples demonstrate that the process of translating a potential biopesticide into a successful field operation requires careful consideration of a complex series of factors associated with application, in addition to those concerned with production and formulation. However, the initial key is having a suitable effective biological agent. This may require a specific subspecies (contrast application of *Bt kurstaki* and *Bt israelensis* depending on the target pest) or strain of the organism if it is to have any chance of being successful.

The use of mycoherbicides has been less successful, due probably to the inherent resistance of plants to invading pathogens. It is possible that in addition to the problems of formulating a suitable pathogen, it may be necessary to devise specific application technology. Thus it may require a two-stage process, one stage of which is to damage the outer plant surface sufficiently for the pathogen to gain access. A sand-blast effect has been demonstrated to reduce the amount of chemical herbicide needed for control; thus a similar approach may increase the effectiveness of a mycoherbicide.

Entomopathogenic nematodes have been mostly applied at high volume as soil treatments, for example to control the vine weevil. Mason *et al.* (1998a,b) have examined the potential for nematodes to control foliar pests with particular reference to the diamond back moth *Plutella xylostella* on crucifers. Using a rotary atomiser, significant numbers of infective juveniles per cm^2 could be deposited, although most droplets contained no nematodes. Deposition was improved by increasing the concentration and/or flow rate. Further studies showed an improvement in deposition with certain adjuvants, including glycerol based and non-ionic surfactants (Mason *et. al.* 1998b). However, nematodes can be separated by centrifugal force from the liquid during application, so the shape of the rotary atomiser is important and needs to be relatively flat, rather than cup- or saucer-shaped. Formulation will undoubtedly also have a major effect on the efficacy of application to ensure survival of the nematodes after deposition on foliage.

Entomophagous insects

There has been increasing interest in mechanical systems for improved field delivery of mass-reared parasitoids and predators. Early attempts included aerial broadcast application of lacewing (*Chrysoperla carnea*) larvae (Kinzer, 1977) and *Trichogramma pretiosum* pupae (Bouse and Morrison, 1985), the latter incorporating a refrigeration system to prevent wasp emergence before delivery. However, extended cold temperatures were said to reduce their vigour (Stinner *et al*., 1974). Giles and Wunderlich (1998) describe apparatus to apply *Chrysoperla* eggs in a liquid suspension. Very large droplets (about 2 mm diameter) were applied using a pulse-width modulated valve to determine the duration and frequency of liquid pulses. The adhesive carrier liquid increased adhesion of eggs on leaves, compared to water, but decreased egg hatch. Hand applied eggs had a higher hatch, but the techniques was very labour intensive and less suitable for adoption by commercial farmers (Wunderlich and Giles, 1999).

A Biosprayer was developed to apply natural enemies directly to foliage (Anon., 1995). Using this equipment, Knutson (1996) reported that application of 240 000–480 000 *Trichogramma* pupae per hectare twice a week did not consistently increase parasitism of bollworm eggs in field plots. Wasp emergence after application through the sprayer was decreased by 22 per cent. Subsequent tests examined the immersion of *Chrysoperla rufilabris* eggs and *Trichogramma pretiosum* pupae for up to 3 h in water and a commercial carrier known as 'BioCarrier', but showed that emergence of the latter was significantly lower (Morrison *et al*., 1998). Similar attempts have been made in Uzbekistan to improve release of *Trichogramma* as aerial dispersal loses many pupae on the soil, but suspended insects in a liquid is clearly undesirable and dry dispersal systems are likely to be more appropriate.

Other application techniques

Apart from using some form of pesticide application equipment it may be more appropriate to develop quite different methods of dispersal of biological agent. In the Pacific area, the baculovirus of the rhinoceros beetle *Oryctes rhinoceros* was collected and beetles were infected in the laboratory by placing about 10 ml of an inoculum in 10% sucrose mixture on their mouthparts (Hunter-Fujita *et al*., 1998); frond damage of coconuts declined over the 24–30 months following their release.

Pheromone traps can be employed to trap and allow automatic release after being coated with a biopesticide. In Egypt, a pheromone mixed with an insecticide as a paste formulation was applied to cotton leaves to check pink bollworm moth populations, but the system was not taken up on a large scale. Isaacs *et al*. (1999) have used a pressurised container with an electrically operated solenoid powered by a 9 V battery to dispense a pheromone. The sprayer has a fuel injector as a spray nozzle and was designed to provide a reliable season-long dispensing device.

18

Equipment for laboratory and field trials

Introduction

The agrochemical industry is screening thousands of new compounds each year to detect new leads in biological activity which can be subsequently be optimised and developed into commercial products. *In vitro* testing enables small quantities of chemicals to be used (Ridley *et al.*, 1998). Predictive models have also been used in an attempt to narrow the number of chemicals that need to be assessed. Nevertheless, detailed tests are required to determine which of the relatively few chemicals selected from these screens shows sufficient activity to justify the enormous investment needed to evaluate the pesticide fully and provide registration authorities with the data required before it can be marketed. Giles (1989) discusses the design of screens to decrease the time, cost and level of risk in the discovery process. Costs of developing new pesticides have escalated, because in addition to activity against pest species, environmental impact studies are increasingly demanded, especially in relation to insecticides and their effect on beneficial and other non-target species (Hassan, 1977, 1985; Carter, 1992; Cooke, 1993; Dohman, 1994; Jensen *et al.*, 1999). In this chapter equipment used to evaluate pesticides is discussed.

Laboratory evaluation

Experimental techniques

Topical tests

A known volume of pesticide formulated as a liquid can be placed precisely on an insect, leaf or other surface, using a precision glass syringe fitted with a very fine hypodermic needle. The syringe is normally attached to a micrometer to control the movement of the plunger, so that repeated regular dosing is possible. Arnold (1967) designed a manually operated micrometer with a cylinder around which are five rings, each having a different number of equally spaced depressions. A spring-loaded ball fits against the appropriate ring that is selected to adjust the amount of pesticide applied from 0.25 up to 5 μl. An electrically operated dispenser was introduced to provide doses in the range 0.1–1.0 μl (Arnold, 1965), and another version is electronically controlled to deliver from 0.1 to 100 μl droplets (Fig. 18.1). In tests to evaluate resistance to insecticides, a small hand-held repeating dispenser with a 50 μl microsyringe was used by Forrester *et al.* (1993) to treat *Helicoverpa armigera*. When treating small insects, an airflow has been used to detach a spray droplet from a needle and deposit on the insect (Hewlett, 1954, 1962; Needham and Devonshire, 1973). MacCuaig and Watts (1968) used small micropipettes to dispense uniform volumes.

The volume applied with a microsyringe is quite large compared with small droplets in a fog or mist, so Johnstone *et al.* (1989a,b) used a vibrating orifice droplet generator (Berglund and Liu, 1973) to deliver an aerosol into a low-speed wind-tunnel so that droplets (10–2 μm diameter), collected on silk threads (about 2 μm diameter) could be transferred to tsetse flies.

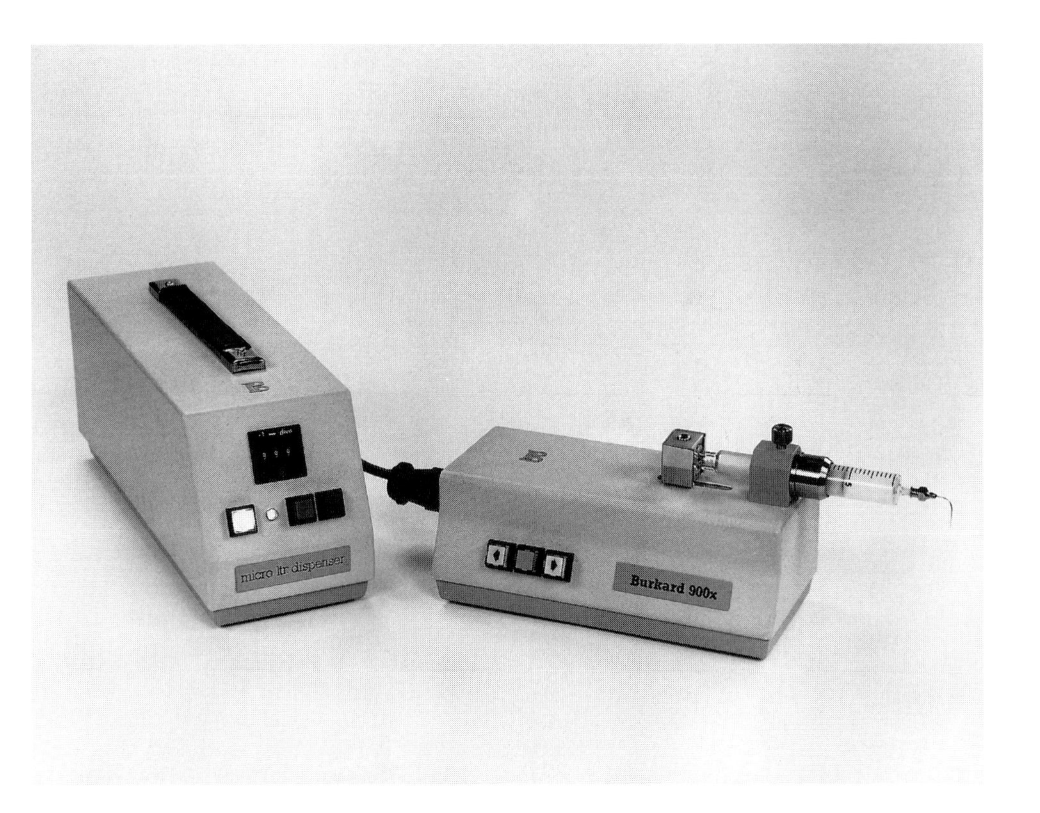

Fig. 18.1 Burkard microapplicator.

Techniques of spraying surfaces

Several methods of treating surfaces with a consistent measured dosage of a pesticide have been devised. Potter (1941,1952) developed a spray tower (Fig. 18.2) to minimise air turbulence and reduce the amount deposited on the sides of the tower. A twin-fluid nozzle was mounted centrally at the top of an open-ended metal tube to spray down onto a horizontal plate. This nozzle produces droplets in the aerosol and mist size range (< 100 μm diameter), quite different from the wider spectrum of droplets produced by a conventional hydraulic nozzle (Matthews, 1994), although the volume of spray can be comparable with field applications of around 200 litres/ha. Recently a similar twin-fluid nozzle has been linked to a computer control operated through a software package, such as Microsoft Paintbrush (Fig. 18.3) (A.C. Arnold, personal communication). Nozzle output can be adjusted by means of a volume control needle and/or the air pressure. The apparatus is supplied as a compact unit, which can be used inside a fume cupboard. Morgan and Pinniger (1987) used a car windscreen-wiper assembly to move the spray nozzle and obtain an even deposit on surfaces up to 27 cm in diameter.

A twin-fluid nozzle was adapted by Coggins and Baker (1983) to produce a very narrow size range of droplets to examine the effect of different droplet sizes (Munthali and Scopes, 1982). A leaf was moved under the nozzle for different periods of time to obtain a range of droplet densities. A fluorescent

Fig. 18.2 Potter tower.

Fig. 18.3 Small laboratory sprayer (Burkard Manufacturing Co.).

tracer was added in the spray to visualize the droplets. A Berglund and Liu droplet generator modified by removing the air column and using orifice plates with larger apertures has also been used to produce uniform droplets in the 50–500 μm diameter range (Reichard *et al.*, 1987). Similarly a piezoelectric disc was used to pulse liquid through a very fine orifice (Young, 1986; Reichard, 1990; Womac *et al.*, 1992) Monosized droplets as small as 60 μm in diameter can be obtained, but well-filtered solutions are needed when the smallest apertures are used. Equipment using a computer is now available to produce either a single or series of droplets on demand (Thacker *et al.*, 1995).

A rotating disc can also be used to apply spray droplets with a narrow size range, provided the flow rate and disc speed are controlled. Droplets larger than 40 μm diameter can be produced, but smaller droplets are more difficult to obtain even at higher disc speeds. Shrouding part of the disc provides a fan-shaped curtain of spray.

Spray chambers

Field trials inevitably take a long time from planning to final analysis, and may not provide significant results due to variability in the populations of pest species. To limit the number of treatments that have to be examined in the field, factors such as dosage rate, formulation and application parameters can be examined partly under glasshouse conditions using plants grown in pots or trays. This section describes a number of spray chambers which have been used to evaluate pesticides. Health and safety regulations require the whole area in which the spray is applied to be enclosed in a chamber so that the operator is not exposed to pesticide and the treated area can subsequently be washed down to remove deposits from the inner surfaces of the chamber.

In the simplest spray chambers, plants placed on a turntable are rotated in front of a fixed nozzle. Plants are often treated until the foliage is completely wetted ('run-off' spray) using large volumes of dilute pesticide. This results in considerable wastage of pesticide and, as noted in earlier chapters, this does simulate field application very accurately. Thus spray chambers are usually made so that volumes similar to those applied in the field can be applied.

One type of spray chamber, the 'Mardrive' track sprayer has a linear transporter, involving a sealed tube in which a small polymer-bonded slug referred to as a 'mole' is pushed to and fro by compressed air. A shuttle mounted on rollers is moved along the tube by a set of permanent magnets, which keep in step with the 'mole' by magnetic coupling. One or more conventional hydraulic nozzles can be mounted on the shuttle to treat plants placed under the centre of the track. The unit can be modified to use a spinning disc atomiser. The sides of the chamber can be washed down by a series of nozzles situated along the length of the chamber after the application of a pesticide and the effluent collected for subsequent disposal. The speed of the mole is normally about 1 m/s, so application is generally equivalent to manually applied sprays.

A larger spray chamber has been used to simulate the higher speeds at which a tractor sprayer operates. Hislop (1989) described a wind tunnel (12.6 m long, 3.6 m wide and 2.7 m high) designed with a ceiling mounted single-axis beam carrying a chain-driven module on which nozzles can be mounted. These nozzles can be transported at speeds between 0.5 and 6.0 m/s in either direction along the track, as the equipment is fitted with a 3 kW electric motor incorporating a 4 kW variable speed controller. The actual speed over a 5 m section in the area used for spray application is detected by sensors which measure to an accuracy of 0.01 m/s. Wind speeds through the chamber can be varied between 1 and 10 m/s so that sprays can be applied either with or against the direction of the wind. This unit and similar wind tunnels have been used for a number of spray application studies, including assessment of airborne drift with different nozzles and airspeeds (Fig. 18.4) (Miller *et al.*, 1993). The nozzle used in the spray chamber is often dictated by the need to apply the same volume rate as that used in the field. In contrast to measurements of spray distribution under static conditions, the pattern across the spray swath will also be influenced by the speed of travel, with the smallest droplets entrained in the air vortices created by the spray (Miller *et al.*, 1993).

Plants grown in a glasshouse have leaves different from those grown outside in fields due to differences in temperature and wind induced leaf movement

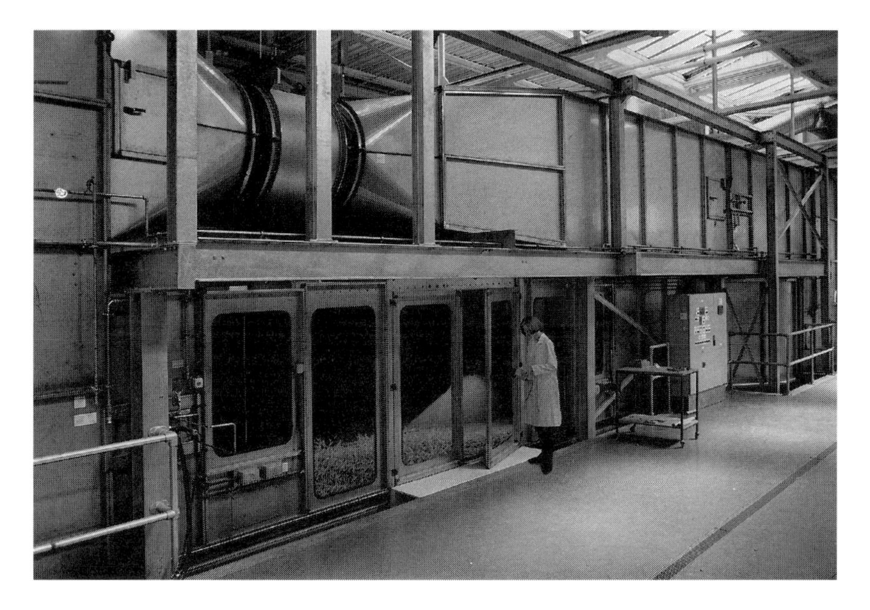

Fig. 18.4 Wind tunnel.

which affect the characteristics of the leaf surfaces. Therefore, to mimic field conditions, plants need to be grown in trays out-of-doors so that the leaf surfaces resemble field plants in terms of their wettability, retention of spray droplets and uptake of a pesticide (Hislop, 1989)

To simulate the effects of an aerial spray on several generations of a population of whiteflies on cotton plants, a more specialised spray cabinet was devised (Rowland *et al.*, 1990). In this cabinet, a spinning disc sprayer was fitted above a cage in which cotton plants were maintained. Separate cabinets were used to compare the response of susceptible and resistant populations to insecticides (Sawicki *et al.*, 1989). Using a spinning disc allowed application of sprays with different droplet sizes. Examination of whiteflies on leaves was achieved with an endoscope so that the cage was not opened to disturb the plants or insects. The equipment also allowed the effect of sprays on mixed populations of whiteflies and natural enemies to be studied.

Special tests have been devised to assess treatments with very small droplets less than 50 μm in diameter used to control flying insects. Thus to evaluate the efficacy of a pressure pack containing an insecticide for fly control, it is discharged while walking down the long axis of a test-room of about $50\,m^3$ capacity (Goodwin-Bailey *et al.*, 1957). The area is uniformly illuminated by fluorescent lights to provide a light intensity of 108 mcd when measured 1 m above the floor. A temperature of $27 \pm 1°C$ and relative humidity of $50 \pm 5\%$ should be maintained and the room ventilated between tests by a fan displacing at least $10\,m^3$ of air per minute. The floor should be covered with a new layer of absorbent paper for each test and deposition on the walls and ceiling avoided, positioning the nozzle at least 1 m from any surface. The pressure pack is weighed before and after spraying to determine the amount discharged. A test may involve the release of 500 flies at floor level. To avoid entering the area treated, some users have a 25 m^3 'room' with transparent panels inside a larger room so that the number of flies 'knocked down' on a grid demarcated on the floor can be assessed at 2 min intervals without entering the room. Later, after adequate ventilation, the flies are collected and placed in clean containers with food so that total mortality can be recorded after 24 h.

Individual clear polythene chambers (2250 × 2000 × 2250 mm^3) with a plastic zip fitted centrally in the front to allow access, have been used to test space sprays (Learmount, 1994). A mesh is fitted in the back wall to allow ventilation, but is covered by a black plastic sheet during the entry of spray from an airbrush nozzle operated at about 2.5 bar pressure and directed through the zipper.

Field trials

Typically, when a particular pesticide, formulation and probable dosage have been selected from the small-scale studies, they are then evaluated in a series of field trials, each with a small number of replicated treatments in a randomised block design. An unsprayed 'check' plot should be included to show the extent to which yields can be improved. However, the usefulness of untreated plots in insecticide trials has been questioned, because they tend to yield more than larger untreated areas not adjacent to treated plots (Reed, 1972). A 'check' with another standard pesticide treatment is often required.

Every effort should be given to collecting adequate pest, plant growth and other crop data throughout the trial to assess the contribution of the application of the pesticide to any improvement in yields. This means that plots need to be sufficiently large so that the sampling area within the peripheral guard section is large enough to allow sampling over an extended period without damage to the plants. Access paths between plots are important to facilitate routine sampling. In insecticide trials, sampling of insect pest species should be supplemented with assessments of natural enemy populations, although detailed sampling of these is often more difficult. Suction sampling equipment has been designed to facilitate assessing insect populations in the field (Arnold, 1994). Such studies may be better on large-scale trials. The operators of the spray equipment should be allocated at random to the sprayers and treatments on each spray date, so that the results are not biased in any way due to the efficiency of a particular person. Another aspect, often neglected, is a check on the distribution of pesticide in relation to the pest and plant growth. Water sensitive cards are a relatively easy way of checking spray distribution within a crop canopy (Cooke and Hislop, 1993).

Plot size is very important; if it is too small, droplets can drift across from adjacent plots, and similarly insect pests and airborne spores of plant pathogens can move with the wind. Sometimes a shield is used along the downwind edge, but this may have little effect unless it is fairly porous and acts as an efficient filter. Otherwise any wind will be deflected over the top of the shield and carry with it the smallest airborne droplets. The method of applying the pesticide will influence the plot size. Relatively small plots are acceptable if there is a placement technique such as granule application or seed treatment. With nozzles moved within the inter-row, plots of 10 m × 10 m were satisfactory for insecticide trials on cotton. If the nozzle is held above the crop canopy some drift is inevitable, so a larger plot is needed with a wider guard area; thus with a spinning disc sprayer, plots at least 30 m × 30 m are required, and aerial spray treatments require much larger plots. Square plots are preferred to long rectangular plots because the effect of spray drift across the latter will be more pronounced if there is a crosswind. Where techniques such as 'lure and kill' and the use of baculoviruses are being examined, area-wide treatments are essential; for example, when treating alternate hosts to reduce the population of the cotton bollworm and tobacco budworm (Bell and Hayes, 1994).

After data has been obtained from field trials, it is important that large-scale farm trials are also conducted. Such trials need to be kept as simple as possible and may only consist of two treatments; the farmer's normal practice and the treatment that research suggests will be an improvement. Plots should be as large as possible (Tunstall and Matthews, 1966; Matthews, 1973). Before embarking on any field trial, its aims and objectives need to be clearly defined and advice on trial design should be sought. Gomez and Gomez (1976) and Pearce (1976) are among many statistical books that can be consulted, but it always advisable to discuss trials with a biometrician at the outset to avoid difficulties with analysis at the end of the trial. In the context of IPM, the efficacy of a pesticide has to be considered in relation to many other factors, such as host plant resistance, the impact on natural enemies and interactions with different cropping practices (Reed *et al.*, 1985).

Spraying equipment

Specialised sprayers have been developed for treating small plots, but some of the manually carried sprayers can also be adapted. Thus a compression sprayer can be used with one or more hydraulic nozzles, preferably mounted on an offset boom so that the operator can walk alongside the plot (Matthews, 1984, 1994). A constant flow valve or pressure regulator is needed to ensure a more uniform application. The pressure at the nozzle needs to be selected in relation to the type of pesticide being applied and the droplet spectrum required. Nozzles should be fitted with a diaphragm check valve to prevent any spray dripping from the lance or boom at the edge of a plot. Some experimenters have used motorised knapsack sprayers (Rutherford, 1985), while Robinson (1985) used a small electrically-driven air compressor to maintain a constant pressure in the pesticide container. J.H. Crabtree (personal communication) has also developed a small plot sprayer using a rechargeable battery to provide power for an air compressor. A constant flow valve is fitted to the nozzle (Fig. 18.5). Where water supplies are scarce, reduced volume spraying has been evaluated initially using hand-carried spinning disc sprayers (Matthews, 1973; Bateman *et al.*, 1994; Fisk *et al.*, 1993), and vehicle mounted equipment (Symmons *et al.*, 1989; Hewitt and Meganasa, 1993).

Fig. 18.5 'Lunch-box' sprayer (photo: J.H. Crabtree).

Mechanised plot sprayers

Small mechanised plot sprayers are used where possible, as walking speed can vary between operators. These are now designed to minimise operator contact with the pesticide to meet more stringent health and safety requirements. Slater *et al.* (1985) described a single-wheeled motorised sprayer, which could also be used as a granule applicator on small plots. An outdoor version of the 'MarDrive' system mounted on a tractor for moving the nozzle across a small plot within a shielded enclosure was described by French (1980), while Skurray (1985) designed a self-propelled gantry which eliminated the need for a passage for the tractor alongside the plots. Crabtree (1993) adapted a hedgecutter arm to mount an offset boom fitted with a separate array of nozzles, protected within a shield, for each treatment (Fig. 18.6). The spray for each treatment was pre-mixed and kept in separate containers, so that several treatments could be applied rapidly in the field. While this equipment has been used primarily for treating cereal crops such as wheat, it could be adapted for other low row crops. In trials with this equipment, plot size, especially its width, is often determined by the mechanised harvesting equipment being used.

Fig. 18.6 Plot sprayer made by adaptation of tractor-mounted hedgecutter. Note individual containers of pesticide for each plot (photo: J.H. Crabtree).

Granule application

Seed treatment and granule application are two other options for more precise placement of insecticides. Precision granule-metering belts were fitted to a precision seed-spacing drill to allow accurate delivery of granules and their incorporation into the soil while sowing (Thompson *et al.*, 1981; Thompson and Wheatley, 1985). Information on techniques of commercial quantities of seed has been published (Jeffs, 1986; Clayton, 1993), but for small samples used in trials, simple mixing of seed with an appropriate formulation of insecticide has usually been carried out in the laboratory. Where solvents have been used, careful volatilisation of the solvent is needed under controlled safe conditions.

General

This chapter has considered only some of the equipment suitable for laboratory, glasshouse and field trials, the choice being very dependent on the ultimate aim of the studies (Krahmer and Russell, 1994). With the escalation of costs, there is a risk of simplified standard tests being used. However, investment in more detailed evaluation may be necessary to understand the more complex interactions that occur, for example, between pests and their natural enemies and host-plant resistance, especially with the introduction of transgenic crops initially affecting herbicide and insecticide use patterns.

19

Selection of application equipment for chemical and biological pesticides

A range of equipment that can be used to apply pesticides is now available, as indicated in the previous chapters. Figure 19.1 shows some of the factors that need to be considered when choosing which equipment should be used on a particular farm or for a specific pest situation. Clearly, sometimes one factor will govern the selection. Too often in the past, the least expensive equipment in terms of capital cost may be selected, especially when the purchase is made by a government or international organisation, without sufficient reference to its long-term operational costs. A compromise between capital cost and technical specifications may be required, but more attention to the latter is needed to ensure greater safety to the user and improve application efficiency to minimise environmental damage due to pesticides. More precise application will be required, especially where the dosage of pesticides has to be reduced as in some Scandinavian countries.

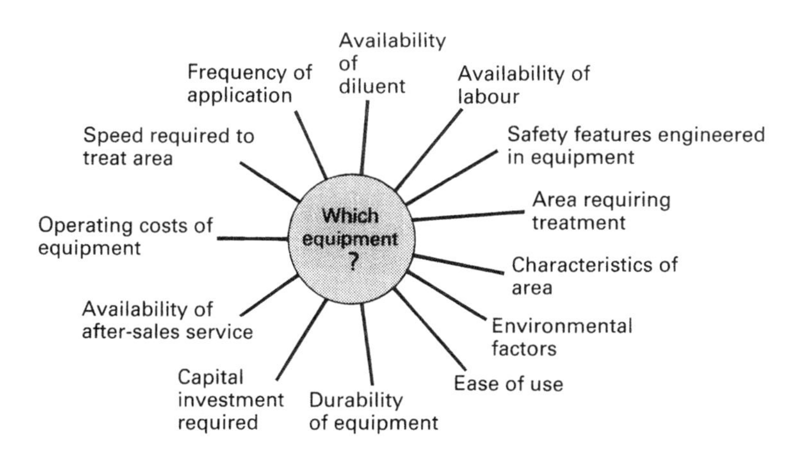

Fig. 19.1 Factors governing the selection of equipment.

Environmental factors

Concern that pesticides can pollute the environment has increased. The Spray Drift Task Force in the USA is one example of research being directed at providing the legislative authorities with sufficient generic information to facilitate registration of pesticides. In the UK a no-spray zone is obligatory for certain pesticides, but this can be amended in some circumstances by carrying out a Local Environmental Risk Assessment for Pesticides (LERAP). Generally, in demonstrating ways of minimising the amount of pesticide spray that can potentially move downwind, the trend has been to advocate coarser sprays as discussed in Chapter 4. However, this increases the risk of less efficient use of pesticides within the area that requires treatment. In the future, there may be more selective treatments within individual fields, thus reducing the total amount of pesticide required. Further research will no doubt build on the initial work of precision agriculture, such as patch spraying using GPS/GIS technology, and develop more sophisticated systems of controlling sprayer output in relation to the spatial distribution of weeds, insect pests and diseases within fields. The time needed to map the distribution of weeds and crop density within fields is currently seen as a problem to the wider adoption of patch spraying, so research is needed to develop on-line systems for crop and weed recognition (Christensen *et al.*, 1997).

The farmer seeks versatile, reliable, cost-effective equipment, so for the foreseeable future more specialised equipment such as the use of CDA equipment will continue to be confined to niche markets. Changes in nozzle selection are very easy without any additional change in the equipment; thus many farmers have adopted air induction nozzles as a drift reduction policy. Similarly, those farmers who have invested in an air-sleeve sprayer can reduce drift during periods when crop growth provides an effective filter to capture the smaller airborne droplets. Adoption of more sophisticated systems will depend not only on their costs, but also whether legislation provides stricter controls on environmental pollution. However, specialised equipment or particular modifications to equipment may be increasingly required if more biopesticides are to be used.

Operational safety factors

Some countries now have some legislation that specifically influences the choice of equipment. In the United Kingdom, the *Food and Environmental Protection Act* (FEPA 1985) provides the framework under which the government can exercise the *Control of Pesticide Regulations* (COPR 1986) and the *Maximum Residue Levels Regulations* (1988). From time to time the regulations are amended. Thus the regulations governing the width of the no-spray zone, i.e. LERAP, were introduced under the umbrella of FEPA. Use of pesticides is also affected by other legislation (Gilbert and Macrory, 1989). In particular the *Health and Safety at Work Act* (1974) with the *Control of*

Substances Hazardous to Health Regulations (COSHH) are aimed at improving the safety of the workers. The *Green Code*, a code of practice published jointly by the Ministry of Agriculture, Fisheries and Food, the Health and Safety Commission and the Department for the Environment, Transport and the Regions, has been updated (MAFF, 1998) to provide guidance to farmers on how to use pesticides. As mentioned in previous chapters, there is also a requirement for obligatory training (Harris, 1988). This training and subsequent tests emphasise the importance of safety and correct calibration of equipment.

A key feature of the Health and Safety requirements is the recognition that the transfer of a pesticide from its original container to the spray tank is potentially the most hazardous task. As indicated in Chapter 7, improvements in sprayer design and in packaging have been made to facilitate safer transfer. Adoption of direct injection systems has not yet been very widespread, despite recognition of the advantages of only mixing pesticide with water as it is applied. However, several closed transfer systems are now available as well as the use of returnable containers that also eliminate the problem of disposal of old containers (Curle *et al.*, 1998).

LinkPak with 5 and 10 litre containers and Ecomatic 25 litre kegs have been introduced (Mills-Thomas *et al.*, 1998), while another development has been the use of water-soluble packs for some pesticide formulations to reduce potential operator contamination (Gilbert, 1998). Use of water dispersible granules (Bell, 1998) also reduces some of the problems associated with disposal of packaging of liquid formulations. In the USA there are similar systems for minimising the problem of disposal of packaging with pesticide residues (Hutton, 1998). Closed systems for delivery of granule formulations and products to be used in seed treaters have also been introduced.

Tractor sprayers

Farmers will choose an item of equipment based on the area of their crops that requires treatment and the time available for each treatment, as well as taking into consideration environmental and safety features. Thus the actual choice will be determined by the work rate for a critical period of the year. When to spray is dictated by when the pest is present, and action is needed without contravening the minimum period before harvest, but weather conditions may not always be suitable. The number of days that are ideal for spraying a crop is often restricted due to the wind, rain or wetness of the soil. Spackmann and Barrie (1982) used threshold values for the United Kingdom (Table 19.1) to calculate the number of days suitable during each month, based on 10 years' weather data from 15 weather stations. Mini-meteorological stations are available for on-farm use to monitor conditions and assist in deciding when to spray, especially in relation to outbreaks of disease such as potato blight. Wider booms can be used if fields are relatively flat, but too wide a boom increases problems of turning at headlands. To allow for rain and other

Table 19.1 Threshold values (Spackmann and Barrie, 1982)

Parameter	Threshold at which spraying is constrained
Hourly mean windspeed*	$\leq$4.6 m/s (hydraulic pressure sprayer) $\leq$6.2 m/s (CDA sprayer with large droplets (250 μm))
Soil moisture	Days when soil moisture deficit <5 mm (conventional tractors), with no constraint on low ground pressure vehicles
Daylight	Not earlier than 06.00 or later than 21.00
Temperature	>1.0°C during spraying, with air temperature >7.0°C some time during the day
Precipitation	None for at least 1 h before or at time of spraying

[a] Measured at a height of 10 metres.

interruptions, it is useful to have sufficient equipment to treat a crop or farm within 3 days, especially if treatments may have to be repeated over several weeks as has been the case on cotton.

Most tractor-mounted booms are in the 18–24 m range of sizes in the United Kingdom. Where tramlines are used, boom width needs to match the width of the seed drill. In practice, spraying can be completed more rapidly by reducing the volume application rate and by shortening the time needed to refill the spray tank. Increasing spray tank capacity can cause soil compaction, although many large farms operate equipment with tanks that contain 2000 litres. Many farmers now apply less than 200 litres/ha on arable crops so that spraying is completed more quickly.

In view of drift reduction requirements, equipment that allows rapid change of nozzles is increasingly important, so that the tractor driver can change spray quality if necessary over part of a field. In some countries spraying is done at speeds greater than 7 km/h, but care is needed to avoid increasing spray drift and ensure that dosage rates are not affected by changes in speed. Smith (1984) developed a model to select ground spraying systems for arable farms (Fig. 19.2).

Equipment needs to have facilities for rinsing containers and sufficient water at the field to clean the pesticide tank and enable the washings to be applied within the treated field as the last swath and avoid overdosing the crop. This eliminates problems of disposal of washings in the farmyard.

Most manufacturers provide a range of machinery, the capital costs and operating costs of which can be estimated (Table 19.2), but these costs are influenced by the actual area that requires treatment each year (Fig. 19.3).

Knapsack and manually–carried sprayers

Many areas requiring pesticide application are too small to justify tractor-mounted or aerial equipment. Sometimes access is difficult, for example around buildings, or terrain may be too hilly, uneven or too wet. Small equipment may be needed in trials, as discussed in Chapter 18, and for treating

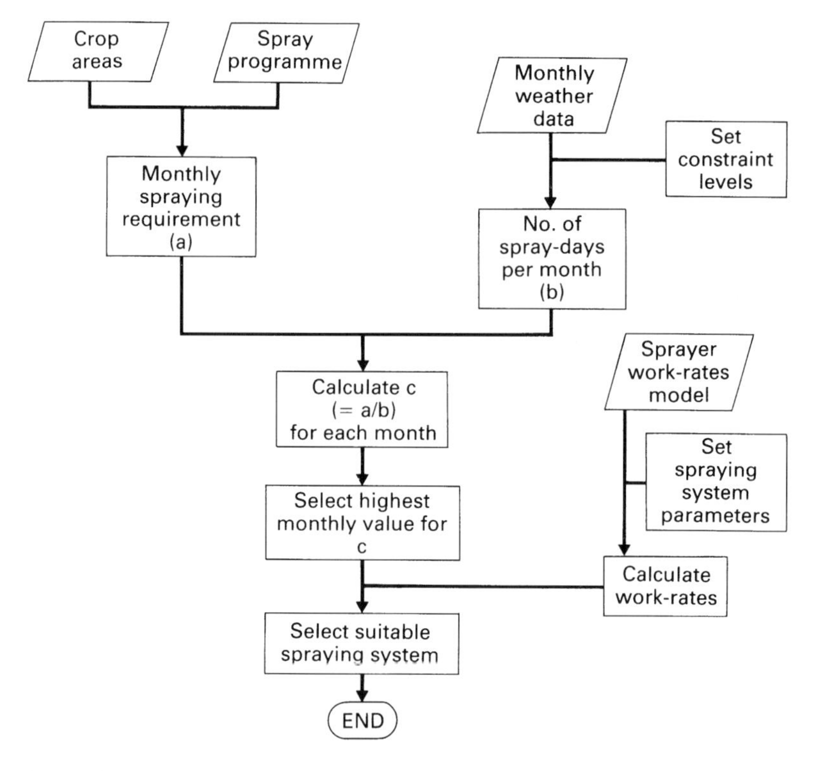

Fig. 19.2 Flow diagram of model suggesting suitable spraying system for a farm (from Smith, 1984).

small areas of infestation to avoid treating whole fields when an early intervention may prevent an infestation spreading more extensively. Under these circumstances manually-carried pesticide application equipment is needed, despite the operator being more exposed to the pesticide.

Many different designs of lever-operated knapsack, compression and other small sprayers are available throughout the world. These vary significantly in price, quality, safety and ease of use. Specifications and buyers' guides have been published by the FAO in an attempt to reduce the number of occasions when equipment has been purchased purely on the basis of the lowest tender price. Cheap equipment often fails due to poor quality, and increases the risk of exposing the operator to pesticides.

Portable equipment must be sufficiently durable so that frequent repairs are not needed and frustrating delays in a spray programme are avoided. It must be comfortable to wear and not too heavy to carry. In some areas, hand-carried battery-operated sprayers have replaced knapsack equipment because of their light weight and the significant reduction in manual effort needed to use them. In particular, the availability of water is often a key factor in favour of CDA/VLV equipment, which allows rapid treatment in response to a pest infestation (Table 19.3).

Table 19.2 Operating costs for four power-operated boom sprayers

Sprayer	Small tractor mounted	Large tractor mounted	Trailed	Self-propelled
Initial capital cost (£)	1 060	3 680	15 000	36 000
Area sprayed annually (ha)	150	600	1 000	2 000
Tank capacity (litres)	320	1 000	2 000	1 600
Boom width (m)	6	12	18	24
Life (years)	8	8	8	8
Hectares/h spraying	3.6	7.2	10.8	14.4
Overall ha/h (50% efficiency)	1.8	3.6	5.4	7.2
Use (h/annum)	83.3	166.7	185.2	277.8
Annual cost of ownership (£)	132.5	460	1 875	4 500
Repairs and maintenance[a] (£)	79.5	276	1 125	2 700
15% interest on half capital (£)	79.5	276	1 125	2 700
Total cost of ownership (£)	291.5	1 012	4 125	9 900
Ownership cost per hour (£)	3.50	6.07	22.27	35.64
Ownership cost per hectare (£)	1.94	1.69	2.06	4.94
Labour costs per hectare (£)	1.39	0.69	0.46	0.35
Tractor cost per hectare (£)	1.67	0.83	0.56	0.42
Total operating costs per hectare[b] (£)	5.0	3.21	3.08	5.71

[a] Based on 7.5% of capital cost, depending on sprayer.
[b] Based on the overall cost of up to £3–6/ha shown above but does not include various overheads, positioning of equipment, secretarial and telephone expenses and other items included in prices charged by farm contractors

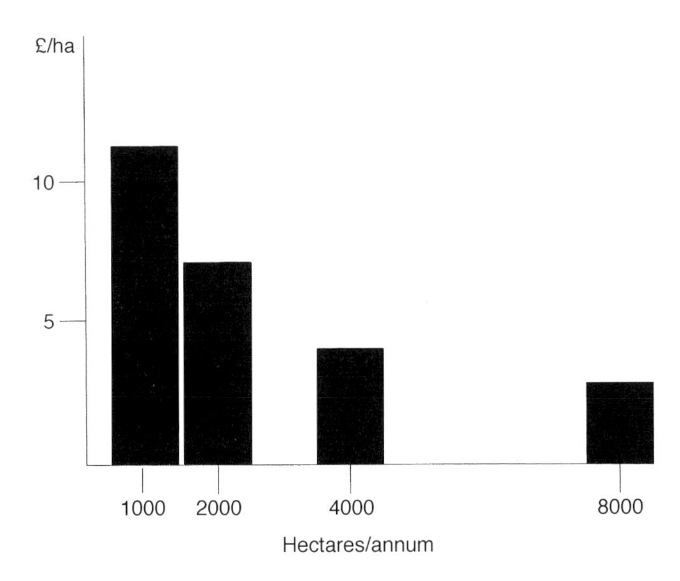

Fig. 19.3 Variation in operating costs with yearly work load, based on 12 m boom (initial cost of tractor-mounted sprayer £4000).

Table 19.3 Comparison of spraying time for a knapsack and battery-operated CDA sprayer[a]

	Knapack sprayer			CDA sprayer		
Application rate (litres/ha)	400	200	100	10	2.5	1
Spray tank (litres)	15	15	15	0.5	0.5	0.5
Mixing and refilling time (h/ha)	2.2	1.1	0.55	0.67	0.17	0.07
Ferry time (h/ha)	14.8	7.4	3.7	0.37	0.08	0.04
Swath width (m)	1	1	1	2.8	1.39	0.56
Spraying time (h/ha)	2.8	2.8	2.8	2.8	1.39	0.56
Turning time (s/row)	2	2	2	2	3	7
Total time (h/ha)	19.9	11.4	7.2	3.95	1.84	0.94
Spraying time as percentage of total	14	25	39	71	76	60

[a] Assuming walking speed of 1 m/s, carrying 15 litres of water from supply 1 km from fields; fields 100 m long, average size 0.5 ha and separated by 150 m.

The cost of operating portable equipment can be calculated using the same basic formula as used for aircraft and tractor equipment, although labour input is proportionally greater, especially when water has to be transported over a long distance (Table 19.4). For the small-scale farmer, the capital cost may be the key factor, although buying the cheapest sprayer is inevitably a false economy when it has to be replaced after very little use. The addition of a pressure or flow control valve is important, as it allows more efficient application of the pesticide.

The need to buy batteries for some equipment is considered a disadvantage, although developments in equipment design have significantly reduced the number of batteries required. The time saved using such equipment also allows the farmer to attend to other work more easily. Air-assisted sprayers, such as knapsack mistblowers are needed when spray has to be projected upwards into trees. Selection from a wide range of mistblowers available should be based on how far the spray has to be projected and the range of droplet sizes that are produced. Other factors that need to be considered are how easily the motor starts, ease of operation and maintenance, and comfort to the operator. Small two-stroke engines need regular maintenance, so suitable facilities need to be available.

Aerial application

Aerial application is generally confined to large areas, especially forests and irrigation schemes, but may be required even when comparatively small areas need treatment because

(1) passage of ground equipment will damage crops, such as cereals unless 'tramlines' are left for access

Table 19.4 Operational costs with knapsack and hand-carried CDA sprayer (actual costs of equipment and labour will depend on local conditions: cost of chemical, which is also affected by choice of formulation is not included)

Sprayer	Manually operated knapsack	Motorised knapsack mistblower	Hand-carried sprayer	
Initial capital cost (£)	60	350	45	45
Area sprayed annually (ha)	20	20	20	20
Tank capacity (litres)	15	10	1	1
Swath width (m)	1	3	1	3
Life (years)	3	5	3	3
Hectares/h spraying[a]	0.36	1.08	0.36	1.08
Overall ha/h (% efficiency)[a]	0.18(50)	0.65(60)	0.31(85)	0.97(90)
Use (h/annum)	111	30.8	64.5	20.6
Annual cost of ownership (£)	20	70	15	15
Repairs and maintenance[b] (£)	6	35	4.5	4.5
15% interest on half capital (£)	4.5	26.3	3.4	3.4
Total cost of ownership (£)	30.5	131.3	22.9	22.9
Ownership cost per hour (£)	0.27	4.26	0.36	1.11
Ownership cost per hectare (£)	0.76	3.95	0.99	1.03
Labour costs per hectare[c] (£)	1.38	0.39	0.80	0.25
Operating cost including batteries[d] (£/ha)	–	0.68	2.2	0.74
Labour costs to collect water[e] (£/ha)	1.38	0.92	0.13	0.04
Total operating costs per hectare (£)	3.52	5.94	4.12	2.06

[a] Assuming walking speed is 1 m/s, actual efficiency will depend on how far water supply is from treated area, application rate and other factors (see Table 19.3).
[b] 10% of capital cost.
[c] Assumes labour in tropical country at £2 per 8 h day.
[d] Assumes batteries cost 50p each and a set of eight will operate for 5 h with a fast disc speed. Fuel for mistblower at 44p/litre/h. Battery consumption is less on some sprayers with a single disc and smaller motors. The 'Electrodyn' sprayer uses only four batteries over 50+ hours, so the costs (£/ha) of batteries on a double row swath is 0.6 instead of 2.2.
[e] Water required for washing, even when special formulations are applied at ULV.

(2) there is a risk of soil compaction, especially on some soils and if repeat sprays are required
(3) crops may be inaccessible to ground equipment at critical periods of pest infestation. This may be due to wet soil, poor drainage or the arrangement of irrigation equipment
(4) pests, including red spider mites, or diseases may be spread by movement of equipment or personnel through the crop
(5) capital investment in equipment is not justified if pest or disease infestations are sporadic
(6) access to certain crops may be difficult or impossible without specialised equipment, for example high-clearance tractors which are needed for late season treatment of tall crops such as maize.

When deciding on aerial application, the higher operating costs and the availability of aircraft must be considered. The higher costs may be offset by less mechanical damage to the crop, but sufficient aircraft may not be available to meet a sudden demand when infestations of a sporadic pest reach the economic threshold over extensive areas simultaneously or in areas remote from the aircraft operator's base.

Aircraft continue to be most important in treating certain crops, such as bananas, locust infestations and pest outbreaks in forests. Public concern about spray drift is imposing more restrictions on treating field crops, but in some countries, e.g. USA, there is still extensive use of aircraft.

General

The availability of spare parts and ease of maintenance are important criteria in selecting equipment. Certain basic spare parts such as nozzle tips, washers, 'O' ring seals and other replaceable components should be purchased wherever possible with the original equipment, to avoid any delay in applying a spray at a crucial period during the crop growing season. The need to stock basic spare parts cannot be overemphasised, particularly when equipment is used in remote areas.

Routine maintenance is strongly advised, so preference should be given to equipment on which components most subject to wear are easily accessible. Some chemicals are particularly corrosive or affect the reliability of a component; for example, certain plastics such as PVC are dissolved by solvents such as isophorone. Elastomers in 'O' ring seals are likely to swell with some solvents, and parts of a sprayer may be abraded by some of the inert fillers used in granular or wettable powder formulations. In general, these problems have decreased as improvements in formulation have been achieved. Manufacturers should be consulted regarding the compatibility of their products with materials used in the construction of application equipment.

The purchase of any application equipment should be preceded by an inspection of the range of the different makes and models of the type of equipment needed. Agricultural shows or exhibitions often provide a suitable venue where equipment is displayed side-by-side, but a more realistic impression of the equipment is provided where the equipment is also operated under field conditions, so that the movement of the spray boom or other features can be assessed.

The ultimate criterion in selecting equipment is whether the pest can be controlled economically. Despite many improvements in equipment, pesticide application still remains one of the most inefficient processes (Graham-Bryce, 1975). This adds to the public perception that chemical pesticides should not be used and crops should be grown organically. That high quality produce is readily available in many countries throughout the year because farmers can protect their crops from pests is too often forgotten. However, the industry should give more attention to improving application. This will require more

Fig. 19.4 Guidebooks on spray application published by the British Crop Protection Council.

training and advisory booklets, such as those published by the British Crop Protection Council (Fig. 19.4), as well as continued research, so that the pesticide whether chemical or biological can be applied more accurately at the most appropriate time. This requires selecting the optimal pesticide formulation, choosing the correct nozzle and delivery system, sometimes with air assistance, to minimise losses in the environment and ensure the correct dose is transferred to the target.

References

Abbott, I.M., Bonsall, J.L., Chester G., Hart, B. and Turnbull, G.J. (1987) Worker exposure to a herbicide applied with ground sprayers in the United Kingdom. *American Industrial Hygiene Association Journal* **48**, 167–175.

Abdalla, M. (1984) *A biological study of the spread of pesticides in small droplets*. PhD thesis, University of London.

Abdelbagi, H.A. and Adams, A.J. (1987) Influence of droplet size, air-assistance and electrostatic charge upon the distribution of ultra-low volume sprays on tomatoes. *Crop Protection* **6**, 226–233.

Adams, A.J. and Lindquist, R.K. (1991) A review of the performance of air-assisted sprayers for use in glasshouses. *British Crop Protection Council Monograph* **46**, 227–235.

Adams, A.J., Abdalla, M.R., Wyatt, I.J. and Palmer, A. (1987) The relative influence of the factors which determine the spray droplets density required to control the glasshouse whitefly, *Trialeurodes vaporarorium*. *Aspects of Applied Biology* **14**, 257–266.

Adams, A.J., Lindquist, R.K., Adams, I.H.H. and Hall, F.R. (1991) Efficacy of bifenthrin against pyrethroid-resistant and -susceptible populations of glasshouse whitefly in bioassays and using three spray application methods. *Crop Protection* **10**, 106–110.

Adams, A.J., Lindquist, R.K., Hall, F.R. and Rolph, I.A. (1989) Application, distribution and efficacy of electrostatically charged sprays on chrysanthemums. *Pesticide Formulation and Application Systems: International Aspects*, Vol. 9, American Society for Testing and Materials, Special Technical Publication. 1036, pp. 179–190. American Society for Testing and Materials, Philadelphia.

Afreh-Nuamah, K. and Matthews, G.A. (1987) Comparative spray distribution in a tree crop with three different spray nozzles. *Aspects of Applied Biology* **14**, 77–83.

Ahmad, S.I., Tate, R.W., Bode, L.E. and Retzer, H.J. (1980) Droplet size characteristics of the by-pass nozzle. *ASAE Paper*, 80–1058.

Ahmad, S.I., Bode, L.E. and Butler, B.J. (1981) A variable-rate pesticide spraying system. *Transactions of the ASAE* **24**, 584–589.

Akesson, N.B. and Yates, W.E. (1974) The use of aircraft in agriculture. *FAO Agricultural Development Paper* 94. Food and Agriculture Organisation, Rome.

Allan, J.R. McB. (1980) Development in monitoring and control systems for greater accuracy in spray application. *British Crop Protection Council Monograph* **24**, 201–213.

Allen, G.G., Chopra, C.G., Friedhoff, J.F., Gara, R.I., Maggi, M.W. Neogi, A.N., Roberts, S.C. and Wilkins, R.M. (1973) Pesticides, pollution and polymers. *Chemical Technology* **4**, 171–178.

Allen, J.G., Smith, A.P., Butt, D.J. and Warman, T.M. (1991) Improved performance of mistblower sprayers by electrostatic charging. *British Crop Protection Council Monograph* **46**, 143–150.

Allsopp, R. (1984) Control of tsetse flies (Diptera: Glossinidae) using insecticides: a review and future prospects. *Bulletin of Entomological Research* **74**, 1–23.

Allsopp, R. (1990) A practical guide to aerial spraying for the control of tsetse flies (*Glossina* spp.). *Aerial Spraying Research and Development Project Final Report*, Vol. 2, Natural Resources Institute, Chatham.

Alves, R.T., Bateman, R.P. and Prior, C. (1998) Performance of *Metarhizium anisopliae* formulations with oil adjuvant on *Tenebrio molitor*. In: *Proceedings of the 5th International Symposium on Adjuvants for Agrochemicals*, Memphis. pp. 170–175.

Amin, M.K., Womac, A.R. Bui, Q.D., Mueller, T.C. and Mulrooney, J.E. (1999) Air sampling or aerosol and gaseous pesticides. *Transactions of the ASAE* **42**, 593–600.

Amsden, R.C. (1959) The Baltin–Amsden formula. *Agricultural Aviation* **2**, 95.

Amsden, R.C. (1962) Reducing the evaporation of sprays. *Agricultural Aviation* **4**, 88–93.

Amsden, R.C. (1970) The metering and dispensing of granules and liquid concentrates. *British Crop Protection Council Monograph* **2**, 124–129.

Amsden, R.C. (1972) Wind velocity in relation to aerial spraying of crops. *Agricultural Aviation* **14**, 103–107.

Amsden, R.C. and Lewins, C.P. (1966) Assessment of wettability of leaves by dipping in crystal violet. *World Review of Pest Control* **5**, 187–194.

Anderson, N.H., Hall D.J. and Seaman, D. (1987) Spray retention: effects of surfactants and plant species. *Aspects of Applied Biology* **14**, 233–244.

Anon. (1971) Application and dispersal of pesticides. *Technical Report Series of the World Health Organisation* no. 465. World Health Organisation, Geneva.

Anon. (1973) *Agricultural Pilots' Handbook*, 1st edn. International Agricultural Aviation Centre, The Hague.

Anon. (1995) 'Spray on' bugs for biocontrol *Agricultural Research* **43**, 23.

Anon. (1998) *Cotton Handbook*, Zimbabwe Commercial Cotton Growers' Association, Harare.

Apt W.J. and Caswell, E.P. (1988) Application of nematicides via drip irrigation. *Annals of Applied Nematology* **2**, 1–10.

Archibald, B.A., Solomon, K.R. and Stephenson, G.R. (1994) A new procedure for calibrating the video imaging technique for assessing dermal exposure to pesticides. *Archives of Environmental Contamination and Toxicology* **26**, 398–402.

Archibald, B.A., Solomon, K.R. and Stephenson, G.R. (1994) Estimates of pesticide exposure to greenhouse applicators using video imaging and other assessment techniques. *American Industrial Hygiene Association Journal* **56**, 226–235.

Arnold, A.C. (1983) Comparative droplet spectra for three different-angled flat fan nozzles. *Crop Protection* **2**, 193–204.

Arnold, A.C. (1987) The dropsize of the spray from agricultural fan spray atomisers as determined by a Malvern and the Particle Measuring System (PM5) instrument. *Atomisation and Spray Technology* **3**, 155–167.

Arnold, A.J. (1965) A high-speed automatic micrometer syringe. *Journal of Scientific Instruments* **42**, 350–351.

Arnold, A.J. (1967) Hand-operated micro-applicator to deliver drops of five sizes. *Laboratory Practice* **16**, 56–57.

Arnold, A.J. (1983) UK patent, 2 119 678.

Arnold, A.J. (1984) Electrostatic application with rotary atomisers. *EPPO Bulletin* **13**, 451–456.

Arnold, A.J. (1994) Insect suction sampling without nets, bags or filters. *Crop Protection* **13**, 73–76.

Arnold, A.J. and Pye, B.J. (1980) Spray application with charged rotary atomisers. *British Crop Protection Council Monograph* **24**, 109–125.

Atias, M. and Weihs, D. (1985) On the motion of spray drops in the wake of an agricultural aircraft. *Atomisation and Spray Technology* **1**, 21–36.
Atkinson, A.M., Filmer, P.J. and Kinsman, K.L. (1968) *The Distribution of Pasture Seed from Light Aircraft*, Commonwealth of Australia.
Avery D.T. (1997) Saving the planet with pesticides, biotechnology and European farm reform. In: *Proceedings of the Brighton Crop Protection Conference – Weeds*, pp. 3–18. BCPC, Farnham.
Babcock, J.M., Brown, J.J. and Tanigoshi, L.K. (1990). Volume and coverage estimation of spray deposition using an amino nitrogen colorimeter reaction. *Journal of Economic Entomology* **83**, 1633–1635.
Bachalo, W.D., Houser, M.J. and Smith, J.N. (1987) Behaviour of sprays produced by pressure atomizers as measured using a phase/doppler instrument. *Atomisation and Spray Technology* **3**, 53–72.
Bache, D.H. (1975) Transport of aerial spray: III. Influence of microclimate on crop spraying. *Agricultural Meteorology* **15**, 379–383.
Bache, D.H. (1994) The trapping of spray droplets by insects. *Pesticide Science* **41**, 351–357.
Bache, D.H. and Johnstone, D.R. (1992) *Microclimate and Spray Dispersion*. Ellis Horwood, Chichester.
Bache, D.H. and Sayer, W.J.D. (1975) Transport of aerial spray: I. A model of aerial dispersion, *Agricultural Meteorology* **15**, 257–271.
Bache, D.H., Lawson, T.J. and Uk, S. (1988) Development of a criterion for defining spray drift. *Atmospheric Environment* **22**, 131–135.
Bailey, A.G. (1986) The theory and practice of electrostatic spraying. *Atomisation and Spray Technology* **2**, 95–134.
Bailey, P.W. (1988) Engineering problems associated with granule application in row crops. *British Crop Protection Council Monograph* **39**, 329–32.
Baldry, D.A.T., Kulzer, H., Bauer, S., Lee, C.W. and Parker, J.D. (1978) The experimental application of insecticide from a helicopter for the control of riverine populations of *Glossina tachinoides* in West Africa. III. Operational aspects and application techniques. *PANS* **24**, 423–434.
Baldry, D.A.T., Zerbo, D.G., Baker, R.H.A., Walsh, J.F. and Pleszak, F.C. (1985) Measures aimed at controlling the invasion of *Simulium damnosum* Theobald, S.I. (Diptera: Simuliidae into the Onchocerciasis Control Programme Area 1. Experimental aerial larviciding in the Upper Sassandra Basin of South-eastern Guinea in 1985. *Tropical Pest Management* **31**, 255–263.
Ball, B.V., Pye, B.J., Carreck, N.L., Moore, D. and Bateman, R.P. (1994) Laboratory testing of a mycopesticide on non-target organisms: the effects of an oil formulation of *Metarhizium flavoviride* applied to *Apis mellifera. Biocontrol, Science and Technology* **4**, 289–296.
Bals, E.J. (1969) The principles of and new developments in ultra-low volume spraying. *Proceedings of the 5th British Insecticide and Fungicide Conference*; pp. 189–193. BCPC, Farnham.
Bals, E.J. (1970) Rotary atomisation. *Agricultural Aviation* **12**, 85–90.
Bals, E.J. (1975a) Development of cda herbicide handsprayer. *PANS* **21**, 345–349.
Bals, E.J. (1975b) The importance of controlled droplet application (CDA) in pesticide applications. In: *Proceedings of the 8th British Insecticide and Fungicide Conference*, pp. 153–160. BCPC, Farnham.
Bals, E.J. (1976) Controlled droplet application of pesticides (CDA). Paper presented at *Symposium on Droplets in Air*. Society of Chemical Industry, London.
Balsari, P. and Airoldi, G. (1998) A survey to determine the amount of unused product and disposal methods used in pesticide application. *British Crop Protection Council Monograph* **70**, 195–202.
Baltin, F. (1959) The Baltin formula, *Agricultural Aviation* **1**, 104; **2**, 6.
Barlow, F. and Hadaway, A.B. (1947) Preliminary notes on the loss of DDT and gammexane by absorption, *Bulletin of Entomological Research* **38**, 335–346.

Barlow, F. and Hadaway, A.B. (1974) Some aspects of the use of solvents in ULV formulations, *British Crop Protection Council Monograph* **11**, 84–93.
Barnett, G. (1990) The increased yield response of winter wheat to low pesticide input programmes with vegetable oil-based carrier adjuvant. *Mededlingen van de Faculteit Landbouwwetenschappen, Rijksuniversiteit Gent* **55**, 1343–1347.
Barnett, G.S. and Matthews, G.A. (1992) Effect of different fan nozzles and spray liquids on droplet spectra with special reference to drift control. *International Pest Control* **34**, 81–83.
Barrett, P.R.F. (1978) Some studies on the use of alginate for the placement and controlled release of diquat on submerged aquatic plants. *Pesticide Science* **9**, 425–433.
Barrett, P.R.F. and Logan, P. (1982) The localised control of submerged aquatic weeds in lakes with diquat alginate. In: *Proceedings of the 6th Symposium of the European Weed Research Society on Aquatic Weeds*, pp. 193–198
Barrett, P.R.F. and Murphy, K.J. (1982) The use of diquat alginate for weed control in flowing waters. In: *Proceedings of the 6th Symposium of the European Weed Research Society on Aquatic Weeds*, pp. 200–208.
Barry, J.W., Ekblad, R.B., Tseke, M.E. and Skyler, P.J. (1990) Technology transfer forest service aerial spray models. *ASAE Paper* 90–1017.
Barry, T., Valcore, D.L. and Hewitt, A.J. (1999) The effect of droplet size and other variables on drift from applications of sprays by ground. *ASAE Paper* 99–1010.
Batchelor, G.K. (1967) *An Introduction to Fluid Dynamics*. Cambridge University Press, Cambridge.
Batel, W. and Hinz, T. (1988) Exposure measurements concerning protective clothing in agriculture. In: *Performance of Protective Clothing: Second Symposium* (eds S.Z. Mansdorf, R. Sager and A.P. Nielsen), American Society for Testing and Materials, Special Technical Publication 989, pp. 584–596. American Society for Testing and Materials, Philadelphia.
Bateman, R.P. (1993) Simple, standardized methods for recording droplet measurements and estimation of deposits from controlled droplet applications. *Crop Protection* **12**, 201–206.
Bateman, R. (1994) Performance of myco-insecticides: importance of formulation and controlled droplet application. *British Crop Protection Council Monograph* **59**, 275–284.
Bateman, R. (1997) Methods of application of microbial pesticide formulations for the control of grasshoppers and locusts. *Memoirs of the Entomological Society of Canada* **171**, 69–81.
Bateman, R. (1999) Delivery systems and protocols for biopesticides. In *Biopesticides: Use and Delivery* (eds F.R. Hall and J.J. Menn) pp. 509–528. Humana Press, Totowa.
Bateman, R.P. and Alves, R.T. (2000) Delivery systems for mycoinsecticides using oil-based formulations. *Aspects of Applied Biology* **57**, 163–170.
Bateman, R.P. and Chapple, A.C. (2000) The spray application of mycopesticide formulations. In: *Fungal Biocontrol Agents: Progress, Problems and Potential* (eds T. Butt, C. Jackson and N. Hagan), Chapter 11. CAB International, Wallingford.
Bateman, R.P., Godonou, I., Kpindu, D., Lomer, C.J. and Paraiso, A. (1992) Development of a novel field bioassay technique for assessing mycoinsecticide ULV formulations. In: *Biological Control of Locusts and Grasshoppers* (eds C.J. Lomer and C. Prior), CABI/IITA, Wallingford.
Bateman, R.P., Price, R.E., Muller, E.J. and Brown, H.D. (1994) Controlling brown locust hopper bands in South Africa with a myco-insecticide spray. In: *Proceedings of the Brighton Crop Protection Conference – Pests and Diseases* pp. 609–616. BCPC, Farnham.
Bateman, R.P., Douro-Kpindou, O.K., Kooyman, C., Lomer, C. and Ouambama, Z. (1998) Some observations on the dose transfer of mycoinsecticide sprays to desert locusts. *Crop Protection* **17**, 151–158.
Bateman, R.P., Matthews, G.A. and Hall, F.R. (2000) Ground-based application equipment. In: *Field Manual of Techniques in Invertebrate Pathology* (eds L. Lacey and H. Kaya), pp. 77–112. Kluwer, Dordrecht.

Beaumont, A. (1947) The dependence on the weather of the dates of outbreaks of potato blight epidemics, *Transactions of the British Mycological Society* **31**, 45–53.
Becher, P. (1973) The emulsifier. In *Pesticide Formulations* (ed. W. Van Valkenburg), pp. 65–92. Marcel Dekker, New York.
Beeden, P. (1972) The pegboard – an aid to cotton pest scouting. *PANS* **18**, 43–45.
Beeden, P. (1975) Service life of batteries used in portable ULV sprayers, *Cotton Growing Review* **52**, 375–385.
Beeden, P. and Matthews, G.A. (1975) Erosion of cone nozzles used for cotton spraying, *Cotton Growing Review* **52**, 62–65.
Beernaerts, S., Debongnie, Ph., Delvaux, A. and Pussemier, L. (1999) Pesticides transport into surface water from a small basin in Belgium. In: *Proceedings of the 51st International Symposium on Crop Protection*, May 1999, Gent, Belgium. University of Gent.
Bell, G.A. (1989) Herbicide granules – review of processes and products. In: *Proceedings of the Brighton Crop Protection Conference – Weeds*, pp. 745–752 BCPC, Farnham.
Bell, G.A. (1998) Water-dispersible granules. In: *Chemistry and Technology of Agrochemical Formulations* (ed. D.A. Knowles), pp. 80–120 Kluwer, Dordrecht.
Bell, M.R. and Hayes, J.L. (1994) Areawide management of cotton bollworm and tobacco budworm (Lepidoptera: Noctuidae) through application of a nuclear polyhedrosis virus on early-season alternate hosts. *Journal of Economic Entomology* **87**, 53–57.
Bennett, L.V. and Symmons, P.M. (1972) A review of the estimates of certain control techniques against the Desert Locust. *Anti-Locust Bulletin* **50**, 1–15.
Bergland, R.N. and Liu, B.Y.H. (1973) Generation of monodisperse aerosol standards. *Environmental Science and Technology* **7**, 147–153.
Beroza, M., Hood, C.S., Trefrey, D., Leonard, D.E., Knipling, E.F., Klassen, W. and Stevens, L.J. (1974) Large field trial with microencapsulated sex pheromone to prevent mating of the gypsy moth. *Journal of Economic Entomology* **67**, 661–664.
Bindra, O.S. and Singh, H. (1971) *Pesticide Application Equipment*, Oxford and IBH, New Delhi.
Birchall, W.J. (1976) Developments in the regulation of agricultural aviation in the United Kingdom, In: *Proceedings of the 5th International Agricultural Aviation Congress*, 22–25 September 1975, Stoneleigh, pp. 190–192. IAAC, The Hague.
Bjugstad N. (1994) Spraying application and deposit of pesticides in orchards. *Acta Horticulturae* **372**, 93–102.
Boatman, N.D. (1998) The value of buffer zones for the conservation of biodiversity. In: *Proceedings of the Brighton Crop Protection Conference – Pests and Disease*, pp. 939–950. BCPC, Farnham.
Bode, L.E., Langley, T.E. and Butler, B.J. (1979) Performance characteristics of by-pass spray nozzles. *Transactions of the ASAE* **22**, 1016–1022.
Bode, L.E., Butler, B.J., Pearson, S.C. and Bouse, L.F. (1983) Characteristics of the Micromax rotary atomiser. *Transactions of the ASAE* **26**, 999–1005.
Boize, L.M. and Dombrowski, N. (1976) The atomization characteristics of a spinning disc ultra low volume applicator. *Journal of Agricultural Engineering Research* **21**, 87–99.
Boize, L., Gudin, C. and Purdue, G. (1976) The influence of leaf surface roughness on the spreading of oil spray drops, *Annals of Applied Biology* **84**, 205–211.
Bond, E.J. (1984) Manual of fumigation for insect control. *FAO Plant Production and Protection*, Vol. 54, Food and Agriculture Organisation, Rome.
Bouse, L.F. (1987) Aerial application. In: *Methods of Applying Herbicides* (eds C.G. McWhorter and M.R. Gebhardt), *Monographs of the Weed Science Society of America*, Vol. 4, pp. 123–136.
Bouse, L.F. and Morrison, R.K. (1985) Transport, storage and release of *Trichogramma pretiosum*. *Southwest Entomologist* Supplement **8**, 36–48.

Bouse, L.F., Carlton, J.B. and Merkle, M.G. (1976) Spray recovery from nozzles designed to reduce drift. *Weed Science* **24**, 361–365.

Bouse, L.F., Carlton, J.B. and Morrison, R.K. (1981) Aerial application of insect egg parasites. *Transactions of the ASAE* **24**, 1093–1098.

Boving, P.A., Winterfeld, R.G. and Stevens, L.E. (1972) A hydraulic drive system for the spray system of agricultural aircraft. *Agricultural Aviation* **14**, 41–45.

Brandenburg, B.C. (1974) Raindrop – drift reduction spray nozzle. *ASAE Paper* 74–1595.

Brazee, R.D., Bukovac, H.J., Cooper, J.A. and Reichard, D.L. (1999) Droplet shape oscillations: effect on spray impaction and retention and fourier analysis of their natural frequencies. *Transactions of the ASAE* **42**, 319–325.

Brazelton, R.W. and Akesson, N.B. (1976) Chemical application equipment regulations in California, In: *Proceedings of the 5th International Agricultural Aviation, Congress*, 22–25 September 1975, Stoneleigh, pp. 206–212. IAAC, The Hague.

Brazelton, R.W. and Akesson, N.B. (1987) Principles of closed systems for handling of agricultural pesticides. *Proceedings of the American Society for Testing and Materials, Philadelphia. 7th Symposium on Pesticide Formulations as Application Systems*, pp. 16–27.

Brazelton, R.W., Roy, S., Akesson, N.B. and Yates, W.E. (1971) Distribution of dry materials, In: *Proceedings of the 4th International Agricultural Aviation Congress* 25–29 September 1969, Kingston, Ontario, pp. 166–174 IAAC, The Hague.

Brooke, J.P., Giglioli, M.E.C. and Invest, J. (1974) Control of *Aedes taeniorhynchus* Wied on Grand Cayman with ULV bioresmethrin, *Mosquito News* **34**, 104–111.

Brown, A.W.A. (1951) *Insect Control by Chemicals*, John Wiley and Sons, New York.

Brown, A.W.A. and Watson, D.L. (1953) Studies on fine spray and aerosol machines for control of adult mosquitoes, *Mosquito News* **13**, 81–95.

Brown, J.R., Zyzak, M.D. Callahan, J.H. and Thomas, G. (1997) A spray management valve for hand-compression sprayers. *Journal of the American Mosquito Control Association* **13**, 84–86.

Bruge, G. (1975) Les appareils localisateurs de microgranules antiparasitaires, *Phytoma* **27**, 9–17.

Bruge, G. (1976) Chlormephos: mutual adaptation of the formulation and the equipment used for its application, *British Crop Protection Council Monograph* **18**, 109–113.

Brunskill, R.T. (1956) Factors affecting the retention of spray droplets on leaves. *Proceedings of the 3rd British Weed Control Conference* **2**, 593–603.

Bryant, J.E. and Yendol, W.G. (1988) Evaluation of the influence of droplet size and density of *Bacillus thuringiensis* against gypsy moth larvae (Lepidoptera: Lymantriidae) *Journal of Economic Entomology* **81**, 130–134.

Bui, Q.D., Womac, A.R., Howard, K.D., Mulrooney, J.E. and Amin, M.K. (1998) *Transactions of the ASAE* **41**, 37–41.

Burges, H.D. and Jarrett, P. (1979) Appliction and distribution of *Bacillus thuringiensis* for control of tomato moth in glasshouses. In: *Proceedings of the British Crop Protection Council Conferences – Pests and Diseases* pp. 433–439. BCPC, Farnham.

Butler Ellis, M.C., and Tuck, C.R. (1999) How adjuvants influence spray formation with different hydraulic nozzles. *Crop Protection* **18**, 101–109.

Butler Ellis, M.C. and Tuck, C.R. (2000) The variation in characteristics of air-included sprays with adjuvants. *Aspects of Applied Biology* **57**, 155–162.

Butler Ellis, M.C., Tuck, C.R. and Miller, P.C.H. (1997a) Emulsions and their effect on spray formation and droplet size with agricultural flat fan nozzles. In: *Proceedings of the Brighton Crop Protection Conference – Weeds*, pp. 561–566 BCPC, Farnham.

Butler Ellis, M.C., Tuck, C.R. and Miller, P.C.H. (1997b) The effect of some adjuvants on sprays produced by agricultural flat fan nozzles. *Crop Protection* **16**, 41–50.

Butt, T.M., Harris, J.G. and Powell, K.A. (1999) Microbial pesticides. In: *Biopesticides: Use and Delivery* (eds F.R. Hall and J.J. Menn), pp. 23–44. Humana Press, Totowa.

Byass, J. and Charlton, G.K. (1968) Equipment and methods for orchard spray application research, *Journal of Agricultural Engineering Research* **13**, 280–289.

Cadogan, B.L., Sundaram, K.M.S., Mickle, R.E., Robinson, A.G., Knowles, K.R. and Scherbach, R.D. (1998) Efficacy of tebufenozide applied by aircraft using only upwind atomizers to control source budworm (Lepidoptera: Tortricidae). *Crop Protection* **17**, 315–321.

Cadou, I. (1959) Une rampe portative individuelle pour la pulverisation a faible volume, *Coton et Fibres Tropicales* **14**, 47–50.

Campion, D.G. (1976) Sex pheromones for the control of Lepidopterous pests using microencapsulation and dispenser techniques. *Pesticide Science* **7**, 636–641.

Campion, D.G., Critchley, B.R. and McVeigh, L.J. (1989) Mating disruption. In: *Insect Pheromones in Plant Protection* (eds A.R. Justum and R.F.S. Garden), pp. 89–119. John Wiley and Sons, Chichester.

Carlton, J.B. (1992) Simple techniques for measuring spray deposit in the field II: dual side leaf washer. *ASAE Paper* 92–1618.

Carlton, J.B. (1999) Technique to reduce chemical usage and concomitant drift from aerial sprays. US Patent, 5 975 425.

Carlton, J.B. and Bouse, L.F. (1980) Electrostatic spinner-nozzle for charging aerial sprays. *Transactions of the ASAE* **23**, 1369–1373, 1378.

Carlton, J.B., Bouse, L.F. and Kirk, I.W. (1995) Electrostatic charging of aerial spray over cotton. *Transactions of the ASAE* **38**, 1641–1645.

Carmen, G.E. (1975) Spraying procedures for pest control on citrus in citrus. *Technical Monograph no.* 4, pp. 28–34. Ciba-Geigy, Basle.

Carroll, M.K. and Bourg, J.A. (1979) Methods of ULV droplet sampling and analysis: effects on the size and spectrum of the droplets collected. *Mosquito News* **39**, 645–655.

Carson, A.G. (1987) Improving weed management in the draft animal-based production of early pearl millet in the Gambia. *Tropical Pest Management* **33**, 359–363.

Carter, N. (1992) A European guideline for testing the effects of plant protection products on arthropod natural enemies. *Aspects of Applied Biology* **31**, 157–163.

Carter, P.L. (1998) The safe disposal of clean agrochemical containers on farm. In: *Proceedings of the Brighton Crop Protection Conference – Pests and Diseases*, pp. 729–736. BCPC, Farnham.

Caudwell, R.W. and Gatehouse, A.G. (1996) Formulation of grasshopper and locust entomopathogens in baits using starch extrusion technology. *Crop Protection* **15**, 33–37.

Cauquil, J. and Vaissayre, M. (1995) Protection phytosanitaire du cotonnier en Afrique tropicale. Constraintes et perspectives des nouveaux programmes. *Agriculture et Developpement* **5**, 17–29.

Cayley, G.R., Etheridge, P., Griffiths, D.C., Phillips, F.T., Pye, B.J. and Scott, G.C. (1984) A review of the performance of electrostatically charged rotary atomisers on different crops. *Annals of Applied Biology* **105**, 279–386.

Cayley, G.R., Griffiths, D.C., Hulme, P.J., Lewthwaite, R.J. and Pye, B.J. (1987) Tracer techniques for the comparison of sprayer performance. *Crop Protection* **6**, 123–129.

Cecil, A.R.G. (1997) Modified spray nozzle design reduces drift whilst maintaining effective. chemical coverage. In: *Proceedings of the Brighton Crop Protection Conference – Weeds*, pp. 543–548 BCPC, Farnham.

Chadwick, P.R. (1975) The activity of some pyrethroids, DDT and lindane in smoke from coils for biting inhibition, knockdown and kill of mosquitoes (Diptera. Culicidae). *Bulletin of Entomological Research* **65**, 97–107.

Chadwick, P.R. and Shaw, R.D. (1974) Cockroach control in sewers in Singapore using bioresmethrin and piperonyl butoxide as a thermal fog. *Pesticide Science* **5**, 691–701.

Chalfont, R.B. and Young, J.R. (1982) Chemigation, or application of insecticide through overhead sprinkler irrigation systems to manage insect pests, infesting vegetable and agronomic crops. *Journal of Economic Entomology* **75**, 237–241.

Chapman, P.J. and Mason, R.D. (1993) British and European Community regulations and registration requirements for non-pesticidal co-formulants in pesticides and adjuvants. *Pesticide Science* **37**, 167–171.

Chapman, P.J., Mattock, S. and Savage, R. (1998) Controls on the authorisation and use of adjuvants with agrochemicals in the UK. In: *Proceedings of the Brighton Crop Protection Conference – Pests and Diseases*. pp. 121–126. BCPC, Farnham.

Chapple A.C. and Bateman, R.P. (1997) Application systems for microbial pesticides: necessity not novelty. *British Crop Protection Council Monograph* **89**, 181–190.

Chapple, A.C., Downer, R.A. and Hall, F.R. (1993a) Effects of spray adjuvants on swath patterns and droplet spectra for a flat-fan hydraulic nozzle. *Crop Protection* **12**, 579–590.

Chapple, A.C., Hall, F.R. and Bishop, B.L. (1993b) Assessment of single-nozzle patternation and extrapolation to moving booms. *Crop Protection* **12**, 207–213.

Chapple, A.C., Downer, R.A., Wolf, T.M., Taylor, R.A.J. and Hall, F.R. (1996) The application of biological pesticides: limitations and a practical solution. *Entomophaga* **41**, 465–474.

Chapple, A.C., Wolf, T.M., Downer, R.A., Taylor, R.A.J. and Hall, F.R. (1997) Use of nozzle-induced air-entrainment to reduce active ingredient requirements for pest control. *Crop Protection* **16**, 323–330.

Chester, G. (1993) Evaluation of agricultural worker exposure to, and absorption of, pesticides. *Annals of Occupational Hygiene* **37**, 509–523.

Chester, G. (1995) Revised guidance document for the conduct of field studies of exposure to pesticides in use. In: *Methods of Pesticide Exposure* (eds P.B. Curry, S. Iyengar, P.A. Maloney and M. Maroni), pp. 179–215. Plenum Press, New York.

Christensen, S., Heisel, T., Secher, B.J.M., Jensen, A. and Haahr, V. (1997) Spatial variation of pesticide doses adjusted to varying canopy density in cereals. *Precision Agriculture* **1**, 211–218.

Clark, C.J. and Dombrowski, N. (1972) On the formation of drops from the rims of fan spray sheets, *Aerosol Science* **3**, 173–183.

Clay, M.M. and Clarke S.W. (1987) Effect of nebulised aerosol size on lung deposition in patients with mild asthma. *Thorax* **42**, 190–194.

Clayphon, J.E. (1971) Comparison trials of various motorised knapsack mistblowers at the Cocoa Research Institute of Ghana. *PANS* **17**, 209–225.

Clayphon, J.E. and Matthews, G.A. (1973) Care and maintenance of spraying equipment in the tropics. *PANS* **19**, 13–23.

Clayton J. (1992) New developments in controlled droplet application (CDA) techniques for small farmers in developing countries – opportunities for formulation and packaging. In: *Proceedings of the Brighton Crop Protection Conference – Pests and Diseases* pp. 333–342. BCPC, Farnham.

Clayton, P.B. (1988) Seed treatment technology – the challenge ahead for the agricultural chemicals industry. *British Crop Protection Council Monograph* **39**, 247–256.

Clayton, P.B. (1993) Seed treatment. In: *Application Technology for Crop Protection* (eds G.A. Matthews and E.C. Hislop), pp. 329–349. CABI, Wallingford.

Coffee, R.A. (1971) Electrostatic crop spraying and experiments with triboelectrogasdynamic generation system. In: *Proceedings of the 3rd Conference on Static Electrification*, Paper 17, pp. 200–211. Institute of Physics.

Coffee, R.A. (1979) Electrodynamic energy – a new approach to pesticide application. In: *Proceedings of the British Crop Protection Council Conference Pests and Diseases* 777–789. BCPC, Farnham.

Coggins, S. and Baker, E.A. (1983) Microsprayers for the laboratory application of pesticides. *Annals of Applied Biology* **102**, 149–154.

Combellack, J.H. and Matthews, G.A. (1981a) Droplet spectra measurements of fan and cone atomisers using a laser diffraction technique. *Journal of Aerosol Science* **12**, 529–540.

Combellack, J.H. and Matthews, G.A. (1981b) The influence of atomizer, pressure and

formulation on the droplet spectra producted by high volume sprayers *Weed Research* **21**, 77–86.
Combellack, J.H., Western, N.M. and Richardson, R.G. (1996) A comparison of the drift potential of a novel twin fluid nozzle with conventional low volume flat fan nozzles when using a range of adjuvants. *Crop Protection* **15**, 147–152.
Combellack, J.H. and Miller, P.C.H. (1999) A new twin fluid nozzle which shows promise for precision agriculture. In: *Proceedings of the Brighton Crop Protection Conference – Weeds*, pp. 473–478. BCPC, Farnham.
Cooke, A.S. (1993) *The Environmental Effects of Pesticide Drift*. English Nature, Peterborough.
Cooke, B.K., Herrington, P.J., Jones, K.G. and Morgan, N.G. (1977) Progress towards economical and precise fruit spraying. In: *Proceedings of the British Crop Protection Council Conference – Pests and Diseases* pp. 323–329. BCPC, Farnham.
Cooke, B.K. and Hislop, E.C. (1987) Novel delivery systems for arable crop spraying – deposit distribution and biological activity. *Aspects of Applied Biology*. **14**, 53–70.
Cooke, B.K. and Hislop, E.C. (1993) Spray tracing techniques. In: *Application Technology for Crop Protection* (eds G.A. Matthews and E.C. Hislop), pp. 85–100. CABI International, Wallingford.
Cooke, B.K., Hislop, E.C., Herrington, P.J., Western, N.M. and Humpherson-Jones, F. (1990) Air-assisted spraying of arable crops in relation to deposition, drift and pesticide performance. *Crop Protection* **9**, 303–311.
Cooke, L.R., Little, G. and Wilson, D.G. (1998) Sensitivity of *Phytophthora infestans* to fluazinam and its use in potato blight control in Northern Ireland. In: *Proceedings of the Brighton Crop Protection Conference – Pests and Diseases* pp. 517–522. BCPC, Farnham.
Cooper, J.F. (1991) Computer programme for the analysis of droplet data. In: *Aerial Spraying Research and Development Project Final Report*, Vol. 2, (ed. R. Allsopp), pp. 37–38. National Resources Institute, Chatham.
Cooper, J.F., Coppen, G.D.A., Dobson, H.M., Rakotonandrasana, A. and Scherer, R. (1995) Sprayed barriers of diflubenzuron (ULV) as a control technique against marching hopper bands of migratory locust *Locusta migratoria capito* (Sauss.) (Orthoptera: Acrididae) in Southern Madagascar. *Crop Protection* **14**, 137–143.
Cooper, J.F., Jones, K.A. and Moawad, G. (1998) Low volume spraying on cotton: a comparison between spray distribution using charged and uncharged droplets applied by two spinning disc sprayers. *Crop Protection* **17**, 711–715.
Cooper, J.F., Smith, D.N. and Dobson, H.M. (1996) An evaluation of two field samplers for monitoring spray drift. *Crop Protection* **15**, 249–257.
Cooper, S.C. and Law, S.E. (1985) *Institute of Electrical and Electronic Engineers (IEEE(IAS)) Conference Record*, 1346–1352.
Cooper, S.C. and Law, S.E. (1987a) Transient characteristics of charged spray deposition occurring under action of induced target coronas: space-charge polarity effect. *Institute of Physics Conference Series No. 85*, Section 1, pp. 21–26.
Cooper, S.C. and Law, S.E. (1987b) Bipolar spray charging for leaf-tip corona reduction by space-charge control. *IEEE Transactions on Industry Applications* **IA-23**, 217–223.
Cooper, S.E. and Taylor, W.A. (1998) A survey of spray operators agrochemical container rinsing skills in June 1997. *British Crop Protection Council Monograph* **70**, 145–148.
Cooper, S.E. and Taylor, B.P. (1999) The distribution and retention of sprays on contrasting targets using air-inducing and conventional nozzles at two wind speeds. In: *Proceedings of the Brighton Crop Protection Conference – Weeds* pp. 461–466. BCPC, Farnham.
Coppen, G.D.A. (1999) A simple model to estimate the optimal separation and swath width of ULV-sprayed barriers of chitin synthesis inhibitors (CSI) to control locust hopper bands. *Crop Protection* **18**, 151–158.

Cornwall, J.E., Ford, M.L., Liyanage, T.S. and Win Kyi Daw D. (1995) Risk assessment and health effects of pesticides used in tobacco farming in Malaysia. *Health Policy and Planning*. **10**, 431–437.
Cory, J.S. and Entwistle, P.F. (1990) The effect of time of spray application on infection of the pine beauty moth, *Panolis flammea* (Den. & Schiff.) (Lep., Noctuida), with nuclear polyhedrosis virus. *Journal of Applied Entomology* **110**, 235–241.
Courshee, R.J. (1959) Drift spraying for vegetation baiting. *Bulletin of Entomological Research* **50**, 355–369.
Courshee, R.J. (1960) Some aspects of the application of insecticides, *Annual Review of Entomology* **5**, 327–352.
Courshee, R.J. (1967) Application and use of foliar fungicides, In: *Fungicides: An Advanced Treatise*, Vol. 1 (ed. D.C. Torgeson), pp. 239–286. Academic Press, New York.
Courshee, R.J. and Ireson, M.J. (1961) Experiments on the subjective assessment of spray deposits. *Journal of Agricultural Engineering Research* **6**, 175–182.
Courshee, R.J. and Ireson, M.J. (1962) Distribution of granulated materials by helicopters – a rotating granule distributor. *Agricultural Aviation* **4**, 131–132.
Courshee, R.J., Daynes, F.K. and Byass, J.B. (1954) A tree-spraying machine. National Institute for Agricultural Engineering Report. Silsoe Research Institute, Bedford. Quoted by Ripper, W.E. *Annals of Applied Biology*, **42**, 288–324.
Coutts, H.H. (1967) Preliminary tests with the UCAR nozzle. *Agricultural Aviation* **9**, 123–124.
Coutts, H.H. and Parish, R.H. (1967) The selection of a solvent for use with low volume aerial spraying of cotton. *Agricultural Aviation*, **9**, 125.
Coutts, H.H. and Yates, W.E. (1968) Analysis of spray droplet distribution from agricultural aircraft. *Transactions of the ASAE* **II**, 25–27.
Cowan, S.L., Tilley, R.C. and Wiczynski, M.E. (1988) Comfort factors of protective clothing: mechanical and transport properties, subjective evaluation of comfort. In: Mansdorf, S.Z., Sager, R. and Nielsen, A.P. (eds) *Performance of Protective Clothing: Second Symposium*. American Society for Testing and Materials, Special Technical Publication 989, pp. 31–42. American Society for Testing and Materials, Philadelphia.
Cowell, C. and Lavers, A. (1988) The flow rate of formulations through some typical hand-held ultra low volume spinning disc atomizers. *Tropical Pest Management* **34**, 150–153.
Crabtree, J.H. (1993) The development of a tractor mounted field trials sprayer. *Proceedings of an International Symposium on Pesticide Application*, pp. 661–668. ANPP/British Crop Protection Council, Strasbourg.
Craig, I.P., Matthews, G.A. and Thornhill, E.W. (1993) Fluid injection metering system for closed pesticide delivery in manually operated sprayers. *Crop Protection* **12**, 549–553.
Craig, I., Woods, N. and Dorr, G. (1998) A simple guide to predicting aircraft spray drift. *Crop Protection* **17**, 475–482.
Crease, G.J., Ford, M.G. and Salt, D.W. (1987) The use of high viscosity carrier oils to enhance the insecticidal efficacy of ULV formulations of cypermethrin. *Aspects of Applied Biology* **114**, 307–322.
Cross J.V. (1988) New trends in orchard spraying. *European Plant Protection Organisation Bulletin* **18**, 587–594.
Cross, J.V. (1991) Patternation of spray mass flux from axial fan air-blast sprayers in the orchard. *British Crop Protection Council Monograph* **46**, 15–22.
Cross J.V. and Berrie, A.M. (1990) Efficacy of reduced volume and reduced dose rate spray programmes in apple orchards. *Crop Protection* **9**, 207–217.
Cross J.V. and Berrie, A.M. (1995) Field evaluation of a tunnel sprayer and effects of spray volume at constant drop size on spray deposits and efficacy of disease control on apple. *Annals of Applied Biology* **127**, 521–532.

Cross, J.V., Ridout, M.S. and Walklate, P.J. (1998) Adjustment of axial fan sprayers to orchard structure. In: *Proceedings of the 4th International Workshop on Integrated Control of Orchard Diseases*, 1996 *International Organisation for Biological Control Bulletin.*

Cross J.V., Berrie, A.M. and Murray, R.A. (2000) Effect of drop size and spray volume on deposits and efficacy of strawberry spraying. *Aspects of Applied Biology* **57**, 313–320.

Croxford, A.C. (1998) Bufferzones to protect the aquatic environment. In: *Proceeding of the Brighton Crop Protection Conference – Pests and Disease*, pp. 923–930. BCPC, Farnham.

Crozier, B. (1976) Specifications for granules. *British Crop Protection Council Monograph* **18**, 98–101.

Cunningham G.P. and Harden, J. (1998) Reducing spray volumes applied to mature citrus trees. *Crop Protection* **17**, 289–292.

Cunningham G.P. and Harden, J. (1999) Sprayers to reduce spray volumes in mature citrus trees. *Crop Protection* **18**, 275–281.

Cunningham, J.C., Brown, K.W., Payne, N.J., Mickle, R.E., Grant, G.G., Fleming, R.A., Robinson, A., Curry, R.D., Langevon, D. and Burns, T. (1997) Aerial spray trials in 1992 and 1993 against gypsy moth, *Lymantria dispar* (Lepidoptera: Lymantriidae), using nuclear polyhedrosis virus with and without an optical brightener compared to *Bacillus thuringiensis. Crop Protection* **16**, 15–23.

Curle, P.D., Emmerson, C.D., Gregory, A.H., Hartmann, J. and Nixon, P. (1998) Packaging of agrochemicals. In: *Chemistry and Technology of Agrochemical Formulations* (ed. D.A. Knowles), pp. 264–301. Kluwer, Dordrecht.

Curran, J. and Patel, V. (1988) Use of a trickle irrigation system to distribute entomopathogenic nematodes (Nematoda: Heterorhabditidae) for the control of weevil pests (Coleoptera: Curculionidae) of strawberries. *Australian Journal of Experimental Agriculture* **28**, 639–643.

Dahl, G.H. and Lowell, J.R. (1984) Microencapsulated pesticides and their effects on non-target insects. In: *Advances in Pesticide Formulation Technology* (ed. H.B. Scher), ACS Symposium Series 254, pp. 141–150. American Chemical Society, Washington DC.

Dai, Y., Cooper, S.C. and Law, S.E. (1992) Effectiveness of electrostatic aerodynamic and hydraulic spraying methods for depositing pesticide sprays onto inner plant regions and leaf undersides. *ASAE Paper* 92–1094.

Dale, J.E. (1979) A non-mechanical system of herbicide application with a rope-wick. *PANS* **25**, 431–436.

David, W.A.L. and Gardiner, B.O.C. (1950) Particle size and adherence of dusts. *Bulletin of Entomological Research* **41**, 1–61.

Davis B.N.K., Brown, M.J. and Frost A.J. (1993) Selection of receptors for measuring spray drift deposition and comparison with bioassay with special reference to the shelter effect of hedges. In: *Proceedings of the Brighton Crop Protection Conference – Weeds* pp. 139–144. BCPC, Farnham.

Davis B.N.K., Frost, A.J. and Yates, T.J. (1994) Bioassays of insecticide drift from air assisted sprayers in an apple orchard. *Journal of Horticultural Science* **69**, 703–708.

De Raat, W.K., van de Gevel, I.A., Houben, G.F. and Hakkert, B.C. (1998) Regulatory requirements in the European Union. In: *Chemistry and Technology of Agrochemical Formulations* (ed. D.A. Knowles), pp. 337–376. Kluwer, Dordrecht.

Dewar, A.M. (1994) The virus yellows warning scheme – an integrated pest management system for beet in the UK. In: *Individuals, Populations and Patterns in Ecology*, (eds S.R. Leather, A.D. Watt, N.J. Mills and K.F.A. Walters), pp. 173–185. Atheneum Press, Newcastle on Tyne.

Dewar, A.M., Westwood, F., Bean, K.M., Haylock, L.A. and Osborne, R. (1997) Relationship between pellet size and the quantity of imidacloprid applied to sugar beet pellets and the consequences for seedling emergence. *Crop Protection* **16**, 187–192.

Doble, S.J., Matthews, G.A., Rutherford, I. and Southcombe, E.S.E. (1985) A system for classifying hydraulic nozzles and other atomisers into categories of spray quality. In: *Proceedings of the Pringhton Crop Protection Conference – Weeds*, pp. 1125–1133. BCPC, Farnham.

Dobson, C.M., Minski, M.J. and Matthews, G.A. (1983) Neutron activation analysis using dysprosium as a tracer to measure spray drift. *Crop Protection* **2**, 345–52.

Dohman, G.P. (1994) The effect of pesticides on beneficial organisms in the laboratory and in the field. *British Crop Protection Council Monograph* **59**, 201–210.

Dohnert, D. (1998) Aspects of modern agrochemical packaging. In: *Proceedings of the Brighton Crop Protection Conference – Pests and Diseases*, pp. 715–722. BCPC, Farnham.

Dombrowski, L.A. and Schieritz, M. (1984) Dispersion and grinding of pesticides. In: *Advances in Pesticide Formulation Technology* (ed. H.B. Scher), ACS Symposium series **254**, pp. 63–73. American Chemical Society, Washington, DC.

Dombrowski, N. and Lloyd, T.L. (1974) Atomisation of liquids by spinning cups, *Chemical Engineering Journal* **8**, 63–81.

Dorow, R. (1976) Red-billed quelea *Quelea quelea* bird control in Nigeria, unpublished report.

Doruchowski, G., Holownicki, R. and Godyn, A. (1996a) Air jet setting effect on spray deposit within apple tree canopy and loss of spray in orchard. In: *Proceedings of AgEng 96*, Madrid. Paper 96A–139.

Doruchowski, G., Svensson, S.A. and Mordmark, L. (1996b) Spray deposit within apple trees of differing sizes and geometry at low, medium and high spray volumes. *International Organisation for Biological Control West Palearctic Regional Section Bulletin* **19**, 289–294.

Downer, R.A., Kirchner, L.M., Hall F.R. and Bishop B.L. (1997) Comparison of droplet spectra of fluorescent tracers commonly used to measure pesticide deposition and drift. In: *Pesticide Formulations and Application Systems*, (eds. G.R. Goss, M.J. Hopkinson and H.M. Collins) Vol. 17, pp. 115–128. American Society for Testing and Materials, Philadelphia.

Downer, R.A., Mack, R.E., Hall, F.R. and Underwood, A.K. (1998a) Roundup ultra with drift management adjuvants: some aspects of performance. In: *Proceedings of the 5th International Symposium on Adjuvants for Agrochemicals*, Memphis. pp. 468–474.

Downer, R.A., Hall, F.R., Thompson, R.S. and Chapple, A.C. (1998b) Temperature effects on atomization by flat-fan nozzles: Implications for drift management and evidence for surfactant concentration gradients. *Atomisation and Sprays* **8**, 241–254.

Doyle, C.J. (1988) Aerial application of mixed virus formulations to control joint infestations of *Panolis flammea* and *Neodiprion sertifer* on lodgepole pine. *Annals of Applied Biology* **113**, 119–127.

Duncombe, W.C. (1973) The acaricide spray rotation for cotton. *Rhodesian Agricultural Journal* **70**, 115–18.

Dunn, P. and Walls, J.M. (1978) An introduction to in-line holography and its application. *British Crop Protection Council Monograph* **22**, 23–34.

Dunning, R.A. and Winder, G.H. (1963) Sugar beet yellows: insecticides. *Report of the Rothamsted Experimental Station for 1962*, pp. 218–219.

Eaton, J.K. (1959) Review of materials intended for aerial application and their properties. *Report of the 1st International Agricultural Aviation Conference*, Cranfield. pp. 92–102. IAAC, The Hague.

Ebert, T.A. and Hall, F.R. (1999) Deposit structure effects on insecticide bioassays. *Journal of Economic Entomology* **92**, 1007–1013.

Ebert, T.A., Taylor, R.A.J., Downer, R.A. and Hall, F.R. (1999) Deposit structure and efficacy of pesticide application 1. Interactions between deposit size, toxicant concentration and deposit number. *Pesticide Science* **55**, 783–792.

Edwards, C.J. and Ripper, W.E. (1953) Droplet size rates of application and the

avoidance of spray drift. In: *Proceedings of the 1953 British Weed Control Conference* pp. 348–367.
Elliott, J.G. and Wilson, B.J. (1983) The influence of weather on the efficiency and safety of pesticide application: the drift of herbicides. *British Crop Protection Council Occasional Publication* no. 3.
Elsworth, J.E. and Harris, D.A. (1973). The 'Rotostat' seed treator – a new application system. In: *Proceedings of the 7th British Insecticide and Fungicide Conference*, pp. 349–356. BCPC, Farnham.
Elvy, J.H. (1976) The laboratory evaluation of potential granular carriers. *British Crop Protection Council Monograph* **18**, 102–108.
Embree, D.C., Dobson, C.M.B. and Kettela, E.G. (1976) Use of radio-controlled model aircraft for ULV insecticide application in Christmas tree stands. *Commonwealth Forestry Review* **55**, 178–181.
Enfalt, P., Bengtsson, P., Engqvist, A., Wretland, P. and Alness, K. (2000) A novel technique for drift reduction. *Aspects of Applied Biology* **57**, 41–47.
Eng, O.K., Omar, D. and McAuliffe, D. (1999) Improving the quality of herbicide applications to oil palm in Malaysia using the CFValve – a constant flow valve. *Crop Protection* **18**, 605–607.
Entwistle, P.F. (1986) Spray droplet deposition patterns and loading of spray droplets with NPV inclusion bodies in the control of *Panolis flammea* in pine forests. In: *Fundamental and Applied Aspects of Invertebrate Pathology* (eds. R.A. Samson, J.M. Vlak, and D. Peters), pp. 613–615.
Entwistle, P.F., Evans, H.F., Cory, J.S. and Doyle, C. (1990) Questions on the aerial application of microbial pesticides to forests. *Proceedings of International Colloquium on Invertebrate pathology and Microbial Control*, Adelaide, Australia. pp. 159–163. Society for Invertebrate Pathology, Wageningen.
Etheridge R.E., Womac, A.R. and Mueller, T.C. (1999) Characterization of the spray droplet spectra and patterns of four venturi-type drift reduction nozzles. *Weed Technology* **13**, 765–770.
Evans, H.F. (1999) Principles of dose acquisition for bioinsecticides. In: *Biopesticides: Use and Delivery* (eds. F.R. Hall and J.J. Menn). pp. 553–574. Humana Press. Totowa.
Evans, W.H. (1984) Development of an aqueous-based controlled/release pheromone-pesticide system. In: *Advances in Pesticide Formulation Technology*. (ed. H.B. Scher), ACS Symposium Series, no. 254, pp. 151–162. American Chemical Society, Washington DC.
Falchieri, D., Mierzejewski, K. and Maczuga, S. (1995) Effects of droplet density and concentration on the efficacy of *Bacillus thuringiensis* and carbaryl against gypsy moth larvae (*Lymantria dispar* L.). *Journal of Environmental Science and Health – Part B, Pesticides, Food Contaminants and Agricultural Wastes* **30**, 535–548.
FAO (1974) The use of aircraft in agriculture, *FAO Agricultural Development Paper no. 94*.
FAO (1998) *Agricultural Pesticide Sprayers*. Food and Agriculture Organisation, Rome.
Farmery, H. (1970) The mechanics of granule application, *British Crop Protection Council Monograph* **2**, 101–106.
Farmery, H. (1976) Granules and their application, *British Crop Protection Council Monograph* **18**, 93–97.
Fee, C.G., Siang, C.S. and Ramalingam B. (1999) Evaluation of 3 types of knapsack equipment for spraying cypermethrin to control *Adoretus compressus* and *Oryctes rhinoceros* in immature palms. *Proceedings of the 5th International Conference on Plant Protection in the Tropics*. Kuala Lumpur, Malaysia. pp. 368–375. Plant Protection Society, Kuala Lumpur.
Felber, H. (1988) Safe and efficient spraying of cereal fungicides with the Ciba-Geigy

Croptilter. In: *Proceedings of the ANPP International Symposium on Pesticide Application*, Paris. pp. 249–258. ANPP, Paris.
Fernando, H.E. (1956) A new design of sprayer for reducing insecticide hazards in treating rice crop, *FAO Plant Protection Bulletin* **4**, 117–120.
Fisk, T., Cooper, J. and Wright, D.J. (1993) Control of *Spodoptera* spp. using ULV formulations of the acylurea insect growth regulator flufenoxuron: field studies with *Spodoptera exempta* and effect of toxicant concentration on contact activity. *Pesticide Science* **39**, 79–83.
Ford, M.G. and Salt, D.W. (1987) Behaviour of insecticide deposits and their transfer from plant to insect surfaces. In: *Pesticides on Plant Surfaces* (ed. H.J. Cottrell), pp. 26–81. Wiley, John Chichester.
Ford, M.G., Reay, R.C. and Watts, W.S. (1977) Laboratory evaluation of the activity of synthetic pyrethroids at ULV against the cotton leafworm *Spodoptera littoralis* Boisd. In: *Crop Protection Agents – Their Biological Evaluation* (ed. N.R. McFarlane), Academic Press, New York.
Forrester, N.W., Cahill, M., Bird, L.J. and Layland, J.K. (1993) Management of pyrethroid and endosulfan resistance in *Helicoverpa armigera* (Lepidoptera: Noctuidae) in Australia. *Bulletin of Entomological Research* Supplement 1. pp. 1–132.
Forster, R. and Rothert, H. (1998) The use of field buffer zones as a regulatory measure to reduce the risk to terrestrial non-target arthropods from pesticide use. In: *Proceedings of the Brighton Crop Protection Conference – Pests and Diseases*, pp. 931–938. BCPC, Farnham.
Foy, C.L. (1992) *Adjuvants for Agrichemicals*. CRC Press, Boca Raton.
Foy, C.L. and Pritchard, D.W. (1996) *Pesticide Formulation and Adjuvant Technology*. CRC Press, Boca Raton.
Fraley, R.W. (1984) The preparation of aqueous-based flowables ranging the sample size from sub-gram to several gallons. In: *Advances in Pesticide Formulation Technology* (ed. H.B. Scher). ACS Symposium series 254, pp. 47–62. American Chemical Society, Washington DC.
Fraser, A.J. and Keeble V.B. (1988) Factors influencing design of protective clothing for pesticide application. In: *Performance of Protective Clothing: Second Symposium* (eds S.Z. Mansdorf, R. Sager and A.P. Nielsen). American Society for Testing and Materials, Special Technical Publication 989, pp. 565–572. American Society for Testing and Materials, Philadelphia.
Fraser, R.P. (1958) The fluid kinetics of application of pesticidal chemicals. In: *Advanced Pest Control Research* Vol. 2, (ed. R. Metcalfe), pp. 1–106. Interscience, New York.
Fraser, R.P., Dombrowski, N. and Routley, J.H. (1963) The production of uniform liquid sheets from spinning cups, *Chemical Engineering Science* **18**, 315–321.
French, P. (1980) A mechanized field sprayer for small plot pesticide trials. In: *Proceedings of the 5th International Conference on Mechanization of Field Experiments*, pp. 135–142. IAMFE, Aas.
Frost, A.R. (1974) Rotary atomisation. *British Crop Protection Council Monograph* **11**, 120–127.
Frost, A.R. (1984) Simulation of an active spray boom suspension. *Journal of Agricultural Engineering Research* **30**, 313–325.
Frost, A.R. (1990) A pesticide injection metering system for use on agricultural spraying machines. *Journal of Agricultural Engineering Research* **46**, 55–70.
Frost, A.R. and Lake, J.R. (1981) The significance of drop velocity to the determination of drop size distribution of agricultural sprays. *Journal of Agricultural Engineering Research* **26**, 367–370.
Frost, A.R. and Law, S.G. (1981) Extended flow characteristics of the embedded electrode spray-charging nozzle. *Journal of Agricultural Engineering Research* **26**, 79–86.

Frost, A.R. and Miller, P.C.H. (1988) Closed chemical transfer system. *Aspects of Applied Biology* **18**, 345–359.

Frost, A.R. and O'Sullivan, J.A. (1988) Verification and use of a mathematical model of an active twin link boom suspension. *Journal of Agricultural Engineering Research* **40**, 259–274.

Fryer, J.D. (1977) Recent developments in the agricultural use of herbicides in relation to ecological effects. In: *Ecological Effects of Pesticides* (eds F.H. Perring, and K. Mellanby), pp. 27–45. Academic Press, London.

Furness, G.O. (1991) A comparison of a simple bluff plate and axial fans for air-assisted, high-speed, low-volume spray application to wheat and sunflower plants. *Journal of Agricultural Engineering Research* **48**, 57–75.

Furness, G.O. (1996) The new 'Hydra' sprayer. *Australian and New Zealand Wine Industry Journal* **11**, 268–270.

Furness, G.O. and Newton, M.R. (1988) A leaf surface scanning technique using a fluorescence spectrophotometer for the measurement of spray deposits. *Pesticide Science* **24**, 123–137.

Furness, G.O., Wicks, T.J. and Campbell, K.N. (1997) Spray coverage on grapevines with the multi-head 'Hydra' sprayer. In: *Proceedings of the 11th Australian Plant Pathology Conference.*

Furness, G.O., Magarey, P.A. Miller, P.H. and Drew, H.J. (1998) Fruit tree and vine sprayer calibration based on canopy size and length of row: unit canopy row (UCR) method. *Crop Protection*, **17**, 639–644.

Galloway, B.T. (1891) The improved Japy knapsack sprayer, *Journal of Mycology* **7**, 39.

Gamliel, A., Grinstein, A., Klein, L., Cohen, Y and Katan, J (1998) Permeability of plastic films to methyl bromide: field study. *Crop Protection* **17**, 241–248.

Gan-Mor, S., Grinstein, A., Riven, Y., Beres, H., Kletter, E., Spenser, J., Forer, G., Tzviele E. and Zur, I. (2000) A new technology for improved pesticide coverage on cotton canopy: Part I. Sprayer development. In: *New Frontiers in Cotton Research. Proceedings of World Cotton Research Conference, Part II.* (ed. F. Gillham), Organising and Scientific Committee, Athens, Greece (in press).

Ganzelmeier, H. and Rautmann, D. (2000) Drift, drift reducing sprayers and sprayer testing. *Aspects of Applied Biology* **57**, 1–10.

Ganzelmeier, H., Rautmann, D., Herbst, A. and Kaul P. (1994) Pflanzenschutzgerate auf dem prufstand. *Forschungsreport* **9**, 15–19.

GCPF (1989) *Codes for Pesticide Formulations.* Technical Bulletin no. 2, Global Crop Protection Federation, Brussels.

Gibbs, J.N. and Dickinson, J. (1975) Fungicide injection for the control of Dutch elm disease. *Forestry*, **48**, 165–176.

Gilbert, A.J. (1989) Reducing operator exposure by the improved design and handling of liquid pesticide containers. In: *Proceedings of the Brighton Crop Protection Conference – Weeds* pp. 593–600. BCPC, Farnham.

Gilbert, A.J. (1995) Analysis of exposure to pesticides applied in a regulated environment. In: *Pesticides – Developments, Impacts and Controls*, (eds G.A. Best and A.D. Ruthen), pp. 28–42. Royal Society of Chemistry, Cambridge.

Gilbert, A.J. (1998) Design guidelines, features and performance characteristics and development of current pesticide containers. *British Crop Protection Council Symposium Series* **70**, 9–16.

Gilbert, A.J. (2000) Local environmental risk assessment for pesticides (LERAP) in the UK. *Aspects of Applied Biology* **57**, 83–90.

Gilbert, A.J. and Bell, G.J. (1990) Test methods and criteria for selection of types of coveralls suitable for certain operations involving handling or applying pesticides. *Journal of Occupational Accidents* **11**, 255–268.

Gilbert, D. and Macrory, R. (1989) *Pesticide Related Law.* British Crop Protection Council, Farnham.

Giles, D.K. (1997) Independent control of liquid flow rate and spray droplet size from hydraulic atomizers. *Atomisation and Sprays* **7**, 161–181.

Giles, D.K. and Ben-Salem E. (1992) Spray droplet velocity and energy in intermittent flow from hydraulic nozzles. *Journal of Agricultural and Engineering Research* **51**, 101–112.

Giles, D.K. and Blewett, T.C. (1991) Effects of conventional and reduced volume charged spray application techniques on dislodgeable foliar residues of captan on strawberries. *Journal of Agricultural and Food Chemistry* **39**, 1646–1651.

Giles, D.K. and Comino, J.A. (1989) Variable flow control for pressure atomization nozzles. *Transactions of The Engineering Society for Advancing Mobility Land, Sea, Air and Space, Journal of Commercial Vehicles* **98**, 257–265.

Giles, D.K. and Law, S.E. (1985) Space charge deposition of pesticide sprays onto cylindrical target arrays. *Transactions of the ASAE* **28**, 658–664.

Giles, D.K. and Slaughter, D.C. (1997) Precision band spraying with machine vision guidance and adjustable yaw nozzles. *Transactions of the ASAE* **40**, 29–36.

Giles, D.K. and Wunderlich, L.R. (1998) Electronically controlled delivery system for beneficial eggs in liquid suspensions. *Transactions of the ASAE* **41**, 839–847.

Giles, D.K. Dai, Y. and Law, S.E. (1991) Enhancement of spray electrodeposition by active precharging of a dielectric boundary. *Institute of Physics Conference Series* **118**, pp. 33–38.

Giles, D.K., Young, B.W., Alexander, P.R. and French, H.M. (1995a) Intermittent control of liquid flow from fan nozzles in concurrent air streams: wind tunnel studies of droplet size effects. *Journal of Agricultural Engineering Research* **62**, 77–84.

Giles, D.K. Welsh, A., Steinke, W.E. and Saiz, S.G. (1995b) Pesticide inhalation exposure, air concentration and droplet size spectra from greenhouse fogging. *Transactions of the ASAE* **38**, 1321–1326.

Giles, D.K., Henderson, G.W. and Funk, K. (1996) Digital control of flow rate and spray droplet size from agricultural nozzles for precision chemical application. In: *Proceedings of the 3rd International Conference on Precision Agriculture*. Minneapolis.

Giles, D.P. (1989) Principles in the design of screens in the process of agrochemical discovery. *Aspects of Applied Biology* **21**, 39–50.

Gledhill, J.A. (1975) A review of ultra-low volume spray usage in Central Africa since 1954 and some recent developments in Rhodesia. In: *Proceedings of the Ist Congress of the Entomological Society of South Africa* pp. 259–267. *Entomological Society of South Africa.*

Godson, T.D., Winfield, E.J. and Davis, R.P. (1999) Registration of mixed formulations of pesticides. *Pesticide Science* **55**, 189–192.

Göhlich, H. (1970) Metering and distribution of coarse and fine granules. *British Crop Protection Council Monograph* **2**, 107–113.

Göhlich, H. (1979) A contribution to the demands of reduced application rates and reduced drift. In: *Proceedings of the British Crop Protection Council Conference – Pests and Diseases* pp. 767–775. BCPC, Farnham.

Göhlich, H. (1985) Deposition and penetration of sprays. *British Crop Protection Council Monograph* **28**, 173–182.

Göhlich, H., Ganzelmeier, H. and Backer, G. (1996) Air-assisted sprayers for application in vine, orchard and similar crops. *European Plant Protection Organisation Bulletin* **26**, 53–58.

Gomes, J., Lloyd, O.L. and Revitt, D.M. (1999) The influence of personal protection, environmental hygiene and exposure to pesticides on the health of immigrant farm workers in a desert country. *International Archives of Occupational and Environmental Health* **72**, 40–45.

Gomez, K.A. and Gomez, A.A. (1976) *Statistical Procedures for Agricultural Research, With Emphasis on Rice*. IRRI, Philippines.

Goodwin-Bailey, K.A., Holborn, J.M. and Davies, M. (1957) A technique for the biological evaluation of insecticide aerosols. *Annals of Applied Biology* **45**, 347–360.

Goose, J. (1991) Mosquito control using mistblower sprayers for residual deposit of bendiocarb ULV in Mexico. *British Crop Protection Council Monograph* **46**, 245–248.

Goss, G.R., Taylor, D.R. and Kallay, W.B. (1996) Granular pesticide formulation. In: *Pesticide Formulation and Application Systems*, Vol. 15 (eds F.R. Hall and H.J. Hopkinson), American Society for Testing and Materials, Special Technical Publication 1268, pp. 114–123. American Society for Testing and Materials, Philadelphia.

Gower, J. and Matthews, G.A. (1971) Cotton development in the southern region of Malawi, *Cotton Growing Review* **48**, 2–8.

Graham-Bryce, I.J. (1975) The future of pesticide technology: opportunities for research. In: *Proceedings of the 8th British Crop Protection Council Insecticide and Fungicide Conference* **3**, pp. 901–914. BCPC, Farnham.

Graham-Bryce, I.J. (1988) Pesticide application to seeds and soil: unrealised potential? *British Crop Protection Council Monograph* **39**, 3–14.

Graham-Bryce, I.J. (1989) Environmental impact: Putting pesticides into perspective. In: *Proceedings of the Brighton Crop Protection Conference – Weeds* pp. 3–20. BCPC, Farnham.

Gratz, N. (1985) Control of dipteran vectors. In: *Pesticide Application: Principles and Practice* (ed. P.T. Haskell), pp. 273–300. Oxford University Press, Oxford.

Gratz, N.G. and Dawson, J.A. (1963) The area distribution of an insecticide (fenthion) sprayed inside the huts of an African village. *Bulletin of the World Health Organisation* **29**, 185–196.

GreatRex, R.M. (1998). Transfer of integrated crop management systems from Northern Europe to other regions. In: *Proceedings of the Ist Transnational Conference on Biological, Integrated and Supervised Controls*, Lille.

Greaves, M.P. and Marshall, E.J.P. (1987) Field margins: definitions and statistics. *British Crop Protection Council Monograph* **35**, 3–10.

Grieshop, J.I. (1988) Protective clothing and equipment: Beliefs and behavior of pesticide users in Ecuador. In: *Performance of Protective Clothing: Second Symposium* (eds S.Z. Mansdorf, R. Sager and A.P. Nielsen), American Society for Testing and Materials, Special Technical Publication 989, pp. 802–809. American Society for Testing and Materials, Philadelphia.

Griffith, J. and Bateman, R. (1997) Evaluation of the Francome MkII exhaust gas nozzle sprayer to apply oil-based formulation of *Metarhizium flavoviride* for locust control. *Pesticide Science* **51**, 176–184.

Grinstein, A., Gan-Mor, S., Kletter, E., Spenser, J., Forer, G., Aharonson, N., Gershon, M., Gerling, D., Navo, D., Riven Y. and Veierov, D. (2000) A new technology for improved pesticide coverage on cotton canopy: Part II. Field efficacy. In: *New Frontiers in Cotton Research*. Proceedings of World Cotton Research Conference, II. (ed. F. Gillham), Organising and Scientific Committee, Athens, Greece (in Press).

Gunn, D.L., Graham, J.F., Jaques, E.C., Perry, F.C., Seymour, W.G., Telford, T.M., Ward, J., Wright, E.N. and Yeo, D. (1948) Aircraft spraying against the desert locust in Kenya, 1945, *Anti-Locust Bulletin* no. 4.

Hadar, E. (1991) Development criteria for an air-assisted ground crop sprayer. *British Crop Protection Council Monograph* **46**, 23–26.

Haggar, R.J. Stent, C.J. and Isaac, S. (1983) A prototype hand-held patch sprayer for killing weeds activated by spectral differences in crop/weed canopies. *Journal of Agricultural and Engineering Research* **28**, 349–358.

Haley, J. (1973) *Expert Flagging*, University of North Dakota Press, Grand Forks, North Dakota.

Hall, L.B. (1955) Suggested techniques, equipment and standards for the testing of hand insecticide spraying equipment. *Bulletin of the World Health Organisation* **12**, 371–400.

Hall, D.W. (1970) Handling and storage of food grains in tropical and subtropical areas, Food and Agriculture Organisation *Agricultural Development Paper* **90**. FAO, Rome.

Hall, D. and Marr, S.G. (1989) Microcapsules. In: *Insect Pheromones in Plant Protection*, (eds A.R. Jutsum, and R.F.S. Gordon), pp. 199–242. John Wiley and Sons, Chichester.

Hall, F.R. and Thacker, J.R.M. (1994) Effects of droplet size on the topical toxicity of two pyrethroids to cabbage looper *Trichoplusia ni*(Hubner). *Crop Protection* **13**, 225–229.

Hall, F.R., Chapple, A.C., Downer, R.A. Kirchner, L.M. and Thacker, J.R.M. (1993) Pesticide application as affected by spray modifiers. *Pesticide Science* **38**, 123–133.

Hall, F.R., Downer, R.A., Wolf, T.M. and Chapple, A.C. (1996) The 'Double Nozzle' – a new way of reducing drift and improving dose transfer? *Pesticide Formulations and Application Systems* Vol. **16**, (eds M.J. Hopkinson, H.M. Collins and G.R. Goss), pp.114–126. American Society for Testing and Materials, Special Technical Publication 1312. American Society for Testing and Materials, Philadelphia.

Hall, F.R., Downer, R.A., Cooper, J.A., Ebert, T.A. and Ferree, D.C. (1997) Changes in spray retention by apple leaves during a growing season. *HortScience* **32**, 858–860.

Hall, F.R., Downer, R.A. and Bagley, W.E. (1998) Laboratory and greenhouse studies with an invert suspension. In: *Proceedings of the 5th International Symposium on Adjuvants for Agrochemicals* Memphis. pp. 407–412.

Halmer, P. (1988) Technical and commercial aspects of seed pelleting and film coating. *British Crop Protection Council Monograph* **39**, 191–204.

Hankawa, Y. and Kohguchi, T. (1989) Improvement in distribution of BPMC deposits and classification of dust particles. *Journal of Pesticide Science* **14**, 443–452.

Hanna, A.D. and Nicol. J. (1954) Application of a systemic insecticide by trunk implantation to control a mealybug vector of the cocoa swollen shoot virus. *Nature* **169**, 120.

Harazny, J. (1976) Agricultural equipment for the M-15 aircraft. In: *Proceedings of the 5th Internatonal Agricultural Aviation Congress*, pp. 378–398. IAAC, The Hague.

Harris, A.G. (1988) The training and certification of pesticide users. *Aspects of Applied Biology* **18**, 311–315.

Harris, W. and Shaw, A.J. (1998) *Cotton Pesticides Guide 1998–99*. NSW Agriculture, Orange.

Harris, D.A., Johnson, K.S. and Ogilvy, J.M.E. (1991) A system for the treatment of waste water from agrochemical production and field use. In: *Proceedings of the Brighton Crop Protection Conference – Weeds* pp. 715–722.

Hart, C.A. (1979) Use of the scanning electron microscope and cathodoluminescence in studying the application of pesticides to plants. *Pesticide Science* **10**, 341–357.

Hart, C.A. and Young, B.W. (1987) Scanning electron microscopy and cathodoluminescence in the study of interactions between spray droplets and leaf surfaces *Aspects of Applied Biology* **14**, 127–140.

Hartley, G.S. and Graham-Bryce, I.J. (1980) *Physical Properties of Pesticide Behavior* – Academic press New York.

Hasegawa, K. (1995) Utilization of remote control helicopters in aerial application of pesticides. *Farming Japan* 35–41.

Hassan, S.A. (1977) Standardized technique for testing side-effects of pesticides on beneficial arthropods in the laboratory. *Zeitschrift für Planzenkrankheiten und Pflanzenenschufz* **84**, 158–163.

Hassan, S.A. (1985) Standard methods to test the side-effects of pesticides on natural enemies of insects and mites developed by the IOBC/WPRS working group 'Pesticides and Beneficial Organisms'. *European Plant Protection Organisation Bulletin* **15**, 214–255.

Heap, I.M. (1997) The occurrence of herbicide resistant weeds worldwide *Pesticide Science* **51**, 235–242.

Heath, D. Knott, R.D., Knowles, D.A. and Tadros, Th.F. (1984) Stabilization of aqueous pesticidal suspensions by graft copolymers and their subsequent weak flocculation by addition of free polymer. In: *Advances in Pesticide Formulation Technology* (ed H.B. Scher), American Chemical Society Symposium Series 254, pp. 11–22. American Chemical Society, Washington DC.

Heestermans, J.M.A.J. (1996) Testing of crop protection equipment in use in Europe. *European Plant Protection Organisation Bulletin* **26**, 43–45.

Heijne, B. (2000) Fruit tree spraying with coarse droplets and adjuvants. *Aspects of Applied Biology* **57**, 279–284.

Heijne, C.G. (1978) A study of the effect of disc speed and flow rate on the performance of the 'Micron Battleship'. In: *Proceedings of the British Crop Protection Council Conference Weeds* pp. 673–679. BCPC, Farnham.

Heijne, B. Hermon, E.A., van Smelt, J.H. and Huijmans, J.F.M. (1993) Biological evaluation of crop protection with tunnel sprayers with reduced emission to the environment in apple growing. In: *Proceedings of the ANPP BCPC 2nd International Symposium on Pesticide Application Techniques*, Strasbourg, Vol. 1, pp. 321–328. ANPP, Paris.

Heinkel, R., Fried, A and Lange, E. (2000) The effect of air injector nozzles on crop penetration and biological performance of fruit sprayers. *Aspects of Applied Biology* **57**, 301–307.

Herbst, A. and Ganzelmeier, H. (2000) Classification of sprayers according to drift risk – a German approach. *Aspects of Applied Biology* **57**, 35–40.

Hewitt, A. (1991) Assessment of rotary atomiser attachments for motorised knapsack mistblowers. *British Crop Protection Council Monograph* **46**, 271.

Hewitt, A.J. (1992) Droplet size spectra produced by the X15 stacked spinning-disc atomizer of the Ulvamast Mark II sprayer. *Crop Protection* **11**, 221–224.

Hewitt, A.J. (1993) Droplet size spectra produced by air-assisted atomizers. *Journal of Aerosol Science* **24**, 155–162.

Hewitt, A.J. and Meganasa, T. (1993) Droplet distribution densities of a pyrethroid insecticide within grass and maize canopies for the control of *Spodoptera exempta* larvae. *Crop Protection* **12**, 59–62.

Hewitt, A.J. and Valcore, D.L. (1998) The measurement, prediction and classification of agricultural sprays. *ASAE Paper* 98–1003.

Hewitt, A.J. and Valcore, D.L. (1999) Drift management: why reading and following pesticide labels is worthwhile. *ASAE Paper* AA99–004.

Hewitt, A.J., Robinson, A.G., Sanderson, R. and Huddleston, E.W. (1994a) Comparison of the droplet size spectra produced by rotary atomizers and hydraulic nozzles under simulated aerial application conditions. *Journal of Environmental Science and Health, Part B* **29**(4), 647–660.

Hewitt, A.J., Sanderson, R., Huddleston, E.W. and Ross, J.B. (1994b) Approaches to measuring atomization droplet size spectra using laser diffraction particle size analyzers in wind tunnels. In: *Proceedings of ILASS-94*. Seattle, WA.

Hewitt, A.J., Valcore, D.L. and Bryant, J.E. (1996) Spray Drift Task Force atomization droplet size spectra measurements. In: *Proceedings of ILASS–Americas* 96. San Francisco, CA.

Hewitt, A.J., Stern, A.J., Bagley, W.E. and Dexter, R. (1999a) The formation of a new ASTM E35.22 task group to address drift management adjuvants. In: *Proceedings of the 19th Symposium on Pesticide Formulations and Application Systems: Global Pest Control Formulations for the New Millennium* (eds R.S. Tann, J.D. Nlaewaja and A Viets), American Society for Testing and Materials, Special Technical Paper. 1373, pp. 135–148. American Society for Testing and Materials, Philadelphia.

Hewitt, A.J., Valcore, D.L. and Young, B. (1999b) Measuring spray characteristics for agricultural orchard airblast sprayers. In: *Proceedings ILASS-Americas*. Indianapolis, IN.

Hewitt, A.J., Valcore, D.L. and Barry, T. (2000) Analyses of equipment, meteorology and other factors affecting drift from applications of sprays by ground rig sprayers. *Pesticide Formulations and Applications Systems*: Vol. 20, American Society for Testing and Materials, Special Technical Paper 1400, (eds A.K. Viets, R.S. Tann and J.C. Mueninghoff), American Society for Testing and Materials, Philadelphia.

Hewlett, P.S. (1954) A micro-drop applicator and its use for the treatment of certain small insects with liquid insecticides. *Annals of Applied Biology* **41**, 45–64.

Hewlett, P.S. (1962) Toxicological studies on a beetle, *Alphitobius laevigatus* (F.) 1. Dose–response relations for topically applied solutions of four toxicants in a non-volatile oil. *Annals of Applied Biology* **50**, 335–349.

Hilder, V.A. and Boulter, D. (1999) Genetic engineering of crop plants for insect resistance – a critical review. *Crop Protection*, **18**, 177–191.

Hill, J.E. (1998) Public concerns over the use of transgenic plants in the protection of crops from pests and diseases and government responses. In: *Biotechnology in Crop Protection Facts and Fallacies. British Crop Protection Council Symposium* **71**, 57–65.

Hill, R.F. (1963) The Beagle Wallis autogyro. *Agricultural Aviation* **5**, 48–51.

Hill, R.F. and Johnstone, D.R. (1962) Tests with a rotary granule dispenser on a dragonfly helicopter. *Agricultural Aviation* **4**, 133–135.

Himel, C.M. (1969a) The fluorescent particle spray droplet tracer method. *Journal of Economic Entomology* **62**, 912–916.

Himel, C.M. (1969b) The optimum size for insecticide spray droplets. *Journal of Economic Entomology* **62**, 919–925.

Himel, C.M. (1974) Analytical methodology in ULV. *British Crop Protection Council Monograph* **11**, 112–119.

Hinze, J.O. and Milborn, H. (1950) Atomisation of liquids by means of a rotating cup. *Journal of Applied Mechanics* **17**, 145.

Hislop, E.C. (1987) Requirements for effective and efficient pesticide application. In: *Rational Pesticide Use* (eds K.J. Brent and R.K. Atkin), pp. 53–71. Cambridge University Press, Cambridge.

Hislop, E.C. (1989) Crop spraying under controlled conditions. *Aspects of Applied Biology* **21**, 119–120.

Hislop, E.C. (1991) Air-assisted crop spraying: an introductory review. *British Crop Protection Council Monograph* **46**, 3–14.

Hislop, E.C. Western, N.M. and Butler, R. (1995) Experimental air-assisted spraying of a maturing cereal crop under controlled conditions. *Crop Protection* **14**, 19–26.

Hobson, P.A., Miller, P.C.H. Walklate, P.J. Tuck, C.R. and Western, N. (1990) Spray drift from hydraulic nozzles: the use of a computer simulation model to examine factors influencing drift. In: *Proceedings of Ag Eng 90* Berlin.

Holden, A.V. and Bevan, D. (1979) *Control of Pine Beauty Moth by Fenitrothion in Scotland*. Forestry Commission, Edinburgh.

Holden, A.V. and Bevan, D. (Eds.) (1981) *Aerial Application of Insecticide against Pine Beauty Moth*. Forestry Commission, Edinburgh.

Holland, J.M. and Jepson, P.C. (1996) Droplet dynamics and initial field tests for microencapsulated pesticide formulations applied at ultra low volume using rotary atomisers for control of locusts and grasshoppers. *Pesticide Science* **48**, 125–134,

Holland, J.M., Jepson, P.C., Jones, E.C. and Turner, C. (1997) A comparison of spinning disc atomisers and flat fan pressure nozzles in terms of pesticide deposition and biological effects within cereal crops. *Crop Protection* **16**, 179–186.

Holloway, P.J. (1970) Surface factors affecting the wetting of leaves. *Pesticide Science* **1**, 156–163.

Holloway, P.J., Butler-Ellis, M.C., Webb, D.A., Western, N.M., Tuck, C.R., Hayes, A.L. and Miller, P.C.H. (2000) Effects of some agricultural tank-mix adjuvants on the deposition of aqueous sprays on foliage. *Crop Protection* **19**, 27–37.

Holownicki, R., Doruchowski, G. and Godyn, A. (1996a) Efficient spray deposition in the orchard using a tunnel sprayer with a new concept of air-jet emission. *Interna-*

tional Organisation for Biological Control/West Palearctic Regional Section Bulletin **19**, 284–288.

Holownicki, R., Goszcynski, W., Doruchowski, G. and Nowacka, H. (1996b) Comparison of apple scab control with traditional and tunnel sprayers at full and reduced chemical rates. *International Organisation for Biological Control/West Palearctic Regional Section Bulletin* **19**, 383–384.

Holownicki, R., Doruchowski, G. and Swiechowski, W. (1997) Low cost tunnel sprayer concept for minimizing pesticide waste and emission to environment. In: *Proceedings of the 5th International Symposium on Fruit, Nut and Vegetable Production Engineering*, Davis, California.

Hooper, G.H.S. and French, H. (1998) Comparison of measured fenitrothion deposits from ULV aerial locust control applications with those predicted by the FSCBG aerial spray model. *Crop Protection* **17**, 515–520.

Hooper, G.H.S. and Spurgin, P.A. (1995) Droplet size spectra produced by the atomization of a ULV formulation of fenitrothion with a Micronair AU5000 rotary atomizer. *Crop Protection* **14**, 27–30.

Hughes, K.L. and Frost, A.R. (1985) A review of agricultural spray metering. *Journal of Agricultural Engineering Research* **32**, 197–207.

Huijmans, J.F.M., Porskamp, H.A.J. and Heijne, B. (1993) Orchard tunnel sprayers with reduced emission to the environment. In: *Proceedings of the ANPP BCPC 2nd International Symposium on Pesticide Application Techniques*, Strasbourg, Vol. 1, pp. 297–304. ANPP, Paris.

Humphrey, S.T. (1998) Agrochemical formulations using natural lignin products. In: (ed. D.A. Knowles), *Chemistry and Technology of Agrochemical Formulations*, pp. 158–178. Kluwer, Dordrecht.

Humphries, A.W. and West, D. (1984) The terramatic boomsprayer – automation in agriculture. In: *Proceedings of the 7th Australian Weeds Conference*, pp. 36–40.

Hunt, G.M. and Baker, E.A. (1987) Application and fluoresence microscopy, autoradiography, and energy dispersive X-ray analysis to the study of pesticide deposits. *Aspects of Applied Biology* **14**, 113–126.

Hunter, D.M., Milner, R.J., Scanlan, J.C. and Spurgin, P.A. (1999) Aerial treatment of the migratory locust *Locusta migratoria* (L.) (Orthoptera: Acrididae) with *Metarhizium anisopliae* (Deuteromycotina: Hyphomycetes) in Australia. *Crop Protection* **18**, 699–704.

Hunter-Fujita, F.R., Entwistle, P.F., Evans, H.F. and Crook, N.E. (1998) *Insect Viruses and Pest Management.* John Wiley and Sons, Chichester.

Hussey, N.W. and Scopes, N.E.A. (1985) *Biological Pest Control – The Glasshouse Experience.* Blandford, Poole.

Hutton, S.A. (1998) Managing pesticide waste and packaging. *British Crop Protection Council Symposium* **70**, 179–186.

Inculet, I.I., Castle, G.S.P., Menzies, D.R. and Frank, R. (1981) Deposition studies with a novel form of electrostatic crop sprayer *Journal of Electrostatics* **10**, 65–72.

Interflug (1975) Method of calculating the optimum work parameters for agricultural flights. *Proceedings of the 5th International Agricultural Aviation Congress*, pp. 213–222. IAAC, The Hague.

Isaacs, R., Ulczynski, M., Wright, B., Gut, L.J. and Miller J.R. (1999) Performance of the microsprayer with application of pheromone-mediated control of insect pests. *Journal of Economic Entomology* **92**, 1157–1164.

Jagers op Akkerhuis, G.A.J.M., Axelsen, J.A. and Kjaer, C. (1998) Towards predicting pesticide deposition from plant physiology: a study in spring barley. *Pesticide Science* **53**, 252–262.

Jeffs, K.A. (1986) *Seed Treatment*, 2nd edn. BCPC, Farnham.

Jeffs, K.A. and Tuppen, R.J. (1986) The application of pesticides to seeds, In: *Seed Treatment* (ed. K.A. Jeffs), *British Crop Protection Council Monograph* 2nd edn. BCPC, Farnham.

Jegatheeswaran, P. (1978) Factors concerning the penetration and distribution of drops in low growing crops. *BCPC Monograph* **22**, 91–99.

Jensen, J.A., Taylor, J.W. and Pearce, G.W. (1969) A standard and rapid method for determining nozzle-tip abrasion. *Bulletin of the World Health Organisation* **41**, 937–940.

Jensen, P.K. (1999) Herbicide performance with low volume low-drift and air-inclusion nozzles. In: *Proceedings of the Brighton Crop Protection Conference – Weeds*, pp. 453–460. BCPC, Farnham.

Jensen, T., Lawler, S.P. and Dritz D.A. (1999) Effects of ultra-low volume pyrethrin, malathion, and permethrin on nontarget invertebrates sentinel mosquitoes, and mosquitofish in seasonally impounded wetlands. *Journal of the American Mosquito Control Association* **15**, 330–338.

Jepson, P.C., Cuthbertson, P.S., Thecker, J.R. and Bowie, M.H. (1987) A computerised droplet size analysis system and the measurement of non-target invertebrate exposure to pesticides. *Aspects of Applied Biology* **14**, 97–112.

Johnson, K.S. (1998) Waste management and disposal of agrochemicals. In: *Chemistry and Technology of Agrochemical Formulations* (ed. D.A. Knowles), pp. 418–434, Kluwer, Dordrecht.

Johnstone, D.R. (1972) A differential thermistor thermometer for measuring temperature gradients in the vicinity of the ground. *East African Agriculture and Forestry Journal* **37**, 300–7.

Johnstone, D.R. (1973a) Spreading and retention of agricultural sprays on foliage. in *Pesticide Formulations* (ed. W. Van Valkenburg), pp. 343–386, Marcel Dekker, New York.

Johnstone, D.R. (1973b) Insecticide concentration for ultra- low-volume crop spray applications, *Pesticide Science* **4**, 77–82.

Johnstone, D.R. (1981) Crop spraying by hovercraft, kites, RPV, ATV and other unconventional devices. *Outlook on Agriculture* **10**, 361–365.

Johnstone, D.R. (1985) Physics and meteorology, In: *Pesticide Application: Principles and Practice* (ed. P.T. Haskell), pp. 35–67. Oxford University Press, Oxford.

Johnstone, D.R. (1991) Variations in insecticide dose received by settled locusts – a computer model for ultra-low volume spraying. *Crop Protection* **10**, 183–194.

Johnstone, D.R. and Huntington, K.A. (1977) Deposition and drift of ULV and VLV insecticide sprays applied to cotton by hand applications in N. Nigeria. *Pesticide Science* **8**, 101–109.

Johnstone, D.R. and Johnstone, K.A. (1977) Aerial spraying of cotton in Swaziland. *PANS* **23**, 13–26.

Johnstone, D.R. and Matthews, G.A. (1965) Evaluation of swath pattern and droplet size provided by a boom and nozzle installation fitted to a Hiller UH-12 helicopter, *Agricultural Aviation*, **7**, 46–50.

Johnstone, D.R. and Watts, W.S. (1966) Physico-chemical assessments of cotton spraying by aeroplane and knapsack ground sprayer, Ilonga. *Tropical Pesticides Research Unit, Porton, Report* **323**.

Johnstone, D.R., Lee, C.W., Hill, R.F., Huntington, K.A. and Coles, J.S. (1971) Ultra-low volume spray gear for installation on an ultra light weight helicopter or autogyro, *Agricultural Aviation* **13**, 57–61.

Johnstone, D.R., Huntington, K.A. and King, W.J. (1972) Tests of ultra-low volume spray gear installed on the light weight Campbell Cricket autogyro. *Agricultural Aciation*, **14**, 82–86.

Johnstone, D.R., Huntington, K.A. and King, W.J. (1974) Micrometerological and operational factors affecting ultra-low volume spray applications of insecticides on to cotton and other crops. *Agricultural Meteorology* **13**, 39–57.

Johnstone, D.R., Huntington, K.A. and King, W.J. (1975) Development of hand spray equipment for applying fungicides to control *Cercospora* disease of groundnuts in Malawi, *Journal of Agricultural Engineering Research* **20**, 379–389.

Johnstone, D.R., Rendell, C.H. and Sutherland, J.A. (1977) The short-term fate of droplets of coarse aerosol size in ultra-low volume insecticide application onto a tropical field crop. *Journal of Aerosol Science* **8**, 395–407.

Johnstone, D.R., Cooper, J.F., Gledhill, J.A. and Jowah, P. (1982) Preliminary trials to examine the drift of charged spray droplets. In: *Proceeding of the British Crop Protection Council Conference – Weeds*, pp. 1025–1032. BCPC, Farnham.

Johnstone, D.R., Cooper, J.F., Dobson, H.M. and Turner, C.R. (1989a) The collection of aerosol droplets by resting tsetse flies, *Glossina morsitans* Westwood (Diptera: Glossinidae). *Bulletin of Entomological Research* **79**, 613–624.

Johnstone, D.R., Cooper, J.F., Flower, L.S., Harris, E.G., Smith, S.C. and Turner, C.R. (1989b) A means of applying mature aerosol drops to insects for screening biocidal activity. *Tropical Pest Management* **35**, 65–66.

Johnstone, D.R., Cooper, J.F., Casci, F. and Dobson, H.M. (1990) The interpretation of spray monitoring data in tsetse control operations using insecticidal aerosols applied from aircraft. *Atmospheric Environment* **24A**, 53–61.

Jollands, P. (1991) Evaluation of knapsack mistblowers for the control of coffee leaf rust in Papua New Guinea. *British Crop Protection Council Monograph* **46**, 177–184.

Jones, K.G., Morgan, N.G. and Cooke, A.T.K. (1974) Experimental application equipment. *Long Ashton Annual Report 1974*, p. 107.

Joyce, R.J.V. (1975) Sequential aerial spraying of cotton at ULV rates in the Sudan Gezira as a contribution to synchronized chemical application over the area occupied by the pest population. In: *Proceedings of the 5th International Agricultural Aviation Congress* pp. 47–54. IAAC, The Hague.

Joyce, R.J.V. (1977) Efficiency in pesticide application with special reference to insect pests of cotton in the Sudan Gezira, Paper presented at a seminar organised by the Rubber Research Institute of Malaysia and Agricultural Institute of Malaysia.

Kao, C., Rafatjah, H. and Kilta, S. (1972) Replacement of spray nozzle tips based on operational considerations. *Bulletin of the World Health Organisation* **46**, 493–501.

Kaul P., Schmidt, K. and Koch, H. (1996) Distribution quality of orchard sprayers. *European Plant Protection Organisation Bulletin* **26**, 69–77.

Keller, A.A. (1971) A note on the development of power-line markers for aerial crop spraying operations in Australia. *Agricultural Aviation* **13**, 81–86.

Kennedy, J.S., Ainsworth, M. and Toms, B.A. (1948) Laboratory studies on the spraying of locusts at rest and in flight. *Anti-Locust Bulletin* **2**.

Khoo, K.C., Ho, C.T., Ng, K.Y. and Lim, T.K. (1983) Pesticide application technology in perennial crops in Malaysia. In: *Pesticide Application Technology* (eds. G.S. Lim and S. Ramasamy), pp. 42–85. Malaysian Plant Protection Society, Kuala Lumpur.

Kinzer, R.E. (1977) *Development and evaluation of techniques for using Chrysopa carnea Stephens to control Heliothis spp. on cotton*. PhD dissertation, Texas A&M University College Station.

Kiritani, K. (1974) The effect of insecticides on natural enemies, particular emphasis on the use of selective and low rates of insecticides. Paper submitted to the *International Rice Research Conference*, April 1974. IRRI, Philippines.

Klein, H.H. (1961) Effects of fungicides, oil and fungicide–oil–water emulsions on the development of *Cercospora* leaf spot of bananas in the field. *Phytopathology*, **51**, 294–297.

Klock, J.W. (1956) An automatic molluscicide dispenser for use in flowing water. *Bulletin of the World Health Organisation* **14**, 639.

Knoche, M. (1994) Effect of droplet size and carrier volume on performance of foliage-applied herbicides. *Crop Protection* **13**, 163–178.

Knoche, M. and Bukovac, M.J. (2000) Spray application factors and plant growth regulator performance. V. Biological response as related to foliar uptake. *Aspects of Applied Biology* **57**, 257–265.

Knoche, M., Noga, G. and Lenz, F. (1992) Surfactant-induced phytotoxicity: evidence for interaction with epicuticular wax fine structure. *Crop Protection* **11**, 51–56.

Knollenburg, R.G. (1976) The use of the low power lasers in particle size spectrometry. *Proceedings in Application of Low Powered Lasers* **92**, 137–152.

Knowles D.A. (1998) (ed.) *Chemistry and Technology of Agrochemical Formulations*. Kluwer, Dordrecht.

Knutson, A. (1996) Evaluation of the Biosprayer for the application of *Trichogramma* to cotton. *Beltwide Cotton Conference* 788–792.

Ko, L.L. and Obendorf, S.K. (1997) Effect of starch on reducing the retention of methyl parathion by cotton and polyester fabric in agricultural protective clothing. *Journal of Environmental Science and Health* B**31**, 283–294.

Koch, H. (1996) Periodic inspection of air-assisted sprayers. *European Plant Protection Organisation Bulletin* **26**, 79–86.

Koo Y.M., Womac, A.R. and Eppstein J.A. (1994) Laser altimetry for low-flying aircraft. *Transactions of the ASAE* **37**, 395–400.

Krahmer, H. and Russell, P.E. (1994) General problems in glasshouse to field transfer of pesticide performance. *British Crop Protection Council Monograph* **59**, 3–16.

Krieger, R.I., Ross, J.H. and Thongsinthusak, T. (1992) Assessing human exposure to pesticides. *Review of Environmental Contamination and Toxicology* **128**, 1–15.

Krishnan, P., Valesco, A., Williams, T.H. and Kemble, L.J. (1989) Spray pattern displacement measurements of TK-SS 2.5 flood tip nozzles. *Transactions of the ASAE* **32**, 1173–1176.

Kruse, C.W., Hess, E.D. and Ludwik, G.F. (1949) The performance of liquid spray nozzles for aircraft insecticide application. *Journal of the National Malaria Society* **8**, 312–334.

Kuhlman, D.K. (1981) Fly-in technology for agricultural aircraft. *World of Agricultural Aviation* **8**, 112–117.

Kuiper, H.A. (1996) The role of toxicology in the evaluation of new agrochemicals. *Journal of Environmental Science and Health* B**31**, 353–363.

Kummel, K., Göhlich, H. and Westpal, O. (1991) Development of practice-oriented control test methods for orchard spray machines by means of a vertical test stand. *British Crop Protection Council Monograph* **46**, 27–33.

Lading, L. and Andersen, K. (1989) A covariance processor for velocity and size measurements. In: *Laser Anemometry in Fluid Mechanics* (eds R.J. Adrian and D.F.C. Durad), Vol. 4, pp. 454–472. Springer-Verlag, Berlin.

Lake, J.R. and Dix, A.J. (1985) Measurement of droplet size with a PMS optical array probe using an X-Y nozzle transporter. *Crop Protection* **4**, 464–472.

Lake, J.R., Frost, A.R. and Lockwood, A. (1980) The flight times of spray drops under the influence of gravitational, aerodynamic and electrostatic forces. *British Crop Protection Council Monograph* **24**, 119–125.

Lake, J.R., Green, R., Tofts, M. and Dix, A.J. (1982) The effect of an aerofoil on the penetration of charged spray into barley. In: *Proceedings of the British Crop Protection Council Conference – Weed*, pp. 1009–1016. BCPC, Farnham.

Lake, J.R. and Marchant, J.A. (1984) Wind tunnel experiments and a mathematical model of electrostatic deposition in barley. *Journal of Agricultural Engineering Research* **30**, 185–195.

Landers, A.J. (1988) Closed system spraying – the Dose 2000. *Aspects of Applied Biology* **18**, 361–369.

Lane, M.D. and Law, S.E. (1982) Transient charge transfer in living plants undergoing electrostatic spraying. *Transactions of the ASAE* **25**, 1148–1153, 1159.

Langenakens, J. and Pieters, M. (1997) The organisation and first results of the mandatory inspection of crop sprayers in Belgium. *Aspects of Applied Biology* **48**, 233–240.

Langenakens, J., De Moor, A., Taylor, W., Cooper, S. and Taylor, B. (2000) The effect of orifice wear on flat fan nozzle performance: using predictive and dynamic techniques to determine quality of liquid distribution. *Aspects of Applied Biology* **57**, 207–217.

Langmuir, I. and Blodgett, K.B. (1946) A mathematical investigation of water droplet trajectories. *USAF Technical Report* 5418.

Langewald, J., Ouambama, Z., Mamadou, A., Peveling, R., Stolz, I., Bateman, R., Attigon, S., Blanford, S., Arthurs, S. and Lomer, C. (1999) Comparison of an organophosphate insecticide with a mycoinsecticide for the control of *Oedaleus senegalensis* (Orthoptera: Acrididae) and other Sahelian grasshoppers at an operational scale. *Biocontrol Science and Technology* **9**, 199–214.

Last, A.J., Parkin, C.S. and Beresford, R.H. (1987) Low-cost digital image analysis for the evaluation of aerially applied pesticide deposits. In *Computers and Electronics in Agriculture*, Vol. 1, Elsevier, Amsterdam.

Law, S.E. (1978) Embedded-electrode electrostatic-induction spray-charging nozzle: theoretical and engineering design. *Transactions of the ASAE* **21**, 1096–1104.

Law, S.E. (1980) Droplet charging and electrostatic depositions of pesticide sprays – research and development in the USA. *British Crop Protection Council Monograph* **24**, 85–94.

Law, S.E. (1986) Charge and mass flux in the radial electric field of an evaporating charged water droplet: an experimental analysis. *IEEE (IAS) Conference Record* **1**, 434–9.

Law, S.E. (1987) Basic phenomena active in electrostatic pesticide spraying. In: *Rational Pesticide Use* (eds K.J. Brent and R.K. Atkin) pp. 83–105. Cambridge University Press, Cambridge.

Law, S.E. (1989) Electrical interaction occurring at electrostatic spraying targets. *Journal of Electrostatics* **23**, 145–156.

Law, S.E. and Bailey, A.G. (1984) Perturbations of charged-droplet trajectories caused by induced target corona-LDA analysis. *Institution of Electrical and Electronic Engineers Transactions on Industrial Applications* **IA-20**, 1613–1622.

Law, S.E. and Cooper, S.C. (1987) Induction charging characteristics of conductivity enhanced vegetable oil sprays. *Transactions of the ASAE* **30**, 75–79.

Law, S.E. and Lane, M.D. (1981) Electrostatic deposition of pesticide spray on to foliar targets of varying morphology. *Transactions of the ASAE* **24**, 1441–1445.

Law, S.E. and Lane, M.D. (1982) Electrostatic deposition of pesticide sprays onto ionizing targets: charge- and masstransfer analysis. *Institution of Electrical and Electronic Engineers Transactions on Industrial Applications* **IA-19**, 673–679.

Law, S.E., Merchant, J.A. and Bailey, A.G. (1985) Charged spray deposition characteristics within cereal crops. *Institution of Electrical and Electronic Engineers Transactions on Industrial Applications* **IA-21**, 685–693.

Learmount, J. (1994) Selection of houseflies (Diptera: Muscidae) with a pyrethroid space spray using a large-scale laboratory method. *Journal of Economic Entomology* **87**, 894–898.

Ledson, T.M., Giles, D.K., Huddlestone, E.W. and Ross, J. (1996) Droplet spectra from hydraulic nozzles using flow control. *ASAE Paper* 96–002.

Lee, C.W. (1974) 'Aerial ULV sprays for Cayman Island mosquitoes', *British Crop Protection Council Monograph* **11**, 190–196.

Lee, K.C. (1976). The design of ducted spreaders for the application of powders and granules. In: *Proceedings of the 5th International Agricultural Aviation Congress* pp. 328–335. IAAC, The Hague.

Lee, K.C. and Stephenson, J. (1969) The distribution of solid materials. In: *Proceedings of the 4th International Agricultural Aviation Congress*, pp. 203–211. IAAC, The Hague.

Lee, C.W., Parker, J.D., Philippon, B. and Baldry, D.A.T. (1975a) Prototype rapid release system for the aerial application of larvicide to control *Simulium damnosum* Theo. *PANS*, **21**, 92–102.

Lee, C.W., Pope, G.G., Kendrick, J.A., Bowles, G. and Wiggett, G. (1975b) Aerosol studies using an Aztec aircraft fitted with Micronair Equipment for tsetse fly control in Botswana. *Centre for Overseas Pest Research Miscellaneous Report* no. 118.

Lee, C.W., Parker, J.D., Baldry, D.A.T. and Molyneux, D.H. (1978) The experimental application of insecticides from a helicopter for the control of riverine populations of *Glossina tachinoides* in West Africa. II. Calibration of equipment and insecticide dispersal. *PANS* **24**, 404–422.

Lee, A.W., Miller, P.C.H. and Power, J.D. (2000) The application of pesticide sprays to tomato crops. *Aspects of Applied Biology* **57**, 383–390.

Lefebvre, A.H. (1989) *Atomization and Sprays*, Hemisphere, Kingston, Ontario.

Lehtinen, J.R., Adams, A.J., Lindquist, R.K., Hell, F.R. and Simmons, H.C. (1989) Use of an air-assisted electrostatic sprayer to increase pesticide efficiency in greenhouses. In: *Pesticide Formulations and Application Systems*, Vol. 9 (eds J.L. Hazen and D.A. Houde), American Society for Testing and Materials, Special Technical Publication 1036, pp. 165–178. American Society for Testing and Materials, Philadelphia.

Lello, E.R., Patel, M.N., Matthews, G.A. and Wright D.J. (1996) Application technology for entomopathogenic nematodes against foliar pests. *Crop Protection* **15**, 567–574.

Leonard, R., Rice, B., Dowley, L.J. and Ward, S. (2000) The effect of air assistance on spray deposition and biological effect in the control of *Phytophthora infestans* in potatoes. *Aspects of Applied Biology* **57**, 243–250.

Le Patourel, G.N.J. (1986) The effect of grain moisture content on the toxicity of a sorptive silica dust to four species of grain beetle. *Journal of Stored Product Research* **22**, 63–69.

Lewis, T. (1965) The effects of an artificial windbreak on the aerial distribution of flying insects. *Annals of Applied Biology* **55**, 503–512.

Lewis, T. (1972) Aerial baiting to control leaf cutting ants. *PANS* **18**, 71–74.

Lim, T.M. (1997) 'TreeJect' – a portable hand-operated pressurised tree injector for pesticide and nutrient application. *Planter* **73**, 353–356.

Lodeman, E.G. (1896) *The Spraying of Plants*. Macmillan, London.

Lofgren, C.S. (1970) Ultra low volume applications of concentrated insecticides in medical and veterinary entomology. *Annual Review of Entomology* **15**, 321–342.

Lofgren, C.S., Ford, H.R., Tonn, R.J. and Jatanasen, S. (1970a) The effectiveness of ultra-low volume applications of malathion at a rate of 6 US fluid ounces per acre in controlling *Aedes aegypti* in a large scale test at Nakhon Sawan, Thailand. *Bulletin of the World Health Organisation* **42**, 15–25.

Longley, M., Cilji, T., Jepson, P.C. and Sotherton, N.W. (1997) Measurement of pesticide spray drift deposition into field boundaries and hedgerows. 1. Summer applications. *Environmental Toxicology and Chemistry* **16**, 165–172.

Longley, M. and Sotherton, N.W. (1997) Measurement of pesticide spray drift deposition into field boundaries and hedgerows. 2. Autumn applications. *Environmental Toxicology and Chemistry* **16**, 173–178.

Lovro, I. (1975) Optimum method of agricultural airstrips planning. In: *Proceedings of the 5th International Agricultural Aviation Congress*, pp. 177–189. IACC, The Hague.

Lunchick, C., Nielsen, A.P. and Reinert, J.C. (1988) Engineering controls and protective clothing in the reduction of pesticide exposure to tractor drivers. In: *Performance of Protective Clothing: Second Symposium* (eds S.Z. Mansdorf, R. Sager and A.P. Nielsen). American Society for Testing and Materials, Special Technical Publication 989. pp. 605–610. American Society for Testing and Materials, Philadelphia.

McAuliffe, D. (1999) Flow control of lever operated knapsack sprayers with CFValve. *International Pest Control* **41**, 21–23, 28.

MacCuaig, R.D. and Watts, W.S. (1968) A simple technique for applying small measured quantities of insecticides to insects. *Bulletin of Entomological Research* **57**, 549–552.

McDaniel, R. and Himel, C.M. (1977) Standardization of field methods for determination of insecticide spray droplet size. *Office of Naval Research Report No. 2*, US Navy.

MacFarlane, R. and Matthews, G.A. (1978) Modifications in knapsack mistblower design to improve spray efficiency on tall tree crops. *British Crop Protection Council Monograph* **22**, 151–155.

MacIver, D.R. (1963) Mosquito coils, Part I. General description of coils, their formulation and manufacture. *Pyrethrum Post* **4**, 22–27.

MacIver, D.R. (1964a) 'Mosquito coils, Part II. Studies on the action of mosquito coil smoke on mosquitoes, *Pyrethrum Post* **7**, 7–14.

MacIver, D.R. (1964b) Mosquito coils, Part III. Factors influencing the release of pyrethrums from coils. *Pyrethrum Post* **7**, 15–17, 19.

Maas, W. (1971) *ULV Application and Formulation Techniques.* NV Philips Gloeilampenfabrieken, Eindhoven.

Mabbett, T.H. and Phelps, R.H. (1974) Low volume and ultra low volume spray systems in the humid tropics. In: *Proceedings of a Symposium on Protecting Horticultural Crops in the Caribbean*, University of the West Indies, St. Augustine, Trinidad. pp. 71–83.

Mabbett, T.H. and Phelps, R.H. (1976) Control of angular leaf spot of cucumber by low and ultra low volume spraying. *Tropical Agriculture (Trinidad)* **53**, 105–110.

Machado-Neto, J.G., Matno, T. and Matou, Y.K. (1998) Efficiency of safety measures applied to a manual knapsack sprayer for paraquat application to maize (*Zea mays* L.). *Archives of Environmental Contamination and Toxicology* **35**, 698–701.

Maczuga, S.A. and Mierzejewski, K.J. (1995) Droplet density and density effects of *Bacillus thuringiensis kurstaki* on gypsy moth (Lepidoptera: Lymantriidae) larvae. *Journal of Economic Entomology* **88**, 1376–1379.

MAFF (1998) *Pesticides: Code of practice for the safe use of pesticides on farms and holdings*, 2nd edn. Ministry of Agriculture, Fisheries and Food, Health and Safety Commission, Department for the Environment, Transport and the Regions, London.

Mahler, D.S. and Magnus, D.E. (1986) Hotwire technique for droplet measurements. *Liquid Pesticide Size Measurement Techniques.* American Society for Testing and Materials, Special Technical Paper 848, pp. 153–165. American Society for Testing and Materials, Philadelphia.

Manor, G., Geva, A. and Tabak, S. (1999) Precision and clean application. Poster presentation at the 14th International Plant Protection Congress, Jerusalem, Israel.

Marchant, J.A. (1980) Electrostatic spraying – some basic principles. In: *Proceedings of the British Crop Protection Council Conference – Weeds* pp. 987–997. BCPC, Farnham.

Marchant, J.A. (1985) An electrostatic spinning disc atomiser. *Transactions of the ASAE* **28**, 386–392.

Marchant, J.A. (1987) Modelling in spray engineering research. In: *Rational Pesticide Use* (eds K.J. Brent and R.K. Atkins), pp. 121–133. Cambridge University Press, Cambridge.

Marchant, J.A. and Frost, A.R. (1989) Simulation of the performance of state feedback controllers for an active boom suspension. *Journal of Agricultural Engineering Research* **43**, 77–91.

Marchant, J.A. and Green, R. (1982) An electrostatic charging system for hydraulic spray nozzles. *Journal of Agricultural Engineering Research* **27**, 309–319.

Marchant, J.A., Dix, A.J. and Wilson, J.M. (1985a) The electrostatic charging of spray produced by hydraulic nozzles. Part I. Theoretical analysis. *Journal of Agricultural Engineering Research*, **31**, 329–344.

Marchant, J.A., Dix, A.J. and Wilson, J.M. (1985b) The electrostatic charging of spray produced by hydraulic nozzles. Part II. Measurements. *Journal of Agricultural Engineering Research* **31**, 345–360.

Markin, G.P., Henderson, J.A. and Collins, H.L. (1972) Aerial application of micro encapsulated insecticide. *Agricultural Aviation* **14**, 70–75.

Marks, R.J. (1976) Field studies with the synthetic sex/pheromone and inhibitor of the red bollworm *Diparopsis castanea* Hmps (Ledpidoptera: Noctiuidae) in Malawi. *Bulletin of Entomological Research* **66**, 243–265.

Marrs, G.J. and Scher, H.B. (1990) Development and uses of microencapsulation. In: *Controlled Delivery of Crop Protection Agent* (ed. R.M. Wilkins), pp. 65–90. Taylor and Francis, London.

Martin, J.T. (1958) The comparison of high and low volume spraying techniques on fruit and ground crops. Levels and distribution of droplets. *Society of Chemistry and Industry Monograph* **2**, 55–68.

Mason, J.M., Matthews, G.A. and Wright, D.J. (1998a) Appraisal of spinning disc technology for the application of entomopathogenic nematodes. *Crop Protection* **17**, 453–461.

Mason, J.M., Matthews, G.A. and Wright, D.J. (1998b) Screening and selection of adjuvants for the spray application of entomopathogenic nematodes against a foliar pest. *Crop Protection* **17**, 463–470.

Matthews, G.A. (1966) Investigations of the chemical control of insect pests of cotton in Central Africa II. Tests of insecticides with larvae and adults. *Bulletin of Entomological Research* **57**, 77–91.

Matthews, G.A. (1969) Performance of some lever-operated knapsack sprayers. *Cotton Growing Review* **46**, 134–142.

Matthews, G.A. (1971) Ultra low volume spraying of cotton – a new application technique. *Cotton Handbook of Malawi*, Amendment 2/71. Agricultural Research Council of Malawi, Zomba.

Matthews, G.A. (1973) Ultra low volume spraying of cotton in Malawi. *Cotton Growing Review* **50**, 242–267.

Matthews, G.A. (1975) Determination of droplet size. *PANS* **21**, 213–225.

Matthews, G.A. (1977) CDA – controlled droplet application. *PANS* **23**, 387–394.

Matthews, G.A. (1984) *Pest Management*. Longman, Harlow.

Matthews, G.A. (1989a) *Cotton Insect Pests and their Management*. Longman, Harlow.

Matthews, G.A. (1989b) Electrostatic spraying of pesticides: a review. *Crop Protection* **8**, 3–15.

Matthews, G.A. (1990) Changes in application technique used by the small scale cotton farmer in Africa. *Tropical Pest Management* **36**, 166–172.

Matthews, G.A. (1994) A comparison of laboratory and field spray systems. *British Crop Protection Council Monograph* **59**, 161–171.

Matthews, G.A. (1996a) The importance of scouting in cotton IPM. *Crop Protection* **15**, 369–374.

Matthews, G.A. (1996b) Pedestrian sprayers: equipment for ultra-low and very-low volume spraying. *European Plant Protection Organisation Bulletin* **26**, 103–111.

Matthews, G.A. (1997) Pesticide application in plasticulture. *Pesticide Outlook* **8**, 16–20.

Matthews, G.A. (2000) A review of the use of air in atomisation of sprays, dispersion of droplets downwind and collection on crop foliage. *Aspects of Applied Biology* **57**, 21–27.

Matthews, G.A. and Clayphon, J.E. (1973) Safety precautions for pesticide application in the tropics. *PANS* **19**, 1–12.

Matthews, G.A. and Johnstone, D.R. (1968) Aircraft and tractor spray deposits on irrigated cotton. *Cotton Growing Review* **45**, 207–218.

Matthews, G.A. and Mowlam, M.D. (1974) Some aspects of the biology of cotton insects and their control with ULV spraying in Malawi. *British Crop Protection Council Monograph* **11**, 44–52.

Matthews, G.A. and Thomas, N. (2000) Working towards more efficient application of pesticides. *Pest Management Science* **56**, 1–3.

Matthews, G.A. and Tunstall, J.P. (1966) Field trials comparing carbaryl DDT and endosulfan on cotton in Central Africa. *Cotton Growing Review* **43**, 230–239.

Matthews, G.A. and Tunstall, J.P. (1968) Scouting for pests and the timing of spray applications. *Cotton Growing Review* **45**, 115–127.

Matthews, G.A., Higgins, A.E.H. and Thornhill, E.W. (1969) Suggested techniques for assessing the durability of lever-operated knapsack sprayers. *Cotton Growing Review* **46**, 143–8.

Matthews, G.A., de Schaetzen, N. and John, F. (1992) Tunnel spraying: an environment-favourable approach for pesticide application in orchards. *Pesticide Outlook* **3**, 13–16.
Maude, R.B. and Suett, D.L. (1986) Application of pesticide to brassica seeds using a film coating technique. In: *Proceedings of the Brighton Crop Protection Conference – Pests and Diseases*, pp. 237–242. BCPC, Farnham.
Mawer, C.J. and Miller, P.C.H. (1989) Effect of roll angle and nozzle spray pattern on the uniformity of spray volume distribution below a boom. *Crop Protection* **8**, 217–222.
May, K.R. (1945) The cascade impactor: an instrument for sampling coarse aerosols. *Journal of Scientific Instruments* **22**, 187–195.
May, K.R. (1950) The measurement of airborne droplets by the magnesium oxide method. *Journal of Scientific Instruments* **27**, 128–130.
May, M.J. (1991) Early studies on spray drift, deposit manipulation and weed control in sugar beet with two air-assisted boom sprayers. *British Crop Protection Council Monograph* **46**, 89–96.
May, K.R. and Clifford, R. (1967) The impaction of aerosol particles, spheres, ribbons and discs. *Annals of Occupational Hygiene* **10**, 83–95.
Mboob, S.S. (1975) Preliminary assessment of the effectiveness of two droplet sizes of insecticide for the control of glasshouse whitefly *Trialeurodes vaporariorum* (Westwood). *Plant Pathology* **24**, 158–162.
Menn, J.J. and Hall, F.R. (1999) Biopesticides – present stautus and future prospects. In: *Methods in Biotechnology 5, Biopesticides: Use and Delivery*, (eds F.R. Hall and J.J. Menn), pp. 1–10. Humana Press, Totowa.
Menzies, D.R., Fisher R.W. and Neff, A.E. (1976) Wear of hollow cone nozzles by suspensions of wettable powders. *Canadian Agricultural Engineering* **18**, 14–15.
Mercer, P.C. (1974) Disease control of groundnuts in Malawi. *World Crops* **26**, 162–164.
Mercer, P.C. (1976) Ultra low volume spraying of fungicides for the control of *Cercospora* leaf spot of groundnuts in Malawi. *PANS* **22**, 57–60.
Merritt, C.R. (1998) The commercialisation of transgenic crops – the Bt experience. In: *Biotechnology in Crop Protection, Facts and Fallacies. British Crop Protection Council Symposium* **71**, 79–86.
Metcalfe, R.J., Shaw, M.W. and Russell, P.E. (1998) Factors affecting strength of selection for resistance to DMI fungicides in *Septoria tritici*. In: *Proceedings of the Brighton Crop Protection Conference – Pests and Diseases*, pp. 535–540. BCPC, Farnham.
Mickle, R.E. (1987) A review of models for ULV spray scenarios. In: *Proceedings of a Symposium on Aerial Application of Pesticides in Forestry*. (ed. G.W. Green), AFA-TN-18, pp. 179–188. National Research Council of Canada, Ottawa.
Mickle, R.E. (1990) Canadian laser mapping technique for aerial spraying. *ASAE Paper* AA90–004.
Middleton, M.R. (1973) Assessment of performance of the 'Rotostat' seed treater. In: *Proceedings of the 7th British Insecticide and Fungicide Conference*, pp. 357–364. BCPC, Farnham.
Miller, A.W.D. and Chadwick, P.R. (1963) Swath marking in aerial spraying. *Agricultural Aviation*, **5**, 114–120.
Miller, D.R. and Stoughton, T.E. (2000) Response of spray draft from aerial applications at a forest edge to atmospheric stability. *Agricultural and Forest Meterorology* **100**, 49–58.
Miller, P.C.H. (1988) Engineering aspects of spray drift control. *Aspects of Applied Biology* **17**, 377–384.
Miller, P.C.H. (1989) The field performance of electrostatically charged hydraulic nozzle sprayers. In: *Proceedings of the 4th European Weed Research Society Mediterranean Symposium*, pp. 324–333. EWRS.
Miller, P.C.H. (1999) Factors affecting the risk of drift into field boundaries. In: *Pro-*

ceedings of the Brighton Crop Protection Conference – Weeds pp. 439–446. BCPC, Farnham.

Miller, P.C.H. and Combellack, J.H. (1997) The performance of an air/liquid nozzle system suitable for applying herbicides in a spatially selective manner. In: *Proceedings of the 1st International Symposium on Precision Agriculture*, pp. 651–659. Warwick, UK.

Miller, P.C.H. and Hobson, P.A. (1991) Methods of creating air-assisting flows for use in conjunction with crop sprayers. *British Crop Protection Council Monograph* **46**, 35–43.

Miller, P.C.H. and Smith, R.W. (1997) The effects of forward speed on the drift from boom sprayers. In: *Proceedings of the Brighton Crop Protection Conference – Weeds*, pp. 399–406. BCPC, Farnham.

Miller, P.C.H., Tuck, C.R., Gilbert, A.J. and Bell, G.J. (1991) The performance characteristics of a twin-fluid nozzle sprayer. *British Crop Protection Council Monograph* **46**, 97–106.

Miller, P.C.H., Hislop, E.C., Parkin, C.S., Matthews, G.A. and Gilbert, A.J. (1993) The classification of spray generator performance based on wind tunnel assessments of spray drift. *ANPP–British Crop Protection Council Second Symposium on Pesticide Application Techniques*, Strasbourg, pp. 109–116. ANPP, Paris.

Miller, P.C.H., Paice, M.E.R. and Ganderton, A.D. (1997) Methods of controlling sprayer output for spatially variable herbicide application. In: *Proceedings of the Brighton Crop Protection Conference – Weeds*, pp. 641–644. BCPC, Farnham.

Miller, P.C.H., Lane, A.G., Walklate, P.J. and Richardson, G.M. (2000) The effect of plant structure on the drift of pesticides at field boundaries. *Aspects of Applied Biology* **57**, 75–82.

Mills-Thomas, G., Piggott, A., Robinson, T. and Watt, A. (1998) Packaging innovation: the development of refillables and its impact on the agrochemical sector. *British Crop Protection Council Symposium* **70**, 137–140.

Mokeba, M.L., Salt, D.W., Lee, B.E. and Ford, M.G. (1998) Computer modelling of the meterological and spraying parameters that influence the aerial dispersion of agrochemical sprays. *International Journal of Biometeorology* **41**, 194–199.

Morel, M. (1985) Field trials with the Girojet. *British Crop Protection Council Monograph* **28**, 107–112.

Morel, M. (1997) Manufacturer's data for 'Turbocoll'. Technoma, Epernay, France.

Morgan, C.P. and Pinniger, D.B. (1987) A sprayer for small scale application of insecticides to test surfaces. *Laboratory Practice* **36**, 68–70.

Morgan, N.G. (1964) Gallons per acre of sprayed area – an alternative standard term for the spraying of plantation crops. *World Crops* **16**, 64–65.

Moreira, J.F., Santos, J. and Glass, C.R. (2000) A comparative study of the potential dermal exposure of an operator with two pesticide application techniques in a tomato greenhouse. *Aspects of Applied Biology* **57**, 399–404.

Morgan, N.G. (1981) Minimizing pesticide waste in orchard spraying. *Outlook on Agriculture* **110**, 342–344.

Morrison, R.K., Rose, M. and Penn, S. (1998) The effect of extended immersion in agitated liquid carriers on the viability of two entomophagous insects. *Southwestern Entomologist* **23**, 131–135.

Morton, N. (1977) The wind leaf orientation and ULV spray coverage on cotton plants, *FAO Plant Protection Bulletin* **25**, 29–37.

Moss, S.R., Clarke, J.H., Blair, A.M., Culley, T.N., Read, M.A., Ryan, P.J. and Turner, M. (1999) The occurrence of herbicide-resistant grass-weeds in the United Kingdom and a new system for designating resistance in screening assays. In: *Proceedings of the Brighton Crop Protection Conference – Weeds*, pp. 179–185. BCPC, Farnham.

Mount, G.A. (1970) Optimum droplet size for adult mosquito control with space sprays or aerosols of insecticides. *Mosquito News* **30**, 70–75.

Mount, G.A. (1998) A critical review of ultralow volume aerosols of insecticide applied with vehicle-mounted generators for adult mosquito control. *Journal of the American Mosquito Control Association* **14**, 305–334.
Mowlam, M.D. (1974) Aerial spraying research on cotton in Malawi. *Agricultural Aviation*, **16**, 36–40.
Mowlam, M.D., Nyirenda, G.K.C. and Tunstall, J.P. (1975) Ultra-low volume application of water-based formulations of insecticides to cotton. *Cotton Growing Review* **52**, 360–370.
Mulqueen, P.J., Patterson, E.S. and Smith, G.W. (1990) Recent developments in suspoemulsions. *Pesticide Science* **29**, 451–465.
Munro, H.A.U. (1961) *Manual of Fumigation for Insect Control.* FAO Agricultural Study No. 56. FAO, Rome.
Munthali, D.C. (1976) Studies on choice of applicator that enables integration of chemical control with *Encarsia formosa* (Gahan) for the control of *Trialeurodes vaporariarum* (Westwood) in glasshouses. Unpublished MSc. thesis.
Munthali, D.C. (1984) Biological efficiency of small dicofol droplets against *Tetranychus urticae* (Koch) eggs, larvae and protonymphs. *Crop Protection* **3**, 327–334.
Munthali, D.C. and Scopes, N.E.A. (1982) A technique for studying the biological efficiency of small droplets of pesticide solutions and a consideration of its implications. *Pesticide Science* **13**, 60–63.
Munthali, D.C. and Wyatt, I.J. (1986) Factors affecting the biological efficiency of small pesticide droplets against *Tetranychus urticae* eggs. *Pesticide Science* **17**, 155–164.
Murray, R.A., Ridout, M.S. and Cross, J.V. (2000) The use of ranked set sampling in spray deposit assessment. *Aspects of Applied Biology* **57**, 141–146.
Nation, H.J. (1982) The dynamic behaviour of field sprayer booms. *Journal of Agricultural Engineering Research* **27**, 61–70.
Nation, H.J. (1985) Construction and evaluation of a universal links spray boom suspension. *National Institute for Agricultural Engineering Divisional Note* DN 1299, NIAE, Silsoe.
Needham, P.H. and Devonshire, A.L. (1973) A technique for applying small drops of insecticide solution to *Myzus persicae* (Sulz). *Pesticide Science* **4**, 107–111.
Nelson, C., Laughlin, J., Kim, C., Rigakis, K., Mastura, R. and Scholten, L. (1992) Laundering as decontamination of apparel fabrics: residues of pesticides from six chemical classes. *Archives of Environmental Contamination and Toxicology* **23**, 85–90.
Nettleton, D.M. (1991) Field experiences with an 'Airtec' twin fluid spraying system. *British Crop Protection Council Monograph* **46**, 107–112.
Ng, K.Y. and Chong, Y.W. (1982) Studies on some aspects of the control of *Darna trima* in oil palm *Malay Peninsula Agricultural Association Year Book, 1982*, pp. 39–45.
Nielsen S.L. and Kirknel, E. (1992) Deposition and distribution pattern of fogs generated by thermal pulse jet applicators throughout glasshouses. *Danish Journal of Plant and Soil Science* **96**, 71–80.
Nillson, U. and Gripwall, E. (1999) Influence of application technique on the viability of the biological control agents *Verticillium lecanii* and *Steinernema feltiae*. *Crop Protection* **18**, 53–59.
Nordbo, E. (1992) Effects of nozzle size, travel speed and air assistance on artificial vertical and horizontal targets in laboratory experiments. *Crop Protection* **11**, 272–277.
Nordbo, E., Steensen, J.K. and Kirknel, E. (1995) Deposition and efficiency of herbicide sprays in sugar beet with twin-fluid, low drift and conventional hydraulic nozzles. *Crop Protection* **14**, 237–240.
Nordby, A. and Skuterud, R. (1975) The effects of boom height, working pressure and wind speed on spray drift. *Weed Research* **4**, 385–395.
Nyirenda, G.K.C. (1991) Effect of swath width, time of application and height on the efficacy of very-low-volume (VLV) water-based insecticides on cotton in Malawi. *Crop Protection* **10**, 111–116.

Ogborn, J. (1977) Herbicides and hoe farmers. *World Crops* **29**, 9–11.

Ogg, A.G. (1986) Applying herbicides in irrigation water – a review. *Crop Protection* **5**, 53–65.

Oke, T.R. (1978) *Boundary Layer Climates*. Methuen, London.

Oliver-Bellasis H.R. and Southerton, N.W. (1986) The cereals and gamebirds research project – an independent viewpoint. In: *Proceedings of the Brighton Crop Protection Conference – Pests and Diseases* **3**, 225–228. BCPC, Farnham.

Omar, D. and Matthews, G.A. (1987) Biological efficiency of spray droplets of permethrin ULV against diamond back moth. *Aspects of Applied Biology* **14**, 173–179.

Onstad, D.W. (1987) Calculation of economic-injury levels and economic thresholds for pest management. *Journal of Economic Entomology* **80**, 299–303.

Orson, J.H. (1998) The role and practical management of buffer strips in crop production. In: *Proceedings of the Brighton Crop Protection Conference – Pests and Diseases*, pp. 951–958. BCPC, Farnham.

Orson, J.H. and Harris, D. (1997) The technical and financial impact of herbicide resistant black-grass (*Alopecurus myosuroides*) on individual farm businesses in England. In: *Proceedings of the Brighton Crop Protection Conference – Weeds*, pp. 1127–1132. BCPC, Farnham.

O'Sullivan, J.A. (1988) Verification of passive and active versions of a mathematical model of a pendulum spray boom suspension. *Journal of Agricultural Engineering Research* **40**, 89–101.

Ozkan, H.E. and Ackerman, K.D. (1992) An automated computerized spray pattern analysis system. *Applied Engineering in Agriculture* **8**, 325–331.

Ozkan, H.E., Miralles, A., Sinfort, C., Zhu, H. and Fox, R.D. (1997) Shields to reduce spray drift. *Journal of Agricultural Engineering Research* **67**, 311–322.

Paice M.E.R., Miller P.C.H. and Bodle, J.D. (1995) An experimental sprayer for the spatially selective application of herbicides. *Journal of Agricultural Engineering Research* **60**, 107–116.

Palmer, J.E. (1970) Dry formulations for selective weed control in cereals. *British Crop Protection Council Monograph* **2**, 114–23.

Park, P.O., Gledhill, J.A., Alsop, N. and Lee, C.W. (1972) A large-scale scheme for the eradication of *Glossina morsitans* Westw. in the Western Province of Zambia by aerial ultra low volume application of endosulfan. *Bulletin of Entomological Research* **6**, 373–84.

Parker, J.D., Collings, B.G.P. and Kahumbura, J.M. (1971) Preliminary tests of a suction spray nozzle for use with aircraft spraying systems, *Agricultural Aviation* **13**, 24–28.

Parkin, C.S. (1979) Rotor induced air movements and their effects on droplet dispersa. *Journal of the Royal Aeronautical Society* (May Issue) 183–187.

Parkin, C.S. (1987) Factors affecting the movement of spray droplets above a forest canopy. In: *Proceedings of a Symposium on the Aerial Application of Pesticides in Forestry* (ed. G.W. Green), pp. 69–79. National Research Council of Canada, Ottawa.

Parkin, C.S. and Newman, B.W. (1977) The bifoil atomiser – a variable geometry venturi atomiser. *Agricultural Aviation*, **18**, 15–23.

Parkin, C.S. and Siddiqui, H.A. (1990) Measurement of drop spectra from rotary cage aerial atomizers. *Crop Protection* **9**, 33–38.

Parkin, C.S. and Spillman, J.J. (1980) The use of wing-tip sails on a spraying aircraft to reduce the amount of material carried off-target by a crosswind. *Journal of Agricultural Engineering Research* **25**, 65–74.

Parkin, C.S. and Wyatt, J.C. (1982) The determination of flight-lane separations for the aerial application of herbicides. *Crop Protection* **1**, 309–321.

Parkin C.S. and Young, P.R. (2000) Measurement and computational fluid dynamic simulations of the capture of drops by spray drift samplers. *Aspects of Applied Biology* **57**, 113–120.

Parkin, C.S., and Wyatt, J.C. and Warner, R. (1980) The measurement of drop spectra in agricultural sprays using a Particle Measuring Systems Optical Array Spectrometer. *British Crop Protection Council Monograph* **24**, 241–249.

Parkin, C.S., Outram, I., Last, A.J. and Thomas, A.P.W. (1985) An evaluation of aerially applied ULV and LV sprays using a double spray system and two tracers. *British Crop Protection Council Monograph* **28**, 211–220.

Parkin, C.S., Brun, L.O. and Suckling, D.M. (1992) Spray deposition in relation to endosulfan resistance in coffee berry borer (*Hypothenemus hampei*) (Coleoptera: Scolytidae) in New Caledonia. *Crop Protection* **11**, 213–220.

Parkin, C.S., Gilbert, A.J., Southcombe, E.S.E. and Marshall, C.J. (1994) British Crop Protection Council scheme for the classification of pesticide application equipment by hazard. *Crop Protection* **13**, 281–285.

Parnell, F.R., King, H.E. and Ruston, D.F. (1949) Jassid resistance and hairiness of the cotton plant. *Bulletin of Entomological Research* **39**, 539–575.

Parnell, M.A., King, W.J., Jones, K.A., Ketunuti, U. and Wetchakit, D. (1999) A comparison of motorised knapsack mistblower medium volume application, and spinning disk, very low volume application, of *Helicoverpa armigera* nuclear polyhedrosis virus on cotton in Thailand. *Crop Protection* **18**, 259–265.

Parr, W.J., Gould, H.J., Jessop, N.H. and Ludlam, F.A.B. (1976) Progress towards a biological control programme for glasshouse whitefly (*Trialeurodes vaporariorum*) on tomatoes. *Annals of Applied Biology* **83**, 349–363.

Parrott, W.N., Jenkins, J.N. and Smith, D.B. (1973) Frego bract cotton and normal bract cotton: how morphology affects control of boll weevils by insecticides. *Journal of Economic Entomology* **66**, 222–225.

Pasian, C., Taylor, R.A.J., McMahon, R.W. and Lindquist, R.K. (2000) New method of acephate application to potted plants for control of *Aphis gossypii, Frankliniella occidentalis* and *Bemisia tabaci. Crop Protection* **19**, 263–271.

Pasquill, F. (1974) *Atmospheric Diffusion*, 2nd edn. Ellis Horwood, Chichester.

Patterson, D.E. (1963) The effect of saturn yellow concentration on the visual inspection and photographic recording of spray droplets. *Journal of Agricultural Engineering Research* **8**, 342–344.

Patty, L., Real, B. and Gril, J.J. (1997) The use of grassed buffer strips to remove pesticide, nitrate and soluble phosphorus compounds from run-off water. *Pesticide Science* **49**, 243–251.

Payne, N.J. (1994) Spray deposits from aerial insecticide spray simulant applications to a coniferous plantation in low and high wind speeds. *Crop Protection* **13**, 121–126.

Payne, N.J. (1998) Developments in aerial pesticide application methods in forestry. *Crop Protection* **17**, 171–180.

Payne, N.J. and Schaefer, G.W. (1986) An experimental quantification of coarse aerosol deposition on wheat. *Atomisation and Spray Technology* **2**, 45–71.

Payne, N.J., Helson, V.B., Sunderam, K.M.S. and Fleming, R.A. (1988) Estimating buffer zone widths for pesticide application. *Pesticide Science* **24**, 147–161.

Payne, N.J., Cunningham, J.C., Curry, R.D., Brown, K.W. and Mickle, R.E. (1996) Spray deposits in a mature oak canopy from aerial applications of nuclear polyhedrosis virus and *Bacillus thuringiensis* to control gypsy moth, *Lymantria dispar* (L.). *Crop Protection* **15**, 425–431.

Payne, N., Retnakaran, A. and Cadogan, B. (1997) Development and evaluation of a method for the design of spray applications: aerial tebufenozide applications to control eastern spruce budworm, *Choristoneura fumiferana* (Clem.). *Crop Protection* **16**, 285–290.

Pearce, S.C. (1976) *Field Experimentation with Fruit Trees and other Perennial Plants.* CAB Technical Communication No. 23 Rev. CAB International, Wallingford.

Pearson, A.J.A. and Masheder, S. (1969) Bi-flon. *Agricultural Aviation* **11**, 126–129.

Pedibhotla, V.K., Hall, F.R. and Holmsen, J. (1999) Deposit characteristics and toxicity of fipronil formulations for tobacco budworm (Lepidoptera: Noctuidae) control on cotton. *Crop Protection* **18**, 493–499.

Pedigo, L.P., Hutchins, S.H. and Hegley, L.G. (1986) Economic injury levels in theory and practice. *Annual Review of Entomology* **31**, 341–68.
Peregrine, D.J. (1973) Toxic baits for the control of pest animals. *PANS* **19**, 523–533.
Pereira, J.L. (1970) Aerial spraying of coffee for disease control. *Agricultural Aviation* **12**, 17–20.
Pereira, J.L. (1972) Modifications of a hydraulic sprayer (Hardi) for improved coffee spraying. *East African Agricultural and Forestry Journal* **37**, 318–324.
Pereira, J.L. and Mapother, H.R. (1972) Overhead application of fungicide for the control of coffee berry disease. *Experimental Agriculture* **8**, 117–122.
Perich, M.J., Tidwell, M.A., Williams, D.C., Sardelis, M.R., Pena, C.J., Mandeville, D. and Boobar, L.R. (1990) Comparison of ground and aerial ultra-low-volume applications of malathion against *Aedes aegypti* in Santo Domingo, Dominican Republic. *Journal of the American Mosquito Control Association* **6**, 1–6.
Perrin, R.M. (2000) Improving insecticides through encapsulation. *Pesticide Outlook* **11**, 68–71.
Perrin, R.M., Wege, P.J., Foster, D.G., Bartley, M.R., Browde, J., Rehmbe, A. and Scher, H. (1998) Fast release capsules: a new formulation of lambda-cyhalothrin. In: *Proceedings of the Brighton Crop Protection Conference – Pests and Diseases*, pp. 43–56. BCPC, Farnham.
Pettifor, M.J. (1988) Practical problems in achieving recommended placement of granules. *British Crop Protection Council Monograph* **39**, 333–5.
Peveling, R., Attignon, S., Langewald, J. and Ouambama, Z. (1999a) An assessment of the impact of biological and chemical grasshopper control agents on ground-dwelling arthropods in Niger, based on presence/absence sampling. *Crop Protection* **18**, 323–339.
Peveling, R., Rafanomezantsoa, J-J., Razafinirina, R. Tovonkery, R. and Zafimaniry, G. (1999b) Environmental impact of the locust control agents fenitrothion, fenitrothion-esfenvalerate and triflumuron on terrestrial arthropods in Madagascar. *Crop Protection* **18**, 659–676.
Phillips, F.T. and Gillham, E.M. (1971) Persistence to rain washing of DDT wettable powders. *Pesticide Science* **2**, 97–100.
Phillips, F.T. and Gillham, E.M. (1973) A comparison of sticker performance against rain washing of microcapsules on leaf surfaces. *Pesticole Science* **4**, 51–57.
Phillips, F.T. and Lewis, T. (1973) Current trends in the development of baits against leaf cutting ants. *PANS* **19**, 483–487.
Phillips, J.C. and Miller, P.C.H. (1999) Field and wind tunnel measurement of the airborne spray volume downwind of a single flat-fan nozzle. *Journal of Agricultural Engineering Research* **72**, 161–170.
Pickler, R.A. (1976) Expanding the applications of firefighting aircraft to improve cost effectiveness. In: *Proceedings of the 5th International Agricultural Aviation Congress*, pp. 79–85. IAAC, The Hague.
Piggott, S and Matthews G.A. (1999) Air induction nozzles: a solution to spray drift? *International Pest Control* **41**, 24–28.
Planas, S. and Pons, L. (1991) Practical considerations concerning pesticide application in intensive apple and pear orchards. *British Crop Protection Council Monograph* **46**, 45–52.
Polles, S.G. and Vinson, S.B. (1969) Effect of droplet size on persistence of ULV malathion and comparison of toxicity of ULV and EC malathion to tobacco budworm larvae. *Journal of Economic Entomology* **62**, 89–94.
Polon, J.A. (1973) Formulation of pesticidal dusts, wettable powders and granules, In: *Pesticide Formulations* (ed. W. Van Valkenburg), pp. 143–234. Marcel Dekker, New York.
Potter, C. (1941) A laboratory spraying apparatus and technique for investigating the action of contact insecticides with some notes on suitable test insects. *Annals of Applied Biology* **28**, 142–169.
Potter, C. (1952) An improved laboratory apparatus for applying direct sprays and

surface films, with data on the electrostatic charge on atomized sprays. *Annals of Applied Biology* **39**, 1–28.
Potts, S.F. (1958) *Concentrated Spray Equipment, Mixtures and Application Methods*. Dorland Books, New Jersey.
Potts, S.F. and German, P. (1950) Concentrated sprays for application by mistblowers for control of forest, shade and fruit tree pests, *Conneticut Agricultural Experimental Station Circular* no. 177.
Povey, G.S., Clayton, J.C. and Bals, T.E. (1996) A portable motorised axial fan air-assisted CDA sprayer: A new approach to insect and disease control in coffee. In: *Proceedings of the Brighton Crop Protection Conference – Pests and Diseases*, pp. 367–372. BCPC, Farnham.
Powell, E.S., Orson, J.H. and Miller, P.C.H. (1999) Guidelines on nozzle selection for conventional sprayers. In: *Proceedings of the Brighton Crop Protection Conference – Weeds*, pp. 467–472. BCPC, Farnham.
Power, J.D. and Miller, P.C.H. (1998) The performance of agricultural chemical induction hoppers and their container rinsing systems. *British Crop Protection Council Symposium* **70**, 113–120.
Prandtl, L. (1952) *The Essentials of Fluid Dynamics*. Blackie, London.
Price, R.E., Bateman, R.P., Brown, H.D., Butler, E.T. and Muller, E.J. (1997) Aerial spray trials against brown locust (*Locustana pardalina*, Walker) nymphs in South Africa using oil-based formulations of *Metarhizium flavoviride*. *Crop Protection* **16**, 341–351.
Quantick, H.A. (1985a) *Aviation in Crop Protection, Pollution and Insect Control*. Collins, London.
Quantick, H.A. (1985b) *Handbook for Agricultural Pilots*. Collins, London.
Radwald, J.D., Shibuya, F., McRae, N. and Platzer, E.G. (1986) A simple, inexpensive portable apparatus for injecting experimental chemicals in drip-irrigation systems. *Journal of Nematology* **18**, 423–425.
Raheja, A.K. (1976) ULV spraying for cowpea in Northern Nigeria. *PANS* **22**, 327–332.
Raisigl, V., Felber, H., Siegfried, W. and Krebs, C. (1991) Comparison of different mistblowers and volume rates for orchard spraying. *British Crop Protection Council Monograph* **46**, 185–196.
Randall, A.P. (1975) Application technology. In: Prebble, M.L. (Ed) *Aerial Control of Forest Insects in Canada*, pp. 34–55. Department of the Environment, Ottawa.
Randall, J.M. (1971) The relationship between air volume and pressure on spray distribution in fruit trees. *Journal of Agricultural Engineering Research* **16**, 1–31.
Ras, M.C.D. (1986) Effective application of chemical agents for the control of pests and diseases by the correct use of spraying machinery in orchards and vineyards. *Deciduous Fruit Grower* (Nov), 467–477.
Rayleigh, Lord (1882) On the equilibrium of liquid conducting masses charged with electricity. *Philosophical Magazine* **14**, 184–186.
Reddy, K.N. and Locke, M.A. (1996) Imazaquin spray retention, foliar washoff and runoff losses under simulated rainfall. *Pesticide Science* **48**, 179–187.
Reed, D.K., Reed, G.L. and Creighton, C.S. (1986) Introduction of entomogenous nematodes into trickle irrigation systems to control striped cucumber beetle (Coleoptera: Chrysomelidae). *Journal of Economic Entomology* **79**, 1330–1333.
Reed, W. (1972) Uses and abuses of unsprayed controls in spraying trials. *Cotton Growing Review* **49**, 67–72.
Reed, W., Davies, J.C. and Green, S. (1985) Field experimentation. In: *Pesticide Application: Principles and Practice* (ed. P.T. Haskell), pp. 153–174. Oxford.
Reichard, D.L. (1990) A system for producing various sizes, numbers and frequencies of uniform-size drops. *Transactions of the ASAE* **33**, 1767–1770.
Reichard, D.L., Retzer, H.J., Liljedahl, L.A. and Hall, F.R. (1977) Spray droplet size distributions delivered by airblast orchard sprayers. *Transactions of the ASAE* **20**, 232–242.

Reichard, D.L., Alm, S.R. and Hall, F.R. (1987) Equipment for studying effects of spray drop size, distribution and dosage on pest control. *Journal of Economic Entomology* **80**, 540–543.
Reichard, D.L., Ozkan, H.E. and Fox R.D. (1990) Nozzle wear rates and test procedures *Transactions of the ASAE* **34**, 2309–2316.
Reichard D.L., Cooper, J.A., Bukovac, M.J. and Fox, R.D. (1998) Using a videographic system to assess spray droplet impaction and reflection from leaf and artificial surfaces. *Pesticide Science* **53**, 291–299.
Rew, L.J., Miller, P.C.H. and Paice, M.E.R. (1997) The importance of patch mapping resolution for sprayer control. *Aspects of Applied Biology* **48**, 49–55.
Rice, B. (1967) Spray distribution from ground crop sprayers. *Journal of Agricultural Engineering Research* **12**, 173.
Rice, B. (1970) A review of procedures and techniques for testing ground crop sprayers. *British Crop Protection Council Monograph* **2**, 1–11.
Richardson, E.G. (1960) Introduction and historical survey. In: *Aerodynamic Capture of Particles* (ed. E.G. Richardson), p. 1. Pergamon Press, Oxford.
Richardson, L.F. (1920) The supply of energy from and to atmospheric eddies. *Proceedings of the Royal Society of London, Series A* **97**, 354–373.
Richardson, R.G., Combellack, J.H. and Andrew, L. (1986) Evaluation of a spray nozzle patternator. *Crop Protection* **5**, 8–11.
Richardson, G.M., Walklate, P.J., Goss, J.V. and Murray, R.A. (2000) Field performance of axial fan orchard sprayers. *Aspects of Applied Biology* **57**, 321–327.
Rickett, F.E. and Chadwick, P.R. (1972) Measurements of temperature and degradation of pyrethroids in two thermal fogging machines, the Swingfog and Tifa. *Pesticide Science* **3**, 263–269.
Ridley, S.M., Elliott, A.C., Young, M. and Youle, D. (1998) High-throughput screening as a tool for agrochemical discovery: automated synthesis, compound input, assay design and process management. *Pesticide Science* **54**, 327–337.
Riley, C.M., Wiesner, C.J. and Ernst, W.R. (1989) Off-target deposition and drift of aerially applied agricultural sprays. *Pesticide Science* **26**, 159–166.
Ripper, W.E. (1955) Application methods for crop protection chemicals. *Annals of Applied Biology* **42**, 288–324.
Robinson, T.H. (1985) A novel sprayer for treatment of small plots. *Aspects of Applied Biology* **10**, 523–528.
Robinson, R.C. and Rutherford, S.J. (1988) A hand-held precision spot applicator for granular insecticide. *British Crop Protection Council Monograph* **39**, 341–347.
Roff, M.W. (1994) A novel lighting system for the measurement of dermal exposure using fluorescent dye and an image processor. *Annals of Occupational Hygiene* **38**, 903–919.
Roger, T.A., Edelson, J.V., Bogle, C.R. and McCrate, S. (1989) Insecticide applicator and insect control using a drip irrigation delivery system. *Pesticide Science* **25**, 231–40.
Rogers, R.B. and Ford, R.J. (1985) The windproof sprayer: its progress and prospects. *Agricultural Engineering* **66**, 11–13.
Rowland, M., Pye, B., Stribley, M., Hackett, B., Denholm, I. and Sawicki, R.M. (1990) Laboratory apparatus and techniques for the rearing and insecticidal treatment of whitefly *Bemisia tabaci* (Homoptera: Aleyrodidae) under simulated field conditions. *Bulletin of Entomological Research* **80**, 209–216.
Rozendaal, J.A. (1989) Self protection and vector control with insecticide-treated mosquito nets. Unpublished document WHO/VBC/89.96. World Health Organisation, Geneva.
Rutherford, I. (1976) An ADAS survey on the utilisation and performance of field crops sprayers. In: *Proceedings of the British Crop Protection Council Conference – Weeds* **2**, 357–361. BCPC, Farnham.
Rutherford, S.J. (1985) Development of equipment and techniques to enable precise

and safe application of pesticides in small and large plot, problem trial situations. *Aspects of Applied Biology* **10**, 487–497.

Rutherford, I., Bell, G.J., Freer, J.B.S., Herrington, P.J. and Miller, P.C.H. (1989) An evaluation of chemical application systems. In: *Proceedings of the Brighton Crop Protection Conference – Weeds* **3**, 601–613. BCPC, Farnham.

Salyani, M. (1999) A technique for stabilizing droplet spots on oil sensitive paper. *Transactions of the ASAE* **42**, 45–48.

Salyani, M. and Fox, R.D. (1999) Evaluation of spray quality by oil and water-sensitive papers. *Transactions of the ASAE* **42**, 37–43.

Salyani, M and Serdynski, J. (1989) Development of a spray sensor for deposition assessment. *ASAE Paper* 89–1526.

Sanderson, R. Hewitt, A.J., Huddlestone, E.W. and Ross, J.B. (1994) Polymer and invert emulsifying oil effects upon droplet size spectra of sprays. *Journal of Environmental Science and Health – B Pesticides, Food Contaminants and Agricultural Wastes* **29**, 815–829.

Sanderson, R., Hewitt, A.J., Huddlestone, E.W. and Ross, J.B. (1997) Relative drift potential and droplet size of aerially applied propanil formulations. *Crop Protection* **16**, 717–721.

Sarker, K.U., Parkin, C.S. and Williams, B.J. (1997) Effect of liquid properties on the potential for spray drift from flat-fan hydraulic nozzles. In: *Proceeding of the Brighton Crop Protection Conference – Weeds*, pp. 555–560. BCPC, Farnham.

Sawicki, R.M., Rowland, M.W., Byrne, F.J., Pye, B.J., Devonshire, A.L., Denholm, I., Hackett B.S., Stribley, M.F. and Dittrich, V. (1989) The tobacco whitefly field control simulator – a bridge between laboratory assays and field evaluation. *Aspects of Applied Biology* **21**, 121–122.

Scher, H.B. (1984) Advances in pesticide formulation technology: an overview. In: *Advances in Pesticide Formulation Technology*, ACS Symposium series 254. pp. 1–7. American Chemical Society, Washington DC.

Scher, H., Rodson, H. and Lee, K.S. (1998) Microencapsulation of pesticides by interfacial polymerisation utilising isocyanate or ammoplast chemistry. *Pesticide Science* **54**, 394–400.

Scherer, T.F., Hofman, V.L. and Albus W.L. (1998). Evaluating the performance of a pivot attached sprayer. *ASAE Paper* 98–2075.

Schmidt, K. (1996) Application of plant protection products by helicopter in Germany (Legislation, requirements, guidelines, use in vineyards and forests, drift). *European Plant Protection Organisation Bulletin* **26**, 117–122.

Schmidt, R.R. (1997) IRAC classification of herbicides according to mode of action. In: *Proceedings of the Brighton Crop Protection Conference – Weeds*, pp. 1133–1140. BCPC, Farnham.

Schuster, W. (1974) Selection standards for agricultural aircraft. *Agricultural Aviation*, **16**, 98–104.

Seaman, D. (1990) Trends in the formulation of pesticides – an overview. *Pesticide Science* **29**, 437–449.

Shang, H. and Li, W. (1990) Study on property of Model 50E electrostatic sprayer and its application. Poster presented at Shenyang Conference on Recent Developments in the Field of Pesticides and their Application to Pest Control, 8–12 October 1990, organised by UNIDO.

Sharkey, A.J., Salt, D.W. and Ford, M.G. (1987) Use of simulation to define an optimum deposit for control of a sedentary pest. *Aspects of Applied Biology* **14**, 267–280.

Sharp, R.B. (1984) Comparison of drift from charged and uncharged hydraulic nozzles. In: *Proceedings of the Brighton Crop Protection Conference – Pests and Diseases*, pp. 1027–1032. BCPC, Farnham.

Shaw, A. Lin, Y.J. and Pfeil, E. (1996) Effect of abrasion on protective properties of polyester and cotton/polyester blend fabrics. *Bulletin of Environmental Contamination and Toxicology* **56**, 935–941.

Shaw, A., Nomula, R. and Patel, B. (1999) Protective clothing and application controls for pesticide application in India: a field study. In: *Performance of protective clothing: Issues and priorities for the 21st Century* (eds. C.N. Nelson and N.W. Henry), Vol. 7, American Society for Testing and Materials, Special Technical Publication 1386. American Society for Testing and Materials, Philadelphia.

Shemanchuk, J.A., Spooner, R.W. and Golsteyn, L.R. (1990) Evaluation of permethrin for the protection of cattle against mosquitoes (Diptera: Culicidae), applied as electrostatic and low pressure sprays. *Pesticide Science* **32**, 253–258.

Siegfried W. and Holliger E (1996) *Application Technology in Fruit Growing and Viticulture.* Swiss Federal Research Station Field Trial Report, December 1996, Wadenswil.

Silvie, P., Le Gall, P. and Sognigbe, B. (1993) Evaluation of a virus–insecticide combination for cotton pest control in Togo. *Crop Protection* **12**, 591–596.

Simard, A.J. (1976) Air tanker utilization and wildland fire management. In: *Proceedings of the 5th International Agricultural Aviation Congress* pp. 7–8. IAAC, The Hague.

Skoog, F.E., Hanson, T.L., Higgins, A.H. and Onsager, J.A. (1976) Ultra low volume spraying: systems evaluation and meteorological data analysis. *Transactions of the ASAE* **19**, 2–6.

Skurray, S.J. (1985) Cereal trials and development of plot machinery as seen by a machinery designer. *Aspects of Applied Biology* **10**, 65–73.

Skuterud, R., Bjugstad, N., Tyldum, A. and Semb Torresen, K. (1998) Effect of herbicides applied at different times of the day. *Crop Protection* **17**, 41–46.

Slater, A.E., Hardisty, J.A. and Yong, K. (1985) Small plot hydraulic sprayer and granule applicator. Aspects of Applied Biology **10**, 477–486.

Slatter, R., Stewart, D.C., Martin, R. and White, A.W.C. (1981) An evaluation of Pestigas, B.B. – a new system for applying synthetic pyrethroids as space sprays using pressurised carbon dioxide. *International Pest Control* **23**, 162–164.

Slaughter, D.C., Giles, D.K. and Tauzer, C. (1999) Precision offset spray system for roadway shoulder weed control. *Journal of Transportation Engineering* **125**, 1–8.

Smith, A.K. (1984) A model to aid decision making in choosing a suitable crop spraying system. In: *Proceedings of the Brighton Crop Protection Conference – Pests and Diseases*, pp. 621–626. BCPC, Farnham.

Smith, R.F. (1970) Pesticides: their use and limitations in pest management. In: *Concepts of Pest Management* (eds. R.L. Rabb and F.E. Guthrie), North Carolina State University, Raleigh.

Smith, R.K. (1988) The 'Electrodyn' sprayer: matching the technology to contrasting areas of smallholder agriculture. *Chemistry and Industry* **6**, 196–199.

Smith, R.K. (1998) Crop protection product (CPP) container management – international experience and manufacturer's strategy. *British Crop Protection Council Monograph* **70**, 21–30.

Snyder, H.E., Senser, D.W. and Lefebvre, A.H. (1989) Mean drop sizes from fan spray atomizers. *Transactions of the ASME* **111**, 342–347.

Sopp, P.I. and Palmer, A. (1990) Deposition patterns and biological effectiveness of spray deposits on pot plants applied by the Ulvafan and three prototype electrostatic sprayers. *Crop Protection* **9**, 295–302.

Sopp, P.I., Gillespie, A.T. and Palmer, A. (1989) Application of *Verticillium lecanii* for the control of *Aphis gossypii* by a low-volume electrostatic rotary atomiser and a high-volume hydraulic sprayer. *Entomophaga* **34**, 417–428.

Southcombe, E.S.E., Miller, P.C.H. Ganzelmeier, H. Miralles, A. and Hewitt, A.J. (1997) The international (BCPC) spray classification system including a drift potential factor. In: *Proceedings of the Brighton Crop Protection Conference – Weeds*, pp. 371–380. BCPC, Farnham.

Southwood, T.R.E. (1977) Entomology and mankind. *American Scientist* **65**, 30–39.

Spackmann, E. and Barrie, I.A. (1982) Spray occasions determined from meterological data during the 1980–81 season at 15 stations in the UK and comparison with 1971–80. *Meterological Office Agricultural Memorandum No. 933*. Meteorological Office, Bracknell.

Spielberger, U. and Abdurrahim, U. (1971) Pilot trial of discriminative aerial application of persistent dieldrin deposits to eradicate *Glossina morsitans submorsitans* in the Anchau and Ikara forest reserves, Nigeria. In: *International Scientific Committee for Trypanosomiasis Research 13th Meeting*, Lagos, Sept. 1971. OAU/STRC publication no. 105, pp. 271–291. Scientific, Technical and Research Commission of the Organisation of African Unity.

Spillman, J.J. (1976) Optimum droplet sizes for spraying against flying targets. *Agricultural Aviation* **17**, 28–32.

Spillman, J.J. (1980) The SB-1 aircraft spreader. In: *Proceedings of the 6th IAAC Congress*, Turin, Italy, 22–26 September. pp. 109–114. IAAC, The Hague.

Spillman, J.J. (1982) Atomizers for the aerial application of herbicides – ideal and available. *Crop Protection* 473–482.

Spillman, J.J. (1984) Evaporation from freely falling droplets. *Journal of the Royal Aeronautical Society* 181–184.

Spillman, J.J. (1987) Improvements required in spray droplet formation to improve application technology. In: *Proceedings of a Symposium on Aerial Application of Pesticides in Forestry, (ed. A.W. Green, Ottawa, October 1987*. National Research Council of Canada, Ottawa.

Spillman, J.J. and Sanderson R. (1983) Design and development of a disc-windmill atomiser for aerial applications. *European Plant Protections Organisation Bulletin* **13**, 265–270.

Staniland, L.N. (1959) Fluorescent tracer techniques for the study of spray and dust deposits. *Journal of Agricultural Engineering Research* **4**, 110–125.

Stent, C.J., Taylor, W.A. and Shaw, G.B. (1981) A method for the production of uniformly sized drops using electrostatic dispersion. *Tropical Pest Management* **27**, 262–264.

Stephenson, J. (1976) Assessment of the performance penalty of spreaders on agricultural aircraft. *Proceedings of the 5th International Agricultural Aviation Congress*, 22–25 September 1975, Stoneleigh, pp. 321–327. IAAC, The Hague.

Stern, V.M. (1966) Significance of the economic threshold in integrated pest control. In: *Proceedings of FAO Symposium on Integrated Pest Control*, Vol. 2, pp. 41–56. Rome, Italy, 11–15 October. Food and Agriculture Organisation.

Stinner, R.E., Ridgway, R.L. and Morrison, R.K. (1974) Longevity, fecundity, and searching ability of *Trichogramma pretiosum* reared by three methods. *Environmental Entomology* **3**, 558–560.

Suett, D.L. (1987) Influence of treatment of soil with carbofuran on the subsequent performance of insecticides against cabbage root fly and carrot fly. *Crop Protection* **6**, 371–378.

Sundaram, K.M.S. (1991) Spray deposit patterns and persistence of diflubenzuron in some terrestrial components of a forest ecosystem after application at three volume rates under field and laboratory conditions. *Pesticide Science* **32**, 275–293.

Sundaram, K.M.S., Milliken, R.L. and Sundaram, A. (1988) Assessment of canopy and ground deposit of fenitrothion following aerial and ground application in a Northern Ontario forest. *Pesticide Science* **25**, 59–69.

Sundaram, A., Sundaram, K.M.S. and Leung, J.W. (1991) Droplet spreading and penetration of non-aqueous pesticide formulations and spray diluents in Kromekote cards. *Transactions of the ASAE* **34**, 1941–1951.

Sundaram, A., Sundaram, K.M.S. and Sloane, L. (1997a) Spray deposition and persistence of a *Bacillus thuringiensis* formulation (Foray(™) 76B) on spruce foliage, following aerial application over a Northern Ontario forest. *Journal of Environmental Science and Health – Part B Pesticides, Food Contaminants and Agricultural Wastes*, **31**, 763–813.

Sundaram, A., Sundaram, K.M.S., Nott, R., Curry, J. and Sloane, L. (1997b) Persistence of *Bacillus thuringiensis* deposits in oak foliage, after aerial application of Foray(™) 48B using rotary and pressure atomizers. *Journal of Environmental Science and Health – Part B Pesticides, Food Contaminants and Agricultural Wastes* **32**, 71–105.

Sutherland, J.A., King, W.J., Dobson, H.M., Ingram, W.R. Attique, M.R. and Sanjani, W. (1990) Effect of application volume and method on spray operator contamination by insecticide during cotton spraying. *Crop Protection* **9**, 343–350.

Sutton, O.G. (1953) *Micrometeorology*. McGraw-Hill, New York.

Sutton, T.B. and Unrath, C.R. (1984) Evaluation of the tree-row-volume concept with density adjustments in relation to spray deposits in apple trees. *Plant Disease* **68**, 480–484.

Svensson, S.A. (1994) Orchard spraying – deposition and air velocities as affected by air-jet qualities. *Acta Horticulturae* **372**, 83–91.

Swaine, G. (1954) A simple and inexpensive insecticide duster. *East African Agricultural Journal* **20**, 38–39.

Swift, D.L. and Proctor, D.F. (1982) Human respiratory deposition of particles during aronasal breathing. *Atmospheric Environment* **16**, 2279–2282.

Swithenbank, J., Beer, J.M., Taylor, D.S., Abbot, D. and McReath, G.C. (1977) Laser diagnostic technique for the measurement of droplet and particle size distribution. *Progress in Astronautics and Aeronautics* **53**.

Symmons, P.M., Boase, C.J., Clayton, J.S. and Gorta, M. (1989) Controlling desert locust nymphs with bendiocarb applied by a vehicle-mounted spinning disc sprayer. *Crop Protection* **8**, 324–331.

Tadros, Th. F. (1989) Colloidal aspects of pesticidal and pharmaceutical formulations – an overview. *Pesticide Science* **26**, 51–77.

Tadros, Th. F. (1998) Suspension concentrates. In: *Pesticide Formulation: Recent Developments and their Applications in Developing Countries* (eds W. Van Valkenburg, B. Sugavanum and S.K. Khetan), pp. 169–202. UNIDO, Vienna/New Age International, New Dehli.

Takenaga, T. (1971) Pesticide applicator used by the granular boom type blow head. *Japan Agricultural Research Quarterly* **6**, 92–96.

Taylor, W.A. and Andersen, P.G. (1991) Enhancing conventional hydraulic nozzle use with the Twin Spray System. *British Crop Protection Council Monograph* **46**, 125–136.

Taylor, W.A. and Cooper, S.E. (1998) Validation of a decontamination method for arable crop sprayers following use with the sulphonyl urea herbicide – amidosulfuron. *British Crop Protection Council Monograph* **70** 189–194.

Taylor, W.A., Andersen, P.G. and Cooper, S. (1989) The use of air assistance in a field crop sprayer to reduce drift and modify drop trajectories. In: *Proceedings of the Brighton Crop Protection Conference – Weeds* **3**, p. 631. BCPC, Farnham.

Taylor, W.A., Pretty, S. and Oliver, R.W. (1988) Some observations quantifying and locating spray remnants within an agricultural field crop sprayer. *Aspects of Applied Biology*, **18**, 385–393.

Taylor, W.A., Cooper, S.E. and Miller, P.C.H. (1999) An appraisal of nozzles and sprayers abilities to meet regulatory demands for reduced airborne drift and downwind fallout from arable crop spraying. In: *Proceedings of the Brighton Crop Protection Conference – Weeds*, pp. 447–452. BCPC, Farnham.

Teske, M.E. and Barry, J.W. (1993) Parametric sensitivity in aerial application. *Transactions of the ASAE* **36**, 27–33.

Teske, M.E. and Thistle, H.W. (1998) Aircraft selection for optimized operation. *ASAE Paper* 98–1012.

Teske, M.E., Barry, J.W. and Ekblad, R.B. (1990) Canopy penetration and deposition in a Douglas fir seed orchard. *ASAE Paper* 90–1019.

Teske, M.E., Bowers, J.F. Rafferty, J.E. and Barry, J.W. (1993) FSCBG: An aerial spray dispersion model for predicting the fate of spray released behind aircraft. *Environmental Toxicology and Chemistry* **12**, 453–464.
Teske, M.E., Thistle, H.W., Barry, J.W. and Eau, B. (1998) Simulation of boom length effects for drift minimization. *Transactions of the ASAE* **41**, 545–551.
Teske, M.E., Thistle, H.W. and Hewitt, A.J. (2000) Conversion of droplet size distributions from PMS Optical Array Probe to Malvern Laser Diffraction. *Proceedings of ICLASS 2000*, Pasadena, CA.
Thacker, J.R.M. and Hall, F.R. (1991) The effects of drop size and formulation upon the spread of pesticide droplets impacting on water-sensitive paper. *Journal of Environmental Sciences and Health* **B26**, 631–651.
Thacker, J.P.M., Young, R.D.F., Allen, I. and Curtis, D.J. (1994) The effect of a polymeric adjuvant on the off-target movement of a pesticide spray. In: *Proceedings of the Brighton Crop Protection Conference – Pests and Diseases*, pp. 1361–1366. BCPC, Farnham.
Thacker, J.R.M., Young, R.D.F., Stevenson, S. and Curtis, D.J. (1995) Microdroplet application to determine the effects of a change in pesticide droplet size on the topical toxicity of chlorpyrifos and deltamethrin to the aphid *Myzus persicae* (Hemiptera: Aphididae) and the ground beetle *Nebria brevicollis* (Coleoptera: Carabidae). *Journal of Economic Entomology* **88**, 1560–1565.
Thomas, M.R., Wood, S.N. and Lomer, C.J. (1995) Biological control of locusts and grasshoppers using fungal pathogen: the importance of secondary cycling. *Proceedings of the Royal Society of London, Series B* **259**, 265–270.
Thompson, A.R., Suett, D.L., Percivall, A.L., Pradbury, C.E., Edmonds, G.H. and Farmer, C.J. (1981) Precision equipment for sowing and treating small plots with granular insecticide *Annual Report, National Vegetable Research Station for 1980*, pp. 34–35 NVRS, Wellesbourne.
Thompson, A.R. and Wheatley, G.A. (1985) Seeder- and planter-mounted attachments for precision evaluation of granule treatments on small plots. *Aspects of Applied Biology* **10**, 465–476.
Thomson, L. (1998) *A Guide to Agricultural Spray Adjuvants used in the United States*. 5th edn, Thomson Publications, Fresno.
Thornhill, E.W. (1974a) The adaptation of a stainless steel container for use as a compression sprayer. *PANS* **20**, 241–245.
Thornhill, E.W. (1982) A summary of methods of testing pesticide application equipment. *Tropical Pest Management* **28**, 335–346.
Thornhill, E.W. (1984) Maintenance and repair of spraying equipment. *Tropical Pest Management* **30**, 266–281.
Thornhill, E.W. (1985) A guide to knapsack sprayer selection. *Tropical Pest Management* **31**, 11–17.
Thornhill, E.W., Matthews G.A. and Clayton, J.C. (1995) Potential operator exposure to herbicides: a comparison between knapsack and CDA hand sprayers. In: *Proceedings of the Brighton Crop Protection Conference – Weeds*, pp. 507–512. BCPC, Farnham.
Thornhill, E.W., Matthews G.A. and Clayton, J.C. (1996) Potential operator exposure to insecticides: a comparison between knapsack and CDA spinning disc sprayers. In: *Proceedings of the Brighton Crop Protection Conference – Pests and Diseases*, pp. 1175–1180. BCPC, Farnham.
Thornton, M.E. and Kibble-White, R. (1974) Apparatus used for spray nozzle evaluation at the Weed Research Organisation. *PANS* **20**, 465–475.
Threadgill, E.D., Eisenhauer, D.E., Young, J.R. and Bar-Yosef, B. (1990) Chemigation. In: *Management of Farm Irrigation Systems* (eds G.J. Hoffman T.A. Howell and K.H. Solomon, pp. 749–780. ASAE, St Joseph.
Tingle, C.C.D. (1996) Sprayed barriers of diflubenzuron for the control of the migratory locust (*Locusta migratoria capito* (Sauss.)) [Orthoptera: Acridiae] in Mada-

gascar: short-term impact on relative abundance of terrestrial non-target invertebrates. *Crop Protection* **15**, 579–592.
Tomlin, C.D.S. (1997) *The Pesticide Manual*, 11th edn., British Crop Protection Council, Farnham.
Trayford, R.S. and Taylor, P.A. (1976) Development of the tetrahedron spreader, *Proceedings of the 5th International Agricultural Aviation Congress*, 22–25 September 1975, Stoneleigh, pp. 294–300. IAAC, The Hague.
Trayford, R.S. and Welch, L.W. (1977) Aerial spraying: a simulation of factors influencing the distribution and recovery of liquid droplets. *Journal of Agricultural Engineering Research* **22**, 183–196.
Tsuji, K. (1990) Preparation of microencapsulated insecticides and their release mechanisms. In: *Controlled Delivery of Crop Protection Agents* (ed. R.M. Wilkins), pp. 99–122. Taylor and Francis, London.
Tsuji, K. (1993) Microcapsules of insecticides for household use. *Pesticide Outlook* **4**, 36.
Tu, Y.Q. (1990) Implications of biological and pesticidal behaviour in chemical control of pests. In: *Proceedings of International Seminar – Recent Developments in the Field of Pesticides and their Application to Pest Control in China and other Developing Countries of the Region*. pp. 155–163. UNIDO, Vienna.
Tuck, C.R., Butler Ellis, M.C. and Miller, P.C.H. (1997) Techniques for measurement of droplet size and velocity distributions in agricultural sprays. *Crop Protection* **16**, 619–628.
Tunstall, J.P. and Matthews, G.A. (1961) Cotton insect control recommendations for 1961–1962 in the Federation of Rhodesia and Nyasaland, *Rhodesia Agricultural Journal* **58**, 289–299.
Tunstall, J.P. and Matthews, G.A. (1965) Contamination hazards in using knapsack sprayers. *Cotton Growing Review* **42**, 193–196.
Tunstall, J.P. and Matthews, G.A. (1966) Large-scale spraying trials for the control of cotton insect pests in Central Africa. *Cotton Growing Review* **43**, 121–139.
Tunstall, J.P., Matthews, G.A. and Rhodes, A.A.K. (1961) A modified knapsack sprayer for the application of insecticide to cotton. *Cotton Growing Review* **38**, 22–26.
Tunstall, J.P., Matthews, G.A. and Rhodes, A.A.K. (1965) Development of cotton spraying equipment in Central Africa, *Cotton Growing Review* **42**, 131–145.
Turner, C.R. and Huntington, K.A. (1970) The use of water sensitive dye for the detection and assessment of small spray droplets. *Journal of Agricultural Engineering Research* **75**, 385–387.
Turner, D.J. and Loader, M.P.C. (1974) Studies with solubilized herbicide formulations. In: *Proceedings of the 12th British Weed Control Conference* pp. 177–184. BCPC, Farnham.
Turner, P.D. (1985) Economics in weed control, costs following changes from conventional knapsack spraying in plantation crops. *BCPC Monograph* **28**, 33–38.
UNEP (1995) *Montreal protocol on substances that deplete the ozone layer*. UNEP 1994 Report of the Methyl Bromide Technical Options Committee. United Nations Environmental Programme, Kenya.
Urech, P.A., Staub, T and Voss, G. (1997) resistance as a concomitant of modern crop protection. *Pesticide Science* **51**, 227–234.
Van de Werken, J. (1991) 'The development of an unmanned air assisted tunnel sprayer for orchards. *British Crop Protection Council Monograph* **46**, 211–217.
Van de Zande, J.C., Porskamp, H.A.J., Michielsen, J.M.G.P., Holterman, H.J. and Huijsmans, J.F.M. (2000) Classification of spray applications for driftability to protect surface water. *Aspects of Applied Biology* **57**, 57–65.
Van Emden, H.F. (1972) Plant resistance to insect pests. Developing 'risk-rating' methods. *Span*, **15**, 71–74.
Van Emden, H.F. and Peakall, D.B. (1996) *Beyond Silent Spring*. Chapman and Hall, London (for UNEP).

Van Hemmen, J.J. and Brouwer, D.H. (1997) Exposure assessment for pesticides: operators and harvesters risk evaluation and risk management. *Mededelingen van de Faculteit Landbouwwetenschappen, Rijksuniversiteit Gent* **62**(2a), 113–130.

Van Valkenburg, W. (1973) The stability of emulsions, In: *Pesticide Formulations*, (ed. W Van Valkenburg). pp. 93–112. Marcel Dekker, New York.

Van Valkenburg, W., Sugavanum, B. and Khetan, S.K. (eds) (1998) *Pesticide Formulation: Recent Developments and their Applications in Developing Countries*, UNIDO, Vienna/New Age International, New Dehli.

Vercruysse, F., Steurbaut, W., Drieghe, S. and Dejonckheere, W. (1999a) Off target ground deposits from spraying a semi-dwarf orchard. *Crop Protection* **18**, 565–570.

Vercruysse, F., Drieghe, S., Steurbaut, W. and Dejonckheere, W. (1999b) Exposure assessment of professional pesticide users during treatment of potato fields. *Pesticide Science* **55**, 467–473.

Vieira, R.F. and Sumner, D.R. (1999) Application of fungicides to foliage through overhead sprinkler irrigation – a review. *Pesticide Science* **55**, 412–422.

Voss, C.M. (1976) The helicopter's contribution to agricultural aviation at present and in the future. *Proceedings of the 5th International Agricultural Aviation Congress*, 22–25 September, Stoneleigh, pp. 223–228. IAAC, The Hague.

Wagner, J.M. (1998) Regulatory requirements in the USA. In: *Chemistry and Technology of Agrochemical Formulations* (ed. D.A. Knowles), pp. 377–417. Kluwer, Dordrecht.

Walker, A., Farrant, D.M., Bryant, J.H. and Brown, P.A. (1976) The efficiency of herbicide incorporation into soil with different implements. *Weed Research* **16**, 391–397.

Walker, D.A. (1973) Agri-Fix, a track guidance system for aerial application. *Agricultural Aviation*, **15**, 99–104.

Walker, P. (1971) The use of granular pesticides from the point of view of residues. *Residue Review* **40**, 65–131.

Walker, P. (1976) Pesticide granules: development overseas, and opportunities for the future. *British Crop Protection Council Monograph*, **18**, 115–122.

Walklate, P.J. and Weiner K.-L. (1994) Engineering models of air assistance orchard sprayers. *Acta Horticulturae* **372**, 75–82.

Walklate, P.J., Richardson, G.M. and Cross, J.V. (1998) Measurement of air volumetric flow rate and sprayer speed on drift and leaf deposit distribution from an air-assisted sprayer in an apple orchard. In: *Proceedings of Agricultural Engineering Symposium*, 1996. Madrid.

Walklate, P.J., Miller, P.C.H. and Gilbert, A.J. (2000a) Drift classification of boom sprayers based on single nozzle measurements in a wind tunnel. *Aspects of Applied Biology* **57**, 49–56.

Walklate, P.J., Richardson, G.M., Cross, J.V. and Murray, R.A. (2000b) Relationship between orchard tree crop structure and performance characteristics of an axial fan sprayer. *Aspects of Applied Biology* **57**, 285–292.

Walton, W.H. and Prewett, W.C. (1949) Atomization by spinning discs, *Proceedings of the Physical Society, London, Section B* **62**, 341–350.

Ware, G.W., Cahill, W.P. and Estesen, B.J. (1975) Pesticide drift: aerial applications comparing conventional flooding vs raindrop nozzles. *Journal of Economic Entomology* **68**, 329–330.

Watkins, T.C. and Norton, L.B. (1955) *Handbook of Insecticide Dust Diluents and Carriers*, 2nd edn. (revised by D.E. Weidhaas and J.L. Brann Jr.), Dorland, Caldwell.

Way, M.J., Bardner, R., van Baer, R. and Aitkenhead, P. (1958) A comparison of high and low volume sprays for control of the bean aphid *Aphis fabae* Scop. on field beans. *Annals of Applied Biology* **46**, 399–410.

Webb, D.A., Western, N.M. and Holloway, P.J. (2000) Modelling the impaction behaviour of agricultural sprays using monosized droplets. *Aspects of Applied Biology* **57**, 147–154.

Werker, A.R., Dewar, A.M. and Harrington, R. (1998) Modelling the incidence of virus yellows in sugar beet in the UK in relation to numbers of migrating *Myzus persicae*. *Journal of Applied Ecology* **35**, 811–818.

Western, N.M., Hislop, E.C., Herrington, P.J. and Jones, E.I. (1989) Comparative drift measurements for BPCP Reference hydraulic nozzles and for an Airtec twin-fluid nozzle under controlled conditions. In *Proceedings of the Brighton Crop Protection Conference – Weeds*, pp. 641–648. BCPC, Farnham.

Western, N.M. and Hislop, E.C. (1991) Drift of charged and uncharged spray droplets from an experimental air-assisted sprayer. *British Crop Protection Council Monograph* **46**, 69–73.

Western, N.M., Hislop, E.C. and Dalton, W.J. (1994) Experimental air-assisted electrohydrodynamic spraying. *Crop Protection* **13**, 179–188.

Western, N.M., Hislop, E.C., Nieswel, M., Holloway, P.J. and Coupland, D. (1999) Drift reduction and droplet size in sprays containing adjuvant oil emulsions. *Pesticide Science* **35**, 640–642.

Wheatley, G.A. (1972) Effects of placement distribution on the performance of granular formulations of insecticides for carrot fly control. *Pesticide Science* **3**, 811–822.

Wheatley, G.A. (1976) Granular pesticides: some developments and opportunities for the future (developed countries). *British Crop Protection Council Monograph* **18**, 131–139.

Whitehead, D. (1976) The formulation and manufacture of granular pesticides. *British Crop Protection Council Monograph* **18**, 81–92.

Whitehead, A.G. (1988) Principles of granular nematicide placement for temperate field crops. *British Crop Protection Council Monograph* **39**, 309–318.

Whittam, D. (1962) Aircraft guidance methods for pest control in the United States, *Agricultural Aviation* **4**, 8–15.

WHO (1970) *Control of Pesticides: A Survey of Existing Legislation*. World Health Organisation, Geneva.

WHO (1973) *Specifications for Pesticides used in Public Health*, 4th edn. World Health Organisation, Geneva.

WHO (1974) *Equipment for Vector Control*. World Health Organisation, Geneva.

WHO (1989) *The use of impregnated bednets and other materials for vector-borne disease control*. WHO/VBC/89.981. World Health Organisation, Geneva.

WHO (1990) *Equipment for Vector Control*, 3rd edn., World Health Organisation, Geneva.

WHO (1998) *The WHO Recommended Classification of Pesticides by Hazard and Guidelines to Classification 1998–1999*, WHO/PCS/98.21/Rev. 1. World Health Organisation, Geneva.

Wicke, H., Backer, G. and Frießleben R. (1999) Comparison of spray operator exposure during orchard spraying with hand-held equipment fitted with standard and air-injector nozzles. *Crop Protection* **18**, 509–516.

Wilce, S.E., Akesson, N.B., Yates, W.E., Christensen, P., Lowden, R.E., Hudson, D.C. and Weigt, G.I. (1974) Drop size control and aircraft spray equipment. *Agricultural Aviation* **16**, 7–16.

Wilkins, R.M. (1990) Biodegradable polymers. In: *Controlled Delivery of Crop Protection Agents* (ed. R.M. Wilkins), pp. 149–168. Taylor and Francis, London.

Wilson, J.D., Hedden, O.K. and Sleesman, J.P. (1963) *Spray droplet size as related to disease and insect control on row crops*. Ohio Agricultural Experiment Station Research Bulletin 945.

Wolf, T.M. (2000) Low-drift nozzle efficacy with respect to herbicide mode of action. *Aspects of Applied Biology* **57**, 29–34.

Wolf, T.M., Grover, R. Wallace, K., Shewchuk, S.R. and Maybank, J. (1993) Effect of protective shields on drift and deposition characteristics of field sprayers. *Canadian Journal of Plant Science* **73**, 1261–1273.

Wolf, R.E., Gardisser, D.R. and Williams, W.L. (1999) Spray droplet analysis of air induction nozzles using WRK DropletScan Technology. *ASAE Paper* 99–1026.

Womac, A.R. (2000) Quality control of standardized Reference Spray Nozzles. *Transactions of the ASAE* **43**, 43–56.

Womac, A.R. and Bui Q.D. (1999) Variable flow control device for precision agriculture. US Patent 5 908 161.

Womac, A.R., Williford, J.R., Weber, B.J., Pearce, K.T. and Reichard, D.L. (1992) Influence of pulse spike and liquid characteristics on the performance of uniform droplet generator. *Transactions of the ASAE* **35**, 71–79.

Womac, A.R., Mulrooney, J.E. and Bouse, L.F. (1993) Spray drift from high-velocity aircraft. *Transactions of the ASAE* **36**, 341–347.

Womac, A.R., Mulrooney, J.E., Young B.W. and Alexander P.R. (1994) Air deflector effects on aerial sprays. *Transactions of the ASAE* **37**, 725–733.

Womac, A.R., Goodwin, J.C. and Hart, W.E. (1997) *Tip selection for precision application of herbicides*. University of Tennessee Agricultural Experiment Station Bulletin 695.

Womac, A.R., Hart, W.E. and Maynard II, R.A. (1998) Drop spectra for pneumatic atomizers at low discharge rates. *Transactions of the ASAE* **41**, 941–949.

Womac, A.R., Maynard II, R.A. and Kirk, I.W. (1999) Measurement variations in reference sprays for nozzle classification. *Transactions of the ASAE* **42**, 609–616.

Wood, B.J., Liau, S.S. and Knecht, J.C.X. (1974) Trunk injection of systemic insecticides against the bagworm *Metisa plana* (Lepidoptera Psychidae) on oil palm *Oleagineux* **29**, 499–505.

Woodford, A.R. (1998) Water-dispersible granules. In: *Pesticide Formulation: Recent Developments and their Applications in Developing Countries*. (eds W. Van Valkenburg, B. Sugavanum, and S.K. Khetan, pp. 203–231. UNIDO, Vienna/New Age International, New Dehli.

Woods, N. (1986) Agricultural aircraft spray performance; calibration for commercial operations. *Crop Protection* **5**, 417–421.

Wooley, D.H. (1963) A note on helicopter spray distribution. *Agricultural Aviation*, **5**, 43–7.

Wraight, S.P. and Carruthers, R.I. (1999) Production, delivery and use of mycoinsecticides for control of insect pests on field crops. In: *Biopesticides: Use and Delivery* F.R. Hall and J.J. Menn, (eds) pp. 233–269. Humana Press, Totowa.

Wright, J.F. and Ibrahim, N.I. (1984) Steps of water dispersible granule development. In: *Advances in Pesticide Formulation Technology* (ed. H.B. Scher). ACS Symposium Series 254, pp. 185–192. American Chemical Society, Washington DC.

Wunderlich, L.R. and Giles, D.K. (1999) Field assessment of adhesion and hatch of *Chrysoperla* eggs mechanically applied in liquid carriers. *Biological Control* **14**, 159–167.

Wyatt, I.J., Abdalla, M.R., Palmer, A. and Munthali, D.C. (1985) Localized activity of ULV pesticide droplets against sedentary pests. *British Crop Protection Council Monograph* **248**, 259–64.

Wygoda, H.J. and Rietz, S. (1996) Plant protection equipment in glasshouses. *European Plant Protection Organisation Bulletin* **26**, 87–93.

Yates, W.E. and Akesson, N.B. (1973) Reducing pesticide chemical drift. In: *Pesticide Formulations* (ed. W. Van Valkenburg), pp. 275–341. Marcel Dekker, New York.

Yates, W.E. and Akesson, N.B. (1976) Systems for reducing airborne spray losses and contamination downwind from aerial pesticide applications. In: *Proceedings of the 5th International Agricultural Aviation Congress* 22–25 September 1975, Stoneleigh, pp. 146–156. IAAC, The Hagne.

Yates, W.E., Cowden, R.E. and Akesson, N.B. (1982) Procedure for determining in situ measurements of pesticide size distribution produced by agricultural aircraft. *Proceedings ICLASS 82* pp. 335–339.

Yeo, D. (1961) Assessment of rotary atomisers fitted to a Cessna aircraft. *Agricultural Aviation* **13**, 131–135.

Younes, M. and Sonich-Mullin, C. (1997) Concepts of the International Programme on chemical safety in the assessment of risks to human health from exposure to chemicals. *International Journal of Toxicology* **16**, 461–476.

Young, B.W. (1986) A device for the controlled production and placement of individual droplets. In: *Proceedings of the 5th Symposium on Pesticide Formulations and Application Systems* pp. 13–22.

Young, B.W. (1991) A method for assessing the drift potential of hydraulic spray clouds and the effect of air assistance. *British Crop Protection Council Monograph* **46**, 77–86.

Young, V.D., Winterfield, R.G., Deonier, C.E. and Getzendaner, C.W. (1965) Spray-distribution patterns from low level applications with a high-wing monoplane. *Agricultural Aviation* **7**, 18–24.

Zhu, H., Dexter, R.W., Fox, R.D., Reichard, D.L. Brazee, R.D. and Ozkan, H.E. (1997) Effects of polymer composition and viscosity on droplet size of recirculated spray solutions. *Journal of Agricultural Engineering Research* **67**, 35–45.

Zhu, H., Fox, R.D., Ozkan, H.E. Brazee, R.D. and Derksen, R.C. (1998) A system to determine lag time and mixture uniformity for inline injection sprayers. *Applied Engineering in Agriculture* **14**, 103–110.

Appendix I

International Standards Relating to Pesticide Application

ISO 4102: 1984	Equipment for crop protection – Sprayers – Connection threading
ISO 5681: 1992	Equipment for crop protection – Vocabulary
ISO 5682–1: 1996	Equipment for crop protection – Spraying equipment – Part 1: Test methods for sprayer nozzles
ISO 5682–2: 1997	Equipment for crop protection – Spraying equipment – Part 2: Test methods for hydraulic sprayers
ISO 5682–3: 1996	Equipment for crop protection – Spraying equipment – Part 3: Test methods for volume/hectare adjustment systems of agricultural hydraulic pressure sprayers
ISO 6686: 1995	Equipment for crop protection – Anti-drip devices – Determination of performance
ISO 8169: 1984	Equipment for crop protection – Sprayers – Connecting dimensions of nozzles and manometers
ISO 9357: 1990	Equipment for crop protection – Agricultural sprayers – Tank nominal volume and filling hole diameter
ISO 10625: 1996 (+ Technical Corrigendum 1: 1998)	Equipment for crop protection – Sprayers nozzles – Colour coding for identification
ISO 10626: 1991	Equipment for crop protection – Sprayers – Connecting dimensions of nozzles with bayonet fixing
ISO 10627–1: 1992	Agricultural sprayers – Data sheets – Part 1: Typical layout
ISO 10627–2: 1996	Hydraulic agricultural sprayers – Data sheets – Part 2: Technical specifications related to components
ISO 13440: 1996	Equipment for crop protection – Agricultural sprayers – Determination of the volume of total residual
ISO 13441–1: 1997	Air-assisted agricultural sprayers – Data sheets – Part 1: Typical layout

ISO 13441–2: 1997	Air-assisted agricultural sprayers – Data sheets – Part 2: Technical specifications related to components
ISO 14710: 1996	Equipment for crop protection – Air-assisted sprayers – Dimensions of swivel nuts of nozzles
ISO 4254–6: 1995	Tractors and machinery for agriculture and forestry – Technical means for ensuring safety – Part 6: Equipment for crop protection
ISO 6720: 1989	Agricultural machinery – Equipment for sowing, planting, distributing fertilisers and spraying – Recommended working widths
ISO 3767–2: 1991 Amendment 1: 1995	Tractors, machinery for agriculture and forestry, powered lawn and garden equipment – Symbols for operator controls and other displays – Part 2: Symbols for agricultural tractors and machinery
ISO 11684: 1995	Tractors, machinery for agriculture and forestry, powered lawn and garden equipment – Safety signs and hazards pictorials – General principles
ISO 1401: 1999	Rubber hoses for agricultural sprayers
EN 907: 1997	Agricultural and forestry machinery – Sprayers and liquid fertiliser distributors – Safety

Index